Geological Controls for Gas Hydrates and Unconventionals

Geological Controls for Gas Hydrates and Unconventionals

Sanjeev Rajput

Baker Hughes, Kuala Lumpur, Malaysia

Naresh Kumar Thakur

Council of Scientific and Industrial Research, New Delhi, India

ELSEVIER

AMSTERDAM • BOSTON • HEIDELBERG • LONDON • NEW YORK • OXFORD • PARIS
SAN DIEGO • SAN FRANCISCO • SINGAPORE • SYDNEY • TOKYO

Elsevier
Radarweg 29, PO Box 211, 1000 AE Amsterdam, Netherlands
The Boulevard, Langford Lane, Kidlington, Oxford OX5 1GB, UK
50 Hampshire Street, 5th Floor, Cambridge, MA 02139, USA

British Library Cataloguing-in-Publication Data
A catalogue record for this book is available from the British Library

Library of Congress Cataloging-in-Publication Data
A catalog record for this book is available from the Library of Congress

ISBN: 978-0-12-802020-3

Publisher: Candice Janco
Acquisition Editor: Amy Shapiro
Editorial Project Manager: Tasha Frank
Production Project Manager: Paul Prasad Chandramohan
Designer: Mathew Limbert

Typeset by TNQ Books and Journals

"The more that you read, the more things you will know. The more that you learn, the more places you'll go."

Dr. Seuss

"To all the Earth Scientists, especially the great ones"

Contents

Authors Biography

Dr. Sanjeev Rajput is a Principal Advisor Geophysics, Oil and Gas Development with Baker Hughes, Malaysia and has 15 years of diverse experience in oil and gas industry. He obtained a B.Sc in Physics and Mathematics in 1998 from M.J.P. Rohilkhand University, India and a Master of Technology (M.Tech) in Applied Geophysics in 2001 from Kurukshetra University, Kurukshetra, India. In 2007, he graduated with a Ph.D. from Kurukshetra University, Kurukshetra, India, specializing in 4C seismic data analysis and quantitative interpretation. Sanjeev began his career as a Marine Geophysicist with CGG International before going for his post-doctorate fellowship with University of Texas at Austin, USA. As a Reservoir Geophysicist with Schlumberger in 2007 he was involved in Full Field Review, Mature Field development, and Quantitative Interpretation of Geophysical data. He joined CSIRO Australia in 2009 as a Senior Research Scientist and moved on as Principal Scientist. His consulting work was focused on geophysical data analysis and quantitative interpretation of oil and gas fields. Sanjeev joined Baker Hughes in 2013 in Kuala Lumpur, Malaysia. His work is primarily on Mature Oil and Gas field development, Unconventional hydrocarbon exploration and development, and Quantitative Interpretation and production enhancement. He is a member of SEG, EAGE, and SPE. He has authored more than 50 international publications.

Dr. Naresh Kumar Thakur Received his Masters in 1970 from Osmania University, Hyderabad, India and Ph.D. from Bombay University, Bombay in 1984. With more than 40 years of research experience in diversified geological and geophysical experimentations, he is presently focusing his research on unconventional hydrocarbon exploration. He has more than 70 peer reviewed papers and two US patents in his credit. He served as a member of United Nations Convention on the Law of the Sea (UNCLOS) and elect representative for five years.

Foreword

The influence of science and technology on our everyday life and on our society as a whole is immense. Technology advancements are also helping us to generate new opportunities for innovation and growth. The petroleum industry has not remained untouched in this regard. Progressive development of energy resources and their conversion technologies has been bringing benefits to mankind since the mid-18th century, longer than almost any other scientific discipline. In the past decade or so, we have noticed some monumental changes to the energy landscape. Our ability to drill, extract, and produce hydrocarbons from different unconventional resources through technological innovations has played a major role in assisting us to access more affordable energy.

Geological Controls for Gas Hydrates and Unconventionals has been written to tell the story of unconventional hydrocarbon resources, especially: gas hydrates, tight gas, shale gas, liquid-rich shale, and shale oil to future generations. It is also a unique tribute to the scientific fraternity engaged in unravelling the intricacies associated with these complex resources. Every chapter abounds with the enthusiasm of the authors: my friend Sanjeev Rajput and Naresh Kumar Thakur, who manage to convey the intellectual fascination of unconventional oil and gas, along with their associated complexity and challenges in a very intriguing and visually appealing way. I strongly believe this book will become a cornerstone in the pursuit to augment our understanding of these unconventional hydrocarbon reservoirs.

Unconventional reservoirs have proven challenging, as these reservoirs are typically composed of complex geological systems of rocks, fluids, and organic material embedded in heterogeneous facies due to changes in stress, pressure, and temperature states over geological time. The authors have done a marvellous job to maintain an approach throughout the book that addresses some of these complex processes through simple descriptions and anecdotes. And we all know: simple is hard.

The book covers not only the geologic processes that give rise to these reservoirs but also describes how to better understand the rock properties through application of multidisciplinary techniques, as well as to address difficulties in understanding fluid flows in these play types. It then elaborates on the critical importance of characterizing the unconventional reservoirs, associated challenges in better understanding the reservoir performances, and how it impacts value creation through cost optimization during development of these resources. Moreover, the authors from their experience and knowledge also define a holistic approach that can be adopted for modeling such reservoirs. The book also highlights the operational part associated with unconventional resource development and ends with an incise insight on the way forward for the industry in this regard.

Blending geology, geophysics, geomechanics, petrophysics, and reservoir engineering, it explains, in simple language, the scientific concepts that are necessary for unconventional hydrocarbon-related studies. Although the emphasis of this book is on geological controls, it contains much that will be of interest to those outside this field and to the students of hydrocarbon resource exploration, development, and production; indeed to anyone with a fascination with the world of "Unconventionals." In their approach, the authors have chosen major stages of an oil and gas field development project that includes discovery, data evaluation, full-field development, production, and abandonment, along with a vivid description on the worldwide occurrence of unconventional hydrocarbons.

I sincerely believe that the authors can be confident that there will be many grateful readers who will gain a broader perspective of the disciplines associated with unconventional hydrocarbon studies, and their book is destined to play an important role in exciting, motivating, and educating the next generation of geoscientists and reservoir engineers from all over the world.

S. Bordoloi
Global Director, Geomechanics, Baker Hughes Incorporated
Houston, USA, 27th January, 2016

Preface

Geological Controls for Gas Hydrates and Unconventionals deciphers the complex processes involved in the generation of unconventional hydrocarbons resources with specifics on gas hydrates, tight gas, shale gas, liquid rich shale and shale oil. It is also a unique tribute to the scientific fraternity engaged in unravelling the intricacies associated with these complex resources. The book first sets the stage by reviewing the structural constituents of continental margins and highlights their role in generating unconventional hydrocarbons particularly gas hydrates, shale gas, and shale oil. The prominent component of gas hydrates, methane, is in turn elaborated on, pointing out the processes favorable for its generation. Kerogen is fully scrutinized sifting the different types of hydrocarbons it produces under thermal processes with different sources. The sedimentation process ultimately giving rise to unconventionals is then explained at depth, from transportation of land materials till deposition on the ocean bottom with a focus on organic matters and associated decomposition. The tectonic settings impacting continental margins, and the resulting influence on the formation of gas hydrate, are illustrated. Channelizing of fluid flow in hydrate formations is analyzed under appropriate petrophysical properties and faulting mechanisms. It then elaborates on the critical importance of characterizing the unconventional reservoirs, associated challenges in better understanding the reservoir performances and how it impacts value creation through cost optimization during development of these resources. *Quantitative seismic interpretation* examples highlighting unconventional plays from different parts of the world are presented. The book reviews the different types of rock properties, elastic, petrophysical, and geophysical along with their estimation methods. Data conditioning, calibration, and processing for rock physics analysis are given in detail. The book covers not only the geologic processes that give rise to these reservoirs but also describes how to better understand the rock properties through application of multidisciplinary techniques, as well as to address difficulties in understanding fluid flows in these play types. Blending geology, geophysics, geomechanics, petrophysics, and reservoir engineering, it explains, in simple language, the scientific concepts that are necessary for unconventional hydrocarbon-related studies. The book concludes with an outlook for the way forward, providing some current facts about the unconventionals including their global distribution and the disputed hydraulic fracturing associated with them. It emphasizes integration of all relevant data for optimal drilling of horizontal wells and for multistage fracture stimulation in tight-oil reservoirs. Finally, key factors for successful exploration and development of unconventional reservoir including discovery, data evaluation, full-field development, production, and abandonment, along with a vivid description on the worldwide occurrence of unconventional hydrocarbons are explained.

AUDIENCE

Geoscientists, physicists, exploration and development geologists/geophysicists, students as researchers, and reservoir engineers with a focus on hydrocarbon research.

Acknowledgments

We derived an unexpected amount of enjoyment in writing this book and we wish the readers to experience the same. We have had many adventures in science and all of them relied on the efforts of many talented people. We have been able to directly benefit from their wisdom, integrity, and creativity, and indirectly from the contributions of the generations of great scientists who came before them. We would like to express our gratitude to the many people who saw us through this book; to all those who provided support, talked things over, read, wrote, offered comments, allowed us to quote their remarks and assisted in the editing, proofreading and design.

This book quite literally would not have been possible without the unparalleled vision, ambition, encouragements and support of Mr. Martin S. Craighead, Chairman, Chief Executive Officer and President of Baker Hughes and Scott Reeves, President - Reservoir Development Services, Baker Hughes. It's a privilege to learn under their leadership.

Very special thanks to Shikha and Arav for their unconditional love, prayers and good wishes throughout this project.

We would like to thank Marisa LaFleur, editorial project manager from Elsevier for design, continuous support at every stage of the project, and many fruitful suggestions. Without her this book would never find its way to the readers.

Similarly, we have once again been aided and abetted by Monica Ronghe, as an external editor, who in addition to her helpful edits contributed some valuable comments.

We thank S. Bordolai, M. Ring, Dr. Mohammed N. Alfaraj, Prof. Anand Joshi, and D. O. Timko for writing foreword for this book.

We express sincere thanks to Professor Dinesh Kumar, Professor S. S. Teotia, Dr. Nader Dutta, Dr. V. K. Gahalaut, John Tinnin, Jonathan E.F. Stearn and Ludovic Le Gurun for coaching and encouragements. We deeply value their guidance.

Although, we have been working on this project for some time, the spark of inspiration that helped us organize our thoughts and drafts into a book was provided by the 'The Black Swan' event of vanishing the concept of 'Peak Oil' and beginning of the era of "unconventional hydrocarbon resources driven economy' in 2014.

Last and not least: we beg forgiveness of all those who have been with us over the course of the years and whose names we have failed to mention."

Dr. Sanjeev Rajput and
Dr. Naresh Kumar Thakur

CHAPTER

CONTINENTAL MARGINS

1

"The sea is only the embodiment of a supernatural and wonderful existence."

Jules Verne

CHAPTER OUTLINE

1.1 INTRODUCTION

The earth is one of the most dynamic elements in our planetary system, where landmasses are created and destroyed. In recent times it has been recognized that geological processes are far more dynamic and on a larger spatial scale than previously thought. Lithological, geochemical and physical properties distinguish the most fundamental elements: the continents and the oceans on the earth's surface. The difference in elevation between the continents and the ocean floor reflects the fundamental difference between the densities (lithological properties) of the continental and oceanic crust and their isostatic response. The ocean floor is divided into three major topographic provinces: the continental margins, the ocean basin floor and the major oceanic ridges. In addition to these major features, secondary topographical features such as rises and plateaus tend to break up the oceans into separate basins. The maximum difference in elevation between the continent and the ocean floor occurs closer to the outer edge of the margin as mid-oceanic ridges occur in the central part of most oceans. The present seas, which cover about 71% of the earth's surface, are relatively young features in the history of the earth. Their evolution is governed by the lithosphere, the outermost rigid part of the earth which is composed of several plates moving relative to each other. When the plates break up, melt from the interior arrives at the surface where it solidifies and gradually moves sideways, cools and subsides, forming new ocean floor by continuous steady state processes of seafloor spreading at the mid-oceanic ridges. The process

Geological Controls for Gas Hydrates and Unconventionals. http://dx.doi.org/10.1016/B978-0-12-802020-3.00001-1

adds new material to the moving outer shell of the earth, or lithosphere. The rigid lithosphere includes both the crust and the uppermost part of the mantle. The production of new lithosphere is compensated at the subduction zone, where the plates underthrust or collide and lithosphere is carried back into the underlying mantle. Geologically, the plate tectonics processes at the plate boundary or plate edges create and destroy plates. The plate boundary consists of mid-oceanic ridges and subduction zones offset by transform faults. Oceans and continents are two primary geomorphic and geological provinces which have distinctly different composition and physical properties. They are separated not by the coastline but by a geological boundary, the continent ocean boundary (COB), which marks the seaward extent of the continent or its outer edge. The COB is commonly found tens to hundreds of kilometers away from the coastline. Refraction and gravity data suggest that the true boundary between the continental crust and the oceanic crust occurs at water depths of 2–3 km.

The continental margin is strictly a geomorphic term. It comprises the continental shelf, slope and rise between the coastline and the deep ocean basin. The margin shows diversified morphological features and structural architecture as its genesis is connected to different geological configurations, tectonic settings and subsequent deposition and erosion history. The continental margins encompass about 21% of the oceanic area. Nearly 71% of the earth's surface is covered with water and only 5.3% of this is less than 200 m deep but this area of shallow water does not represent all the continental margins. Models for the development of continental margins have been based on the morphological features, patterns of rock emplacement, structural features associated with the margins and sedimentation and erosion history. The continental margins include the continental shelf, slope and rise.

1.2 MORPHOLOGY

Suess (1885) first introduced the terms "Atlantic margin" and "Pacific margin" to emphasize what he saw as the fundamental difference between the wide coastal plains, steady subsidence and thick layer cake sediments of the Atlantic Ocean and the folded mountain belts, island arc and volcanism on the margins of the Pacific Ocean. The classical configuration of the continental margin that was established by Heezan et al. (1959) was based on the detailed bathymetry that summarizes the North Atlantic Ocean. Three significant marine areas were recognized, including shelf, continental slope and continental rise based on major changes in the inclination of the bathymetric surface. Continental margins can be characterized depending on whether they have experienced a relatively long period of stability (Atlantic type) or whether they have suffered active tectonism during later geological times (Pacific type). Topographically the Atlantic-type margin is typified by a relatively wide continental shelf, which may vary from 30 to 400 km in width, and a relatively dissected continental slope indented by submarine canyons.

A continental margin that has a broad continental shelf, a gentle continental slope and a pronounced continental rise is known as a passive continental margin (Fig. 1.1a). This type of margin experiences little, if any, volcanic or earthquake activity and the primary activity that affects it is the build up of sediment. As these margins are found along the east coasts of North and South America and the west coasts of Europe and Africa, they are also known as "Atlantic-type" or "divergent-type" margins. The continental margins along India and Antarctica are other examples of passive margins.

A continental margin that has a very narrow or almost nonexistent continental shelf and a narrow, steep continental slope is known as an active continental margin. Instead of ending in a continental rise, the continental slope of this type of margin often plunges into a deep ocean trench, which may be filled with sediment (Fig. 1.1b). Earthquake and volcanic activity are prominent here. Since active

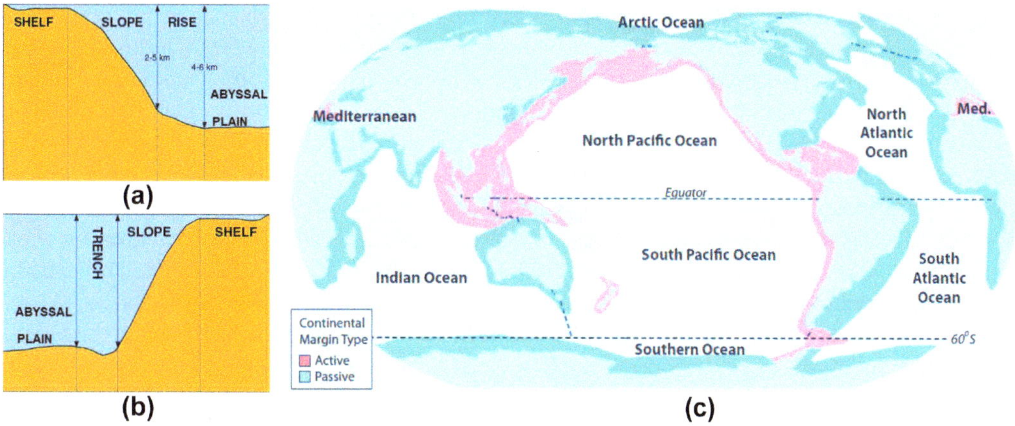

FIGURE 1.1

Morphology of passive continental margin and convergent margin. (a) The passive margins are characterized by a broad shelf, well-defined slope and continental rise. (b) The convergent marines have a narrow to broad shelf, a steep slope merging with the nearby trench and do not have a conspicuous rise. (c) Map showing the locations of active and passive continental margins and the eight ocean regions. (From http://www.bluehabitats.org/)

continental margins occur along many coasts of the Pacific Ocean, these types of margins are also known as "Pacific-type" or "convergent-type" margins.

Usually continental shelves slope gently with an average seaward slope of about 0.1°. They tend to have the same topography or surface features that dominate the adjacent land. Where a coastal area is mountainous, dominated by low hills or flat plains, the continental shelf next to it is similarly shaped. The average depth of the continental shelf at the shelf break is about 130 m meters below the ocean surface. The continental shelf generally slopes gently (gradient less than 1:1000) away from the shoreline to a relatively well-defined shelf edge or shelf break at an average depth of about 130 m. Its relief is generally subdued (less than 20 m), the width ranging from a few kilometers to 400 km (average 78 km). The extent of the continental shelf varies from a few kilometers in western North and South America to 700 km off the Russian Siberian coast.

The continental shelves are underlain by a thick sequence of sedimentary rocks whose structural pattern may be very complex. Tectonic, depositional and erosional modifications can lead to a variety of margin types. As shallow water environments are in a state of disequilibrium and sediments are trapped in the near shore areas, most continental shelves are covered with relict sediments and are still dominated by morphology related to Quaternary low stands of sea levels. It is estimated that about 70% of the continental shelf area is covered with relict sediments (Emery, 1968).

The continental margin is the submerged shelf and slope forming the outer edge of a major landmass. The differences between the continental and oceanic crust composition are fundamental in margin development. The transition between the two crustal types lies under the area of the continental slope, with continental crust under the shelf and oceanic crust under the continental rise. The difference in densities and isostatic balance of the continental and oceanic crustal blocks is expressed dramatically by the presence of the continental slope, which is a great declivity from the shelf break downwards to the deep ocean depths. It is a continuous and impressive geomorphic feature – an escarpment 3 km high, 10 km wide and extending more than 350,000 km.

In contrast to the continental shelf, the continental slope is generally narrow in width, ranging from 10 to 100 km. Relatively steeper than the shelf, the slope angles down to the ocean basin at an average of 4°; in some areas, the angle may be as much as 25°. The steepness of a slope often reflects the steepness of the nearby coastal area. Slopes along mountainous coasts are steeper than those along flat coasts. In general, the steepest slopes tend to be found in the Pacific Ocean and the least steep ones in the Atlantic and Indian Oceans. Continental shelves and slopes, although covered by water, are part of the continental crust.

At the base of some continental slopes is a gently sloping, smooth-surfaced thick accumulation of sediment from land that has been transported to the shelf and then down the slope. This transition between the continental slope and the ocean basin is known as the continental rise and can be 100–1000 km wide. Some sources include the continental rise as being a major section of the continental margin, along with the shelf and the slope. Since it does not occur on all continental margins around the planet, it is best considered a characteristic of the margin in some areas. Continental rises are well developed around Antarctica and in the Atlantic and Indian Oceans but are hardly present in the Pacific Ocean.

The continental rise lies between the slope and deep ocean basin and varies in width from 100 to 1000 km with gentle gradients (1:100 to 1:700) and low local relief. Many margins do not have a continental rise for a variety of tectonic and depositional reasons. Although the continental rise is a geomorphic term it is really used to describe the depositional feature caused by the accumulation of sediments largely derived from continents and transported both down and along the slope. It also represents the geophysical characteristic of being the sediment apron at the base of a slope. The same processes that create the rise also create the abyssal plain seawards of the continental rise. Abyssal plains result from the pounding of turbidity current deposits in the last stages of deposition; only a small fraction of clastic sediments that reach the continental rise are carried forward and deposited on the adjacent abyssal plain. Even on the Atlantic-type margin, the simple configuration shelf slope and rise is relatively uncommon due to the presence of complex slope-associated areas of low gradient, which are referred to as "terraces" or "plateaus." Also, margins may have poorly developed rises and sometimes are not characterized as having a rise. The total area of the continental rise is difficult to estimate but the area of this ocean margin may be equal to the total area of the continental slope. The sediment thickness of the continental rise is often so great that seismic systems are incapable of recording reflections from areas of the greatest sediment accumulations.

1.3 EVOLUTION OF CONTINENTAL MARGINS

During the entire evolutionary process of the earth, continents have been evolving through geological time periods; five major episodes of clustering and fragmentation have been observed over this period. Continents episodically cluster together into a supercontinent and eventually break up with intense magmatic activity. Random motion of the continental blocks probably led to the cyclic formation of supercontinents every 350–400 Ma. During the present cycle, the breaking apart of the supercontinent of Pangaea was initiated during the Mesozoic Era (Early-Middle Jurassic) due to the episodic geodynamic event that began as early as 200 Ma. The emplacement of the largest known continental basalts accompanied the break up of Pangaea. Initially the supercontinent contained segments of passive margin from Northwest Africa to Northern Australia on a small segment on the eastern flank of the active oceanic subduction margin circumventing the remaining periphery of the continent (Fig. 1.2).

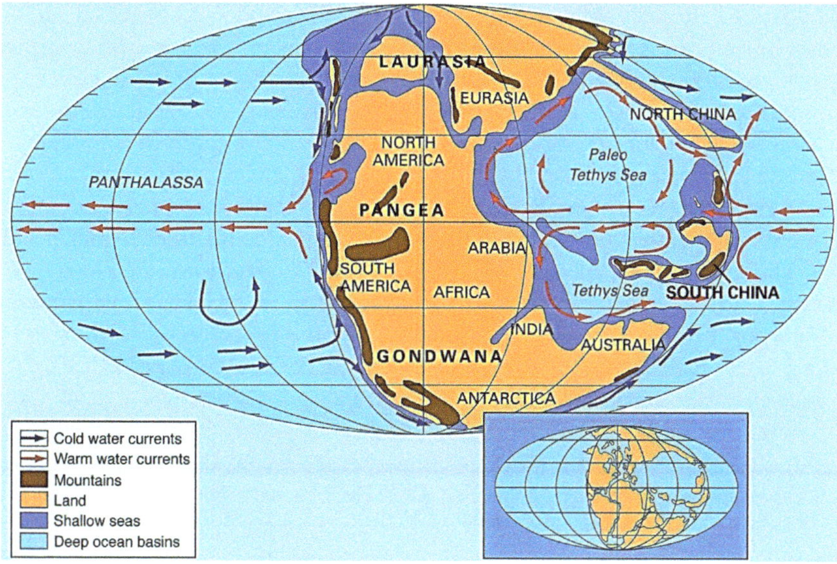

FIGURE 1.2

Distribution of landmasses, mountainous regions, shallow seas, and deep ocean basins during the early Triassic Period. Included in the paleogeographic reconstruction are cold and warm ocean currents. The present day coastlines and tectonic boundaries of the configured continents are shown in the inset at the lower right. (Adapted from Scotese CR. The University of Texas at Arlington. http://www.britannica.com/media/full/604667/146753)

The history of the breaking apart of Pangaea is well documented. However, the mechanism of the breaking up of the supercontinent is still being debated. Most of the divergent continental margins evolved during the break up of Pangaea in the early Mesozoic Era (Dietz and Holden, 1970). It is not fully understood whether the mantle convection source (Anderson, 1994; Coltice et al., 2007) or mantle plume heat source (Morgan, 1983) was responsible for separating the constituents of the supercontinent or whether the excessive thickening of the upper rigid part of the lithosphere was responsible for its instability and subsequent breaking into different segments along the weak zones. Vink et al. (1984) suggested that the lithosphere becomes progressively weaker as the crust thickens and some rifting is undoubtedly associated with a thick crust (Morgan and Burke, 1985). Thermomechanical processes govern rifting and subsequent breaking up of continents. Owing to the differences in rheological and thermal properties in the earth's subsurface, the continental and oceanic lithospheres exhibit different deformational responses to applied stresses. The continental parts, being more rigid, are not easily deformed. Owing to their thickness continents preclude heat diffusion thus making the continental lithosphere weaker and susceptible to rupture. The mobile belts (young orogeny/suture zones) surrounding the rigid continental part have lower mechanical strength and are likely to deform when stresses are applied. These zones retain the memory of old active faults for longer periods of time and are most suitable for deformation and breaking apart. Weakened mechanical strength over the preexisting crustal discontinuities plays a major role in the localization and distribution of crustal strain, leading to deformation (Corti et al., 2003). Preexisting mechanically weak zones in the lithosphere may emerge as a "diffuse plate boundary," to constitute zones of deformation with lateral dimensions on the order of hundreds and thousands of kilometers (Gordon and Stein, 1992; Zatman et al., 2005). Geological records suggest that rifting processes often localize along ancient orogenic belts and suture zones (Dunbar and Sawyer, 1988). Numerical models have shown that mantle convection can account for the present-day continental drift and the continents can fragment (rifting) over weak zones (low viscosity). Suture zone–circumventing continental fragments work as a buffer to prevent the cratonic roots from undergoing deformation over a longer period of time (Yoshida, 2010).

Segments of the former Gondwanaland exhibit variations in the lithospheric thicknesses, eg, the thickness of the African lithosphere is ~200–300 km in contrast to the thinner Indian lithosphere of ~100–150 km. The variations in the lithospheric thickness due to thermal erosion or metamorphism of the lower part of the lithosphere and the continental aerial extent of the plate result in varying mobility of different segments of Gondwanaland. In the earlier part of the 20th century Wegner (1924) proposed that lithospheric plates have always been in motion, sliding past over the more dense oceanic lithosphere. The concept of plate tectonics emerged strongly during the 1960s when geophysical measurements taken mainly over the oceans demonstrated that fragmented segments of the earth's lithosphere move relative to each other in complex ways, pulling apart at the spreading ridges, subducting or sliding past one another over the transform faults (Fig. 1.3). The plates are constantly moving, crashing into each other and subducting; they are pushed by convection currents in the mantle. In the distant past there have been several supercontinents where all the continents had come together to from one. Past supercontinents have names such as Rodinia, Godwana and Pangaea. It is predicted that a new supercontinent will occur in the next 250 Ma.

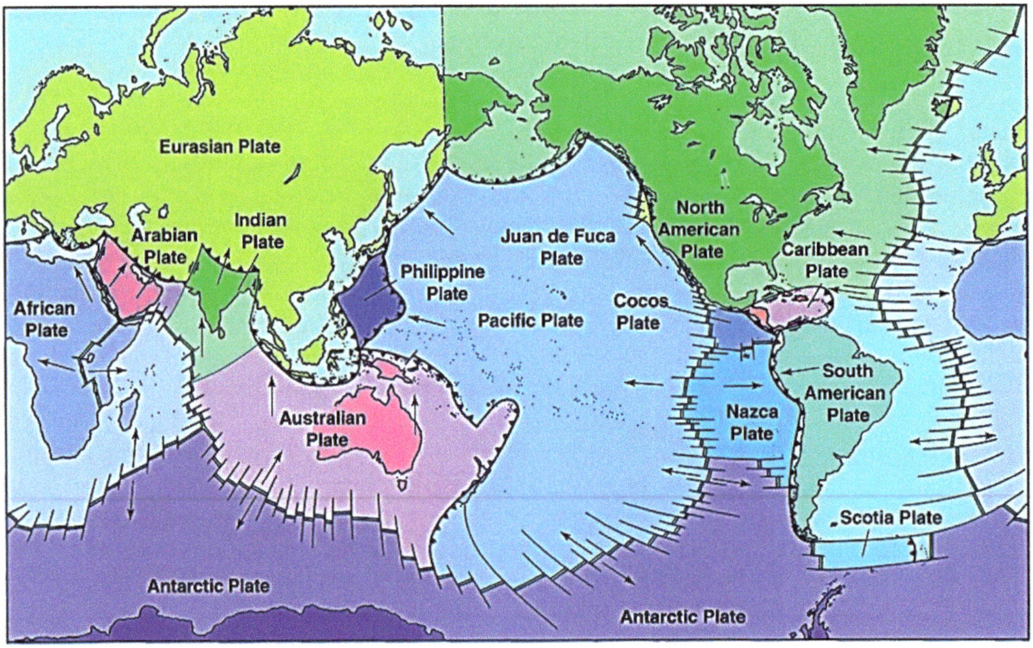

FIGURE 1.3

Plate movement for different parts of the world. (From http://www.westcoastplacer.com/)

Studies in the past (Hales, 1969; Chapple and Tullis, 1977) have suggested that ridge push, slab pull and trench pull (suction) forces are the main factors contributing to plate mobility, slab pull force being almost two to three times higher than the ridge push force. Trench suction adds lesser magnitude towards plate mobility due to a smaller segment length of the trench. Continental plates surrounded by spreading ridges and trenches have moderate average plate velocities (such as the Indian and Arabian plates). Recent studies have suggested that the mobility of the Indian plate is the fastest and the African plate slowest since their (Indian and African plates) separation during the late Cretaceous period (Prakash Kumar et al., 2007). In this regard the contribution of different factors relating to the ridge push, slab pull and trench suction pull forces to the mobility of different plates still remains uncertain.

The world's deep oceans and adjacent continental margins can be considered a part of the dynamic system moving with the plates, modified by fundamental geological processes such as uplift and subsidence, sedimentation and erosion, volcanism and rupturing (Eldholm and Tsikalas, 2004). Geologically the plate tectonic processes at the plate boundary or plate edges create and destroy plates. As the plates move, they interact with each other to produce different scenarios: add mass at the peripheries of the plate (divergent), consume land mass at the plate boundary (convergent) or different segments of the plate slide past each other (transform). Transform margins reflect the relative plate motion directions of the newly formed crust and occur at offsets of the normal faulted median rift valley that marks the axis of the ridge. Continental margins can be characterized depending on the period of stability (Atlantic, passive) or exposure to active tectonics during the later geological period (Pacific, active; Burke and Dewey, 1973). Passive margins like the Atlantic create landmass by the accretion of oceanic basin sediments or crust at the plate edge (Fig. 1.4).

The active margins are located at the plate boundary whereas the passive margins are located within the rigid moving plate. Two plates of different characteristics converge over the Pacific margins making the denser plate subduct and friction between the two plates and detachment of the subducting plate result in immense seismic activity (Fig. 1.5).

Plate movements, both in passive and active types of margins, normally occur in preexisting weak zones such as rifts/trenches in the region. Plates may be offset by transform (sheared) margins where plates move parallel to fractures or faults. About 10% of the plate boundaries neither diverge nor converge but instead transform or slide past one another. On land, such a boundary is called a slip-strike boundary. A classic example of this is the San Andreas Fault system, an area of cracks in the earth's surface that occurs at the boundary between two tectonic plates which extend down the coast of California. The active transforms are regions of decreased magma generation resulting in the transform zone being starved of volcanism and expressed as a deep trough in the oceanic crust.

Evolution of continental margins and their structural and morphological pattern are controlled by many factors such as the geological setting before the deformation, the tectono-magmatic events (rupturing, rifting, breakup) and erosion depositional history since seafloor spreading, reflecting the complex sequences of geological events resulting in the great variety in the continental margin's dimensions and style (Kennettt, 1982). Most of the present-day divergent margins were formed within the last 200 Ma, subsequent to the break up of Pangaea. The strength of the continental lithosphere controls the initiation and subsequent evolution of extensional deformation and is primarily controlled by three important factors, ie, rheology, geothermal gradient and thickness of the crust. Different components of the lithosphere (the upper crust, lower crustal and upper mantle part) react differently to variations in the crustal thickness (lithostatic pressure) and geothermal gradients. Increase in the geothermal gradient during extension serves to weaken the lithosphere, whereas thinning due to extension increases its

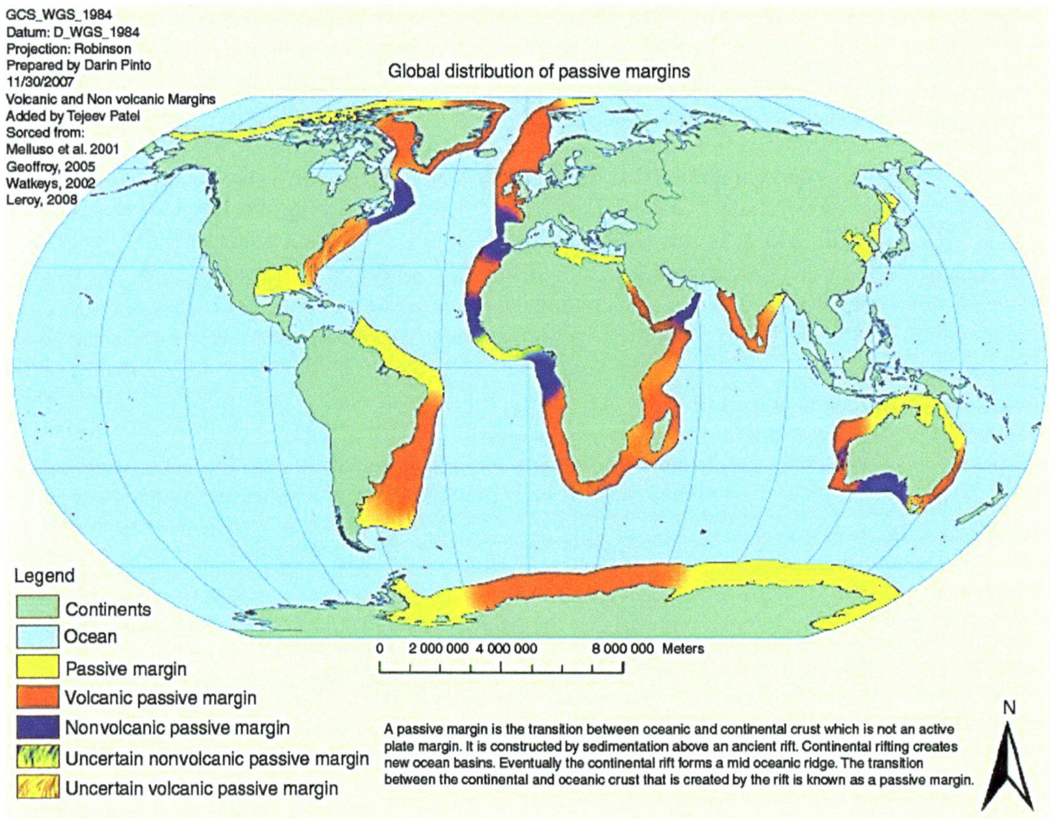

FIGURE 1.4

Global distribution of nonvolcanic and volcanic passive margins of the world. About 70% of the world's passive margins are characterized as volcanic passive margins. (From http://commons.wikimedia.org/wiki/File:Globald.png)

strength (Kusznir and Park, 1987). The strength of the continental lithosphere decreases in the presence of weak zones (old faulting) and increases in volatile content in the mantle and increased lateral heterogeneity within the crust (Meissner, 1986; Pollack, 1986; Dewey, 1986).

1.4 DIVERGENT PASSIVE (ATLANTIC) MARGINS

Intracontinental rifts evolve into continental margins when two segments across the rift are subjected to extensional forces leading to seafloor spreading and the creation of new oceanic floor and a new plate boundary (Symonds et al., 2002). The continental margin marks the zone where active divergent plate tectonics commences and the ocean begins to grow. Evolution of continental margins can be classified into three primary stages sometimes referred to as the initial (preexisting weak zone, rifting, subduction), youth (rupture and breakage) and maturity (subsidence, erosion, deposition) stages (Kennettt, 1982). Thus the margins may reflect the complex sequence of geological events caused by the

FIGURE 1.5

Map showing active global subduction zones and the ancient arc systems referenced in the text. Accretionary plate margins are shown with solid triangular marks along the plate boundary, and tectonically erosive margins are indicated by open triangles. (From Draut and Clift, 2013.)

interacting magmatic, tectonic and depositional processes (Eldholm and Tsikalas, 2004). Divergent continental margins undergo predominantly vertical tectonic development. Plate tectonics emphasizes horizontal movement of the lithosphere which also induce vertical movement due to changes in thickness and thermal character (geothermal gradients) of the crust.

The continental rifting process is usually the first event (initial stage) in the formation of a major ocean basin. Lithospheric extension due to the extensional stresses at the plate edges precedes rifting in about 20–25 Ma. Preexisting weak zones and episodes of previous extensions undergone by the plate are characteristic of each continental margin. Two types of models have been proposed to explain the rifting process. In passive rifting, the extension forces at the plate edges result in thinning of the lithosphere and passive upwelling of the asthenosphere (Fig. 1.6). Passive upwelling brings mantle material into a lower temperature or/ pressure regime resulting in partial melting (Foucher et al., 1982). The melt is spread under the stretched continental crust (underplating) and preserved as thin plutonic crust. New ocean floor is generated at the diverging boundary, and the edges of the now-split continents get farther apart. The now fully developed mid-ocean ridge is the site of hot new ocean floor, and so it floats high on the asthenosphere beneath. Divergent margins include both the shallow water deposition on the continental shelf and the deep water deposition on the continental rise below.

In the active mode of rifting, the thermal upwelling of the asthenosphere induces tensional stresses in the crust leading to rifting (Spohn and Schubert, 1983). In fact, both types of margins appear similar

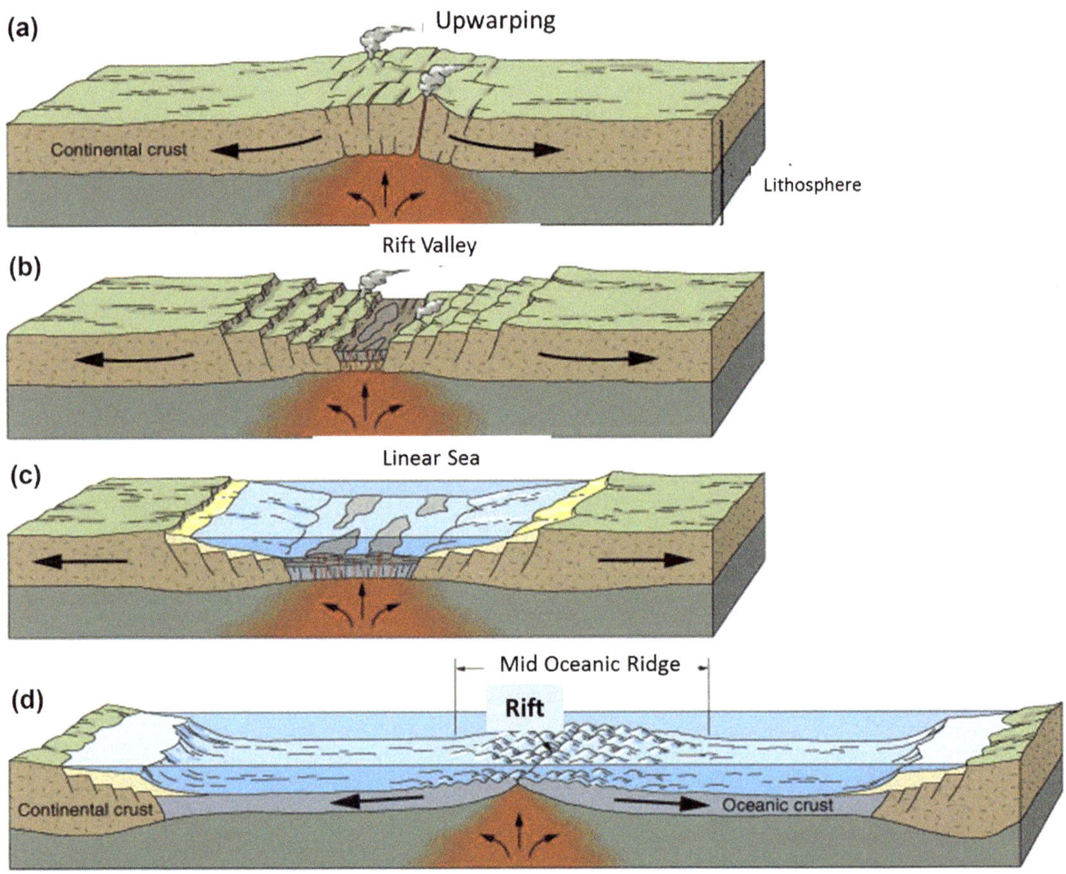

FIGURE 1.6

(a–d) The formation of divergent margins, with the development of rift and uplifted rift shoulders due to extension forces at the plate edges. (From http://www.csus.edu/indiv/k/kusnickj/Geology12/tectonicbasins.html)

in structural style and deformation history for much of their evolution until the time of breakup and the start of seafloor spreading (Eldholm et al., 1995). The nonvolcanic rifts are associated with plate–tectonic-driven horizontal stresses acting on the lithosphere leading to crustal thinning (passive model) rifting and subsidence that predates and accompanies the continental break up and development of passive continental margins (McKenzie, 1978). Brittle failure of the upper crust and ductile stretching in the lower crust and upper mantle causes thinning of the lithosphere prior to rifting. Extensional forces at the plate edges (closed system) impose divergent velocity to the lithosphere leading to the formation of rifts. Mass from the mantle moves upwards to replace the mass that has been removed, creating upwarp (necking) in the lower part of the lithosphere (Armitage et al., 2010). Mechanically or thermally induced weaknesses in the upper mantle, possibly resulting from prerift underplating, control the initial location of upper mantle necking that can be offset from crustal deformation, as seen in the Iberian region (Whitmarsh et al., 2001). The width of the rift is controlled by the magnitude of the

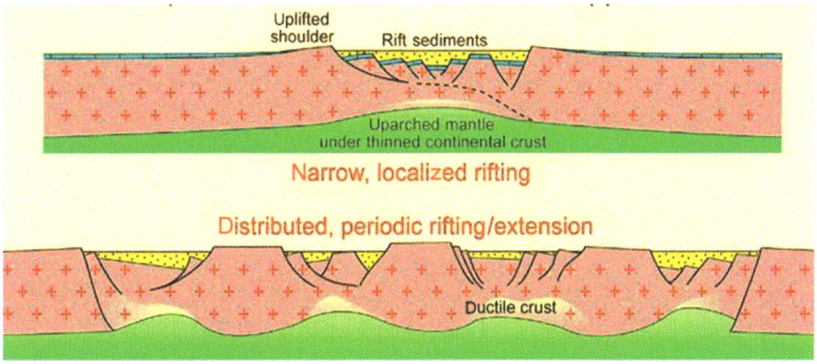

FIGURE 1.7

Two modes of continental extension. (Adapted from Brun and Choukroune, 1983.)

extension as well as temperature gradients in the horizontal (lateral) direction. Over time, reduction of tensile forces at the edges leads to local extension resulting in a narrow rift. Lateral variation in temperature may also lead to formation of faults and narrower rifts (Behn, 2002).

Early graben formation, uplifted shoulder, rotated fault blocks, subsidence, absence of extensive volcanic activity and neighboring thinner oceanic crust (3–4 km) such as the Exmouth plateau characterize nonvolcanic (Fig. 1.7), crevice or sedimentary passive margins. The replacement of denser lithosphere by lighter asthenosphere makes the mass imbalanced and isostatic compensation makes the rift valley subside and the rift shoulders elevate immediately after the extension (Hellinger and Sclater, 1983). Following this, subsidence may occur due to the removal of thermal anomaly as the margin moves away from the spreading center and sediment loading. The shape of the divergent margin is related to the trend of initial rifting although subsequent modifications take place by subsidence, erosion and deposition.

About 90% of the passive margins are volcanic whereas the remaining 10% are nonvolcanic with only limited amounts of volcanic rocks (Menzies et al., 2002). Active rifting margins are characterized by immense magnetism (open system) and extend laterally over distances of hundreds of kilometers. In contrast to the closed system, the asthenosphere can initially rise independent of the magnitude of lithospheric extension. Production of melt is not due to the extension but is produced during the asthenospheric upwelling or is advected in the system atop a rising diapiric plume (Leeder, 1995). Volcanic rifted margins generally show a short-lived pulse of intense magmatism and quickly abate to a steady-state midoceanic ridge (Nielsen and Hooper, 2004). Rifting associated with immense volcanism over hotter than normal mantle (plume or convection) originates from thermally driven domal uplift under the thinned lithosphere (McKenzie and Bickle, 1988; Dunbar and Sawyer, 1988). Convected thinning of the lithosphere may be due to the upward movement of the solidus because of the increase in heat flux or mechanically removed material from the base (erosion) of the lithosphere (Ingersoll and Busby, 1995). Adiabatic mantle melting occurs due to the penetration of the low-density mantle in the lithosphere with rates increasing with time due to the continuous decrease in mantle pressure as the lithosphere thins. Plume-driven or convection-driven heat is considered to be the main source for heating the asthenosphere resulting in domal uplift (Sengor and Burke, 1978; Morgan, 1983). However, the existence of volcanic margins away from the known hotspot regions has prompted an alternative mechanism for the formation of volcanic margins. Such margins can be formed by intrinsic dynamic rifting processes (Holbrook and Keleman, 1993; King and Anderson, 1995). The amount of melt generated by passive stretching and thinning of the lithosphere during the rifting is significantly lower when compared to models for volcanic margins. Seismic studies suggest that transition from continental to oceanic crust is abrupt in the case of volcanic margins. Lateral temperature differences in the lithosphere evoked by the rifting and upwelling of the asthenosphere drive convection in the upper mantle (Mutter et al., 1988), which transports a large volume of igneous material into the crust making the igneous crust thicker, as seen in many volcanic margins like Iceland.

Rifts are considered unequally deep all along their length as continental crust gets thinner on a series of subplanar listric faults with a throw of several kilometers. Active rift zones are initially elevated above the sea level; syn-rift sediments are continental or lagoonal and strongly influenced by syn-rift volcanism.

Volcanic arc volcanic margins are associated with high-velocity (7.2–7.7 km/s), thicker, lower crust, seaward dipping reflectors and volcanic activity (Vøring Plateau). The extrusive magmatic bodies overlying intrusive bodies are identified in the geophysical signature as seaward dipping reflectors and are suggestive of extensive volcanism. In some regions this geophysical signature is found to be missing, even in cases of extensive magmatism. The Great African Rift, the Rhine Valley Graben and the Rio Grande Rift are believed to exemplify this basic rifting process, which sometimes leads to formation of an ocean.

Continental rifting is followed by plate separation (break up) and onset of seafloor spreading. The youthful stages of margin formation relate to the breaking up of the continent and spreading of the continental blocks, presently preserved in the Red Sea, and last about 50 Ma (Kennett, 1982). Little is known about the character of changes from rifting to spreading phases. The timing of transition from the rifting phase to spreading phase is not spontaneous, may take millions of years and may differ from margin to margin. The most recent period of rifting and the break up event is accompanied by magmatic activity manifested by intrusion or extrusion in the crust and volcanic eruption at the surface. The intensity and volume of magma production may vary between margins and individual margin segments. The volume of magmatism varies from a thin plutonic layer to a 20-km-thick layer. Following the rifting, mafic magma intrudes the continental crust as dyke swarms; sills and oceanic crust begin to accrete at the edges of the separate blocks. The magma penetration and thermal erosion of the lower part of the adjoining continental lithosphere results in its thinning and attenuation (Fig. 1.8).

Overlapping and underlapping (intermixing) of the continental crust and the oceanic crust causes their physical characteristics to change and makes them hybrids, leading to the formation of quasi-continental crust. The modified continental crust constitutes a high-velocity underplated magmatic layer at its base. With intense melt production in the process of volcanic rifting, the underplated magmatic layer is considerably thicker when compared to nonvolcanic margins where melt generation is substantially lower.

Kinsman (1975) suggested that the width of the attenuated continental crust seaward of the rift shoulder (rim) is relatively narrow (60–80 km) and others have suggested that larger attenuated margins are common (Lister et al., 1986). Further seaward, the intrusion rather than the extrusion of basaltic magma under the high rate of sedimentation results in the quasi-oceanic layer (Keen and Beaumont, 1990). It is suggested that this part of the crust is formed by subaerial lava flow on the thinned continental crust and might have evolved during the early phase of seafloor spreading (Mutter et al., 1982). The region is characterized by lower crustal thickness with thinning of the magmatic layer and seaward dipping reflectors in the upper part. This region is limited on the seaward side by uncontaminated

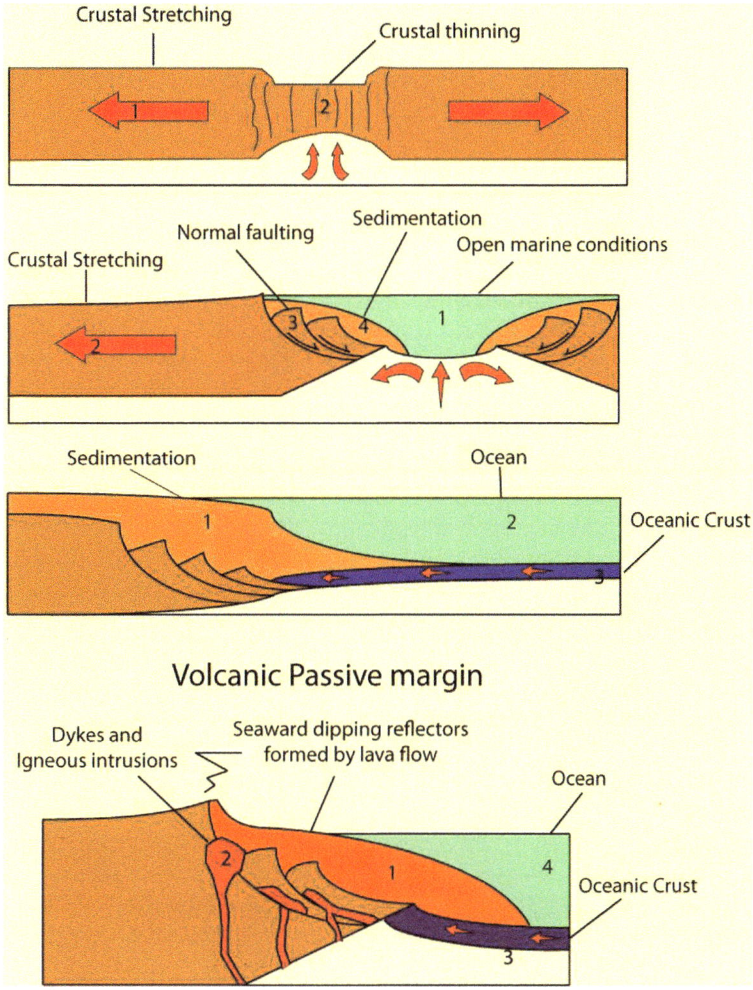

FIGURE 1.8

Development of passive continental margin after the initial phase of rifting. (1) In the first stage a continental rift is established due to stretching and thinning of the crust and lithosphere by plate movement. (2) Second stage leads to the formation of an oceanic basin. (3) The last stage of formation happens only when crustal stretching ceases and transitional crust and lithosphere subsides as a result of thermal subsidence. (From https://en.wikipedia.org/wiki/Passive_margin)

oceanic crust. Rift shoulders persist until the decay of diffused extension of the lithosphere and the shoulders subside below sea level once extension ceases; this is the culmination of the proto-oceanic phase. Accretion of oceanic crust to the continental crust may form the initial phase of the continent breaking and spreading. The imprint of break-up volcanism is presently buried by the thick sedimentary sequence and observed along many passive margins.

The change from rifting to accretion marks a major change in the thermal regime of the margin. Thermal contraction of the lithosphere leads to its subsidence. During rifting, the heat source remains beneath the rift axis but the continental margin becomes successively remote from the heat source allowing it to cool and subside (maturity stage). The initial line of spreading is not necessarily located within the zone of initial rifting. The continuous thermal subsidence increases the rigidity of the oceanic crust resulting in flexural bending of the continental and young oceanic crust. The onset of breaking coincides with combining of the quasi-continental and quasi-oceanic to regulate formation of the transition crust underlying the subsiding continental crust. Erosion of the uplifted shoulder leads to sedimentation at the flanks of the rift valley. A zone of syn-rift sediments are deposited into the rift system prior to its break up and is overlain by a broad zone of post rift and post break-up sediments deposited following the initiation of seafloor spreading. The rate of thermal subsidence declines with time as the margins become remote from the spreading ridge. The formation of continental terraces and rises are direct consequences of continental rifting (Bond et al., 1995). The modified shape of the continental margin following the rifting and breakup is primarily controlled by sedimentation and sea level oscillations. Sedimentation of the terrace and rise begins after rifting and commencement of seafloor spreading. Pitman (1978) modeled the seawards thickening of the sedimentary wedge of the shelf and slope (terrace) forms in response to the combined effect of tilting along the hinge line (slope–slope transition). Repeated sea-level advances and retreats or transgression and regression have been instrumental in shaping continental shelves especially during the low-sea-level stands when the shelf is exposed to subaerial processes. The shelf would be closer to the shoreline during periods of sea level fall (glaciations) or in the area of high sedimentation. These cyclic patterns have been very effective in producing flat surfaces on the shelves coinciding with deposition of alluvium, and this coincided with the late Cenozoic glacial maximum epoch. Erosion became the dominant process over the surface of the shelf. The broad flat characteristic of the shelf has resulted from erosional activity related to transgression and regression. Both erosion and sedimentation are concentrated in the near shore areas and repeated sea level migrations have leveled the margin forming the present-day shelf surface.

As sedimentation proceeds the troughs and tectonic basins formed as a result of the faulting fill up. The onward distribution of sediments is restricted by barriers created by the marginal basement high, coral reefs and diapirs. The progressive seaward tilting of the continental edge and flexural bending of the lithosphere under the growing sedimentary load may lead to further subsidence and sediment deposition. Further seaward with the increase in relief due to tilting, most of the sediments are deposited in the slope, ie, towards the seaward side of the terrace, where the sediment deposition could be the thickest. A continental rise is a broad wedge-shaped apron where the sediment thickness tapers off towards the abyssal plane. Most of the rise is covered with sediment deposition due to turbidity currents but final distribution deposition may be due to contour currents (Kennett, 1982). At this stage, terrace (shelf slope) and rise morphology is mostly developed.

The sediment deposition pattern on the rifted continental margin depends on the geology of the hinterland and the erosion processes. Large river systems transport sediment loads several kilometers into the ocean, leading to the formation of submarine levees. When the sediment deposition is faster and overcome subsidence and the removal of deposits by offshore drift leads to formation of progradation deltas front occurs. Delta formation is more common or well developed in semienclosed or enclosed sea regions (Kennett, 1982). During the low sea level stand, rivers will incise the convex up topography creating channels through which the turbidity currents feed voluminous sediment to the submarine fan seawards over the oceanic crust and older pelagic sediments. Most of the major rivers such as Ganges, Amazon erode elevated continental landmass and transport the sediments loads to deeper parts of the sea, thereby producing submarine fans, which are characterized by high sediment thicknesses.

1.5 CONVERGENT ACTIVE (PACIFIC) MARGINS

Plate tectonic forces drive continental plates to diverge or converge leading to creation or destruction of land masses at the plate edges or boundaries. Divergent margins develop as consequences of the extensional stresses whereas compression dominates the processes in the formation of convergent margins. When two plates converge, the surface expression on the ocean floor is deep-sea trenches. Unlike divergent margins, the convergent margins and adjacent seafloor belong to different tectonic plates. When two plates with different physical or geological characteristics interact the result is their deformation and consumption by convergent processes. The various types of convergent plate boundaries are sometimes defined as active or seismic margins because they are often marked by shallow to deep earthquake as well as volcanism, crustal deformation and metamorphism. They form much of the margin around the Pacific Ocean as well as parts of the southern Atlantic Ocean, the Northeast Indian Ocean and the Southwest Pacific region. Convergent margins are sites of subduction (the consumption of one segment of lithosphere beneath another adjacent segment) of the ocean and continental crust characterized by deep-sea trenches at their seaward (forearc) margin. Most of them have a volcanic arc and many have rifted backarc basins behind the volcanic arc. A variety of crustal interactions may occur resulting in the complex, structural configuration of the convergent margin (Fig. 1.9).

Depending on the density contrast and morphological configuration, one of the incoming plates is tectonically inserted or subducted under the other plate, which mostly has a terrestrial crust.

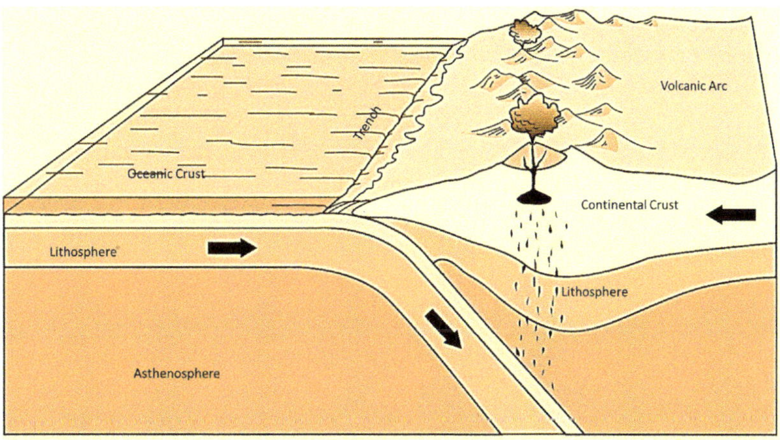

FIGURE 1.9

Figure shows that oceanic lithosphere is being subducted beneath the edge of a continent. The sediments from the ocean floor and pieces of the oceanic crust are scraped from the descending oceanic plate and plastered against the edge of the overriding continent. (From United States Geological Survey.)

The morphological as well as structural features of the subducting plate prior to the subduction is inherited from the constructive phase of the extensional regime, volcanic activity and sediment deposition on the abyssal plain (Aubouin, 1989) and the basement relief of the plate is controlled by the rate of sediment accumulation on the abyssal plain (Jarrad, 1989). The overriding terrestrial plate is 20–40 km thick. In most places the plate of oceanic lithosphere descends into the mantle along the Benioff–Wadati zone. Such subduction can occur under the oceanic (Mariana type) or continental (Andean type); in both cases the outer trench slope (seaward slope) is formed in response to the downward flexure of the subducting plate. This provides a mechanism for the downward displacement of deep-sea sediments several kilometers beneath the convergent margin. It is also widely recognized that tectonic processes at the convergent margins constantly fashion new regions of continental crust from fragments of the older or preexisting crustal material. A juvenile mass of terrestrial crust is added to the overriding crust through arc volcanism or magmatism or accretion of sedimentary strata (accrectionary prism). Arc magmatism is probably the most important process sustaining earth stock of terrestrial matter (Reymer and Schubert, 1984). At the convergent ocean margin the lower plate is the principal contributor to the input material. The lower plate is constructed of igneous rocks of the ocean crust with an overlying cover of ocean basin sediments. Ocean basin sediments primarily constitute pelagic ooze, hemipelagic mud and clayey far-traveled terrigenous or continental detritus contributed by continental erosion; it may even contain turbidites. The incoming oceanic sediments are scraped off through skimming of the upper part of the sedimentary layer during subduction and accreted to the overriding plate by a mechanical transfer process. Sediments may also underthrust the seaward position of resistive consolidated sediments and rock framework (buttress) of overriding plate (von Huene and Scholl, 1991). Under certain conditions the sedimentary deposits and oceanic crust may bypass the buttress and subduct, resulting in the consumption of incoming masses. Tectonic removal into the subduction zone of the down-going slab including the sediments and oceanic crust is called consumption (Kennett, 1982). The convergence process also may lead to loss of landmass by erosion of the lower part of the upper terrestrial crust and the eroded material may be taken by the subducting plate into deeper parts of the earth (Underwood and Moore, 1995). The main control on the morphology and geological structure of the convergent margin depends on the amount of sediment entering the trench or backarc region. Sediment input is quite variable, depending on the distance of the trench or backarc basin from major river systems. Moreover, the variation depends on whether the continental or oceanic crust is subducted in the forearc or rifted in the backarc region. The observed seismic activity over Chile and its absence in the Mariana may be ascribed to prevailing compressional and tensional stresses produce the pattern of backarc basin opening in these regions (Uyeda and Kanamori, 1979). Interaction of converging plates at the convergent margin produces complex structural patterns and their surface expressions on the seafloor.

1.6 STRUCTURAL CONFIGURATION

Convergent margins are subject to a complex interrelationship of structural, sedimentary and volcanic processes that produce a great range of margin types, features and characteristics. The convergent margin may constitute the following units from the seaward side (front) to the rear towards the terrestrial lithosphere, trench, subduction complex (part of the accrectionary prism) with slope basement high, forearc basin, frontal arc, volcanic chain, backarc basin and remnant arc (Fig. 1.9). The forearc region includes all the features lying on the trench side of the volcanic arc and constitutes the subduction complex, basement high, forearc basin and trench slope basin, extending to at least 100 km. The outer or the seaward slope of the trench may be quite variable in relief.

The outer (seaward) slope develops in response to the downward flexing of the subducting plate. The seaward swell is quite observable in the case of subduction of the young oceanic lithosphere without much sediment deposition (Mariana trench) on the abyssal plain. The outer swell may cover a region of 120–150 km from the trench axis and may attain a maximum uplift of the order of 300–500 m (Forsyth, 1980). The structural pattern with the presence of horst and graben of the outward slope may also exhibit variability depending on the extensional rifting and the processes of continental breaking (Fig. 1.10). The deformation on the outer slope may also result from a crustal response as the subducting plate passes through the outer swell or bulge (Hilde, 1983).

Deep-sea trenches may vary greatly in their depth, width and type of sedimentary succession (Uyeda, 1982). The trench floor basin sediments forming a wedge overlay the oceanic basement. The depth of the trench axis and wedge width is controlled to an extent by the age of the subducting lithosphere and subduction rate (Hilde and Uyeda, 1983). In general, trenches with thick sediment accumulation display nearly flat surface topography with a gentle gradient along the strike. The sediment thicknesses on the abyssal plain are of the order of tens of meters as in the case of intraoceanic trenches (Mariana, Tonga trenches). In the area of terrigenous sedimentations fed by channel and levee systems in submarine fans (Curray and Moore, 1974; Stevenson and Embley, 1987), such as the Makran and Sunda trenches, the sedimentations could be of the order of a few kilometers.

Trenches with lower sediment thicknesses exhibit pronounced topographical changes. The amount of sediment fill is perhaps the most important factor to shape the overall geometry of the trench wedge and primarily depends on whether the forearc is accretionary or erosional. Erosional trenches, such as the Mariana and Tonga trenches, are typically isolated from large amounts of terrigenous sediments and are deep (>9000 m) and narrow (<5 km). In contrast, accretionary trenches, such as Eastern Aleutian,

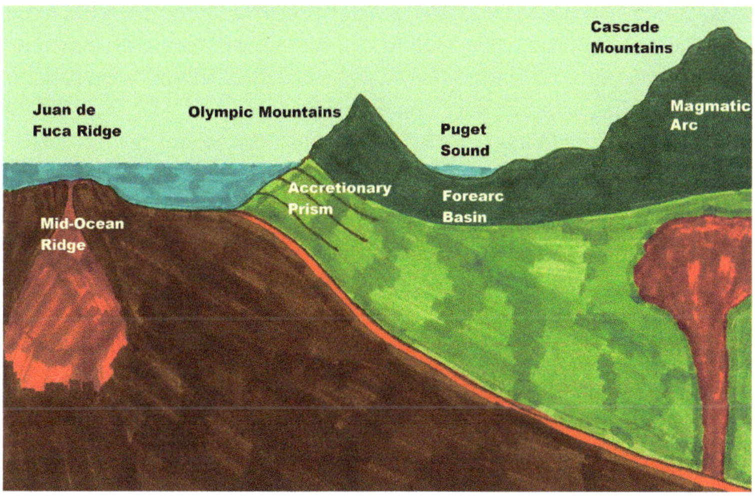

FIGURE 1.10

Schematic presentation of accretionary prism and underlying basement. (From http://blogs.scientificamerican.com/)

Nankai and Makran, are choked with terrigenous sediments, and are shallow (3000–5000 m) and wide (>10 km). Accretionary trenches with less terrigenous sediment deposition, such as the Northern Chilean and Peru trenches, have intermediate depths of 5000 to 7000 m and widths of 5–10 km.

Forearc regions adjacent to the continents generally have significant sediment fill in the trench or in the incoming oceanic plate and exhibit large, wedge-shaped bodies that have been accreted. The overall width of forearc regions between the volcanic arc and the trench can be greater than 300 km. The entire forearc basin should be considered part of the continental shelf (or slope or rise) on the basis of sediment thickness, basement type and morphology. Although most of the oceanic lithosphere is subducted into the mantle, large components of sedimentary cover and fragments of the basement high in the down-going plate are retained in the crust by mechanical transfer to the overriding plate (Underwood and Moore, 1995). In most cases, the ocean plate sediments are dominated by pelagic ooze and hemipelagic mud of the abyssal plain, although turbidites and submarine fan deposits are also possible. Normally the abyssal plain deposits are subsequentlyburied by the considerable thickness of terrigenous trench wedge turbidites as the down-going plate enters the subduction zone.

Accretion is the process in which material from the outer plate and trench (during the periods of discontinuous subduction) is removed and added to the outer continental margin or by other mechanisms such as imbricate thrusting or a combination of folding and thrusting (Karig, 1974; Karig and Sherman, 1975). Between the zone of uplift (frontal arc) and the trench axis is a 50- to 400-km-wide accretionary prism or the arc trench gap. The accretionary prism is made up of arcward-dipping, wedge-shaped, thrust-bound packets characterized internally by complex, trench-verging folds (Kennett, 1982). A well-developed accretionary prism is called an accretionary trench system, whereas those that lack a prism are called nonaccretionary trench systems. Davis and Hyndman (1989) infer that some of the truly massive accretionary bodies may have achieved their exceptional size because of factors that favor the efficient scrapping off of the subducting plate operating at these margins, for example, slow rate of convergence speeds and thick (2–6 km plus) incoming section of relatively coarse-grained (silty and sandy) terrigenous beds. The prism typically grows in width as successive increments are added by underthrusting at the trench, but it also increases in height because of compression to become a positive feature such as an island chain or uplands such as the Mentawai Islands off Sumatra and Middleton Island off Aleutian. Uyeda and Kanamori (1979) suggested that tectonic-controlled prisms are well developed when the trench lies in the vicinity of a continent (such as Chile) providing proper mechanical coupling between the upper and lower plates. In contrast, the prisms are not well developed when the trench interacts with the island arcs (Mariana).

In some cases, such as eastern Japan, where there is subsidence of the accretionary prism rather than uplift, the operating mechanism for such a configuration may be quite different. The accretionary prism displays rugged and irregular seafloor morphology due to the complex pattern of folding and faulting creating tectonic ridge-like features. The slopes of the accretionary prism (Fig. 1.10) are usually low and contain small basins near the toe of the slope (trench slope basin) that are filled with slope sediments. The inner slope (landward trench slope) of the trench is a major feature of the forearc system and its morphology reflects the subduction and accretionary processes. The shape and internal structure of the prism are also influenced by the geometry of its backstop (buttress), which is formed by the edge of the rifted continent or the older compacted accretionary body (Byrne et al., 1993) dipping towards the trench (Silver et al., 1985; Westbrook et al., 1988). The prism may be simple as in the case of oceanic regions of low sedimentation, such as the Mariana and Tonga arc, or may be subdivided into the upper and seaward lower (inner slope) trench slope. These two slopes are separated by the slope break. Much of the structural information at the trench slope break is obtained from the observation made from the nearby island. In the Indonesian arc system the exposed ridges over the sea level confine the seaward limits of the upper slope and are sometimes called a tectonic arc (Vening Meinesz, 1964). The trench slope break expressed as a bathymetric high and is termed the "outer ridge" and "outer arc high", which marks the boundary between the highly deformed strata of the lower slope and the open folded less-deformed sediments of the upper slope. The zone of active deformation between the trench axis and the trench slope break defines the surface expression of the subduction zone (Karig and Sherman, 1975). This region is sometimes referred to as the subduction complex.

The areal extent between the trench slope break and the upper slope discontinuity in most of the sediment-rich accretionary prisms is marked by a depression, which is filled mostly with terrigenous sediments and often called "the forearc basin." Forearc basins may occur near the intraoceanic island on the fringes of the marginal seas or adjoining the inland volcanic arcs on the active continental margins and along offshore island chains near the continental edge (Dickinson, 1995). Forearc basins are generally wide (>75–100 km) and filled with a large volume of sediments that have been delivered from the terrain (Sunda Trench, Nankai Trough, Eastern Aleutian). The basement of these forearc basins is usually older accreted sediment on the trenchward margin and volcanic arc terrain on the arcward margin (Sunda, Makran).

A zone of uplift immediately in front of the volcanic arc is called the frontal arc and forms the upper trench slope of the accretionary prism. The upper slope discontinuity marks the landward limit of the forearc basin and seaward termination of the frontal arc. The upper slope discontinuity in younger trenches may correspond to the upper part of the continental or insular slope that existed before the subduction process began (Karig and Sherman 1975; Underwood and Moore 1995). The morphological relief between the frontal arc platform and the upper slope indicates the pattern of sedimentation in the accretionary prism. In the case of young subduction zones the relief of the upper plate is significant and in older subduction zones with appreciable sedimentation, such as the Aleutian and Kermadec, the upper slope is characterized by low relief (Karig and Sherman, 1975).The upper slope discontinuity or outer ridge demarcates the landward side of the accretionary prism upper slope and determines the extent of the frontal arc (Fig. 1.9).

At the beginning of the frontal accretion, thrust slices that detach the upper part of the incoming sedimentary section have dimensions determined by sediment strength and its thickness. Stacking and upward rotation of the thrust slices thicken and shorten the accreted material. Once the maximum thickening and shortening is attained, the accretionary prism further maintains its seaward taper by subsurface underplating or by new thrust faults that cut the accretionary body at a low angle (out of sequence thrust; Platt, 1989). The seismic reflectors clearly show the reverse and thrust faults within the accretionary complex as well as the top of the oceanic crust underneath it. A model of converging plates explains these compressional features. A model of an earth where there is no convergence must account for these observations (both conceptually and specifically right here on this spot on the earth) (Fig. 1.11).

The term decollment refers to a fault along the bedding plane. Decollment marks the structural boundary (Fig. 1.11) between sediments that are scraped off and added to the toe of the accretionary prism and material that is carried towards the arc and underthrust beyond the toe of the prism. The underplating process adds incoming sedimentary material to the bottom of the accretionary prism at an intermediate position between the volcanic arc and the trench, leading to thickening and uplift of the prism without additional shortening (Cloos and Shreve, 1988). The depth of decollment serves as an important boundary to determine the amount of sediments to be accreted as opposed to the material that

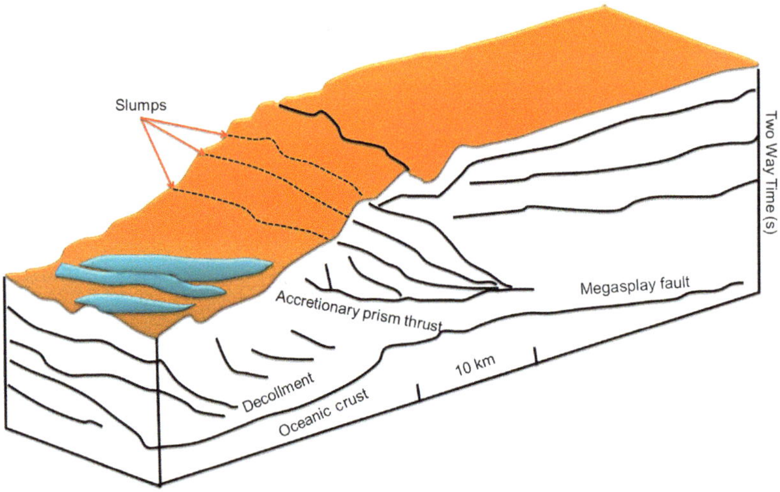

FIGURE 1.11

Structural representation of accretionary wedge (offshore Japan's southeastern margin). (Modified from Moore GF, Bangs NL, Taira A et al. Three-dimensional splay fault geometry and implications for tsunami generation. Science 2007;318:1128–1131. http://dx.doi.org/10.1126/science.1147195.)

is being underplated or subducted. Trenches wedges without sediments filling the decollment boundary may intersect at the seafloor itself. Most of the incoming sediments subduct at the base of the trench wedge and may lead to accretion of trench fill sediments, and make the incoming abyssal basin sediment subduct. Downward movement of decollment may promote the transfer of material from the lower to the upper plate, thus leading to underplating and prism growth. In contrast, upward movement of the decollment boundary may lead to subduction of most of the incoming sediments.

Volcanism always encompasses convergent margins, located either on the edge of the continent or island arc. The persistence of volcanism (in spurts) over a wide zone (~50–250 km) all along the subduction zone is called the volcanic arc. Towards the trenchward side intense volcanism is confined to the narrow zone and represents the volcanic front, the axis of which lies over the zone where the subducting block is at a depth of 100–150 km (Gill, 1981). The arc massif or the arc basement (Fig. 1.10) constitutes the crustal part that owes its genesis to arc magmatism (Dickinson, 1974) and may form the basement over a wider range covering the forearc basin to trench wall in some cases (Hussong and Uyeda, 1981). Earthquakes in the convergent plate margins mainly occur due to thrusting, fault failures and detachment of part of the subducting block in the deeper part of the earth. Subduction processes in which the Benioff zone is only confined to the lithosphere may produce earthquakes without much volcanism.

The occurrence of deeper earthquakes at depths on the order of 100 km associated with volcanism has led to the suggestion that they are caused by steep dips in the Benioff zone where the subducting plate penetrates deep into the low-velocity medium (asthenosphere) in the vicinity of volcanic arcs (Barazangi and Isacks, 1976). The mechanism of magma formation is still not clear. The magma generated at depths of 80–1000 km is mostly hydrous (tholeiitic) in nature; as its propagates upwards it fractionates into different compositions ranging from basaltic andesites to andesitic magmas. At shallower

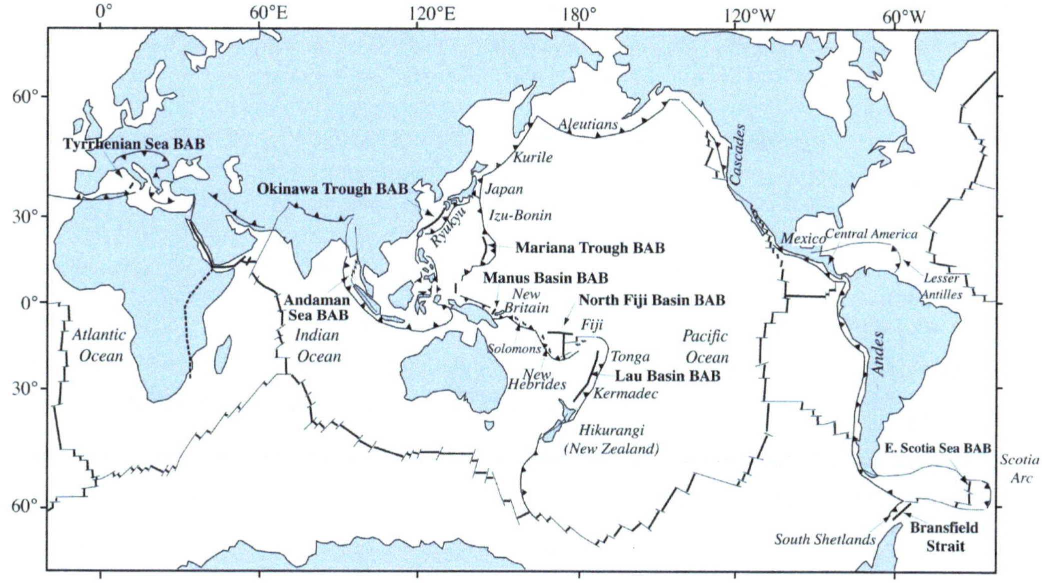

FIGURE 1.12

Active backarc basins (marginal seas) of the world. Most of the basins are confined to the Western Pacific. (From http://en.wikipedia.org/wiki/Back-arc_basin)

depths further fractionation may lead to the separation of amphibole, pyroxene and plagioclase producing dacites and rhyolites (Ringwood, 1977). It has been observed that the axis of volcanism shifts, moving randomly away from the trench or towards the trench (Williams and McBirney, 1979; Ingersoll 1988). Over the long-lived volcanic arcs intraarc basins may form where thick volcanic and mostly volcaniclastic sedimentary material is found (Smith and Landis, 1995).

Backarc basins occur behind the volcanic island arc and commonly along the continental margins and the convergent plate margins, hence they are often called "marginal" basins (Karig, 1974). Not all the trenches have backarc basins or marginal seas (Peruvian and Chilean trenches; Uyeda and Kanamori, 1979). Most of the world's backarc basins are located on the western edge of the Pacific Ocean (Fig. 1.12).

The genesis of backarc basins is not fully understood due to their submarine occurrence and subaerial exposure of ancient basins is difficult to recognize (Marsaglia, 1992). Different modes of formation of backarc basins have been proposed to account for the different characteristics associated with each region. Such models include extensive magma intrusion (Hasbe et al., 1970), mantle convection by mantle-wedge flow induced by the subducting slab (McKenzie, 1969; Hsui and Toksoz, 1981), trench migration due to sinking of the subducting plate (Forsyth and Uyeda, 1975) or ridge push associated with spreading (Hsui and Toksoz, 1981).

Based on the mode of formation they have been categorized as extensional (associated rifting seafloor spreading, eg, Sea of Japan), intraoceanic (rifting producing basin flanked by remnant arc, e.g, Mariana, Lau Basin) and nonextensional back basins that are (a) trapped during the plate reorganizations (Bering Sea) or (b) developed on the continental crust (Sunda Shelf and Java Sea; Hamilton, 1979). Active backarc regions are marked by high heat flow and development of marginal seas, with subdued magnetic lineations (Mariana Trough and Lau Basin). Extension in the crust could lead to rifting and subsequent seafloor spreading may initiate the formation of backarc basins marked by faulted boundaries and high heat flow (Karig, 1971; Taylor and Karner, 1983; Tamaki and Honza, 1991). These basins are characterized by the presence of thin sediment cover and a rugged basement. Uyeda and Kanamori (1979) and Brooks et al. (1984) propose that the degree of coupling of the overriding plate with the subducting plate determines whether a backarc basin develops as an active or nonactive type.

Notwithstanding the attention given to the process of juvenile crustal growth and terrain accumulation, it has been recognized that other subduction processes act at a convergent margin to remove or fail to store rock and sedimentary masses (Gilluly, 1963). These processes that respectively involve sediment subduction and subduction erosion can retard sediment accumulation or thwart continental growth (Scholl et al., 1980). Subduction erosion can also truncate the seaward wedges of the accretionary bodies and ancient continental masses (Scholl et al., 1977; Scholl and Vallier, 1981). The absence of marine pelagic sediments in the exposed subduction complexes and the small size or virtual absence of accretionary wedges along the continental margins has led to the proposal that part of the sedimentary matter is supposed to be consumed or taken into the mantle. Such observations in the initial stages was proposed by Coats (1962) who attempted to correlate the geochemistry of the sediments injected into the mantle by underthrusting and exposed volcanic rocks on the Aleutian Island arc. This has led to the proposition that on a convergent margin all the incoming sediments lying over the oceanic crust are not accreted to the toe of the accretionary prism (basement, rock framework or consolidated sediments accreted earlier) but some of the sediments bypass the toe (back stop, buttress) and either underthrust or subduct. In front of the buttress the upper part of the sediments is scraped off by the process of frontal accretion and the lower part is underthrusted. In the accretionary margins the buttress is buried by sediment thickness. Once the sediments bypass the toe of the accretionary prism they either accrete to the base of the accretion prism by underplating or are transported on the lower plate to deeper parts of the earth (Fig. 1.13).

It is presumed that sediment subduction has occurred where the volume of sediments accreted frontally to a buttress is less than what could have been accreted (von Huene and Scholl, 1991). Decollement is the principal boundary that separates the sediments into the upper and lower parts. Sediments above the decollement are accreted and those below are underplated; further landward in the subdecollement the sediments are subducted.

Subduction erosion involves the removal of sediment or rock mass that was supposed to have been part of the upper plate, which may constitute material that was attached to the seaward wedge (frontal) of the upper plate accretionary process as well as the landward part of the rock (basal) framework (von Huene and Scholl, 1991). Both the frontal and basal erosion reduce the quantum of material in the crust, making the upper plate thinner and thereby leading to landward migration of the trench (Von Huene and Lallemand, 1990). Murauchi and Ludwig (1980) proposed that a low-friction mechanism between the overriding and subducting plates might lead to subduction erosion. Conversely, Hilde (1983) proposed that the abrasive or scraping off mechanism of the incoming undulating topography subducting plate accounted for the basal erosion. However, identification of horst and graben way below the decollement in the seismic section in the region of Japan (von Huene and Scholl, 1991) and over the Middle America Trench (Shipley and Moore, 1986) suggests that abrasive interaction may not be a requirement for basal erosion.

The total amount of accreted sediments can be estimated as the product of (a) average time over which the sediments have accreted, (b) the average rate of orthogonal convergence, and (c) the average

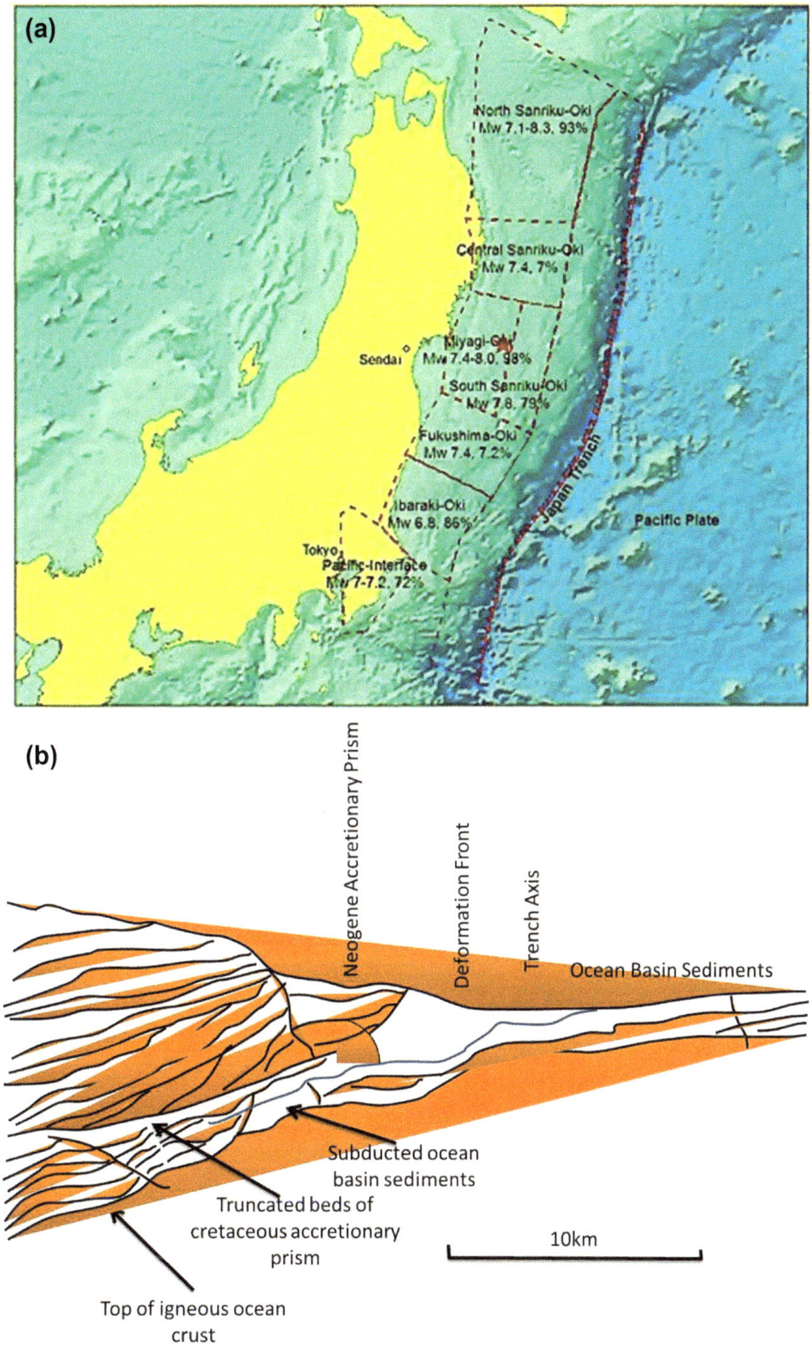

FIGURE 1.13

Japan Trench and structure interpreted from seismic data. (a) Japan Trench seismic source zones according to the latest national earthquake hazard models for Japan. The label in each zone shows the characteristic magnitude and its probability of rupture within the next 30 years. (Adapted from Headquarters of Earthquake Research Program, Japan, 2011.) (b) Interpreted seismic depth section across Japan Trench, indicating the truncation of older sediments of the overriding plate. Truncation is ascribed to subduction erosion along the base of the upper plate. (Modified from von Huene R, Scholl DW. Observations at convergent margins concerning sediment subduction, subduction erosion, and the growth of the continental crust. Rev Geophys 1991;3:279–316.)

thickness of sediments entering the subduction zone over time (von Huene and Scholl, 1991). Based on the difference between the estimated quantum of sediments accreted and observed in the field, mass balance can be calculated. Mass balance calculations from margins having different accretionary prism dimensions ranging from small to medium sized (40–150 km, covering 16,000 km convergent margin), well-constrained or not well-constrained by data sets, have suggested that in most of these regions only 20% of incoming sediments have accreted and about 80% have subducted below and landward of the active margin buttress in these regions. These calculations suggest that over the large accretionary prism (such as the Barbados region) about 70% of the incoming material in the subduction zone is subducted landward of the active buttress (von Huene and Scholl, 1991).

1.7 GLOBAL HYDROCARBON RESOURCE ESTIMATES

Continental margins with suitable sediment deposition, sediment accretion, biogenic content and geological properties, such as porosity, permeability, pressure–temperature regimes and tectonic conditions, provide prime locations for hydrocarbon accumulations. Continental margins hold most of the vital mineral deposits and potential avenues for the occurrence of hydrocarbons reservoirs in the world. With growing understanding of the genesis of deep-water hydrocarbon reservoirs and developments in the drilling technology, reservoirs earlier thought to be inaccessible are becoming economically viable and are being extensively explored and drilled to overcome the present oil and gas shortfalls (Fig. 1.14).

Passive margins are petroleum storehouses because these are associated with favorable conditions for accumulation and maturation of organic matter. Early continental rifting conditions led to the development of anoxic basins, large sediment and organic flux, and the preservation of organic matter that led to oil and gas deposits. Mann (2004) classified 592 giant oil fields into 6 basin and tectonic-setting

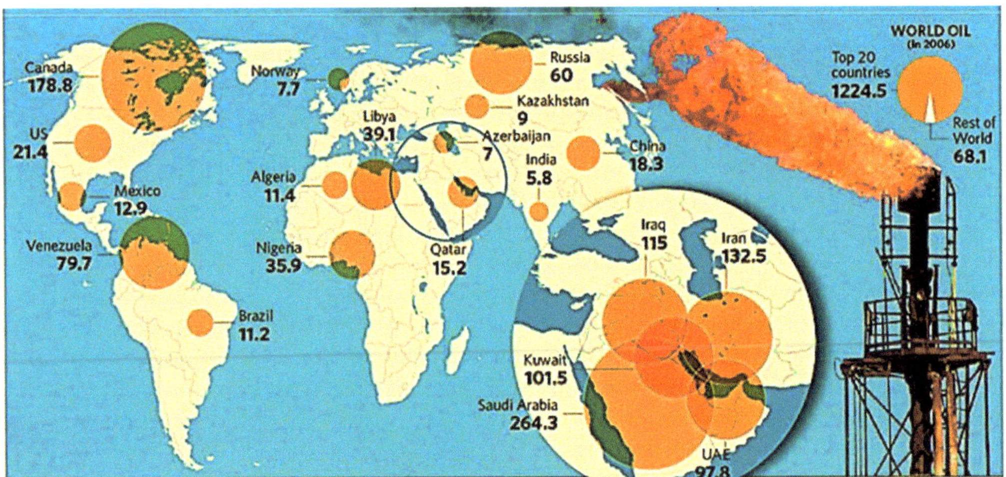

FIGURE 1.14

Proven oil reserves in billions of barrels. (From http://www.endofcrudeoil.com/2011/04/world-oil-reserves.html)

categories, and noted that continental passive margins account for 31% of giant oil fields. Continental rifts (which are likely to evolve into passive margins with time) contain another 30% of the world's giant oil fields. Basins associated with collision zones and subduction zones are where most of the remaining giant oil fields are found. These basins put together contain proved reserves of >500 million barrels of oil or >3 trillion cubic feet of natural gas.

Continental margins also provide suitable biological, physical and chemical conditions for the formation of gas hydrates. The shape and relief of the different elements of the continental margins constrain the physical, ie, temperature, pressure, and geological conditions (methane generation, porosity and permeability, water content in the sediments) necessary for the stability and formation of gas hydrates. Gas hydrates crystallize under moderate pressures and low temperatures when the lattice structure of water molecules holds molecules of methane and other hydrocarbon gases, such as ethane and pentane. Sufficient concentration of gas, far exceeding the solubility of gas in water and ionic strength of water, is important for formation of gas hydrates (Kvenvolden, 1993). A specific pressure–temperature (SPT) regime governs the stability of gas hydrates. In most parts of the world the necessary pressure–temperature conditions are met but gas hydrates are likely to form only on the continental margins where high inputs of organic carbon, focusing of methane-bearing fluids and the supply of methane are sufficient for the formation of gas hydrates (Davie et al., 2004).

Most of the countries with large oil reserves, notably, Iran, Iraq, Kuwait, Saudi Arabia and Venezuela, are known to report up to five times more reserves to sustain a similar level of production capacity when compared with other major oil-producing countries such as Mexico, Norway, Russia, the United Kingdom and the United States. Five are members of OPEC – Saudi Arabia, Kuwait, Iran, Venezuela and Nigeria – and together with the United States, Russia, Mexico, Norway and the United Kingdom account for almost two-thirds of the reported world's crude oil reserves; all are net exporters of oil with the exception of the United States.

1.7.1 GAS HYDRATES

Methane gas in the shallower regions of the earth gets trapped in the voids (cages) present in the crystal lattice of water molecules under moderate pressure and lower temperatures. Such conditions are prevalent in the polar regions where the lower temperature on the surface and lithostatic pressure (created by the weight of the sediments) provide the proper conditions for the formation of gas hydrates in the subsurface earth. In the tropical regions temperatures at the bottom of the sea at depths of tens of meters are appreciably lower than the observed temperatures at the sea surface. The thickness of the water column provides necessary pressures for the formation of gas hydrates in the subsurface earth. In both the environments sufficient availability of methane far exceeding the solubility of methane in water is a prerequisite for the formation of gas hydrates. The occurrence of gas hydrates has been extensively reported from the continental margin and permafrost regions reflecting the relics of an ancient oceanic environment. Gas hydrates can occur in both divergent and convergent margins. However, inference of the presence of gas hydrates on the continental margins by the identification of a characteristic seismic marker, bottom-simulating reflector (BSR), is rather simplistic. Some characteristic signatures of BSR may be missing in regions of high sedimentation on the divergent margins, thus making it difficult to identify hydrate-bearing locations. However, other geological and geophysical proxies can be utilized to identify the presence of gas hydrate in the regions of study. The presence of gas hydrate has been inferred by drilling in most continental margins around the world (Kvenvolden, 1993). Gas hydrates

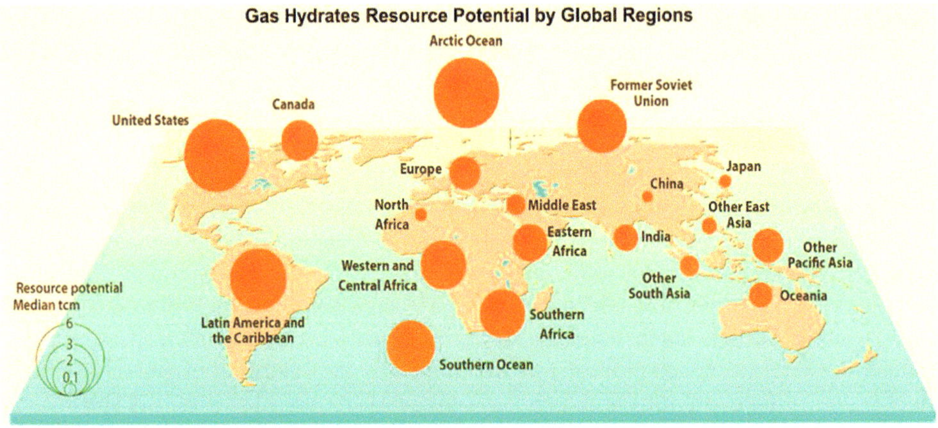

FIGURE 1.15

Distribution of known methane hydrate accumulations (courtesy of United Nations Envionmental Programme). This figure includes only that subset of global in-place gas hydrates that appear to occur at high concentrations in sand-rich reservoirs. (From Johnson A. Global resource potential of gas hydrate: a new calculation. Proceedings of the 7th International Conference on Gas Hydrates (ICGH 2011), Edinburgh, Scotland, 2011)

occur worldwide but because of the pressure–temperature and gas volume requirements, they are restricted to two regions: polar and deep oceanic. In polar regions gas hydrates are normally associated with permafrost, both onshore in continental sediment and offshore in sediment of the continental shelves. In deep oceanic regions gas hydrates are found in the outer continental margins in the sediment of slopes and rises where cold bottom water is present. The worldwide occurrence of known and inferred gas hydrates is shown on Fig. 1.15.

Although total in-place natural gas resources represented globally by methane hydrates are enormous, those resources are not all created equal. Instead they occur in a wide range of accumulation types. As with other petroleum resources, the accumulation types that are the most favorable for production are the least abundant, while the most challenging are the most abundant. This common attribute of natural resources creates a pyramidal distribution (Boswell and Collett, 2006). A generalized resource pyramid (Fig. 1.16) for gas hydrates (right) is shown in relation to a resource pyramid for all gas resources (left). Society continues to progress down through the global gas pyramid (left), aided by occasional technological breakthroughs that enable significant access to previously unrecoverable resources. Gas hydrates (right) may experience a similar progression with initial production most likely to occur within marine or arctic sands. Substantial new technological breakthroughs will be needed to access the large resources at the base of the hydrate pyramid. However, given the vast scale of hydrate resources, potential volumes even at the apex of the hydrate pyramid are significant.

1.7.2 SHALE GAS

In recent times shale gas and oil have emerged as viable unconventional resources to fulfill the global energy demand. Shale gas is considered a so-called "unconventional gas," together with "tight gas" from sandstones or limestones with low permeability and coal bed methane. As per the nomenclature,

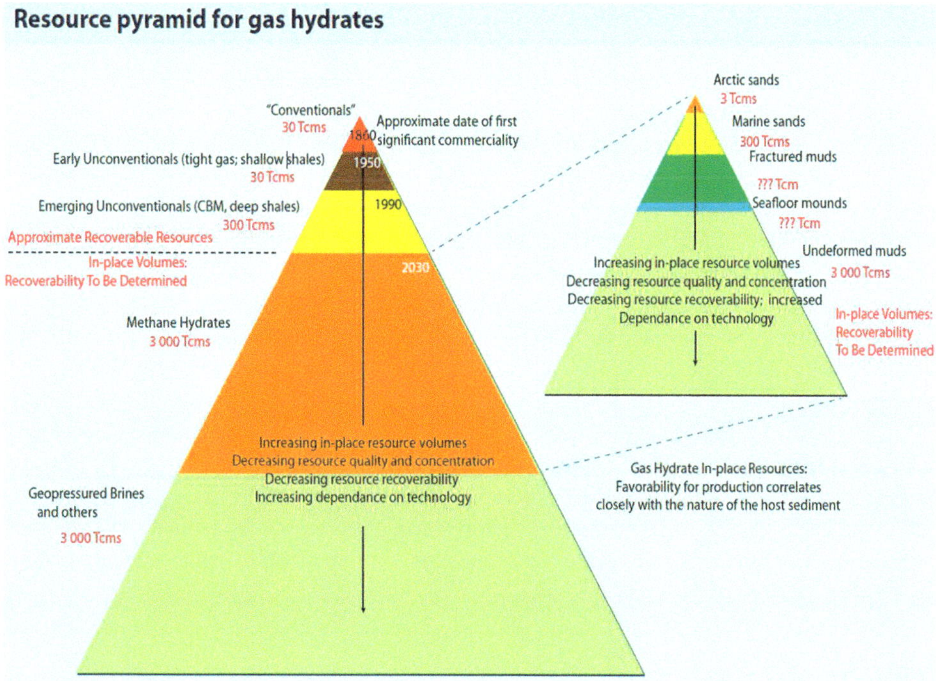

FIGURE 1.16

Pyramidal distribution of gas hydrate resources. A generalized resource pyramid for gas hydrates (right) is shown in relation to a resource pyramid for all gas resources (left). (From Boswell, R., Collett, T., 2006. DOE-NETL Fire in the Ice. Methane Hydrate Newsletter. Summer. 5. www.netl.doe.gov/mh/FITI06_Pyramid/.)

gas and oil that is preserved in rock where the primary component is shale is termed as shale gas. Unconventional gas reservoirs refer to low to ultra-low permeability, which mainly produces dry gas. There is no clear-cut scientific definition to distinguish between conventional and unconventional reservoirs. However, studies have characterized reservoirs with permeability >0.1 DM as conventional and those with permeability <0.1 DM as unconventional (Boyer et al., 2006). Globally, 32% of the total estimated natural gas resources are in shale formations, while 10% of estimated oil resources are in shale or tight formations. Fine-grained shale rock mainly formed due to the compaction of silt- and clay-sized particles and constitutes a major portion of the world's sedimentary rock deposit. However, gas occurs only in some localities where shales have rich organic content and organic matter is exposed to low oxygen levels. Such conditions prevail in many marine and lacustrine environments. Most of the organic matter, such as trees, plants, soil from the continental land mass (terrigenous) and animal and plant life in the oceans (pelagic), perished and deposited along with other sediments (see Chapter 2 for details) in the oceanic basins and major lakes. Depleted oxygen conditions in deeper parts of the ocean where anaerobic bacteria are found to occur (bacteria are not efficient in the consumption of organic matter) and less consumption of organic matter due to smaller food chain life in the shallower parts of the basin allow the organic matter to accumulate in sediments, making organic-rich sediments. Organic matter preservation in sediments seems to result from complex interdependent parameters and may be controlled by the site-specific bio-physicochemical conditions. Organic matter is preserved in ocean sediments after it meets its fate of degradation/respiration/remineralization due to several physical, chemical and biological factors operating in the ocean water column and near the seafloor environment. Under normal circumstances it seems that the organic matter that does not suffer remineralization is preserved. Most shales around the world were deposited in marine or restricted basin settings under conditions that allowed for the preservation of organic matter in the sedimentary rocks. Generation of methane and other gases and oil from the constituent organic matter is a complex process and is linked to physical and chemical conditions and organisms present at the site of deposition of organic-rich sediments (see Chapter 3 for details). Sedimentation has been a continuous process since the Precambrian period to recent times. Breaking up of continental landmasses during different geological periods leading to the formation of continental margins and major basins around the world provided a suitable environment for the deposition of sedimentary rocks in the oceanic basins. Events of warm and cold climatic conditions modified the erosional and depositional patterns over the entire globe, resulting in varying thicknesses of the sedimentary column and proportions of its constituent minerals.

Shale gas is found in many parts of the world in old reworked basins (Fig. 1.17). These reserves have been formed throughout the geological periods. However, prominent shale gas reserves are mainly confined to the Devonian Period. During this period warm climatic conditions led to melting of ice and an enormous amount of water flowed into the oceans resulting in a rise in sea levels across much of the

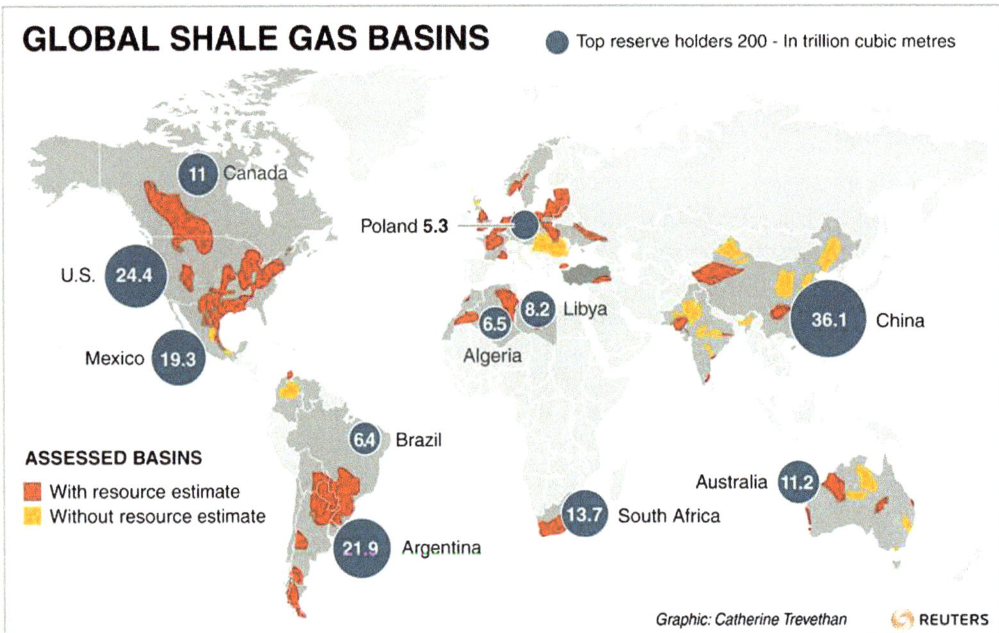

FIGURE 1.17

Worldwide distribution of shale gas resources. Present day mapped shale gas basins and probable global resource estimates of shale gas. With regard to the basins, they are differentiated by "with resource" and "without resource" estimates. (From http://blog.thomsonreuters.com/)

earth. Natural gas stored in shale gas reservoirs is found to exist either as free gas in the pores and fractures, sorbed gas in organic matter and on inorganic minerals, or dissolved gas in oil and water. Because of a lower permeability than tight gas and coal bed methane, the stratum becomes a seal for many conventional oil and gas reservoirs. Shale formations act both as source and reservoir for natural gas. The gas in shale is stored in pore spaces, in the fractures and absorbed gas. The presence of gas in different settings makes extraction of gas from the shale difficult. Abundant occurrences of shale gas have been reported from the United States, Canada and Asia Pacific (Perry et al., 1984).

It is estimated that about 23,000 TcF of natural gas may exist in the shale formations worldwide. Geochemical techniques such as total organic carbon (TOC), gas volume and capacity are being used extensively to assess the shale reservoir characteristics. TOC values show significant variation (2–10%) from one region to another. TOC values >4% are considered to be a good prospect (Alexander et al., 2011). Petrophysical properties generally used for the exploration of conventional gas reservoirs, such as gamma ray, resistivity, porosity and acoustic measurement, are also utilized for shale gas exploratory studies to assess porosity and gas concentration in pore spaces (increase in resistivity), fracture distribution and fluid migration.

Unconventional reservoirs require hydraulic fracturing, horizontal drilling and other techniques to make the gas flow feasible. Horizontal drilling is a technique that allows the wellbore to come into contact with significantly larger areas of hydrocarbon-bearing rock than in a vertical well. As a result of this increased contact, production rates and recovery factors can be increased. Hydraulic fracture stimulation, or "fracking", is a process through which a large number of fractures are created mechanically in the rock, thus allowing the natural gas and/or crude oil trapped in subsurface formations to move through those fractures to the wellbore from where it can then flow to the surface.

1.7.3 TIGHT OIL AND GAS

Unconventional resource plays introduce challenges that are not present for the development of conventional reservoir systems. These plays are driving evolutionary technological advances necessary to develop the resource base economically, particularly with respect to drilling and completing wells. The ability to efficiently exploit ultralow-permeability plays has invigorated the oil and gas industry around the globe. The transition from vertical to horizontal wells was spurred by the development of revolutionary techniques for drilling and completion. A worldwide unconventional resource map is shown in Fig. 1.18.

Tight sands are characterized by their low permeability (less than 0.1 mD) because of compaction, the nature of the fine sediments or pore space infilling. A generally accepted industry definition of "tight" is also those reservoirs that do not produce economic volumes of natural gas without assistance from massive stimulation treatments or special recovery processes and technologies. The tight sands reservoirs are mainly associated with compact rock types such as sandstones and limestones. It is assumed that tight sands occur in vast quantities worldwide but their quantum is not well defined. It is estimated that gas from tight sands contributes to about 30% of total domestic supply in the United States. The flow rates of tight sands are uneconomical and it takes sophisticated technologies eg, multistage hydraulic fracturing, to produce this unconventional resource. Higher commodity prices and development techniques including horizontal and directional drilling, underbalanced drilling, directional fracturing and carbon dioxide and nitrogen fracturing fluids have made tight sands exploration targets.

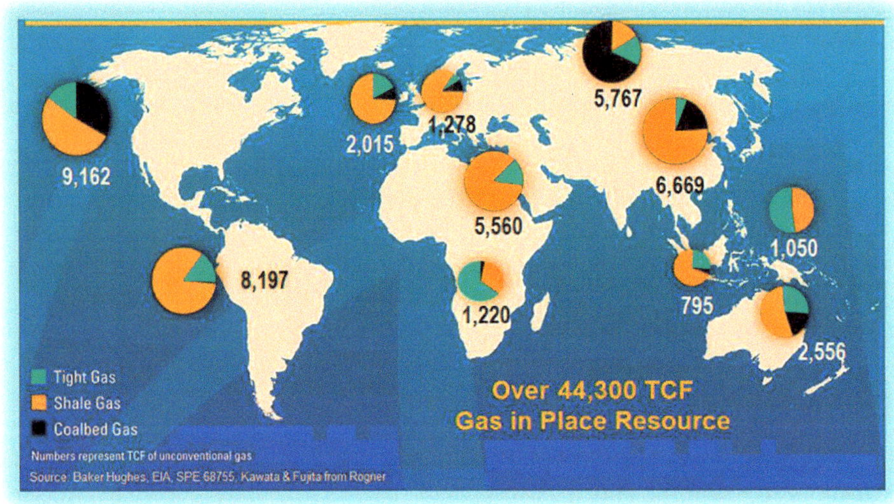

FIGURE 1.18

Worldwide distribution of tight gas, shale gas and coal bed methane. (From Baker Hughes, EIA, SPE 68755 Kawata & Fujita from Rogner.)

For tight oil/gas and shallow depth coalbed methane plays, a large number of wells are needed, and thus the environmental footprint is larger. To improve production from individual wells, a good understanding of the reservoir is needed so the wells can be optimally fractured and completed. Tight oil generally refers to the oil resources that have experienced short distance migration, are gathered in either tight sandstone, tight limestone or tight carbonate reservoir layers, that may be interbedded with or in close proximity to the source bed, where the overburden pressure matrix permeability is less than or equal to 0.1 mD, and where reservoir porosity is less than 10%. The oil contained within these reservoir rocks typically does not flow to the wellbore at economic rates without assistance from technologically advanced drilling and completion techniques.

Hydraulic fracturing of oil and gas reservoirs has been an accepted technique for enhancing production since its commercial introduction in 1949, and plays a major role in the development of many oil and gas fields globally. The combination of fracturing technology and horizontal drilling has made exploiting source rock reservoirs a thriving endeavor in North America. As a result of the explosion in demand for petroleum worldwide, the horizontal well technique is being increasingly applied in the international oil and gas industry. If the existing horizontal well data could be combined with seismic inversion on reservoir characterization, it would be an innovative approach.

REFERENCES

Alexander, T., Baihly, J., Boyer, C., Clark, B., Waters, G., Jochen, V., Le Calvez, J., Lewis, R., Miller, C.K., Thaeler, J., 2011. Shale gas revolution. Oilfield Review 23 (3), 40–55.

Anderson, D.L., 1994. Superplumes or supercontinents? Geology 29, 39–42.

Armitage, J.J., Collier, J.S., Minshull, T.A., 2010. The importance of rift history for volcanic margin formation. Nature. http://dx.doi.org/10.1038/nature09063.

Aubouin, J., 1989. Some aspects of the tectonics of subduction zones. Tectonophysics 160, 1–21.

Barazangi, M., Isacks, B.L., 1976. Spatial distribution of earthquakes and subduction on the Nazca plate beneath South America. Geology 4, 686–692.

Behn, M.D., 2002. The Evolution of Lithospheric Deformation and Crustal Structure from the Continental Margin to Oceanic Spreading Center (Ph.D. dissertation), p. 243.

Bond, G.C., Kominz, M.A., Sheridan, R.E., 1995. Continental terraces and rises. In: Busby, C.J., Ingersoll, R.V. (Eds.), Tectonics of Sedimentary Basins. Blackwell Science, Cambridge, MA, pp. 149–178.

Boswell, R., Collett, T., 2006. DOE-NETL Fire in the Ice. Methane Hydrate Newsletter. Summer. 5. www.netl.doe.gov/mh/FITI06_Pyramid/.

Boyer, C., Kieschnick, K., Suarez-Rivera, R., Lewis, R.E., Waters, G., 2006. Producing gas from its source. Oilfield Review. 36–49. http://geoworld.geoservices.com.

Brooks, D.A., Carlson, R.L., Harry, P.J., et al., 1984. Characteristics of back arc regions. Tectonophysics 102, 1–6.

Brun, J.P., Choukroune, P., 1983. Normal faulting, block tilting and decollement in a stretched crust. Tectonics 2, 345–356.

Burke, K., Dewey, J.F., 1973. Plume generated triple junctions: key indicators in applying plate tectonics to old rocks. J. Geol. 81, 406–433.

Byrne, D.E., Wang, W.H., Davis, D.M., 1993. Mechanical role of backstop in the growth of backarcs. Tectonics 12, 123–144.

Chapple, W.M., Tullis, T.E., 1977. Evolution of the forces which drive the plate. J. Geophys. Res. 82, 1967–1984.

Cloos, M., Shreve, R.L., 1988. Subduction-channel model of prism accretion, mélange formation, sediment subduction, and sediment erosion at convergent plate margins: implications and discussions. Pure Appl. Geophys. 128, 501–545.

Coats, R.R., 1962. Magma type and crustal structure in Aleutian Arc. In: Macdonald, G.A., Kuno, H. (Eds.), Crust of the Pacific Basin. Geophysical Monograph Series, Vol. 6. AGU Publications, Washington DC, pp. 92–109.

Coltice, N., Phillips, B.R., Bertrand, H., Ricard, Y., Rey, P., 2007. Global warming of the mantle at the origin of flood basalts over supercontinents. Geology 35, 391–394.

Corti, G., Wijk, J.W., Bonini, M., et al., 2003. Transition from continental break-up to punctiform seafloor spreading: how fast, symmetric and magmatic. Geophys. Res. Lett. http://dx.doi.org/10.1029/2003GL017374.

Curray, J.R., Moore, D.G., 1974. Sedimentary and tectonic process in the Bengal deep-sea fan and geosyncline. In: Burke, C.A., Drake, C.L. (Eds.), The Geology of the Continental Margins. Springer Verlag, Berlin, p. 1009.

Davie, M.K., Zatsepina, O.Y., Buffet, B.A., 2004. Methane solubility in marine hydrate environment. Marine Geol. 203, 117–184.

Davis, E.E., Hyndman, R.D., 1989. Accretion and recent deformation of sediments along northern Cascadia subduction zone. Geol. Soc. Am. Bull. 101, 1465–1480.

Dewey, J.F., 1986. Diversity in the lower continental crust. In: Dawson, J.B., Carswell, D.A., Hall, J., Wedpohl, K.H. (Eds.), The Nature of the Lower Continental Crust. Geological Soc. London, pp. 71–78.

Dickinson, W.R., 1995. Forearc basins. In: Busby, C.J., Ingersoll, R.V. (Eds.), Tectonics of Sedimentary Basins. Blackwell Science, (University of California, California) USA, pp. 221–261.

Dickinson, W.R., 1974. Plate tectonics and sedimentation. Soc. Econ. Paleontol. Mineral. Spl. Publ. 22, 1–17.

Dietz, R.S., Holden, J.C., 1970. Reconstruction of Pangaea. Breakup and dispersion of continents: Permian to present. J. Geophys. Res. 75, 4939–4956.

Draut, A.E., Clift, P.D., 2013. Differential preservation in the geologic record of island-arc sedimentary and tectonic processes: Earth-Science Reviews 116, 57–84.

Dunbar, J.A., Sawyer, D.S., 1988. Continental rifting at pre-existing lithospheric. Nature Weakness 333, 450–452.

Eldholm, O., Tsikalas, F., 2004. Scientific aspects of the continental shelf. In: Nordquist, M.H., Moore, J.H., Heidar, T.H. (Eds.), Legal and Scientific Aspects of Continental Shelf Limits. Kninklijke Brill, Leiden, The Netherlands, p. 467.

Eldholm, O., Skogseid, J., Planke, S., Gladczenko, T.P., 1995. Volcanic margin concepts. In: Banda, E., Talwani, M., Torne, M. (Eds.), Rifted Ocean-continent Boundaries. Kluwer Academic Publishers, London, pp. 1–16.

Emery, K.O., 1968. Relict sediments on the continental shelf of the world. Am. Asso. Petro. Geol. Bull. 52, 445–464.
Forsyth, D., Uyeda, S., 1975. On the relative importance of the driving forces of plate motion. Geophys. J. Roy. Ast. S. 43, 163–200.
Forsyth, D.W., 1980. Comparison of mechanical models of the oceanic lithosphere. J. Geophys. Res. 85, 6364–6368.
Foucher, J.P., LePichon, X., Sibuet, J.C., 1982. The ocean-continent transition in the uniform stretching model: role of partial melting in the mantle. Philos. Trans. Roy. Soc. A. 305, 27–43.
Gill, J.B., 1981. Orogenic Andesites and Plate Tectonics. Springer Verlag, Berlin, p. 390.
Gilluly, J., 1963. Tectonic evolution of western United States. Geol. Soc. Lond. Quat. 119, 133–174.
Gordon, R.G., Stein, S., 1992. Global tectonics and space geodesy. Science 256, 333–342.
Hales, A.L., 1969. Gravitational sliding and continental drift. Earth Planet Sci. Lett. 6, 31–34.
Hamilton, W., 1979. Tectonics of the Indonesian Region. Professional Paper 1078. USGS Publications, USA, p. 345.
Hasbe, K., Fujii, N., Uyeda, S., 1970. Thermal processes under island arcs. Tectonophysics 10, 335–355.
Heezan, B.C., Tharp, M., Ewing, M., 1959. The floor of oceans: 1 the North Atlantic. Geol. Soc. Am. Spec. Paper 65, 1–122.
Hellinger, S.J., Sclater, J.G., 1983. Some comments on two-layer extensional models for the evolution of sedimentary basins. J. Geophys. Res. 88, 8251–8269.
Hilde, T.W.C., Uyeda, S., 1983. Trench depth: variation and significance. In: Hilde, T., Uyeda, S. (Eds.), Geodynamics of the Western Pacific-Indonesian Region. Geodyn. Series, Vol. 11. American Geophysical Union and Geological Society of America, Washington DC, pp. 75–89.
Hilde, T.W.C., 1983. Sediment subduction versus accretion around the Pacific. Tectonophysics 93, 381–397.
Holbrook, W.S., Kelemen, P.B., 1993. Large igneous province on the U.S. Atlantic margin and implications for magmatism during continental breakup. Nature 364, 433–436.
Hsui, A.T., Toksoz, M.N., 1981. Back-arc spreading: trench migration, continental pull, or induced convection? Tectonophysics 74, 89–98.
Hussong, D.M., Uyeda, S., 1981. Tectonic processes and history of Mariana arc: a synthesis of the results of the DSDP Leg 60. Initial Rep. DSDP 60, 909–929.
von Huene, R., Lallemand, S., 1990. Tectonic erosion along the Japan and Peru convergent margins. Geol. Soc. Am. Bull. 102, 704–720.
von Huene, R., Scholl, D.W., 1991. Observations at convergent margins concerning sediment subduction, subduction erosion, and the growth of the continental crust. Rev. Geophys. 3, 279–316.
Ingersoll, R.V., Busby, C.J., 1995. Tectonics of sedimentary basins. In: Busby, C.J., Ingersoll, R.V. (Eds.), Tectonics of Sedimentary Basins. Blackwell Science, USA, pp. 1–51.
Ingersoll, R.V., 1988. Tectonics of sedimentary basins. Geol. Soc. Am. Bull. 100, 1704–1719.
Jarrad, R.D., 1989. Relation among subduction parameters. Rev. Geophys. 24, 217–284.
Karig, D.E., Sherman III, G.F., 1975. Subduction and accretion in trenches. Geol. Soc. Am. Bull. 86, 377–389.
Karig, D.E., 1971. Origin and development of marginal basins in the western Pacific. J. Geophys. Res. 76, 2542–2561.
Karig, D.E., 1974. Tectonics erosion of trenches. Earth Planet Sci. Lett. 21, 209–212.
Keen, C.E., Beaumont, C., 1990. Geodynamics of rifted continental margins. In: Keen, M.J., Williams, G.L. (Eds.), Geology of the Continental Margin of Eastern Canada. The Geology of North America, Vol. I-1. Geological Society of America, pp. 391–472. Published Prentis-Hall, Eaglewood Cliff, NJ.
Kennett, J.P., 1982. Marine Geology. Prentic Hall, Englewood Cliff, NJ, p. 813.
King, S.D., Anderson, D.L., 1995. An alternative mechanism for flood basalt formation. Earth Planet Sci. Lett. 36, 269–279.
Kinsman, D.J.J., 1975. Rift valley basins and sedimentary history of trailing continental margins. In: Fisher, A.G., Judson, S. (Eds.), Petroleum and Global Tectonics. Princeton University Press, Princeton, pp. 83–126.

Kusznir, N.J., Park, R.G., 1987. The extensional strength of continental lithosphere: its dependence on geothermal gradient and crustal composition and thickness. Geol. Soc. Lond. Spec. Pub. 28, 35–52.

Kvenvolden, K.A., 1993. Gas hydrate as potential energy resource – a review of their methane content. In: Howell, D.G. (Ed.), The Future of Energy Gases. Professional Paper 1570. USGS Publications.

Kumar, P., Yuan, X., Ravi Kumar, M., et al., 2007. The Rapid Drift of the Indian Plate. http://dx.doi.org/10/nature06214.

Leeder, M.R., 1995. Continental rifts and proto-oceanic rift troughs. In: Busby, C.J., Ingersoll, R.V. (Eds.), Tectonics of Sedimentary Basins. Blackwell, Oxford, UK. p. 199–148.

Lister, G.S., Etheridge, M.A., Symonds, P.A., 1986. Application of the detachment fault model to the formation of passive continental margins. Geology 14, 246–250.

Mann P. Tectonics setting of world's giant oil and gas fields, HGS Northsiders Committee: Invited talk, 19 October 2004. http://www.hgs.org/articles. The article was accessed in January–February, 2015.

Marsaglia, K.M., 1992. Petrography and provenance of volcanoclastic sands recovered from the Izu-Bonin arc. In: Taylor, B., Fujioka, K., , et al. (Eds.), Proceedings of the Ocean Drilling Program, Scientific Results. Ocean Drilling Program, College Station, TX, Vol. 126, pp. 139–154.

McKenzie, D., 1978. Some remarks on the development of sedimentary basins. Earth Planet Sci. Lett. 40, 25–32.

McKenzie, D.P., Bickle, M.J., 1988. The volume and composition of melt generated by extension of lithosphere. J. Petrol. 29, 625–679.

McKenzie, D.P., 1969. Speculation on the consequences of plate motion. Geophys. J. Roy. Astr. Soc. 18, 1–32.

Meissner, R., 1986. The Continental Crust. Academic Press, New York, pp. 162–187.

Menzies, M.A., Klemperer, S.L., Ebinger, C.J., et al., 2002. Characteristics of volcanic rifted margins. In: Menzies, M.A., Klemperer, S.L., Ebinger, C.J., Baker, J. (Eds.), Volcanic MarginsSpec. Pap. Geol. Soc. Am., Vol. 362, pp. 1–14.

Morgan, P., Burke, K., 1985. Continental plateaus. Tectonophysics 119, 137–151.

Morgan, W.J., 1983. Hotspot tracks and the early rifting of the Atlantic. Tectonophysics 94, 123–139.

Murauchi, S., Ludwig, W.J., 1980. Crustal structure of the Japan trench: the effect of subduction of the ocean crust. Initial Rep. DSDP 56/57, 463–470.

Mutter, J.C., Buck, W.R., Zehnder, C.M., 1988. Convective partial melting: a model for the formation of thick basaltic sequences during the initiation of spreading. J. Geophys. Res. 93, 1031–1048.

Mutter, J.C., Talwani, M., Stoffa, P.L., 1982. Origin of the seaward-dipping reflectors in the ocean crust off the Norwegian margin by "subaerial sea-floor spreading". Geology 10, 353–357.

Nielsen, T.K., Hooper, J.R., 2004. From rift to drift: mantle melting during continental breakup. Geochem. Geophys. Geosyst. 5. http://dx.doi.org/10:1029/2003GC000662.

Perry, R.H., Green, D.W., Maloney, J.O., 1984. Perry's Chemical Engineer's Handbook. McGraw-Hill Boo.

Pitman, W.C., 1978. Relationship between eustacy and stratigraphic sequences of passive margins. Geol. Soc. Am. Bull. 89, 1389–1403.

Platt, J.P., 1989. The mechanics of frontal imbrication: first order analysis. Geol. Rundsch. 77, 577–589.

Pollack, H.N., 1986. Cratonization and thermal evolution of the mantle. Earth Planet Sci. Lett. 80, 175–182.

Reymer, A., Schubert, G., 1984. Phanerozoic addition rates to the continental crust and crustal growth. Tectonics 3, 63–77.

Ringwood, A.E., 1977. Petrogenesis in the Island arc systems. In: Talwani, M., Pitman III, W.C. (Eds.), Maurice Ewing Series, Vol. 1. American Geophysical Union, Washington DC, pp. 311–324.

Scholl, D.W., Marlow, M.S., Cooper, A.K., 1977. Sediment subduction and offscraping at Pacific margins. In: Talwani, M., Pitman III, W.C. (Eds.), Island Arcs Deep Sea Trenches and Back-arc Basins. Maurice Ewing Ser 1. AGU, Washington DC, pp. 199–210.

Scholl, D.W., Vallier, T.L., 1981. Subduction and rock record of Pacific margins. In: Carey, S.W. (Ed.), Expanding Earth Sym., 1981. University of Tasmania, Hobart, Tasmania, Australia, pp. 235–245.

Scholl, D.W., von Huene, R., Vallier, T.L., Howell, D.G., 1980. Sedimentary masses and concepts about tectonic processes at underthrust ocean margins. Geology 8, 564–568.
Sengor, A.M., Burke, K., 1978. Relative timing of rifting and volcanism on Earth and its tectonic implications. Geophys. Res. Lett. 5, 419–421.
Shipley, T.H., Moore, G.H., 1986. Sediment accretion, subduction and dewatering at the base of the trench slope off Costa Rica: a seismic reflection view of decollement. J. Geophys. Res. 91, 2019–2028.
Silver, E.A., Ellis, M.J., Breen, N.A., Shipley, T.H., 1985. Comments on the growth of accretionary wedges. Geology 13, 6–9.
Smith, G.A., Landis, C.A., 1995. Intra-arc basins. In: Busby, C.J., Ingersoll, R.V. (Eds.), Tectonics of Sedimentary Basins. Tectenophysics, Netherlands, pp. 263–298.
Spohn, T., Schubert, G., 1983. Convective thinning of lithosphere: a mechanism for rifting and mid-plate volcanism on Earth, Venus and Mars. Tectonophysics 94, 67–90.
Stevenson, A.J., Embley, R., 1987. Deep-sea fan bodies, terrigenous turbidite sedimentation and petroleum geology, Gulf of Alaska. Circum-Pacific Counc. Energy Mineral Resour. Earth Sci. Ser. 6, 503–522.
Suess, E., 1885. Das Antlitz der Erde, 1A. Prague.
Symonds, P.A., Eldholm, O., Mascle, J., Moore, G.F., 2002. Characteristics of continental margins. In: Cook, P.J., Carleton, C.M. (Eds.), Continental Shelf Limits: The Scientific and Legal Interface. Oxford University Press, New York, pp. 25–63.
Tamaki, K., Honza, E., 1991. Global tectonics and formation of marginal basins: role of western Pacific. Episodes 14, 224–230.
Taylor, B., Karner, G.D., 1983. On the evolution of marginal basins. Rev. Geophys. Space Phys. 21, 1727–1741.
Underwood, M.B., Moore, G.F., 1995. Trenches and trench-slope basins. In: Busby, C.J., Ingersoll, R.V. (Eds.), Tectonics of Sedimentary Basins. Blackwell, Oxford, UK, pp. 179–219.
Uyeda, S., Kanamori, H., 1979. Back-arc opening and the mode of subduction. J. Geophys. Res. 84, 1049–1061.
Uyeda, S., 1982. Subduction zones: an introduction to comparative subductology. Tectonophysics 81, 133–159.
Vening Meinesz, F.A., 1964. The Earth's Crust and Mantle. Elsevier, Amsterdam, p. 124.
Vink, G.E., Morgan, W.J., Zhao, W.L., 1984. Preferential rifting of the continents: a source of displaced terranes. J. Geophys. Res. 89, 10072–10076.
Westbrook, G.K., Ladd, J.W., Bruhl, P., Bangs, N., Tiley, G.J., 1988. Cross section of the accretionary wedge: Barbados Ridge complex. Geology 16, 631–635.
Wegener, A., 1924. The Origin of Continents and Oceans (Entstehung der Kontinente und Ozeane). Methuen & Co.
Whitmarsh, R.B., Manatschal, G., Minshull, T.A., 2001. Evolution of magma-poor continental margin from rifting to seafloor spreading. Nature 413, 150–154.
Williams, H., McBirney, A.R., 1979. Volcanology. Freeman, Cooper & Co., San Francisco, p. 397.
Yoshida, M., 2010. Preliminary three-dimensional model of mantle convection with deformable mobile continental lithosphere. Earth Planet Sci. Lett. 295, 205–218.
Zatman, S., Gordon, R.G., Mutnuri, K., 2005. Dynamic of diffuse oceanic plate boundaries: insensitivity to rheology. Geophys. J. Int. 162, 239–248.

CHAPTER

GENERATION OF METHANE IN EARTH

2

"Nothing in life is to be feared, it is only to be understood. Now is the time to understand more, so that we may fear less."

Marie Curie

CHAPTER OUTLINE

2.1 INTRODUCTION

Methane is a naturally occurring organic compound and can be considered as a future energy source. It is an essential element of the earth's atmosphere. Human activities, including hydroelectric dams, composting, cultivation of rice, and semiconductors in computer and mobile devices, have increased the amount of methane released into the atmosphere. Consequently, methane is of considerable interest as a potential energy source and because of its possible role in global climate change. A lack of state of the art technologies and the low quantum of energy generation from renewable resources make these unviable alternatives to the present energy scenario. The uneconomical production of hydrogen fuels makes the use of hydrogen as an energy carrier a remote possibility in the near future. Methane hydrates are expected to become an important source of energy in bridging carbon-intensive fossil energies. Of all the hydrocarbons being utilized today, methane stands out as the cleanest energy with the least emissions. The potential use of gas hydrates as a viable alternative to declining or stagnant fossil-fuel energy resources has provided much impetus towards its exploration and technological development for its exploitation, particularly in energy-starved countries such as Japan and India. In recent times, gas

Geological Controls for Gas Hydrates and Unconventionals. http://dx.doi.org/10.1016/B978-0-12-802020-3.00002-3

hydrates with their immense resource gas potential are being considered a viable alternative to overcome the present energy crisis.

The term "natural gas hydrates" indicates that methane gas occupies most of the cavities in the hydrates. The nomenclature is in accordance with the petroleum industry, where methane is formed under subsurface earth by natural processes (Hunt, 1996). The methane gas contained in the gas hydrates burns more efficiently with fewer emissions of hazardous gases when compared to conventional petroleum and coal. On the utilities front, methane gas is replacing conventional petroleum as feedstock in power plants as well as in the running of cars and other conveyance modes (Sloan, 2003). Gas hydrates are stable in shallow regions of the earth where pressure–temperature regimes are favorable for their formation.

2.2 WHAT ARE GAS HYDRATES?

Gas hydrates are ice-like crystalline materials that have a structured water molecular framework linked together to create cages, in which guest gas molecules can be trapped. They are considered to be a class of clathrates (clathratus – meaning cages or gratings; Brown, 1962). Clathrates are inclusion chemical compounds in which molecules of chemical components get trapped in the cavities of the host molecule (Webster, 1994). In terms of inorganic chemistry a "hydrate" is a stoichiometric compound where water is one of the integral parts of the crystal structure and the guest molecule may be any other chemical compound. At the standard atmospheric pressure freezing point of water (0°C), the molecules of water are arranged in an ordered (hydrogen bonding) manner and occupy more spaces creating a hexagonal structure. At higher pressures, during cooling the water molecules get solidified with complex structures at temperatures above the freezing point. These complex structures unlike ice are inherently unstable, and have a network of small and large cavities that develops in the crystal lattice. The presence of certain other compounds causes the crystalline structure to stabilize. Molecules of water are the host and those of other compounds are the guest molecules. The guest molecules are considered hydrate formers. Guest molecules in the hydrate fill the cavities without chemically bonding with the crystal structure. The structure is held together not by the electrostatic forces of hydrogen bonding but by other weak forces called van der Waals forces and the guest molecules are free to rotate in the cages (Carroll, 2002). When the guest molecules occupy sufficient numbers of cavities the crystal structure becomes stable. The size and the geometry of the cages (voids) and the availability of gases determine the formation of hydrates. Hydrates are formed under suitably high pressure and subzero temperature regimes when sufficient quantities of low molecular weight gases (formers) and water are present. Formers are low molecular weight gases, such as hydrogen and helium, which are smaller than the small cavities (3.8 Å), whereas high molecular weight gases, such as pentane, hexane, and paraffin hydrocarbons, are bigger than the large cavities (>9 Å), which prevents the formation of hydrates under natural conditions. However, in recent times attempts have been made to synthesize hydrates with low molecular weight gases in laboratories. Most low molecular weight hydrocarbon and nonhydrocarbon gases, such as oxygen (O_2), hydrogen (H_2), nitrogen (N_2), carbon dioxide (CO_2), methane (CH_4), hydrogen sulfide (H_2S), argon (Ar), and krypton (Kr), as well as some higher hydrocarbons and freons are trapped in the cavities and form hydrates at suitable temperatures and pressures. Without the support of the trapped molecules, the lattice structure of hydrate clathrates would collapse into a conventional ice crystal structure or liquid water. When the cavities are fully occupied by guest molecules of single chemical components then the composition can be expressed in terms of definite formulae. However, in nature, depending upon the physical conditions of pressure, temperature, and availability of suitable gases, only parts of the void are filled with guest molecules; this is called the degree of saturation. The maximum number of gas molecules that can be trapped depends on the geometry of the crystal (free space available in the lattice of water molecules), pressure, and temperature. The hydrate sample can hold more gases under low temperatures and/or higher pressures. Based on the degree of saturation, the water molecules can hold 70–160 volumes of gases. The terms "clathrate hydrates" or "gas clathrates" generally refer to crystalline structures such as ice whereas gas hydrates are usually crystalline structures where low molecular weight hydrocarbon gases such as methane (CH_4), ethane (C_2H_6), propane (C_3H_8), and pentane (C_5H_{12}) are trapped in the water molecules (Hardage and Roberts, 2006). Gas hydrates of low order molecular hydrocarbons (<9 Å), such as methane, ethane, and others, are formed under low temperatures (<300 K) and moderate pressures (>0.6 MPa) (Sloan, 2003). In the case of gas hydrates, most of the cavities are occupied by methane gas molecules but some cavities are filled with ethane and other gases. As a consequence, these hydrates are considered nonstoichiometric compounds, which means they cannot be expressed with a simple formula.

2.3 WHERE ARE GAS HYDRATES FOUND?

Gas hydrates in subsurface earth may exist as individual grains or particles distributed throughout the sediments but they may also cement sediment grains together forming nodules or form laminate layers of different shapes and sizes with dimensions of just a few centimeters or less (Collett, 2003). Two-dimensional samples such as laminate or layers were associated with fine-grained sediments. Three-dimensional samples such as granules and nodules also occurred in fine-grained sediments and with mixed grain sizes. Hydrate cementing was more prevalent in coarse-grained samples. In many cases the difference between such shapes is based only on the orientation of the hydrate within the sedimentary bedding. Gas hydrates primarily occupy the pore spaces in sediments and cement them. They also form in secondary voids, in fractures and joints. Hydrates during growth create their own space by disturbing the sediment matrix (Suess et al., 1999). Gas hydrate deposits occur in different shapes and sizes, such as: veins and layers of comparable size (Fig. 2.1a,b); those with spherical to oblate features, sometimes with edges, which are considered to be nodules (Fig. 2.1c); and massive hydrates (Fig. 2.1d), which on a larger scale probably look like a lens, an example of which was obtained during drilling over the Blake Ridge (Site 570: Shipboard Scientific Party, 1985).

2.4 METHANE GENERATION IN SUBSURFACE EARTH

Carbon is the fourth most abundant element on the earth's surface and is most stable individually or in combination with other elements forming covalent bonds, i.e. sharing electrons with itself and other elements. It is readily reduced with hydrogen or oxidized with oxygen to form the most common carbon compounds in the earth's crust: methane and carbon dioxide (Hunt, 1996). Carbon is unique as it is the basic element in all life-forms, with its ability to form long carbon chains, rings, and complex bridged structures. Most of the carbon on earth is concentrated in sedimentary rocks in the form of organic carbon formed by biological processes over geological time (reduced carbon 18%), and a major portion as carbonate (oxidized carbon 82%) according to Tissot and Welte (1978). Primary organic matter is formed directly from the atmospheric reservoir by terrestrial plants or photosynthesis of marine plants from dissolved carbon dioxide in the hydrosphere. Photosynthesis is the preliminary stage for the production of organic matter and indirectly plays a significant role in determining the geochemical characteristics of petroleum. The simple organic molecule formed by photosynthesis is

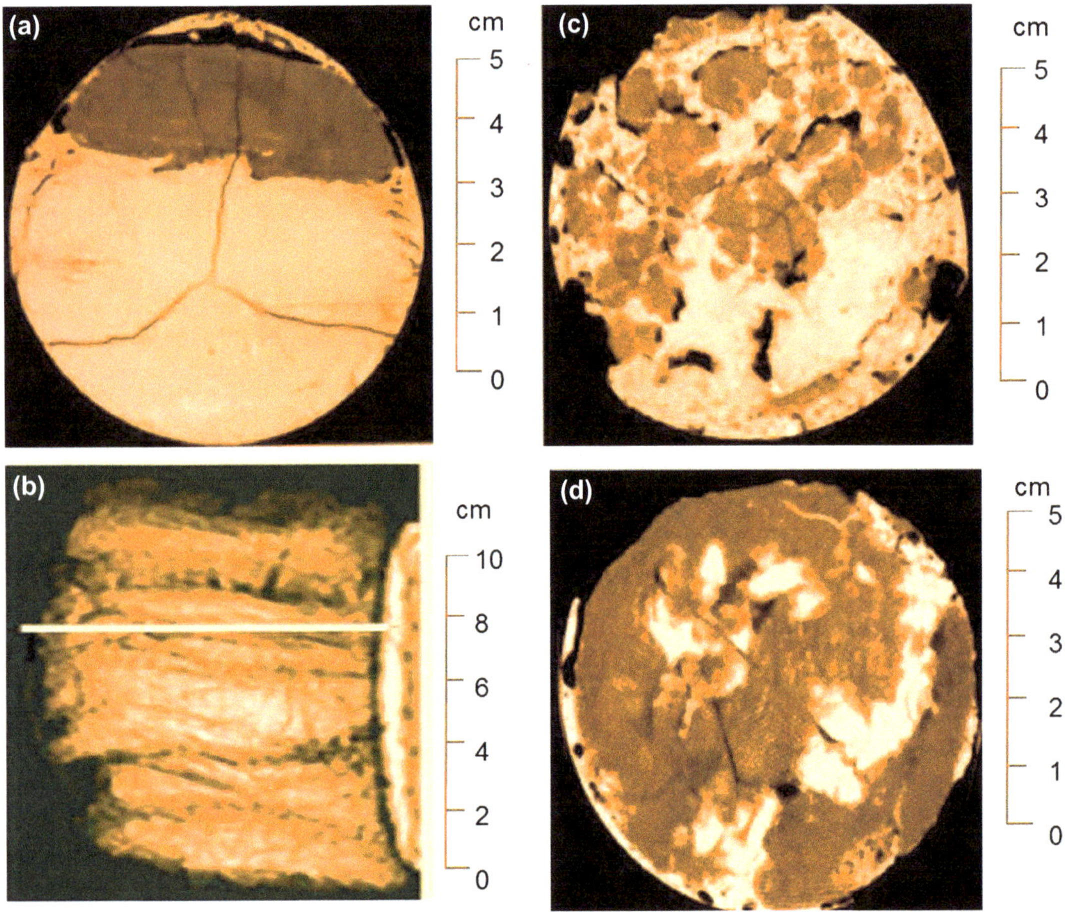

FIGURE 2.1

Various gas hydrate deposits of different shapes and sizes, such as veins and layers of comparable size (a, b), spherical to oblate features, sometimes with edges, which are considered as nodules (c), and massive hydrates (d), which on a larger scale probably look like a lens, were obtained during drilling over the Blake Ridge (Site 570: Shipboard Scientific Party, 1985).

glucose. Plants on land and phytoplankton in the oceans are primary producers, which photosynthesize to provide a source of energy to secondary consumers for their survival and growth through biosynthesis of organic compounds. Ultimately all living creatures, both primary producers and primary and secondary consumers, die and may be deposited in some aquatic environment. In the aquatic environment, the precipitation or deposition of the remains of dead organisms (shells and skeletons) leads to the formation of carbonates. Phytoplanktons (plants) utilize carbon dioxide in the form of carbon for cell growth, whereas zooplanktons (animals) eat plants and excrete excess carbon as carbon dioxide. Dead organisms are oxidized to carbon dioxide either by the metabolic processes of microbes or by chemical reactions. During its journey from the sea surface to its burial in deep sediment, organic

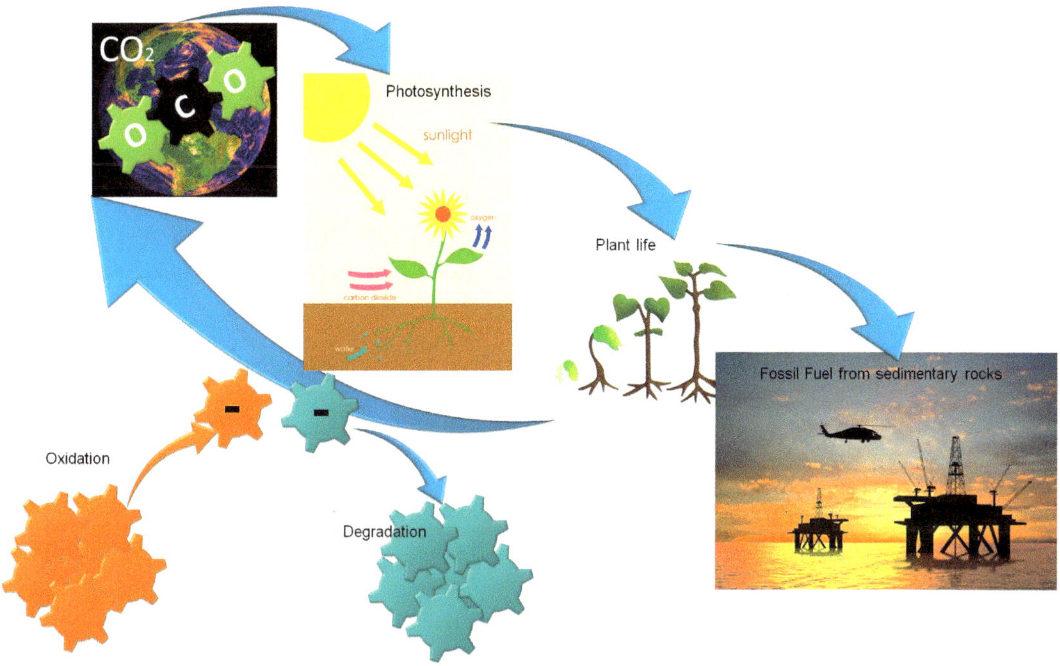

FIGURE 2.2

Carbon cycle in the biosphere; the main source of carbon is by means of photosynthesis and oxidation. Only a small fraction of the carbon is carried in the subsurface earth.

material undergoes considerable alteration. Only a small portion (<0.1%) of organic carbon from the photosynthesis cycle is preserved in sedimentary rocks and an even smaller percentage of this is ultimately converted to oil and gas; the majority of carbon is released into the atmosphere in the form of carbon dioxide (Fig. 2.2), maintaining the equilibrium in the atmosphere and hydrosphere (Hunt, 1996).

Organic carbon found on land is mainly the result of decomposition of plants; major carbon accumulations are found in surface detritus and peat. The subsurface layer under the landmass also contains vast accumulations of organic carbon and these are associated with fossil fuels (oil, coal, and natural gas reservoirs). In marine regions, organic carbon occurs in surface biota and is dissolved in surface waters, intermediate water depths, and deep waters. The largest pool of organic carbon occurs as dispersed organic matter, terrestrial sediments, and sedimentary rocks. Gas hydrates constitute an important component of marine resources. Natural gas is the gaseous phase of petroleum, which is a naturally occurring mixture of hydrocarbon gases that is highly compressible and expansible. Typically a natural gas reservoir contains methane (CH_4) as the predominant constituent (70–100%) with smaller contributions from ethane and lower concentrations of higher hydrocarbon gases such as propane, butane (C_4H_{10}), pentane, and traces of some other constituent gases (Tissot and Welte, 1978). The reservoir may also hold some nonhydrocarbon gases such as carbon dioxide, hydrogen sulfide, nitrogen, and helium (He). Methane gas is the potential primary resource which occurs as natural gas in the reservoir, in its wet form in association with oil, or as the main constituent in gas hydrates. It is generated in subsurface earth as a result of complex biogeochemical and physical processes. It is formed as an end product of alteration of organic carbon due to metabolic activity associated with microorganisms for their survival and cell growth, and alteration of organic matter due to thermal cracking.

2.4.1 TERMINOLOGY

Methane in subsurface earth is produced as the end product of biological, chemical, physical, and geological processes and reactions by different agencies. It is found in association with coal beds, petroleum reservoirs, gas hydrates, and as reservoirs of natural gas (Rice, 1993). Although the mechanics of methane production are not completely understood, it has received a great deal of attention by virtue of its importance in the anaerobic digestion process as practiced in sewage treatment, decomposition of plant material and other organic wastes, and in the proper understanding of deep subsurface microbial processes and the biogeochemical role of biota (Kotelnikova, 2002).

The first few meters of sediment just below the water-sediment contact zone represent the interface through which organic matter passes from the biosphere to the geosphere. During the sedimentation processes and later in young sediments, organic material is subjected to alteration by various degrees of microbial and chemical actions. Depending on the activity of subsurface agencies such as microorganisms, the depth of the buried organic matter-rich sediments, and chemical reactions taking place in deeper parts of the earth at elevated temperatures, different nomenclatures (terminology) have been proposed to distinguish the elements and processes that govern the formation of methane in subsurface earth (Fig. 2.3). Methane derived from decomposition of organic matter is biogenic as opposed to the abiogenic methane formed by chemical processes that take place in a deeper part of the earth in the absence of organic matter. Biogenic methane has been referred to in the literature as methane derived from bacterial origin (Rice and Claypool, 1981; Martens et al., 1986). A much better term, microbial methane, is advocated (Coleman et al., 1988) to distinguish methane derived from the anaerobic bacterial activity at shallow depths from other processes operating in subsurface earth for methane generation. Methane produced by thermochemical degradation, i.e., thermal degradation and cracking processes at elevated temperatures (50–150°C) and extended periods of heating (Rice, 1993), is termed as thermogenic or thermal process in accordance with the terminology adopted in the oil industry for natural gases (Schoell, 1988; Wiese and Kvenvolden, 1993). It is estimated that microbial methane contributes 20% and thermogenic methane about 80% to the world's natural gas reserves. Abiogenic methane refers to methane which has its origin in the recombining of organic carbon and hydrogen-rich gases during chemical reactions and involves no biological assistance. Carbon and hydrogen may originate from primordial material deep in the earth's crust or mantle. Chemical reactions such as that of graphite with water under suitable temperature conditions may lead to the formation of methane (Howell et al., 1993). The genesis of abiogenic methane is widely debated (Hunt, 1996; Gold, 1993). In the absence of clear signatures in the identification of subsurface abiogenic methane, the quantum of methane from such sources remains speculative.

2.4.2 BIOLOGICAL PROCESSES

The study of methane on earth reveals that 1700 parts per billion (ppb) of this greenhouse gas in the earth's atmosphere is produced by biological processes. Nonbiological processes such as volcanism contribute to less than 1% of methane generation. Inside the earth, insoluble parts of decomposing

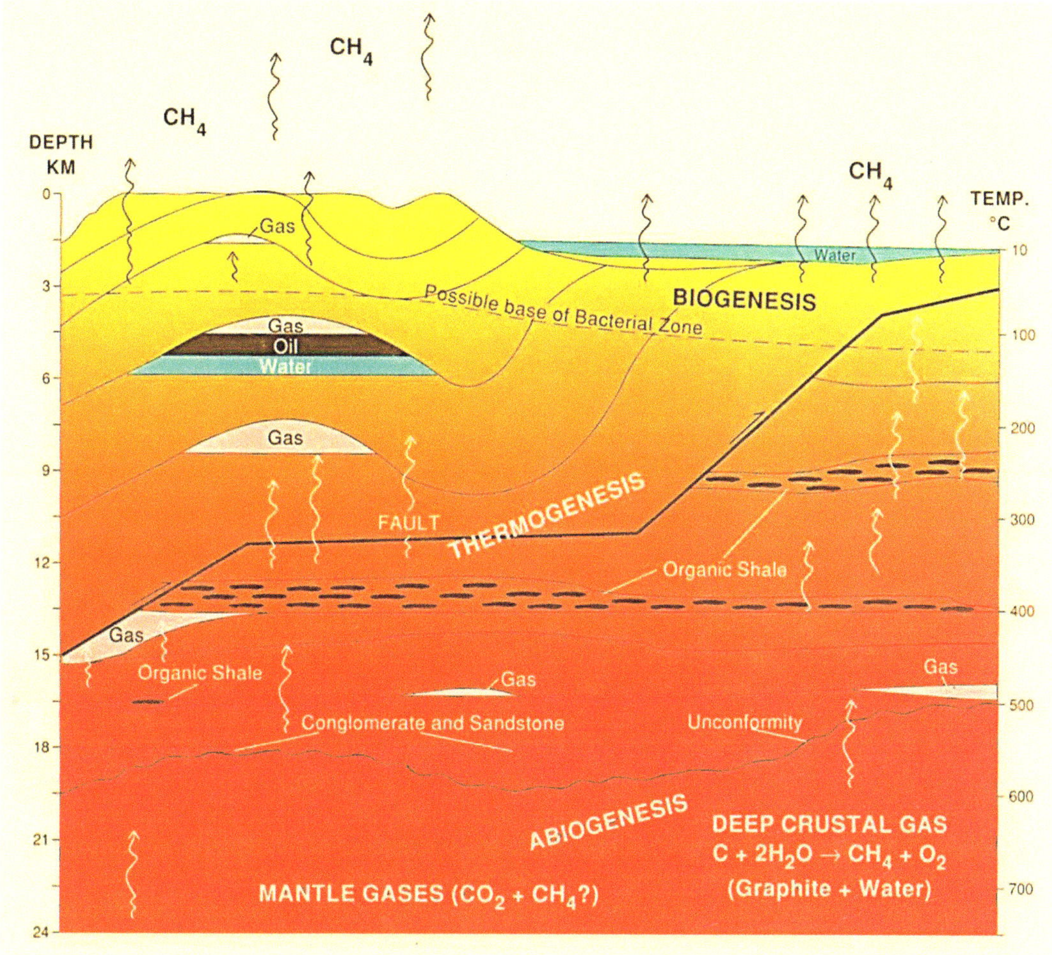

FIGURE 2.3

Different processes operate on organic matter at different levels for the generation of methane in the shallow regions of the earth, which is termed as biogenic/microbial/ bacterial and thermogenic methane. Methane in deeper parts of the earth is generated by the chemical reaction of water with graphite. (From Howell et al., 1993.)

buried plants convert into kerogen, which then breaks down through thermal "cracking" to generate methane together with longer-chain hydrocarbons (Tenenbaum, 2005) such as ethane, propane, and butane. A significant quantity of methane is generated by anaerobic microbes that are tolerant of extremely high temperatures called methanogens, which are also called "extremophiles" due to their tendency to prosper under extreme acidity, alkalinity, or saltiness. Methanogens or extremophiles can be found anywhere on earth in the absence of oxygen. Researchers at the University of Michigan recently concluded that methanogens living on wetlands produce about 21% of the methane in the earth's atmosphere, and those in the guts of cow together with other ruminants produce about 20%. Microbes in termites and similar organisms produce 15% of atmospheric methane, and those in rice paddies, about 12%. Other major sources of methane include natural gas releases and biomass burning.

Zeeya Merali (2004) proposed that this untapped fossil-fuel reserve could be hidden deep within mother earth due to its production by nonbiological processes but the idea remains controversial. Indiana University researcher Henry Scott suggested that the hydrocarbons can form from nonbiological material and be stable, deep within the earth. Because the mantle makes up 80% of the planet's volume, "much more carbon in the Earth could be in the form of hydrocarbons than we previously imagined." The announcement of the discovery of methane on Mars came in 2003 (Atreya, 2007), when researchers proposed two sources of methane generation: the first being "Martians" – specifically, methane-belching bacteria like those in the guts of cows on earth; and the second being a rock–water reaction called serpentinization, which occurs in black smokers on the earth's seafloor.

2.4.2.1 Microbial (Bacterial) Methane

Organic matter (biomass) is derived from terrestrial, marine, and anthropogenic sources and is deposited in lakes, basins, and marine environments by transport mechanisms such as wind, rivers, and other agencies; it is a continuous process. Hydrogen and carbon are the main constituents for the formation of hydrocarbons. In view of the retentivity of these elements in the source and their relative enhancement during the physical and chemical processes involved, they play a pivotal role in the formation of petroleum products. Organic material derived from terrestrial vegetation (humic) primarily contains organic compounds with different chemical structures such as carbohydrates, lignins, and some proteins derived from the cell and walls of plants rich in resins and waxes. The transportation of humic organic material through rivers and other modes of transportation to its final destination exposes it to oxygenation, thus reducing the hydrogen content in the organic debris. Some organic components derived from aquatic life (algae, phytoplankton, zooplankton) are added to the land-derived organic matter before they are all deposited in the marine environment. Sapropelic organic material is comprised of structureless vegetable (algae) mud or slime and the remains of phytoplankton and zooplankton containing polymerized material lipids and proteins rich in hydrogen content. Lipids are derived from pollen, spores, cuticles, and similar plant constituents. Most organic matter is made up of large complex organic chemical structures (polymers, such as carbohydrates, proteins, lipids, lignin, etc.) which are formed from interconnecting chains of simple basic carbon structures (CH_2O – monomers). All organic compounds play a specific role in plants as well as other living organisms, forming the building blocks for their survival and growth. Higher plants contain more lignin to provide rigidity and keep them upright. Marine organisms contain more lipid to maintain buoyancy and provide a source of energy (Philp, 2005). So that microorganisms can utilize the energy potential of organic material, these chains must first be broken down into their smaller constituent parts (glucose). The deposited organic material is decomposed or degraded by microbial communities for their survival and cell growth. Biological processes involve microorganisms ranging from single-celled bacteria to higher order microorganisms. Biodegradation is the process by which organic substances are broken down into smaller compounds by enzymes produced by living microbial organisms (in biochemical processes, enzymes are organic substances that cause chemical reactions without being consumed in the reaction; they are called organic catalysts). The microbial organisms transform the organic matter through metabolic or enzymatic processes. Depending on the mechanism of metabolism the microorganisms utilize different energy resources (nutrients). Organisms are divided into autotrophs and heterotrophs according to their energy pathways. Autotrophs are organisms that are able to make energy-containing organic molecules from inorganic raw material by using basic energy sources such as sunlight. Plants are prime examples of autotrophs, because of their ability to photosynthesize. Organisms that must make use of food that comes from other organisms in the form of liquids, carbohydrates, and proteins are called heterotrophs. Some microorganisms have the microbial catabolic diversity to degrade, transform, or accumulate a huge range of compounds including hydrocarbons. By decomposing organic material for energy these microorganisms help recycle nutrients like nitrogen and carbon back into the environment.

The physiological and ecological constraints restrict microorganisms from adapting to different modes of metabolic processes. Bacteria obtain their energy for growth and cell maintenance from a series of dehydrogenations or coupled oxidation–reduction reactions. In nature, metabolism is governed by two basic processes, i.e., carbon dioxide reduction and fermentation of acetates (CH_3COO-) (Whiticar et al, 1986). The former process is more prevalent in a shallow marine environment whereas conditions in freshwater lakes favor microbial fermentation (Whiticar, 1994).

Reduction–oxidation reactions (redox) are central to all these energy transformations and the resulting flow of electrons manifests in the form of biogeochemical cycles which are kept in motion with bacterial activity. Respiration is a series of coupled oxidation and reduction reactions that result in the transfer of electrons from an appropriate electron donor (such as a reduced coenzyme) to an appropriate acceptor of electrons. Bacterial growth occurs when energy is released by the flow of electrons from donors to acceptors. Typical electron acceptors for bacterially mediated reactions are oxygen, nitrate, sulfate, and carbon dioxide. When oxygen is the electron acceptor, the highest amount of energy is released, whereas the lowest is released with carbon dioxide. These reactions are mediated biologically and the chemical reaction sequence is paralleled by an ecological succession of microorganisms: aerobic heterotrophs, denitrifiers [loss or removal of nitrogen or nitrogen compounds, specifically reduction of nitrates or nitrites commonly by bacteria (as in soil), which usually results in the escape of nitrogen into the air], fermenting bacteria, sulfate reducers (which can obtain energy by oxidizing organic compounds or molecular hydrogen while reducing sulfate to hydrogen sulfide), and methane producers (Cappenberg, 1975).

Respiration serves two purposes in a living organism: the harvest of electrons from organic compounds such as glucose generated during catabolism; and the production of adenosine triphosphate (ATP), the energy storage molecule, and free carbon molecules. Chemical changes occur during cellular respiration as cellular structures are affected by the transfer of electrons. Inorganic compounds such as oxygen, nitrates, sulfates, and carbonates are also used as electron acceptors during the respiratory metabolism process.

Microorganisms seldom occur naturally as a single species. Usually there are many different metabolic types present, both aerobes and anaerobes, each group occupying a special ecological niche depending on substrate availability and restrictions of the physical environment. The species best able to adapt to new conditions are favored and become dominant. This activity results in the succession of a metabolic type that is manifested as a temporal or a spatial progression.

Aerobic biodegradation is the breakdown of organic contaminants by microorganisms in the presence of oxygen. The chemistry of the system, environment, or organism is characterized by oxidative conditions and specifically refers to living only in the presence of oxygen. Many organic contaminants are rapidly degraded under aerobic conditions by aerobic bacteria (aerobes) that have an oxygen-based metabolism. During cellular respiration aerobes use oxygen to oxidize substrates such as sugars and fats to obtain energy. Before cellular respiration begins, glucose molecules are broken down into two smaller molecules in the cytoplasm of the aerobes. The smaller molecules then enter a mitochondrion, where aerobic respiration takes place. In this process, the electron transfer is to oxygen or the dioxygen molecule, which is reduced to water with concomitant coupled ion translocation and generation of an electrochemical gradient. Oxygen is used in the chemical reactions that break down the small molecules into water and carbon dioxide and release energy.

The physiological top zone in contact with the atmosphere (Fig. 2.4) has an appreciable quantum of dissolved oxygen in water, and microorganisms in this zone draw most of their energy from aerobic

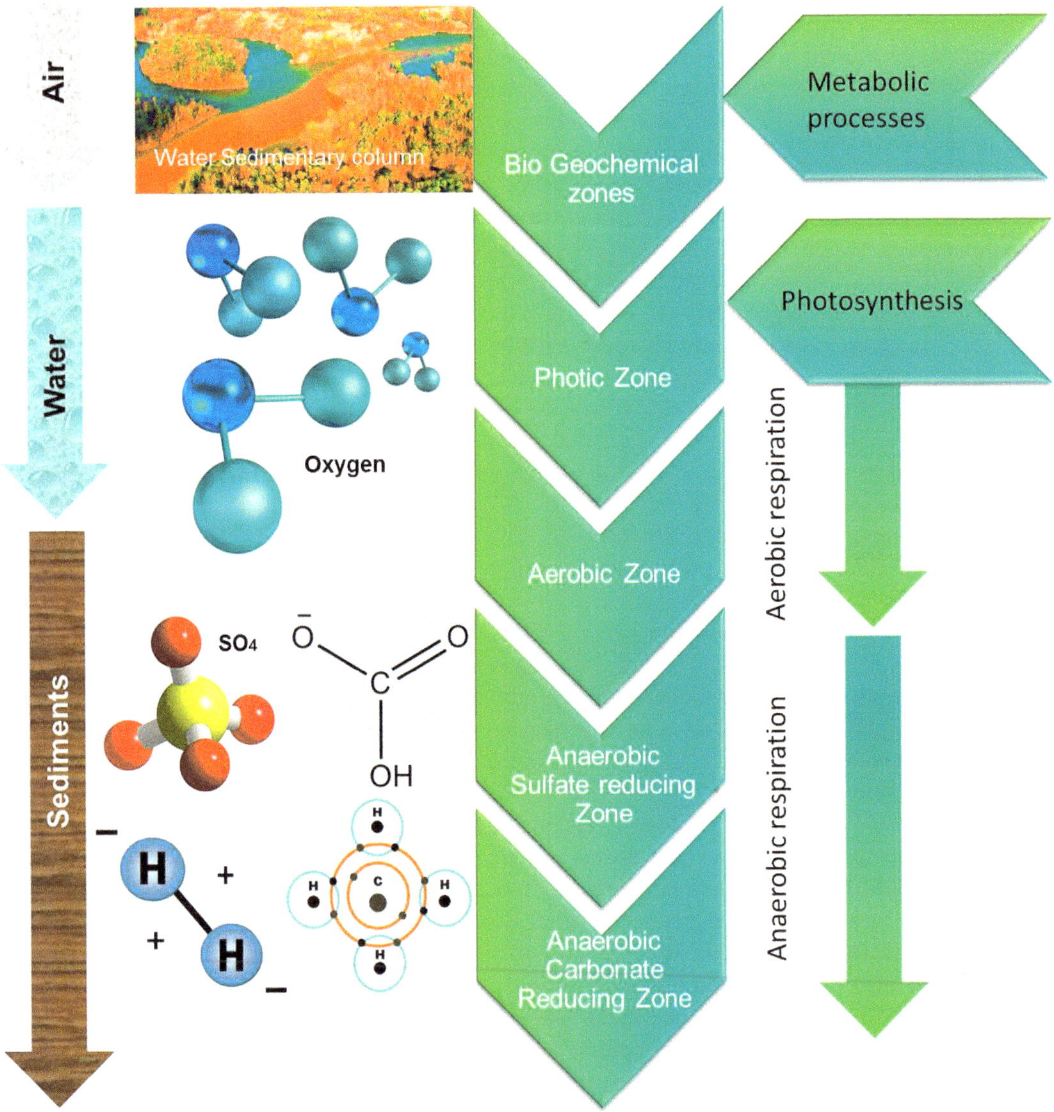

FIGURE 2.4

Cross-section of an open marine organic-rich sedimentary environment. The biological zones are the consequence of ecological succession. (Modified from Claypool and Kaplan, 1974.)

respiration. The sedimentary environment in which organic matter is deposited at a rate exceeding the supply of dissolved oxygen is characterized by the appearance of some anaerobic or anoxic conditions. The depth below which there is no dissolved oxygen marks the boundary between aerobic and anaerobic metabolism (Hunt, 1996). The aerobic–anaerobic boundary may be above but usually is below or at the sediment–water interface.

The basic terminal acceptor is of course oxygen. During aerobic assimilation, oxygen is rapidly utilized by microorganisms, and carbon is converted to carbon dioxide, sulfur to sulfate (SO_4), nitrogen to nitrate (NO_3), and phosphorous to phosphate (PO_4) until energy is no longer available from the molecule to support the biological process. As the dissolved oxygen concentration in seawater is in equilibrium with the atmosphere, a point is reached when the demand for oxygen exceeds the rate at which the gas can be introduced from the atmosphere. This creates oxygen-depleted zones in various areas of the sea.

Anaerobic mineralization is a process by which the organic matter in the absence of molecular oxygen is converted to low molecular weight substances such as carbon dioxide and methane. The complete anaerobic degradation of complex organic matter requires a diversity of enzymatic reactions and involves a variety of bacteria in large numbers. Extracellular enzymes, apparently essential in the initial liquefaction stage, have been identified in anaerobic sewage digesters.

The types of microbial metabolism that are feasible in anoxic sediments are determined by the types of organic substrates and the ultimate electron acceptors that are available. These organisms use nitrogen, phosphorus, and other nutrients to live and to develop cell protoplasm, but they reduce the organic nitrogen to organic acids and ammonia. In anaerobes, the chain still functions, but oxygen is not used as the final electron acceptor. Other less oxidizing substances such as sulfate, nitrate, or sulfur (S) are used. Anaerobic respiration is mainly used by prokaryotes that live in environments devoid of oxygen. Many anaerobic organisms are obligate anaerobes, meaning they can respire using only anaerobic compounds and will die in the presence of oxygen.

Once the level of dissolved oxygen in the water column decreases appreciably, the other groups of microorganisms utilize converted nitrates for their own respiration process (Claypool and Kaplan, 1974). Several groups of bacteria share the ability to utilize nitrates for coenzyme oxidation. This respiratory process reduces oxidized forms of nitrogen in response to the oxidation of an electron donor such as organic matter. In general, it occurs where oxygen, a more energetically favorable electron acceptor, is depleted, and bacterial respiration uses nitrate as a substitute terminal electron acceptor (Mechalas, 1974). The preferred nitrogen electron acceptors in descending order of thermodynamically favorable include nitrate, nitrite (NO_2), nitric oxide (NO), and nitrous oxide (N_2O). Several species of bacteria are involved in the complete reduction of nitrate to molecular nitrogen and more than one enzymatic pathway has been identified in the reduction process, which is performed primarily by heterotrophic bacteria, e.g., *Paracoccus denitrificans*, and various pseudomonads (Carlson and Ingraham, 1983). Denitrification is a microbially facilitated process of nitrate reduction that may ultimately produce molecular nitrogen (N_2) through a series of intermediate gaseous nitrogen oxide products. In terms of the general nitrogen cycle, denitrification completes the cycle by returning nitrogen to the atmosphere.

In a given water body below the photic zone, devoid of oxygen and nitrate, the sulfate ion becomes the principal oxidant for respiration (Fig. 2.4). The sulfate-reducing bacteria is capable of reducing the sulfate to sulfide, using sulfate as the terminal electron acceptor for respiration. Sulfate reduction is carried out by an unusual group of bacteria represented by the type species *Desulfovibrio desulfurican* (Tissot and Welte, 1984). These organisms are obligate anaerobes and require a reducing environment for active growth. They are generally regarded as heterotrophs that require the presence of organic carbon for the synthesis of cell material. *D. desulfuricans* is not solely dependent on organic materials for its energy metabolism. These sulfate-reducing bacteria contain a hydrogenase enzyme system that enables them to produce energy by coupling the reduction of sulfate using hydrogen gas. In addition to the utilization of selected compounds as substrate for both cell synthesis and energy, *D. desulfuricans* can carry out a second set of reactions. It can derive energy from the oxidation of nonassimilable organic molecules and uses this energy in turn to derive cell carbon from organic molecules which are unable to act as energy sources. The carbon metabolism of sulfate-reducing bacteria is incomplete yielding acetate and carbon dioxide from substrate as lactate, malate, or ethyl alcohol (Cappenberg, 1975). Doose (1980) observed that methane generation becomes important once the content of sulfate has been largely diminished by anaerobic sulfate-reducing bacteria. The observation is interpreted as resulting from competition for hydrogen (issued from organic matter) between sulfate-reducing bacteria and methane producers.

During the decay process, electron acceptors (such as oxygen, ferric iron, sulfate, nitrate, and manganese) become depleted, while hydrogen and carbon dioxide accumulate along with light organics produced by fermentation. During advanced stages of organic decay, all electron acceptors become depleted except carbon dioxide. Carbon dioxide is the product of most catabolic processes so it is not depleted like other potential electron acceptors.

All methanogens form methane by reducing a methyl group. The major energy-yielding reactions used by methanogens utilize substrates such as hydrogen, formic acid, methanol, acetic acid, and methylamine to produce methane. Methanogenesis is the final step in the decay of organic matter. Methane is generated biologically by methanogenic bacteria (*Methanobaterium*, a major division of the Archaea kingdom) largely from acetate dissimilation and carbon dioxide. Methanogens are found in freshwater sediments where the dominant metabolic process is fermentation of acetate and sulfate-reducing marine sediments where the reduction of carbon dioxide is the predominant metabolic process. In the freshwater sediments, hydrogen and acetate are important precursors for the formation of methane and a part of methane produced by reaction of hydrogen with carbon dioxide. In marine environments, competition for acetate and hydrogen by sulfate reducers makes the methane-producing bacteria highly dependent on sulfur concentrations (Marty, 1992).

In fermentative metabolic processes, energy is derived from oxidation of organic compounds such as carbohydrates usually from intermediates produced in the dissimilation of the substrate itself as the electron acceptors (Claypool and Kaplan, 1974). Fermentation can take place in aerobic or anaerobic conditions and only allows the breakdown of large organic compounds to produce smaller organic compounds.

In the anaerobic fermentation process, bacteria use oxidized forms of organic matter (carbohydrates) instead of using molecular oxygen as in respiration. Anaerobic digestion is considered a two stage process consisting of a nonmethanogenic phase (liquefaction) and a methanogenic phase (gas formation). In the first phase, a heterogeneous group of microorganisms converts proteins, polysaccharides, and lipids into short chain alcohols, fatty acids, carbon dioxide, and hydrogen gas by the combined processes of liquefaction and fermentation. In the second stage, some of these organic acids and carbon dioxide are converted to methane by a unique and metabolically limited group of anaerobic bacteria. Cellulose is degraded enzymatically by bacteria like *Clostridia* producing low molecular weight precursors for methane formation (acetate, bicarbonate, etc.; Fig. 2.5) and the final stage is reduction of carbon dioxide or acetate by methane-generating bacteria such as *Methanobacteria* (Doose, 1980). In the fermentation process, methane is derived from the methyl group of acetate (Smith and Mah, 1980). The presence of about 70% methane in freshwater is due to its genesis in the fermentation process (Cappenberg and Jongejan, 1978). Possible methane sources and sinks are shown in Fig. 2.5.

Living organisms, especially microorganisms carry out metabolic processes that affect the amounts of oxygen, carbon dioxide, nitrogen, hydrogen, methane, hydrogen sulfide, and other volatile products in the water. The residual oxidant is required to maintain the sequence of coenzyme reoxidations, which provide controlled release of energy from the substrate molecules. The chemical pathways involved in utilizing the biomolecules by different respiration processes produce varied amounts of energy. Aerobic respiration is more efficient than anaerobic respiration and generates more energy for the same amount of glucose utilized. The energy availability varies according to the electron acceptors utilized and the maximum amount of energy is generated with oxygen (Fig. 2.6).

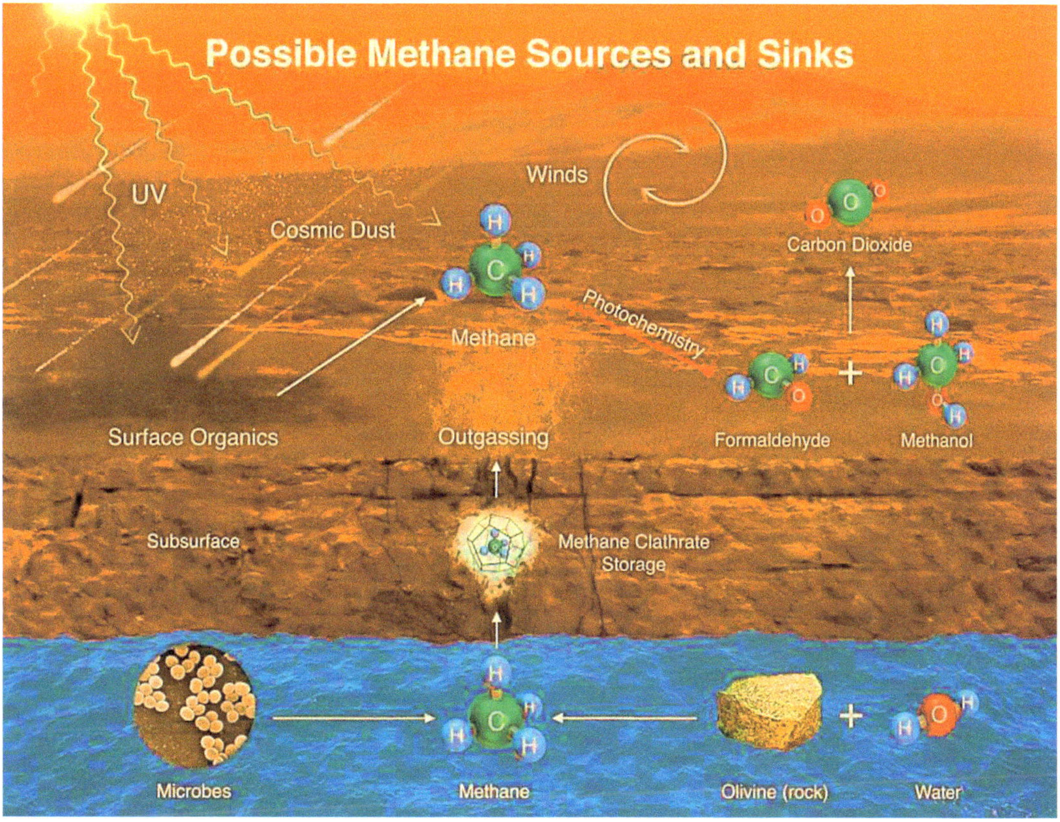

FIGURE 2.5

Sources and sinks of methane gas. (Modified figure from http://www.valuewalk.com/2014/12/nasas-curiosity-rover-finds-organic-matter-mars/.)

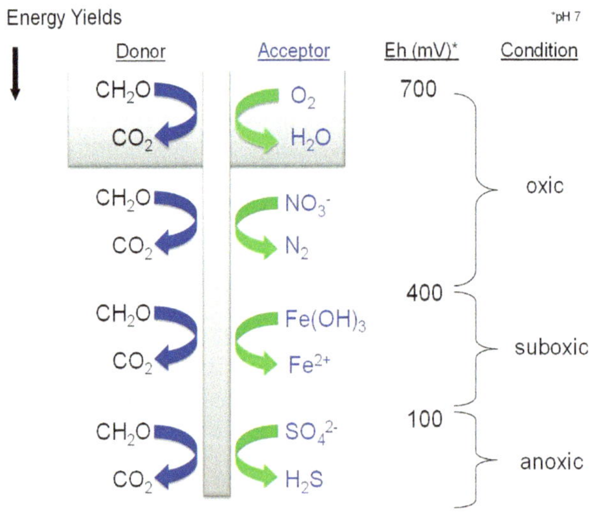

FIGURE 2.6

Quantum of energy produced during the respiration processes. Aerobic respiration emerges as more efficient compared to anaerobic respiration as it generates a higher quantum of energy for the amount of glucose utilized.

Seasonal variation, temperature, organic substrate, and age may be the controlling factors for the relative importance of the two pathways of reduction and fermentation (Schoell, 1988). Fermentation is more prevalent during winter and in low temperature regimes, whereas carbon dioxide reduction is confined to summer and warmer climates. However, both pathways may operate simultaneously with fermentation preceding the reduction. Methane generated by the fermentation process is mostly confined to younger sediments and may eventually escape into the atmosphere in the absence of proper traps for holding such a gas.

Bacterial processes do not produce significant quantities of higher order hydrocarbons such as ethane and propane except methane. However, Deep Sea Drilling Project (DSDP) results have indicated that the ratio of ethane to propane increases in the depth range from 1000 to 1500 m. This progressive increase in ratio was attributed to the mixing of biogenic methane with early products of thermal evolution from kerogen (Rice and Claypool, 1981). Biogenic gas occurs in younger sediments (Cretaceous to Pleistocene) which were not exposed to higher temperature regimes (maturation).

The activity of microorganisms decreases with depth due to hostile ecological conditions and microbial activity terminated due to an increase in temperature. Their presence in different geological environments has been identified to a depth of about 1000 m (Hunt, 1996; Kotelnikova, 2002). Microbial methane has also been detected in oil reservoirs (Whiticar, 1994).

2.4.2.2 Thermogenic or Thermal Methane

Microbial activity is confined to shallow layers of crust (<1000 m) as the bacterial population decreases due to an increase in temperature (>70°C) and paucity of nutrients. Productivity and preservation of organic matter are two driving forces behind the formation of rocks that are rich in organic matter, which ultimately become the source for oil and gas, as preservation primarily depends on the physical state of the environment. In a well-oxidized environment, the preservation factor is minimal but in a reducing environment, preservation is appreciably higher. Only a small percentage of organic matter reaches the sediments following its decomposition or degradation and consumption by organisms in the water column and scavengers just at and below the water–sediment interface. Methane is formed by microbial processes during diagenesis (<1000 m) in which the organic matter is altered mainly as a result of biological and chemical changes but with no pronounced effects due to temperature. Methane formed under diagenetic conditions is contained in nonreservoir rocks and about 20% of the world estimate is preserved in reservoir rocks. Because methane is formed in shallower depths and without proper geological structural configuration (traps) to entrap the gases, most of it generated by microbial processes escapes into the atmosphere. With rapid rates of sediment deposition, early formation of traps and seals and slowing down of the diffusion gas due to thickening of sediments may facilitate gas accumulation in reservoirs (Rice, 1993).

The volume of methane more than its solubility in water causes microbial gas to be trapped in cages of water molecules at low temperatures and higher pressures in shallow subsurface earth leading to the formation of gas hydrates on the continental margins and providing a valuble energy resource.

About 80% of natural gas (methane and other higher order hydrocarbons) in the reservoirs has its origin in thermogenic formation. In this mode, altered substrates during their burial in subsurface earth undergo extensive chemical structural changes resulting in the formation of gas. Organic compounds such as carbohydrates with their higher oxygen content are more susceptible to degradation and initially degraded products are easily soluble in water. Proteins are degraded to peptides and amino acids which can dissolve in water easily. Other organic compounds of living systems are recalcitrant and selectively preserved polymers to form the basic building blocks of kerogen (Philp, 2005).

During microbial activity, organic matter is decomposed and broken down into lower order molecules (monomers). The surviving organic matter in the early stages of diagenesis contains fractions of fulvic acid, humic acid, and humin, which are diluted during the progressive stages of diagenesis. Some biomonomers do not undergo further reactions, but others are condensed to form complex high molecular weight geopolymers, which along with ungraded biopolymers become the precursors for kerogen (Hunt, 1996). Kerogen is the portion of naturally occurring poly-condensed structural organic matter preserved at moderate temperature and pressure. Kerogen in the present from is to not extractable by using non-oxidizing acids and bases having organic solvents (Durrand, 1980; Tissot and Welte, 1984). Typical organic constituents of kerogen are algae, phytoplankton remnants of organisms, and woody outer coverings of plants. Besides high molecular weight organic material, sediments also hold low molecular weight substances called bitumen. The formation of kerogen concurrently involves the removal of low molecular weight substances such as water, carbon dioxide, ammonia, and sulfur dioxide. With changes in the physical environment due to its burial and exposure to higher temperature over longer periods of time (maturity), kerogen is no longer in equilibrium; it is altered in its structural configuration and attains a more stable degree of ordering due to thermal cracking (thermochemical reactions), resulting in the formation of bitumen. The conversion of kerogen to bitumen leads to the enrichment of hydrogen and carbon and depletion of oxygen, nitrogen, and sulfur (Huc and Durrand, 1977; Schulz and Zabel, 2006). As the depth of burial increases, the residual kerogen becomes more mature, enhancing the degree of bitumen formation, which may subsequently lead to the formation of oil and gas (Tissot and Welte, 1984).

Methane in the subsurface earth can be generated by:

- Methanogenic bacteria (as detailed above)
- All types of kerogen
- Coal
- Oil

Thermal cracking of kerogen and coal is known as primary cracking and that of oil is considered to be secondary cracking (Hunt, 1996). When physical processes such as temperature and pressure act on organic matter over time at different stages it leads to the formation of hydrocarbon gases. The formation of oil and gas are dependent on the type of the kerogen, burial history (temperature and time), and the geothermal gradient.

Various types of organic matter are defined according to their chemical structure (e.g. carbohydrates, lipids, proteins, etc.) but an alternate classification scheme widely used by petroleum geoscientists is based on the chemical composition and structure of kerogen. Studies of extraction of pollen and spores from young sediment deposits indicate different patterns in hydrogen to carbon ratios (H/C) depending on the genesis of organic material (Bernard et al., 1978):

- <0.8 predominantly humic
- Between 0.8 and 1.0 mixed humic and sapropelic
- >1.0 predominantly sapropelic

Van Krevelen (1950) proposed a simple and rapid technique of plotting the hydrogen carbon ratio versus the oxygen carbon ratios for tracing the processes for coal maturation. The approach of Van Krevelen has been modified to incorporate and interpret the formation of other petroleum products. Kerogen is categorized according to its genesis and content of carbon, hydrogen, and oxygen in the residue of the organic substrate (Tissot and Welte, 1984). Type I kerogen results from the predominant accumulation of algal material and severe biodegradation of organic material during deposition results in a high proportion of lipids in the substrate. In Type II kerogens, the organic content is mainly derived from phytoplanktons, zooplanktons, and microorganisms in the reducing marine environment. Type I and II kerogens together are called sapropelic. Type III kerogen is mainly derived from continental plants. More recently Type IV has been added, which includes kerogen with a low hydrogen to carbon and high oxygen to carbon ratio. The genesis of this kerogen could be of mixed mode and organic matter representing highly oxidized and recycled organic matter during deposition. The types of kerogen are illustrated in Fig. 2.7a together with the classic Van Krevelen diagram. The evolutionary paths of kerogens are shown in Fig. 2.7. In conjunction with the pattern kerogen track, vitrinite reflectance of kerogen is utilized to determine the maturity (temperature) history of the sediments and the pattern of thermal control for generation of the hydrocarbons.

Commercial quantities of oil are formed from kerogen Types I and II (sapropelic) whereas gas is formed from kerogen Types III and IV (humic) (Fig. 2.7b). Type III kerogen is different from sapropelic Type I and II kerogens. Sapropelic kerogens during maturation generate a significant volume of higher order hydrocarbons. Humic kerogens result in high amounts of thermogenic methane relative to higher order components, yielding mostly carbon dioxide at low thermal maturity levels (Fig. 2.7b).

The formation of different types of petroleum products depends on the types of kerogen, thermal regimes, and the duration of exposure of organic material. During their maturation history, sapropelic kerogens generate a significant volume of C_2^+ hydrocarbons (Hunt, 1996). At low thermal maturity

(a)

Types of Kerogen and Their Hydrocarbon Potential				
Environment	Kerogen Type	Kerogen Form	Origin	HC Potential
Aquatic	I	Alginite	Algal bodies	
		Amorphous Kerogen	Structureless debris of algal origin	Oil
	II		Structureless planktonic material, primarily of marine origin	
Terrestrial		Exinite	Skins of spores and pollen, cuticle of leaves and herbaceous plants	
	III	Vitrinite	Fibrous and woody plant fragments and structureless, colloidal humic matter	Gas, some oil
				Mainly gas
	IV	Inertinite	Oxidized, recycled woody debris	None

(b)

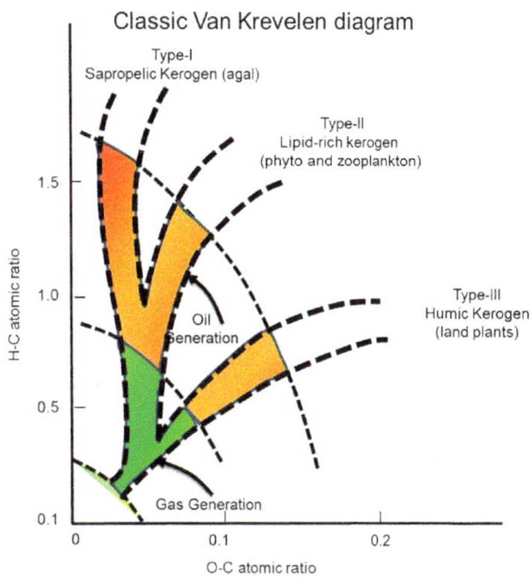

FIGURE 2.7

Types of kerogen and their classification and the Van Krevelen diagram. Pattern of oil and gas generation in the subsurface earth controlled by the type and degree of maturity the organic matter (kerogen) undergoes during its burial. Types I and II kerogen (sapropelic) with a lower degree of maturity mainly produce oil whereas humic kerogen (Type III and IV) is responsible for the generation of dry gas at deeper levels. (a) Origin, type, source, and hydrocarbon potential of different kerogens. Organic content in gas shales is usually Type II, as opposed to coals, which contain mostly Type III. (b) The classic Van Krevelen diagram. (From Crain's Petrophysical Handbook: https://spec2000.net/11-vshtoc.htm.)

levels, humic kerogens result in high amounts of thermogenic methane relative to ethane, propane, and other higher components yielding mostly carbon dioxide. The successive formation of different components of petroleum under different thermal regimes as a function of depth is given in Fig. 2.8. With increase in the depth of burial (maturity regime) there is a decrease in the amount of liquid hydrocarbons and an increase in the proportion of methane compared to higher alkanes (propane and others). During thermal cracking the molecular structure of kerogen is broken down resulting in lower order organic compounds and the production of methane. Thermogenic gas may be formed by thermal degradation of kerogen or thermal cracking of crude oil with increasing maturity (Hunt, 1996; Whiticar, 1999). Exposure of kerogen to further increases in temperature causes dehydration and decarboxylation resulting in the rapid loss of oxygen and enhancement of carbon and nitrogen in the organic residue. Increases in carbon lead to a reduction of the hydrogen/carbon ratio. In this context, depending on the extent of higher temperature regimes, alteration of kerogen may lead to two fractions (North, 1985):

1. A fluid product high in hydrogen leading to the formation of oil and gas
2. A residue high in carbon such as bituminous coal

The quantity of petroleum generated and expelled increases as the atomic hydrogen/carbon ratio increases. The pattern of formation of methane gas at various depths shows dry gas at shallower levels followed by wet gas in an oil window, and mostly dry gas at the very bottom (Fig. 2.8). The dry gas in the

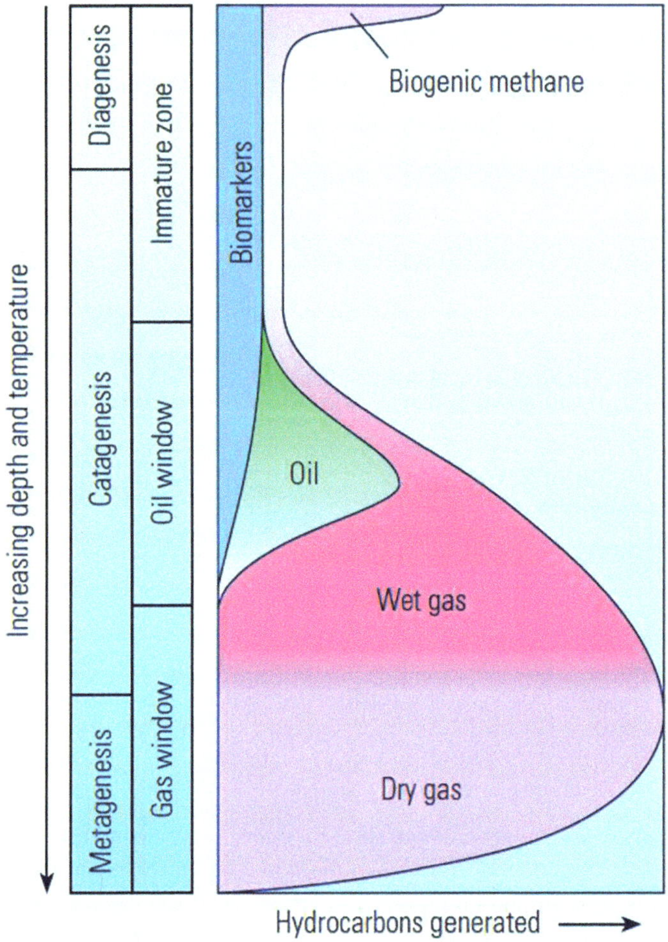

FIGURE 2.8

Thermal transformation of kerogen. Pattern of generation of oil and different types of gases produced in subsurface earth and its dependence on depth of burial. The generation of hydrocarbon in source rock is controlled primarily by temperature as the kerogen content evolves from reactive carbon to dead carbon. Gas is given off during early diagenesis primarily through biologic activity. Catagenesis takes place with further burial, during which oil and gas are given off. With increasing depth and temperature, any remaining oil is cracked during metagenesis, initially giving off gas followed by a simpler form of dry gas. (Modified from Tissot et al., 1974.)

zone closest to the earth's surface (<75°C) is mainly derived from microbial processes either by reduction of carbon dioxide or fermentation of acetates with methane, the most prominent constituent (>90%). At temperatures of 100–150°C gases can be generated either directly from kerogen or through thermal cracking of oil at 150–180°C. Gases generated directly by kerogen have low-order components (C_1–C_5) whereas gases produced by thermal cracking of oil constitute higher order gas components (C_6–C_{40}; Fig. 2.9) (Clayton, 1991). Depending on the operating processes at different stages of gas formation,

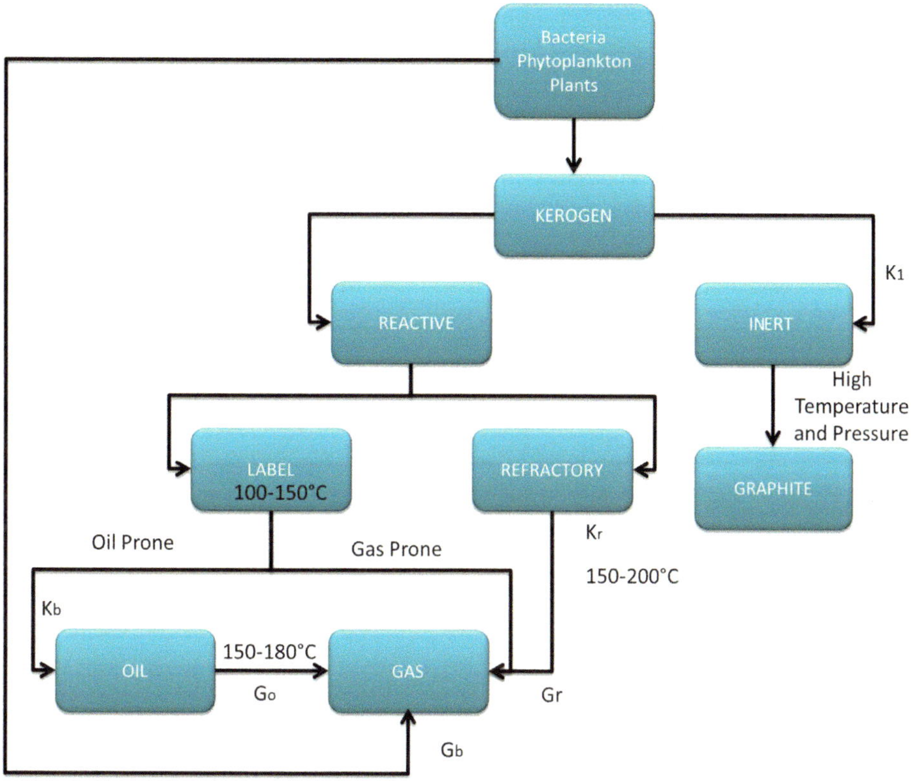

FIGURE 2.9

Processes governing generation of methane by different mechanisms. Low temperature gas (biogenic gas, G_b) is produced in shallow levels of the earth by bacterial degradation of organic matter. In the higher temperature regime, methane is produced by thermal degradation of different types of kerogen. Labile kerogen (K_l) produces both oil and gas. Refractory kerogen (K_r) generates gas (G_r) mainly from the .humic type of kerogen at higher temperatures. Thermal cracking of oil results in wet gas G_O. (Modified from Clayton, 1991.)

different percentages of condensate liquid are entrapped in methane. Natural gases (methane) are classified as dry gas (1.3 L/100 m^3) and wet gas (>4 L/100 m^3) of liquid condensate. Wet thermogenic gases are the main products of the mature stage (70–150°C) with generation at ~120°C (Hunt, 1996).They are formed by sapropelic (Type I and II) decomposition of kerogen and thermal cracking of oil (Tissot and Welte, 1978; Hunt, 1996). Dry gas in deeper levels of the earth with a higher order of maturity (150–200°C) primarily originates from humic constituents of kerogen derived from terrestrial plants and other organic matter. Under this regime the production of wet thermogenic gases diminishes to a minimum.

2.4.3 GEOCHEMICAL CHARACTERISTICS

Only 3% of the advective methane in gas hydrate-bearing sediments is released into the atmosphere as a result of the anaerobic oxidation of methane (AOM), which is a specific microbial process (methanotroph) occurring in marine sediments. The distinct geochemical signature in methane evolves through the processes of biochemical reduction of sulfate, enrichment of bicarbonate, and precipitation of calcium, and magnesium's geochemical signatures provide vital information in understanding the genesis of different constituent hydrocarbon gases and source rock characteristics in sedimentary basins. These geochemical fingerprints have been used in determining the source potential of gases (Sassen and Chinn, 1990), their maturity profiles (Clayton, 1991), migration pathways, and secondary alterations (Sassen et al., 1993). In the shallower subsurface regimes, organic matter is subjected to bacterial activity resulting in the generation of methane and other hydrocarbon components of gases. In deeper parts of the earth, organic matter preserved as kerogen is subjected to complex processes at higher temperature and pressure conditions leading to thermal degradation of kerogen and generation of hydrocarbon gases. Gases are also generated due to thermal cracking of oil at higher temperatures. All these processes produce different geochemical signatures. There are significant components in geochemical studies together with compositional variations among different gas components. Molecular and isotopic signatures are primarily utilized to study the complex processes and mechanisms by which organic matter leads to the generation of microbial gases, the formation of kerogen, and subsequent generation of gaseous components.

The seven primary saturated aliphatic compounds (C_1–C_5), ranging from methane to normal pentane, retain important information about the mode of their formation. The relative proportion of C_1–C_5 components in gas samples may provide information about the processes by which they are formed (microbial versus thermogenic; Bernard et al., 1978; Schoell, 1983). Microbial processes generate a very high percentage of methane, where gases that are emanated due to thermogenic processes have a significant contribution from higher order (C_2–C_5) constituent gases. In this respect, molecular composition studies have provided simple methods for distinguishing between microbial and thermogenic origins of methane. Earlier studies have utilized different configurations of molecular ratios such as C_1/C_2 (Claypool, 1974), C_1/C_2+C3 (Bernard et al., 1978), and $1/\sum C_2+$ (Faber and Stahl, 1984) to distinguish microbial and thermogenic components in hydrocarbon gases.

There are three naturally occurring stable isotopes of carbon namely carbon 12 (^{12}C), carbon 13 (^{13}C), and carbon 14 (^{14}C) in subsurface earth. Each is distinguished by the number of protons in the nucleus with respect to the fixed number of neutrons (six). Carbon ^{12}C and ^{13}C are the original forms of organic carbon preserved in the earth. ^{14}C is produced by the bombardment of neutrons produced by cosmic radiation on atmospheric nitrogen which subsequently enters the biosphere as carbon dioxide (Hunt, 1996). The predominant isotopic component is ^{12}C (99%), with 1% of isotope ^{13}C and traces of ^{14}C. However, ^{14}C is not as stable as ^{12}C or ^{13}C, as it decays into nitrogen with a half-life of 5530 ± 40 years. There are no processes that can alter the ratio of ^{12}C and ^{13}C significantly, but small variations can occur. These small variations are not due to changes in the nuclei themselves but to processes that show slight preferences for either light or heavy isotopes. The differences in mass due to a difference in the number of neutrons results in partial separation of the light isotopes from the heavy isotopes during chemical reactions and physical processes such as diffusion and vaporization. A selection process which enriches one or the other isotope is called fractionation. The resulting fractionated material is called isotopically heavier or lighter (Gold, 1993). In nature, the $^{12\text{-}13}$carbon content in the hydrocarbons has contributions from the source, from the fractionated part of the gas, and from the altered hydrocarbons (Huex, 1977).

Variations in the atomic mass bring about changes in the chemical and physical properties; this is called an isotopic effect, which leads to isotopic fractionation. Two types of isotopic effects have been utilized to draw inferences on the source characteristics: the reactionary processes and the resulting final product, i.e., kinetic isotopic effect and equilibrium isotopic effect. The kinetic isotope effect is measured as ratio of rate constants, i.e. rate constant of abundant isotope to that of altered isotope, with the rate constant for the reaction of the altered isotope, such as substitution of hydrogen with deuterium. Breaking bonds of the isotopes of lower atomic mass gases such as ^{12}C-^{12}C, results in isotopically lighter gases (smaller molecular size) with depletion of higher molecular (^{13}C) content in the gas. The residual organic matter is enriched with ^{13}C with a decrease in size (Fig. 2.10; Silverman, 1971). Breakage of ^{12}C-^{12}C bonds has a higher rate constant than that of ^{13}C-^{13}C bonds.

When the interchange of isotopes alters the equilibrium it is called the equilibrium isotopic effect. It is best to express kinetically (Stahl, 1973; Chung et al., 1988) and thermodynamically controlled fractionation as an equivalent of kinetic effect (Gilmov, 1985) in terms of isotopic ratios (Hayes, 1993). Frequent breaking of bonds results in a gaseous product that is depleted in isotopic properties and a residue that is enriched in isotopic properties. Isotopic bonds can be broken when the microbes have preferential consumption of ^{12}C over ^{13}C resulting in depletion of ^{13}C (Clayton, 1991). This type of bond breakage can also be accomplished as a result of thermal cracking, when the organic matter (kerogen) and oil is exposed to higher temperatures over longer periods of time.

The molecular and stable isotopic composition of natural gas is of particular importance for genetic identification of gas and gas-source rock correlation because they reflect the thermal maturity of source rocks and the isotopic composition of source organic matter (Wang et al., 2005). Measurement of small-scale variations in the carbon isotope ratio in different samples is usually not done in absolute terms but by minor deviation from the selected standard rock type. The selected standard for this purpose is carbon dioxide prepared from belemnites collected from the Upper Cretaceous Peedee formations of South Carolina, a marine carbonate rock. The standard is called Peedee belemnite or PDB and it has a carbon isotope value that represents the most likely median value of distribution of all marine carbonates. ^{13}C forms a very small percentage (0.11%) in naturally occurring carbon and varies from about 1.02% in most depleted carbon to the slightly higher figure of 1.13% in enriched carbon. As this

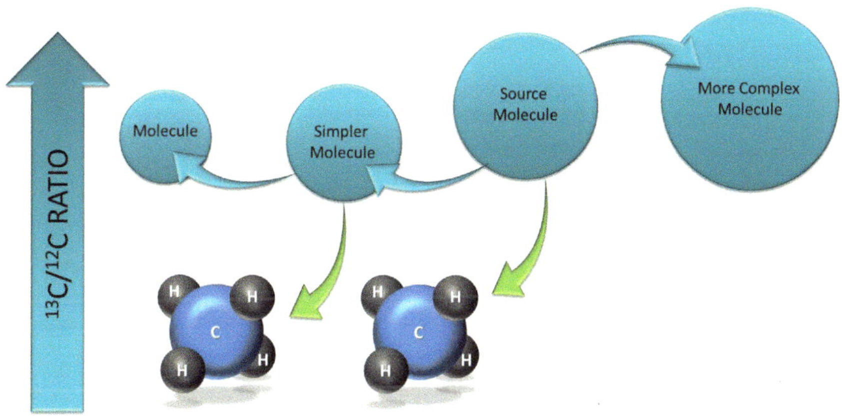

FIGURE 2.10

The change molecular structure of kerogen during successive fractionations. With continuous fractionations the kerogen gets enriched in ^{13}C with increase in molecular size as shown by bigger rectangular size. (Reinterpreted and modified from Silverman, 1971.)

amount is so small, it is necessary to measure its absolute abundance with very high precision (Fuex, 1977). The measurements are then quoted as the departure of ^{13}C content of the sample from that of the standard, and the figure is usually given in parts per thousand (per mil) and referred to as the $\partial^{13}C$ value of the sample.

The $\partial^{13}C$ value for carbon is defined as (Craig, 1953, 1957):

$$\partial^{13}C = [(R_{sample}/R_{pdb}) - 1] \, x1000$$

where $R_{sample} = {}^{13}C/{}^{12}C$ and the R_{pdb} PDB standard is 0.0112372 having biological origin.

The $\partial^{13}C$ values versus the PDB reference show a large spread (Fig. 2.11a). Since the $\partial^{13}C$ varia at PDB standard are very small they are usually expressed in parts per thousand (per mil).

Unoxidized carbon in plants is derived from atmospheric carbon dioxide by the process of photosynthesis. In this process, the light isotope ^{12}C is slightly favored. As a result, the carbon is slightly depleted in ^{13}C relative to atmospheric carbon dioxide. This effect is larger than that produced by any other single nonbiological chemical process occurring in nature. There is a marked depletion of ^{13}C in most of the deposits – unoxidized carbon such as petroleum, methane, coal, and kerogen.

$\partial^{13}C$ for plant organic carbon is generally in the range of −8 to −35 per mil (PDB standard). The atmospheric carbon dioxide from which carbon is derived has a $\partial^{13}C$ value of −6 per mil showing a large fractionation effect. Marine carbonate laid down from atmospheric carbon dioxide dissolved in ocean waters has $\partial^{13}C$ values ranging from +5 to −5 per mil (average 0), and thus evidently a fractionation averaging 6 per mil occurs in favor of the heavy isotope during that process.

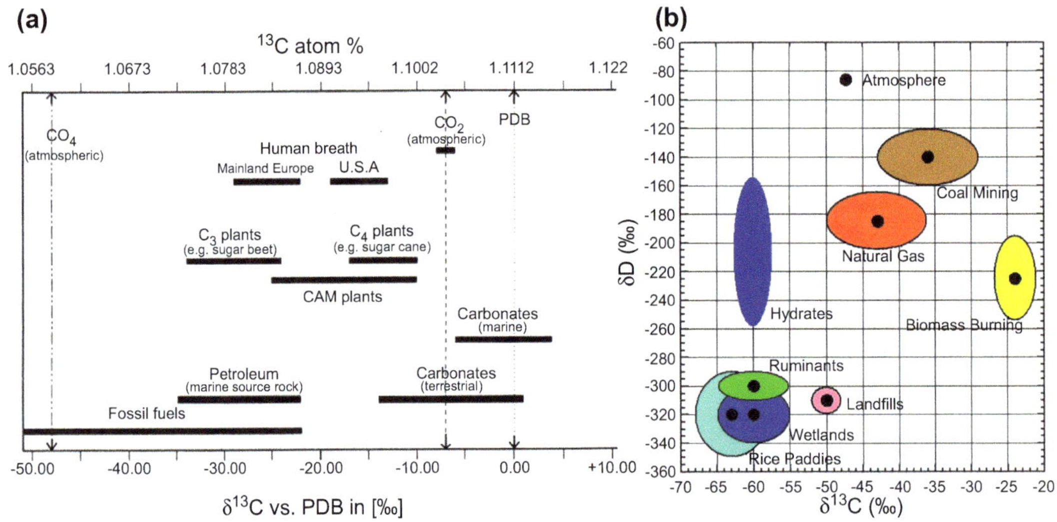

FIGURE 2.11

Delta ^{13}C values compared to the PDB reference. (a) Values of $\partial^{13}C$ over the PDB reference values. The values of $\partial^{13}C$ are estimated from the equation given by Craig (1957). (b) The isotopic composition of methane CH_4 from different sources. $\delta^{13}C$ indicates the deviation in per mil of the $^{13}C/^{12}C$ ratio to a defined standard (PDB). Data from Quay et al. (1999) and Whiticar (1999).

In the production of methane from plant debris, a further fractionation takes place that once again favors the light isotope, and plant-derived methane isotopically is even lighter with $\partial^{13}C$ values ranging from −50 to −80 per mil. Owing to the large variations in $\partial^{13}C$ values the world over, an arbitrary division has generally been assumed, so that carbon with $\partial^{13}C$ values lower than −30 per mil are considered to be of biological origin, whereas heavier carbon is assumed to be from other sources. Methane is the only carbon-bearing molecule that is light enough to suffer significant isotopic fractionation and shows the largest spread of values. The isotopic composition of methane from different sources is shown in Fig. 2.11b. $\partial^{13}C$ indicates the deviation in per mil of the $^{13}C/^{12}C$ ratio to a defined standard (PDB).

2.4.3.1 Microbial (Bacterial) Gases

Molecular studies have clearly distinguished different modes of formation of methane in subsurface earth (Bernard et al., 1978; Faber and Stahl, 1984). Fig. 2.12 shows a plot of molecular ratios against the $\partial^{13}C$ of methane gas (Bernard et al., 1978). The mode of formation of microbial and thermogenic methane gas give quite distinctive plots, as shown in Fig. 2.12. Microbial (bacterial) gas obtained from most of the regions studied exhibited very high molecular ratios with a higher order of depletion of $\partial^{13}C$ for methane gas. However, the lower molecular ratios of thermogenic gases indicated a higher content of higher order molecular gases such as C_2 and C_3 with distinctly lower values of $\partial^{13}C$.

Carbon isotopic studies have been extensively carried out to isolate the origin and the processes responsible for the generation of methane in different environments (Stahl, 1973; Schoell, 1980, 1988). The stable isotope bonds between carbon and hydrogen of the microbial (bacterial) methane may be fractioned by consumption of ^{12}C carbon or breaking of bonds due to higher temperatures. The fractionation results in depletion of ^{13}C. The observed $\partial^{13}C$ as estimated from the above definition shows a variation from −60‰ to −110‰ over the PDB values recorded from marine and freshwater sediments. The former has a mean value of about −68‰ and the latter a value of −59‰, the transition from marine to freshwater occurring at around 60‰. Owing to such a small variation over the different regimes it becomes difficult to isolate processes for the generation of methane either by reduction of carbon dioxide or by acetate fermentation (Schoell, 1980). Methane produced through carbon dioxide reduction (mostly marine environments) compared with acetate fermentation (freshwater sediments) has a different isotopic composition of carbon (Whiticar et al., 1986; Whiticar, 1994).

In order to resolve the detailed mode of formation of microbial methane, carbon isotopic fractionation in conjunction with hydrogen fractionation has been attempted (Schoell, 1980, 1988; Whiticar et al., 1986; Hunt, 1996). Microbes also consume lighter hydrogen compared to heavier (deuterium) hydrogen, thereby depleting deuterium content in the gas. The hydrogen isotopic composition of methane is also the diagnostic parameter in classifying the type of gas. Deviation of deuterium from the standard mean ocean water (SMOW) is defined and has been done in the case of carbon isotopes (Craig, 1961):

$$\partial D = (R_{sample} - R_{standard}) / R_{standard} * 1000$$

Hydrogen in the methane molecule is derived primarily from two sources: organic matter and the formation of water. The isotopic deuterium concentration of organic matter in the sedimentary rocks is significantly lower than its concentration in natural waters. Laboratory studies (Daniels et al., 1980) have shown that all hydrogen incorporated into methane by carbon dioxide reduction pathways was deuterium and originated from the formation of water. Bacterial methane in marine environment has ∂D value averaging above −192‰ from the SMOW with small variation over the mean. However,

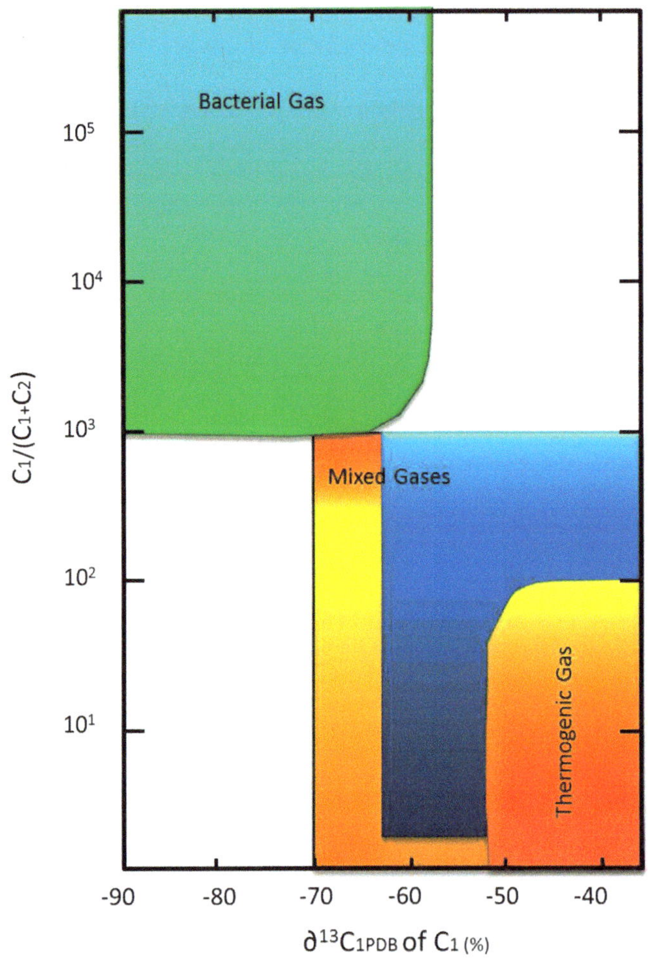

FIGURE 2.12

Revised Bernard plot after Hunt (1979) comparing gas wetness and $\delta^{13}C$ of methane to characterize origin of gas.

methane formed in freshwater environments is strongly depleted with a mean value of −318‰ (Whiticar et al., 1986; Woltemate et al., 1984).

Based on the pattern of depletion of deuterium, initial studies have indicated that fermentation is the predominant pathway in the freshwater environment (Woltemate et al., 1984; Whiticar et al., 1986) but later studies (Jenden and Kaplan, 1986; Burke et al., 1988) showed that such depletion also occurs in marine sediments, when methane is formed by carbon dioxide reduction. These studies indicate that the methanogenic process is independent of the depositional environment (Schoell, 1988). Analysis of the isotope data (Woltemate, 1982) indicated that the fermentation process shows an "aging" effect.' Based on carbon and hydrogen data, Schoell (1988) presented the generalized concept of fermentation versus

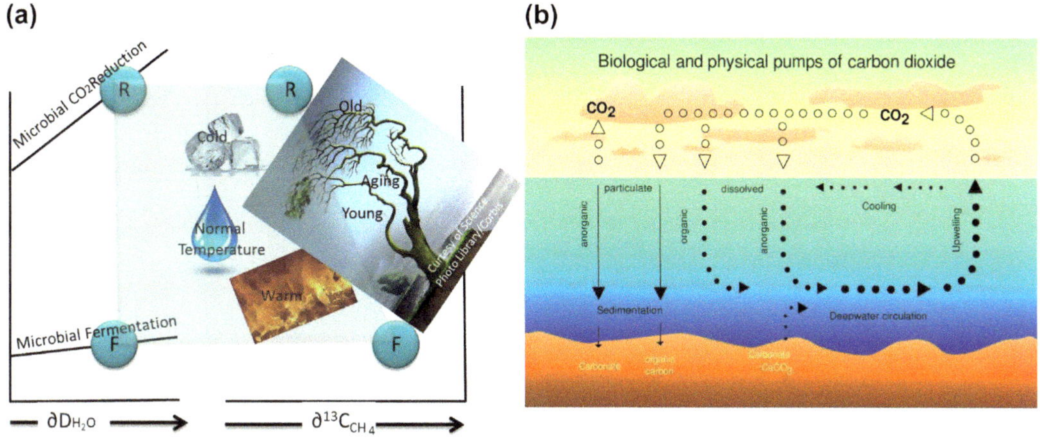

FIGURE 2.13

Bacterial processes models for CO_2 reduction (R) and fermentation (F). (a) Model provides the plot between deuterium concentration in water and methane and (b) gives the plot between deuterium and $\partial^{13}C$ methane. Methane derived by fermentation is enriched in ^{13}C compared to methane generated by CO_2 reduction. Younger sediments can hold methane formed by fermentation and older sediments predominantly hold methane formed by process of reduction of CO_2. (Modified from Schoell, 1988.)

the carbon dioxide reduction process (Fig. 2.13a), indicating that fermentation-derived methane is generally enriched in ^{13}C in comparison to methane derived from the carbon dioxide reduction process. However, methane formed by the fermentation process is much more depleted in deuterium than methane derived from carbon dioxide reduction. With increasing age methane deuterium gets enriched and the carbon dioxide reduction process becomes more prominent in the generation of methane.

Conversion of carbon dioxide by bacteria to methane results in fractionation into carbon isotopes depleting $\partial^{13}C$ by about −75‰ and with this $\partial^{13}C$ of carbon dioxide is depleted by order of −15‰; leading to net depletion of the order of −15‰ in the methane (Rosenfeld and Silverman, 1959). Analysis of carbon isotope $\partial^{13}C$ data of CO_2 by Whiticar et al. in 1986 indicated that freshwater carbon dioxide has more negative value (mean value −12‰) compared to CO_2 data from marine environments (mean value −6‰). This suggests that $\partial^{13}C$ values of interstitial waters are controlled by a combination of equilibrium and kinetic isotope effects and are related to the atmospheric exchange, precipitation–dissolution, organic matter decomposition, and other digenetic processes, which in turn may affect the carbon isotope ratio of biogenic methane.

The isotopic effects of microbial metabolism and other processes should be reported in terms of α, the fractionation factor, rather than the $\partial^{13}C$ of a product resulting from the fractionation. This is important because the $\partial^{13}C$ of product depends on the $\partial^{13}C$ of the source. To segregate the marine and freshwater carbon isotope values, Whiticar et al. (1986) cross-plotted the $\partial^{13}CO_2$ and $\partial^{13}CH_4$ and estimated the fractionation factor by the following equation:

$$\alpha_c = \frac{\left(\partial^{13}CO_2 + 10^3\right)}{\left(\partial^{13}CH_4 + 10^3\right)}$$

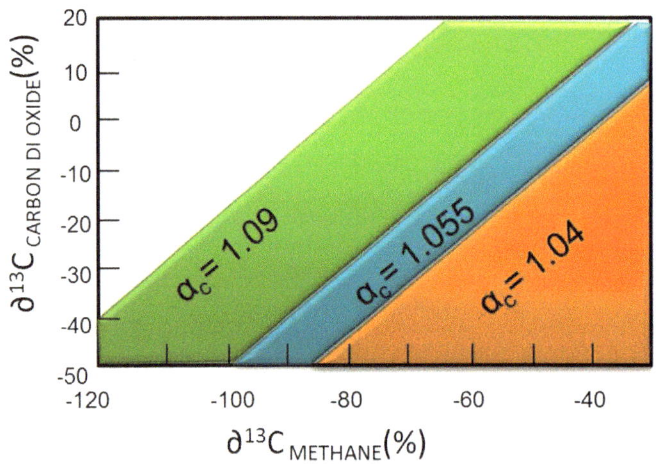

FIGURE 2.14

Cross-plot of $\partial^{13}C_{methane}$ and $\partial^{13}C_{carbon\ dioxide}$ for estimating the fractionation of sedimentary organic matter for different environments. Lines of equal fractionation (α_c) are drawn in the range 1.04 to 1.09 to differentiate the fractionation in marine and freshwater environments. α_c values of 1.06 demarcate the boundary between marine and freshwater processes. (Modified from Whiticar et al., 1986.)

Lines of equal α_c values ranging from 1.04 to 1.09 are drawn on the cross-plot to ascertain the probable zone of marine and freshwater distribution on the cross-plot (Fig. 2.14). A wider zone of α_c values ranging from 1.05 to 1.09 represents the carbon isotopic ratios. The freshwater carbon isotope data are clustered in the α_c value range. The line with a α_c value of 1.55 may represent the region of marine and nonmarine separation (Whiticar et al., 1986).

2.4.3.2 Thermogenic Gases

Thermal alteration of organic matter is a gradual and complex process which produces liquid and gaseous compounds continuously changing composition (Schoell, 1980).Thermogenic gases produced by the fractionation of kerogen have complex structure and carbon distribution because they originate from different types of organic matter in various depositional environments (Chung et al., 1988). The carbon content of kerogen is already depleted owing to microbial processes in shallower parts of the earth (<1000 m and temperatures <75°C). The organic matter (kerogen) which sustains the biogeochemical reaction at lower depth levels is subjected to thermal degradation and fractionation at higher temperatures over long periods of time (maturity). Apart from methane, gases generated at deeper levels have significant contributions from higher hydrocarbon gas components such as C_2 to C_5. Molecular and isotopic characteristics of carbon and hydrogen provide reliable information about the sources and kerogen types such as sapropelic and humic (Prinzhofer and Huc, 1995). Gases generated by bacterial activity are known to be highly depleted in $\partial^{13}C$ and spread over larger values. The $\partial^{13}C$ of the carbon isotopes estimated for thermogenic gases overlap with bacterial gases in the range −60‰ to −50‰. In this regard the upper limit of $\partial^{13}C$ for bacterial gases (−50‰) is strictly empirical and there is no theoretical explanation for this upper limit (Huex, 1977).

Thermogenic gases have a wide range of molecular and isotopic properties related to the type, amount of organic matter (kerogen), and the extent of their exposure to higher temperature and maturity (Tissot et al., 1974). Thermal degradation or fractionation of enriched kerogen, however, still depleted when compared to residual kerogen and oil in the deeper part of earth, results in the generation of methane and thermogenic higher molecular hydrocarbon gases. In subsurface earth (Fig. 2.9) the thermal generation of gaseous hydrocarbons takes place in the maturation of organic matter as well as the thermal cracking of liquid petroleum. The gaseous hydrocarbons are more depleted than the petroleum products with respect to parent source material and are also dependent on the type of organic matter (Fuex, 1977). In gases formed by terrestrial organic matter, the ∂C1 values are consistently heavier when compared to gases originating in the sapropelic organic matter of marine environments (Stahl, 1975). Gases associated with oil generation have a significant quantity of C_{2+} (5% to <15%) components and are wetter (Rice et al., 1989). Thermogenic gases derived from oil have a ∂^{13}C1 range in the order of −58‰ to −40‰ indicating the effect of thermal maturation (Fuex, 1977).

Deep dry gas is obtained from source rock that has gone past the oil-generating phase, i.e., more mature stage, or coaly source rock with any level of maturity and has a lower percentage of C_{2+} gases (Fuex, 1977). Immature gas is typically depleted in ^{13}C (<5% C_{2+}) (Rice et al., 1989; Hunt, 1996).The dry gases are less depleted in $\partial^{13}C_1$ and cover a probable range of −40‰ to −25‰ (Schoell, 1983; Whiticar, 1994).

The isotopic properties of wet thermogenic gas range from −30‰ to 60‰ PDB and −120‰ to −300‰ SMOW; for dry gas it ranges from −15‰ to −40‰ PDB and −70‰ to −150‰ SMOW (Hunt, 1996).

To differentiate between the components of bacterial and thermogenic gases, Schoell (1983) plotted the hydrogen and carbon isotopic ratios (Fig. 2.15). Methane gas produced by bacterial activity has two components, i.e., continental (∂D~ −250 to −200 ppt) and marine (∂D~ −200 to −150 ppt), and $\partial^{13}C_1$ covers a range from ~ −75 to −60 ppt. The differentiation between the various components of thermogenic gas was proposed by Fuex (1977) and Stahl (1977) and is also incorporated in Fig. 2.15. Thermogenic methane gas associated with crude oil (T_o) or associated with condensate (T_c) is depleted in ∂D compared to dry methane gas. The gas derived from thermal cracking of oil is depleted in $\partial^{13}C_1$ −50 to −60 ppt over PDB and ∂D −245 to −340 ppt over SMOW. The condensate associated with thermogenic wet gas is enriched $\partial^{13}C_1$ and has a range of −40 to −30 ppt (Schoell, 1983). Methane in the deeper parts of the earth shows an appreciable decrease in ∂D as a consequence of high maturity conditions. However, $\partial^{13}C_1$ shows a distinct pattern depending on whether methane is derived from marine (sapropelic, $TT_{(m)}$) or humic sources (continental, $TT_{(h)}$). Mixing of bacterial and thermogenic gases occurs at shallower levels (M) and in deeper parts of the earth between $TT_{(m)}$ and $TT_{(h)}$.

As the pattern of thermogenic gases is mostly dependent on the maturity level that the organic matter or oil is subjected to, James (1983) plotted the difference between the isotopic composition of individual gases (methane to n-pentane) versus the ∂^{13}C as the calculated difference of gas components. Thermogenic gases data from different regions of the world suggested that the separation of carbon isotopes between the normal alkane components of the natural gas generally depends on the maturity at which the gas was generated, and the difference between the ∂^{13}C values of different alkane components is inversely proportional to maturity (Fig. 2.16, adopted from James, 1983).

Empirical relations between molecular and isotopic values of different gases have been used to generate cross-plots. Heating of oil from different locations was utilized (Chung et al., 1988) to ascertain the pattern and quantum contribution from each thermogenic component (Fig. 2.17b). The carbon isotope

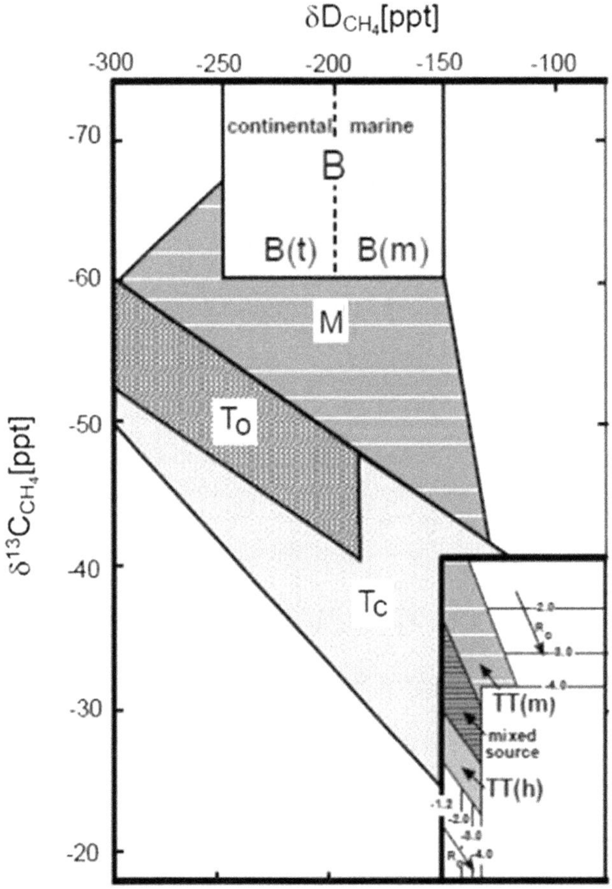

FIGURE 2.15

Formation mechanism for methane based on the pattern of deuterium $\partial D_{methane}$ and $\partial^{13}C_{methane}$. Methane formed at low temperatures by bacterial processes in the terrestrial B(t) and marine environment B(m). Methane formed due to thermogenic processes, thermal cracking of oil, thermogenic wet gas (To), and condensate-associated thermogenic gas (Tc). Thermogenic dry gas derived from humic organic matter [TT(h)] and dry gas derived from liptinitic organic matter [TT(m)]. (Adopted from Schoell, 1983.)

ratio was plotted against inverse carbon number (i.e., 1/n, n going from 1 to 5 for all main constituent components of gas) and called a "natural gas plot." The plots for oils from two different regions, a Miocene oilfield (Monterey, California, USA) and a Jurassic oilfield (Statfjord, Offshore Norway), indicate different characteristics. The Monterey oil shows a linear trend, whereas the Statfjord oil shows that n-butane and n-pentane are isotopically heavy with methane being more light (Chung et al., 1988).

Such plots have also been utilized to study the effect of mixing microbial and thermogenic gases. Fig. 2.18 shows the natural gas plot of data from the Gulf of Mexico and natural gas plots of gaseous hydrocarbons generated from laboratory heating experiments for the Statfjord and Monterey oilfields.

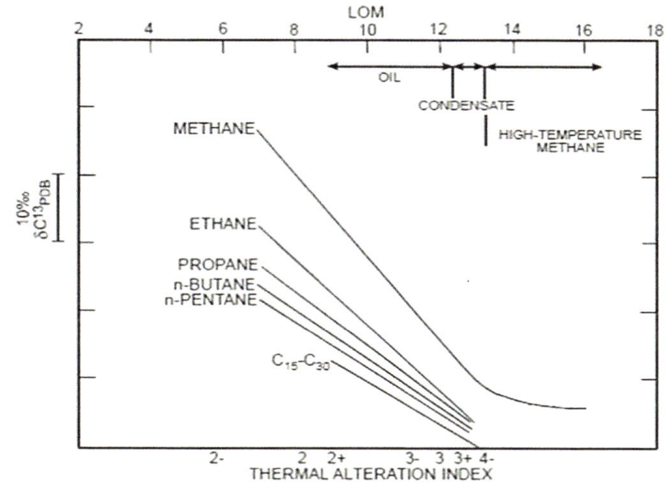

FIGURE 2.16

Plot of the difference between the isotopic composition of individual gases (methane to n-pentane) and the $\partial^{13}C$ as the calculated difference of gas components. Separation of carbon isotopes between the normal alkane components of the natural gas generally depends on the maturity. (From James, 1983.)

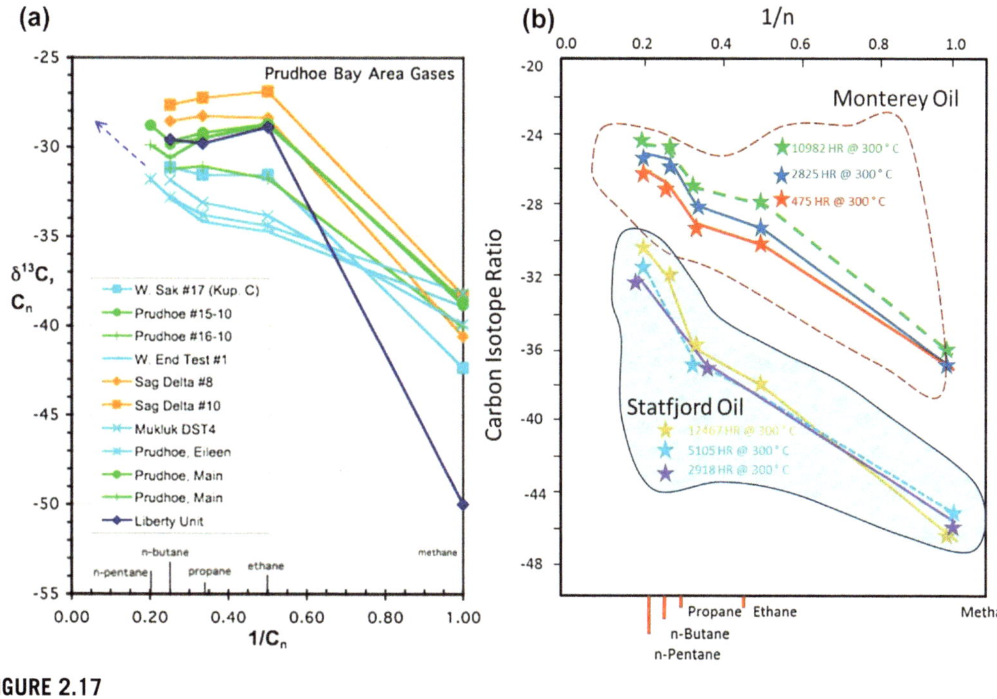

FIGURE 2.17

Natural gas plot (Chung et al., 1988) showing the carbon isotopic composition of individual hydrocarbons as a function of the carbon number (C_n) of the hydrocarbon. Ranges of the carbon isotopic composition of kerogen from the Shublik Formation, Kingak Shale, Torok Formation Shales, and the Pebble Shale unit are shown adjacent to the y-axis. Data are from Burwood et al. (1985) and Schoell et al. (1985). (a) Gases from the greater Prudhoe Bay area are shown (From US Geological Survey Open-File Report 03-041. http://pubs.usgs.gov/of/2003/of03-041/fig10c.htm). (b) Natural gas plots of gaseous hydrocarbons generated from laboratory heating experiments for Statfjord and Monterey oilfields. The Monterey oilfield shows a linear trend; the Statfjord oilfield shows that n-butane and n-pentane are isotopically heavy with methane being more light. (Modified from Chung et al., 1988.)

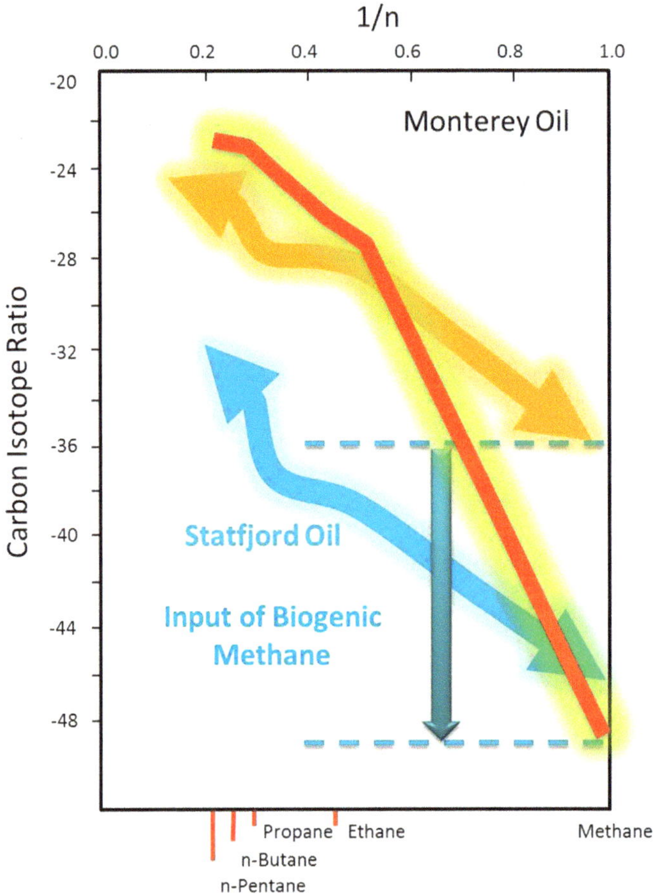

FIGURE 2.18

Procedure followed for estimating the percentage of biogenic methane in the mixed gas samples from the Gulf of Mexico based on the carbon isotopic ratio. (Modified from Chung et al., 1988.)

Studies suggest that the carbon isotope ratio of the obtained sample is very light (−58‰) indicating that the methane is of microbial (bacterial) origin. Extrapolation of the trend in carbon isotopic ratios for ethane and propane suggests that thermogenic methane has a carbon ratio of the order of −40‰. The studies suggest that the microbial contribution has ~43% of methane present in the reservoir (Chung et al., 1988).

2.4.4 SOLUBILITY OF METHANE

The solubility of methane is appreciably low – 0.045 volumes per unit volume of water. Methane dissolved in water can coexist with free gas at high temperature and with solid hydrates at high

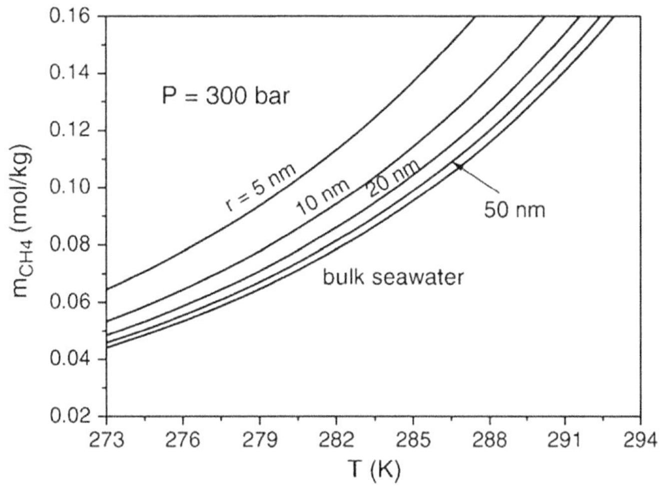

FIGURE 2.19

The variation of methane solubility in seawater at hydrate–liquid equilibrium (H–L) with pore sizes. (From Sun and Duan, 2007.)

pressures and low temperatures. In the absence of gas hydrates, the solubility increases with decrease in temperature and increase in pressure as per Henry's law (Osegovic et al., 2006). The equilibrium of methane solubility is important, because it determines the minimum quantum of methane required for the formation of hydrates. This decrease is attributed to the hydrate formed directly from aqueous solution without the need for free gas (Zatsepina and Buffet, 1997). The presence of hydrates facilitates further hydrate formation, hence the equilibrium concentration of gas in water decreases and gas solubility increases with increase in temperature (Duan and Mao, 2006; Davie et al., 2004). In this regard the numerical models for gas hydrate formation show that the shape of the methane solubility profile with depth has importantance for hydrate formation and has a bearing on the vertical distribution of hydrates in the sediments (Davie and Buffet, 2001). The peak solubility of methane occurs at the three phase equilibrium and coincides with the base of the hydrate stability zone. Within the hydrate stability zone the solubility is insensitive to pressure; it is only a function of temperature and increases with temperatures. At temperatures above the equilibrium phase curve, methane and water are in stable phases and solubility is a function of both temperature and pressure. Below the hydrate stability zone, pressure and temperature both increase with depth and affect solubility in different ways. The solubility increases with pressure but decreases with temperature. Below the hydrate stability zone there are hardly any appreciable changes in solubility due to the counteracting effects of pressure and temperature (Zatsepina and Buffet, 1997). The stability of methane hydrate and CH_4 solubility in the marine environment is illustrated in Fig. 2.19, which shows the variation of CH_4 with pore size at hydrate–liquid equilibrium (H-L). Methane solubility in porous media is greater than that in a bulk system for a given temperature and pressure since the capillary forces increase the pressure (Sun and Duan, 2007).

REFERENCES

Atreya, S.K., 2007. The mystery of methane on Mars and Titan. Sci. Am. 42–51.
Bernard, B.B., Brooks, J.M., Sackett, W.M., 1978. Light hydrocarbons in recent Texas continental shelf and slope sediments. J. Geophys. Res. 83, 4053–4061.
Brown, J.F., 1962. Inclusion compounds. Sci. Am. 207, 82–92.
Burke Jr., Martens, R.A.C.S., Sackett, W.M., 1988. Seasonal variation of D/H and 13C/12C ratios of biogenic methane in surface sediments. Nature 1988 (332), 829–831.
Burwood, R., Cole, G.A., Brozd, R.J., et al., 1985. Carbon isotope characteristic of some North Alaska petroleum and potential rocks, kerogen assemblages: basic source rock evaluation and carbon isotope, in Alaska North slope crude. AAPG Studies in Geology, USA 123–137.
Cappenberg, T.E., Jongejan, E., 1978. Microenvironment for sulfate reduction and methane production in freshwater sediments. In: Krumbein, W.E. (Ed.), Environmental Biogeochemistry and Geomicrobiology, Ann Arbor Science, Michigan, USA, pp. 129–138.
Cappenberg, T.E., 1975. Some aspects on the role of bacteria in deep water ecosystems. Plant Soil 43, 125–139.
Carlson, C.A., Ingraham, J.L., 1983. Comparison of denitrification by Pseudomonas stutzeri, Pseudomonas aeruginosa and Paracoccus denitrificans. Appl. Environ. Microbiol. 45, 1247–1253.
Carroll, J.J., 2002. Natural Gas Hydrates, a Guide to Engineers. Elsevier Publishing, Washington DC, USA p. 274.
Chung, M., Gormly, J., Squires, R., 1988. Origin of gaseous hydrocarbons in subsurface environment: theoretical consideration of carbon isotope distribution. Chem. Geol. 71, 97–103.
Claypool, G.E., 1974. Anoxic diagenesis and bacterial methane production in deep sea sediments. Ph.D Thesis. UCLA, USA.
Claypool, G.E., Kaplan, I.R., 1974. The origin and distribution of methane in marine sediments. In: Kaplan, I.R. (Ed.), Gas in Marine Sediments, Elsevier Netherland, pp. 99–116.
Clayton, C., 1991. Carbon isotopic fractionation during natural gas generation from kerogen. Mar. Petrol. Geol. 8, 232–240.
Coleman, D.D., Chao, L.L., Riley, K.M., 1988. Microbial methane in the shallow Paleozoic sediments and glacial deposits of Illinois, USA. Chem. Geol. 71, 23–40.
Collett, T.S., 2003. Natural gas hydrate as a potential energy resource. In: Max, M.D. (Ed.), Natural Gas Hydrate in Oceanic and Permafrost Environment. Kluwer Academic Publishers, London, pp. 123–136.
Craig, H., 1957. Isotopes standards for carbon and oxygen and correction factor for mass spectrometric analysis of carbon dioxide. Geochim. Cosmochim. Acta 12, 133–149.
Craig, H., 1953. The geochemistry of the stable carbon isotopes. Geochim. Cosmochim. Acta 3, 53–92.
Craig, H., 1961. Standard for reporting concentration of deuterium and oxygen-18 in natural waters. Science 133, 1833.
Daniels, L., Fulton, G., Spencer, R.W., Orme-Johnson, W.H., 1980. Origin of hydrogen in methane produced by methanobacterium thermoautotrophicum. J. Bacteriol. 141, 694–698.
Davie, M.K., Buffet, B.A., 2001. A numerical model for the formation of gas hydrates below the seafloor. J. Geophy. Res. 106, 497–514.
Davie, M.K., Zatsepina, O.Y., Buffet, B.A., 2004. Methane solubility in marine hydrate environment. Mar. Geol. 203, 117–184.
Doose, R.R., 1980. The Bacterial Production of Methane in Marine Sediments PhD Thesis. UCLA, p. 240.
Duan, Z.H., Mao, S.D., 2006. A thermodynamic model for calculating methane solubility, density and gas phase composition of methane-bearing aqueous fluids from 273 to 523 K and from 1 to 2000 bar. Geochim. Cosmochim. Acta 70, 3369–3386.
Durrand, B., 1980. Kerogen: Insoluble Organic Matter from Sedimentary Rocks. France, Paris, p. 519. Edition Technique.
Faber, E., Stahl, W., 1984. Geochemical surface exploration for hydrocarbons in the North Sea. AAPG Bull. 68, 363–386.
Fuex, A.N., 1977. The use of stable isotopes in hydrocarbon exploration. J. Geochem. Explo. 7, 155–188.

Gilmov, E.M., 1985. The Biological Fractionation of Isotopes. Academic Press, New York, p. 261.
Gold, T., 1993. The origin of methane in the crust of the earth. In: Howell, D.G. (Ed.), The Future of Energy Gases. US Govt Printing Office, Washington, pp. 57–81.
Hardage, B.A., Roberts, H.H., 2006. Gas hydrate in the Gulf of Mexico: what and where is the seismic target? Leading Edge 25, 566–571.
Hayes, J.M., 1993. Factors controlling ^{13}C contents of sedimentary organic compounds: principles and evidence. Mar. Geol. 113, 111–125.
Howell, D.G., Cole, F., Fanelli, M., Weise, K., 1993. An introduction to 'The Future of Natural Gas'. In: Howell, D.G. (Ed.), The Future of Energy Gases. USGS Professional Paper1570, pp. 1–20.
Huc, A.Y., Durrand, B.M., 1977. Occurrence and significance of humic acids in ancient sediments. Fuel 56, 73–80.
Huex, A.N., 1977. The use of stable carbon isotopes in hydrocarbon exploration. J. Geochem. Explor. 7, 155–188.
Hunt, J.M., 1996. Petroleum Geochemistry and Geology, second ed. W. H. Freeman, San Francisco, CA. 743 p.
Hunt, J.M., 1979. Petroleum Geochemistry and Geology, second ed. W.H. Freeman and Co, New York, p. 189.
James, A.T., 1983. Correlation of natural gas by use of carbon isotope distribution between hydrocarbon components. AAPG Bull. 67, 1176–1191.
Jenden, P.D., Kaplan, I.R., 1986. Comparison of microbial gases from Middle American trench and Scripps submarine canyon: implications for the origin of natural gas. Appl. Geochem. 1, 631–646.
Kotelnikova, S., 2002. Microbial production and oxidation of methane in deep subsurface. Earth-Sci. Rev. 58, 367–395.
Martens, C.S., Balir, N.E., Green, C.D., Des Marais, D.J., 1986. Seasonal variations in the stable carbon isotopic signature of biogenic methane in coastal sediments. Science 233, 100–1303.
Marty, D.G., 1992. Ecology and metabolism of methanogens. In: Vailley, R. (Ed.), Bacterial Gas. Editions Technip, Paris, p. 1324.
Mechalas, B.J., 1974. A guide to marine pollution. Marine Chem. 2, 153–154.
Merali, Z., 2004. Earth's mantle can generate methane. Nat. News. http://dx.doi.org/10.1038/news040913-5.
North, F.K., 1985. Petroleum Geology. Allen and Unwin, Boston, p. 607.
Osegovic, J.P., Tatro, S.R., Homan, S.A., 2006. Physical chemical characteristics of natural gas hydrates. In: Max, M.D., Johnson, A.H., Dillon, W.P. (Eds.), Economic Geology of Natural Hydrates, Springer, Berlin, pp. 45–104.
Philp, R.P., 2005. Formation and geochemistry of oil and gas. In: Mackenzie, F.T. (Ed.), Sediments, Diagenesis and Sedimentary Rocks, Elsevier, Netherland, pp. 223–251.
Prinzhofer, A., Huc, A.Y., 1995. Genetic and post-genetic molecular and isotopic fractionation in natural gases. In: Rice, D.D., Scoell, M. (Eds.), Sources of Natural Gas, 126. Chemical Geology Special Publication, pp. 281–290.
Quay, P., Stutsman, J., Wilbur, D., et al., 1999. The isotopic composition of atmospheric methane. Glob. Biogeochem. Cycles 13, 445–461. http://dx.doi.org/10.1029/1998GB900006.
Rice, D.D., Claypool, G.E., 1981. Generation, accumulation and resource potential of biogenic gas. AAPG Bull. 65, 5–25.
Rice, D.D., Clayton, J.L., Pawlewicz, M.J., 1989. Characterization of coal derived hydrocarbons and source rock potential of coal beds, San Juan Basin, New Mexico and Colorado, Usa. Int. J. Coal Geol. 13, 597–626.
Rice, D.D., 1993. Biogenic gas: controls, habitats and resource potential. In: Howell, D.G. (Ed.), The Future of Energy GasesUSGS Professional Paper, 1570, pp. 583–606.
Rosenfeld, W.M., Silverman, S.R., 1959. Carbone isotope fractionation in the bacterial production of methane. Science 130, 1658–1659.
Sassen, R., Chinn, E.W., 1990. Implication of Lower Tertiary source rocks in south Louisiana to the origin of crude oil, offshore Louisiana. In: Schumacher, D., Perkin, B.F. (Eds.), Gulf Coast Oil and Gases. GCSSEPM Foundation 9th Annual Research Conference Proceedings, pp. 175–179.
Sassen, R., Roberts, H.H., Aharon, P., Chinn, E.W., Carney, R., 1993. Chemosynthetic bacterial mats at cold hydrocarbon seeps, Gulf of Mexico continental slope. Org. Geochem. 20, 77–89.
Schoell, M., 1983. Genetic characterization of natural gases. AAPG Bull. 67, 2225–2238.

Schoell, M., 1988. Multiple origins of methane in the earth. Chem. Geol. 71, 1–10.

Schoell, M., 1980. The hydrogen and carbon isotopic composition of methane from natural gases of various origins. Geochim. Cosmochim. Acta 44, 649–661.

Schoell, M., Durrant, B., Ondiv, T.L., 1985. Migration of oil and gas in isotope and biomarker studies. Indonesia Petro Asso. In: Proceedings of 14th Annual Convention, Jakarta, pp. 49–56.

Schulz, H.D., Zabel, D., 2006. Marine Geochemistry. Springer Verlag, Berlin, p. 579.

Shipboard Scientific Party, 1985. Site 570 (Leg 84). In: Huene, R., et al. (Ed.), Proceedings of the Deep Sea Drilling Project. Initial Reports, 9. US Government Printing Office, Washington DC, pp. 3–424.

Silverman, S.R., 1971. Influence of petroleum origin and transformation on its distribution in sedimentary rocks. Proc. 8th World Pet. Congr. 2, 47–54.

Sloan, E.D., 2003. Review article: fundamental principles and applications of natural gas hydrates. Nature 426, 353–363.

Smith, M.R., Mah, R.A., 1980. Acetate as a sole carbon and energy source for the growth of Methanosarcina strain 227. Appl. Envion. Microbiol. 39, 939–999.

Stahl, W., 1973. Carbon isotope ratio of German natural gases in comparison with isotopic data of gaseous hydrocarbons from other parts of the World. In: Tissot, B., Bienner, F. (Eds.), Advances in Organic Geochemistry. Editions Technip, Paris, pp. 453–462.

Stahl, W.J., 1975. Kohlenstoff isotopenverthaltnise von erdgasen: reifekenzeichen ihrer muttersubstanzen. Erdoll Kohle 28, 188–191.

Stahl, W.J., 1977. Carbon and nitrogen isotopes in hydrocarbon research and exploration. Chem. Geol. 20, 121–149.

Suess, E., Bohrmann, G., Lausch, E., 1999. Flammable ice. Sci. Am. 281, 76–83.

Sun, R., Duan, Z., 2007. An accurate model to predict the thermodynamic stability of methane hydrate and methane solubility in marine environments. Chemi. Geol. 244, 248–262.

Tenenbaum, D., 2005. Methane on Earth. Astrobiological Magazine. http://www.astrobio.net/news-exclusive/methane-on-earth/.

Tissot, B., Durand, B., Espitalie, J., Combaz, A., 1974. Influence of nature and diagenesis of organic matter in formation of petroleum. AAPG Bull. 58, 499–506.

Tissot, B.P., Welte, D.H., 1978. Petroleum Formation and Occurrence. Springer Verlag, Berlin, p. 538.

Tissot, B.P., Welte, D.H., 1984. Petroleum Formation and Occurrence, second ed. Springer Verlag, New York, p. 699.

Van Krevelen, D.W., 1950. Graphical-statistical method for the study of structure and reaction processes of coal. Fuel 29, 269.

Wang, W.C., Zhan, L.Y., Liu, W.H., Kang, Y., Ren, J.H., 2005. Effects of biodegradation on carbon isotope composition of natural gas: a case study in the Bamianhe oil field of the Jiyang Depression. Geochem. J. 39, 301–309.

Webster, M., 1994. Webster Dictionary. Houghton Miffin Co., Boston, p. 1536.

Whiticar, M.J., Faber, E., Schoell, M., 1986. Biogenic methane formation in marine and freshwater environment: CO_2 reduction vs acetate fermentation – isotope evidence. Geochim. Cosmochin. Acta 50, 693–709.

Whiticar, M.J., 1994. Correlation of natural gases with their sources. In: Magoon, L.B., Dow, W.G. (Eds.), The Petroleum System: From Source to Trap, 60. AAPG Memoir, Tulsa, USA pp. 261–284.

Whiticar, M.J., 1999. Carbon and hydrogen isotope systematics of bacterial formation and oxidation of methane. Chem. Geol. 161, 291–314. http://dx.doi.org/10.1016/S0009-2541(99)00092-3.

Wiese, K., Kvenvolden, K.A., 1993. Introduction to microbial and thermal methane. In: Howell, D.G. (Ed.), The Future of Energy Gases. USGS Professional Paper1570, pp. 13–20.

Woltemate, I., Whiticar, M.J., Schoell, M., 1984. Carbon and hydrogen isotopic composition of bacterial methane in shallow fresh water lake. Limno. Oceanogr. 29, 985–992.

Woltemate, I., 1982. Isotopische Untersuxhungen zur bakteriellen Gasbildung in Einem Susswassersee, Diplomarbiet. Technical University Clausthal, Clausthal Zellerfeld, p. 90.

Zatsepina, O.Y., Buffet, B.A., 1997. Phase equilibrium of gas hydrate: implication for the formation of hydrate in the deep sea floor. Geophys. Res. Lett. 24, 1567–1570.

CHAPTER

SEDIMENTATION PATTERN

3

"If you wish to make an apple pie from scratch, you must first invent the universe."

Carl Sagan, Cosmos

CHAPTER OUTLINE

3.1 INTRODUCTION

Since the formation of the earth, over long periods of time running into millions of years, geodynamic forces (such as plate movement, tectonic cycles, large-scale denudation, and sedimentation) have been active resulting in the formation of continental margins. These mechanisms also control the shaping of continental margins, their configuration and landscape. Since the time of earth's formation, physical, chemical, and biological factors have acted upon continental and oceanic crust in an effort to smoothen the elevated regions on land and deepen and fill the oceanic basins in order to maintain a natural balance between land and the marine environment. The exposed rock caps on the land are subjected to low-temperature weathering processes and high-temperature volcanic activity (Chester, 1990). Additionally, anthropological activities over long periods of time have acted as weathering agents in their attempt to reshape the landscape. The physical and chemical processes operating on the surface of the earth involve three distinct mechanisms: weathering, transportation, and deposition. Weathering and erosion are terms frequently used for the characteristic changes that the rock undergoes due to the physical and chemical processes operating on the surface of the earth. However, the process of weathering involves both mechanical and chemical activity in shaping rock properties. Rocks on the surface of earth are either disintegrated (mechanical breaking down) or decomposed (chemical reaction). Disintegration of rock primarily results from thermal heating, variation in temperature, the effect of plant roots, and to an extent burrowing animals. The process only involves the breakdown of rock without altering the chemical properties of the host bedrock. Disintegration reduces the thickness of the crust

Geological Controls for Gas Hydrates and Unconventionals. http://dx.doi.org/10.1016/B978-0-12-802020-3.00003-5

on which different physical processes operate. In the process of chemical weathering, the host rock undergoes chemical change due to alteration in its mineral properties and this brings about distinct changes in the chemical constituents of the host rock. Certain constituents are more soluble and more readily attacked and hence, become part of the solutions carried away from the site of weathering. Water percolating through fractures plays an important role in terms of disintegration and decomposition. At low temperatures, water which percolates into fractures solidifies and expands, making the rock more susceptible to disintegration, and with an increase in temperature, water may flow out of the fracture zone carrying dissolved minerals. This chemical action does not significantly reduce the layer thickness on which it operates. Chemical weathering is primarily controlled by the chemical activity of water on the exposed surface area of the rock, the susceptibility of the rock (chemical composition), and the prevailing climatic conditions. The weathering process also depends upon the surface of rock exposed and therefore to a large degree upon the amount of disintegration, which will increase the area of exposed rock surface (Twenhofel, 1939). Only a small portion of the bedrock on the surface of the earth is exposed because the majority of the land surface is covered by vegetation. The process of disintegration promotes chemical weathering by exposing the bedrock to different agencies. Both processes are confined to the surface of the earth. Climatic conditions near the source rock and agencies acting upon it determine the extent of disintegration or decomposition of the rock.

3.2 EROSION, TRANSPORTATION, AND DEPOSITION

Erosion refers to the movement of weathered material from the site of denudation. Loosened rock particles and degraded organic matter are carried away from the source area by natural agencies like rivers, glaciers, winds. These particles on their onwards journey are subjected different physical conditions and chemical alteration before being deposited in oceanic basins and lakes. The two processes of weathering and erosion operate concurrently. Erosion is also closely related to transportation because individual particles often come to rest between the time they are loosened from the bedrock until they reach the place of accumulation (Hjulström, 1955). Generally, erosive agents remove sediments from the site of weathering in one of the following ways: impact of the agent, abrasion (mechanical erosion or corrosion), or corrosion (chemical erosion). Eroded material from the land is transported to the sea by different agencies. There are three categories of load that may be transported by an erosion agent: dissolved load, suspended load, and bedload. Rivers carry particulate (suspended), dissolved matter, and bedload. Bedload refers to heavy sediment particles that are in constant contact with the bedrock and move forward by rolling, skipping, or sliding along the bottom. Glaciers and sea ice, winds, and volcanic and biological activity all contribute to the transport of eroded materials. Material transported by rivers and erosion of the shoreline by waves deposits most of the material closer to the shore. However, wind, volcanic activity, and glaciers or icebergs carry the material load over long distances from the coastline. Sediments are most often transported by water (fluvial processes), wind (aeolian processes), and glaciers. The movement of the water column under the wave and tidal action results in the deposition of coarse material closer to the coastline. Beach sands and river channel deposits are examples of fluvial transport and deposition, though sediment also often settles out of slow-moving or standing water in lakes and oceans. Desert sand dunes and loess are examples of aeolian transport and deposition. Glacial moraine deposits and till are ice-transported sediments. The entire process of erosion, transportation, and deposition of grains and decomposed organic materials depends on the velocity of water, climatic conditions at the erosion and deposition sites, as well as the nature of the rock being eroded. Vegetation, cultivation of land, and climatic (precipitation, glaciation) and geological or tectonic (modifying the terrain configuration) factors have a great influence on the process of erosion.

Running water (runoff) not only facilitates erosion of the land but also transports materials to the sea for deposition on the seafloor. Gravity is the main force for the movement of particles (grains) and it works in concert with water and glaciers. Erosion by rivers takes place in two forms, i.e., interfluvial and by scouring the river channel. Rivers primarily transport rock particles derived from erosion of interfluvial areas. The movement of water is the main cause for erosion or decomposition of the rock and organic matter (OM) on land in the form of rivers or ice, over the sea surface, or deep at the seafloor. Climatic conditions favoring higher precipitations (rainfall) provide a higher rate of runoff and feed a substantial amount of eroded materials to the rivers. The eroded material (soil) is transported either as suspended particles or dissolved matter in water. The net quantum of eroded land surface is estimated either by the mass that is carried by the rivers (Holeman, 1968) or from the surface denudation of continents (Schumm, 1963). However, all the rock mass that is eroded may not reach the ocean as part of the eroded soil gets deposited in the river basins or valleys, making estimates derived from surface denudation of land appreciably higher (Milliman and Meade, 1983).

Rivers not only transport rock material from the site of weathering but during their journey towards the deposition site, they also erode the bedrock (scouring of river valleys). Rivers and streams even take the grains covering bedrock into suspension and carry them forward during the onward journey. Scouring of the river valley requires an appreciable velocity of running water to overcome the cohesiveness of particles in the riverbed for their erosion and transportation.

Wind-driven mechanisms are some of the most important factors involved in transporting fine-grained particles over large distances on land (in arid desert regions, volcanic ejected ash, gases emitted through anthropogenic activity, and biomass particles from constituents of aerosols) and over sea. Erosion and transportation by the wind is more prevalent in regions of high wind velocities and in areas which are not covered with thick vegetation. In mountainous, desert, and semiarid regions where the wind velocities are appreciably higher, the transportation of matter is promoted by wind energy. In order for particles from the surface of land to be transported large distances over the sea, they need a correct trajectory to carry the matter to a higher altitude where wind velocities are appreciably higher. However, most of the continentally derived matter is initially ejected into the planetary boundary layer of the atmosphere. This layer exists at about 1000–1500 m over land and about 300–600 m over sea (Hasse, 1983). The marine boundary layer is responsible for transporting most of the continental fine particles in the vicinity of the coast for short distances over the sea. If the particles from the ground are to be carried for great distances then they need to be in the tropospheric heights. Long-distance particle transportation requires dry surface conditions, strong surface winds, small particle size, appropriate trajectory into the troposphere, and vertical mixing in the troposphere (Kennet, 1982). Retention time in the order of days in the stratosphere, prevailing wind direction, and high velocities may result in transport of particles for long distances over the sea. In some localities, the amount of material transported in the air over the sea is sufficient to form dust clouds. It has been reported that matter originating in dust storms over the African Sahara desert has been found a long distance away in the deep-sea sediments of the Atlantic Ocean (Folger, 1970). During volcanic activity ash is ejected to great heights and can enter the troposphere, which then transports the particles for long distances over the sea. In recent times, eruption of the volcano near the Eyjafjallajökull glacier in Iceland caused a vast cloud of volcanic ash to be swept across many parts of northern Europe and the ash was carried more than 1000 km by winds high up in the atmosphere. Sands transported by wind tend to be well sorted and rounded compared to those transported by other agencies.

On earth, 99% of glacial ice is contained within vast ice sheets in the polar regions. The transportation of sedimentary grains and sedimentary debris by ice has been and still is extremely important in high latitudes, specifically in the Arctic and Antarctic regions. Glaciers carry large amounts of material that they erode from the land surface and embedded fine rock particles and OM for long distances to the sea. Owing to the high viscosity of ice, sediments deposited directly from the melting of a glacier can range from very poorly sorted to better sorted and may undergo little rounding. Colluvial sediments from the adjoining valley walls are incorporated in the ice and they move down with gravity. The textural pattern of sediments depends on the extent of water transport after the ice melts. As the ice moves, material from underlying bedrock or sediment is picked up by the glacier and "held" in the ice. Material falling onto the surface (often the result of freeze-thaw activity or frost shattering of surrounding rock walls) is also transported and often finds its way down through crevasses to the base of the glacier. Basal debris is also incorporated as the constituents of glacial sedimentation. Material held within the glacier is called "englacial moraine." It is this material trapped in the ice that allows the glacier to erode its surroundings. The nature of the rock that the glacier has passed over defines the character of debris. Cold glaciers and those under transition from warm to cold hold a thick layer containing sediments (Shaw, 1977; Weertman, 1961). Glaciers sculpt the underlying landscape by abrasion and the rate of abrasion depends on the thickness of ice, the presence of water, and the hardness of basal debris relative to the bedrock. During the thawing period, melting takes place at the surface; ice below the water starts rising due to buoyancy and in the process also moves partially settled sediment particles. The number of particle sizes that can be carried by a glacier is tremendous, ranging from very fine ground rock termed "rock flour" to massive boulders. Icebergs also carry with them the enclosed rock particles, shells, and remains of organic material as they float great distances from the source, and pebbles and boulders that may be found in otherwise pelagic deposits far from land. In addition to subsurface alterations, glacier movement produces a number of distinct erosional and depositional landforms and structures. The fjords of Norway stand out as examples of bedrock erosion by moving glaciers. The erosion and transportation of sediments by glacial and iceberg agencies as input to marine sedimentation over the entire globe is low as this activity is confined to a small region.

Most of the natural processes are affected by the biggest force of universe which is gravity. Mass movement or mass wasting is one of the natural processes that wears away at mountainous slopes and delivers fine and coarse sediment to waterways (Jones, 1995). This is a movement of soil or rock mass (regolith soil, loose stones and rocks) downslope under the influence of gravity. It differs from mass transport, which is described as a movement caused by wind water or ice. Mass wasting includes all processes by which soil and rock materials are dislodged and transported downslope under the direct application of gravitational stresses. It includes slow displacement such as creep and earthflows, as well as rapid movements including rockfall and debris slides. Generally mass wasting can be classified into three categories;

- Flow: material moves as an unconsolidated fluid
- Fall: material moves as a consolidated mass along a surface of slip
- Slide: free fall

Flows can be divided into five categories including debris flow, debris avalanche, earth flow, mudflow and creep. These differ from one another in fundamental ways. Falls are the abrupt movements of mass of geological material such as rocks and bounders that becomes detached from steep slopes or cliffs. These are strongly influenced by gravity, mechanical weathering and presence of interstitial water. The term "landslide" describes many processes that result in the movement of slope forming material such as rock mass, soil, and artificial fill. Slides can be rotational or translational. Typically, a number of elements will contribute to a landslide, but often there is one which triggers the movement of material. Natural causes include earthquake vibration, volcanic eruptions, undercutting of cliffs and elevation of pore water pressure by saturation of slope material from both intense or prolonged rainfall and seepage. Human causes also contribute in occurrence of land slides. These include vegetation removal, interference with natural drainage system, and modification of slope by construction of structures, overloading the slopes, mining, heavy traffic vibration etc.

Turbidity currents are one of the most important modes of transporting the suspended particles from continental shelf to Deep Ocean. Turbidity currents mainly arise from density difference between two mediums. The steep gradients in continental slope gravity must act on the sediments to pull them down the slope and along with it pulls down the slope, and in process pushes dense water aside creating turbidity current conditions. During motion the currents become self accelerating as its scours the sea bottom putting more sediment in suspension and increasing density configuration and thereby increasing the speed of turbidity currents. In deeper part of ocean owing smaller gradients in seafloor relief the currents weaken out depositing sediments in deep sea floor (Pinet, 2009, 2014).

A less significant amount of debris may be transported in the sea by the buoyant organic material from land and marine environments. Trees and clumps of vegetation eroded during floods or by wave action may float over great distances before they decompose or settle. Leaves, branches, and other organic constituents from land are carried for long distances over the sea before they coagulate and mix with fine sediments and sink to the bottom.

The particulate matter that is being transported for deposition on the seafloor is further subjected to different physical, chemical, and biological processes and is always in motion until it becomes compact and resists further degradation. Many of the sediment particles which reach the oceans are dispersed by currents before they settle down on the ocean floor. The particles oscillate and their movement is controlled by the waves, tidal forces, and the strength of the current. Particles subjected to higher current velocity and turbulent conditions always keep the fine particles in suspension, re-suspending finer particles which have settled momentarily. Particles under suspension are always acted up on by tidal and ocean currents. Depending on the magnitude and strength of these two the sediment load starts settles down on the ocean floor as these forces weaken segregating coarse and fine grained sediments from coast. Such conditions in the deep oceans do not receive a supply of coarse-grained sediments. They are deposited under high energy conditions and confined to regions where the river load is dumped in the ocean.

Particulate matter in the oceanic water column is confined to two distinct depths. In the shallower parts above the thermocline, particulate matter is appreciably higher due to the influx of matter from the top layers; in the deeper depths, just above the seafloor, the particulate concentration increases again due to suspended particles resulting from vertical mixing and advection of sediments from the seafloor (Biscay and Eittreim, 1977). The water column trapped between two particulate high concentrations is quite clean. These two distinct layers with turbid (cloudy) particle matter are termed "nepheloid layers." The top layer mainly contains OM and is mainly composed of surface plants (phytoplankton) undergoing photosynthesis. The deeper layer primarily consists of suspended particles as the result of high currents operating at the seafloor. Intermediate nepheloid layers have also been identified in various parts of the world (Pak et al., 1980).

Under normal sea floor relief conditions the sorting of sediment grain size distribution is governed by the strength of the oceanic currents. Grain size characterizes type of sediments. In normal sea floor relief, turbulent (higher current strength) water conditions fine grained particles are constantly in agitated sated and remain suspended in water. Suspended finer grains can easily be carried to large distances from coast in with average current strength. Quieter current conditions in deeper part of the sea over large areal extent make the finer sediment grain well sorted and favourable regime for settlement on the sea floor. Sediment like clays with fine grained particle can remain in suspension over long period of time and can be carried to long distances by average ocean currents. On the contrary near coast turbulent condition make the coarse grain size not well sorted and to settle over short distances. Owing to pattern of grain size in the sediment sample fare estimate of current energy environment and probable time of its deposition can be ascertained (Pinet 2009, 2014).

The strength of ocean currents required for erosion, transportation, and deposition in deep waters is not precisely known (Kennet, 1982). Laboratory experiments have provided a working relationship between the grain size and current velocity that specifies whether sediments of a particular size will be eroded, transported, or deposited under a given set of energy conditions. The general rules derived from these experiments are summarized in the Hjulström diagrams (Fig. 3.1). This chart plots the average current velocity versus particle diameter on a log scale. Based on the laboratory results Hjulström diagram provide the information on the strength of the current required for different size of the particles to be either eroded from the sea floor transported or deposited. The pattern in general indicates that dependence on grain size with current strength (energy). However in some cases minerology of the rock type plays important role in characterizing the energy-grain size dependent relationship. Large diameter grain sizes such as gravel require higher energy to erode but it cannot transported and deposited to large distance owing to absence high energy requirement in deeper part of ocean. Sand particle at the sea floor comparatively require lower energy of the currents to erode and its transportation and deposition. However, due to the mineral content silt and clay irrespective of their finer grain size require higher energy environment to erode the particles at sea floor but require lower energy requirement for their transportation and deposition (Hjulström 1939; Sundborg 1956). These studies provide basic information about the strength of current required for erosion, transportation, and deposition in streams. Erosion of unconsolidated clay and silt requires bottom currents of the order of 10–20 cm/s whereas velocities of 20–40 cm/s are required for the erosion of sand size particles (Flood, 1978). Simulation deepwater environments in the laboratory for pelagic, biogenic sediments have suggested that critical velocities of the order of 15–35 cm/s are required for erosion of carbonate oozes. But higher current strength is required for eroding consolidated sediments (Gardner, 1978; McCave, 1975). Once the sediments are brought into suspension they require lower velocities to transport them over greater distances. The particle matrix influences transport dynamics because of the dependencies of suspension and settling on particle density, size, and shape (Paphitis et al., 2002). The consequence of water velocity and particle characteristic dependencies is that the hydrodynamic sorting of fine-grained (clays and/or low density small plant fragment materials from coarse or denser material) sediments will be most pronounced in the low gradient systems or during low discharge events. Upland regions of high gradients will mobilize larger aggregates of sediment that will include biogenic debris (logs) and rock fragments, ranging from sand to boulders (Blair and Aller, 2012).

The rate of sedimentation is also one of the major factors for the sorting of grains. With a slower rate of erosion, the sediment supply to the ocean slows down providing more time for water to sort grains by their shape, size, and density. A faster rate of erosion leads to a higher quantum of sediment and may not allow water much time for sorting, resulting in sediment that is poorly sorted and heterogeneous, having coarse-grained sediments such as gravel sediment mixed with coarse-grained sand and sand mixed with mud.

Sediments are the ultimate product of weathering, erosion, transport, and chemical reactions of the particles (mineral grains) and the OM within and under the water column of the sea and near seafloor environment. They are part of the early diagenetic process in shallower parts of subsurface earth,

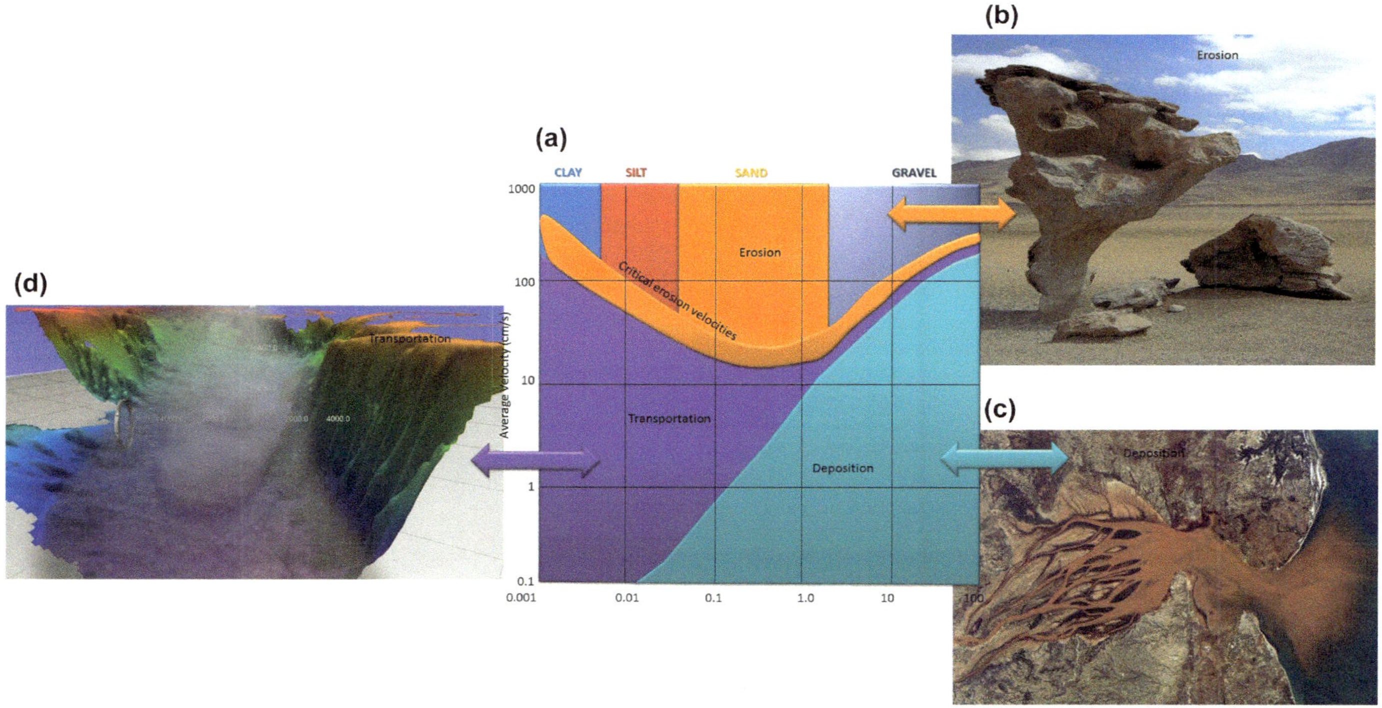

FIGURE 3.1

Average current velocities required for erosion, transportation, and deposition. Velocities determined based on the laboratory experimental data following the procedure adapted by Hjulström (Hjulström diagrams; from Hjulström, 1939; Sundborg, 1956). (a) Hjulström–Sundborg diagram showing sediment particle size and water velocity. (b) A rock sculpted by wind erosion in the Altiplano region of Bolivia. (c) Aerial photograph showing the deposition of reddish brown sediment. The sediment is carried by water and deposited forming a delta (group of islands) at the mouth of the river. The river flows from left to right. Note the sediment being carried into the sea (blue) on the right. (Source: http://mamontoff.org/weathering%20-%20erosion%20-%20deposition.htm). (d) A submarine canyon, a big indention in the seafloor, transports sediments from the nearshore to deeper water. Sediment from the Himalayas is ultimately deposited in the Bengal Fan, where scientists aboard the JOIDES Resolution drill sediment cores to study the mountain range and past climates. (Credit: Lisa Strong: http://blogs.agu.org/geospace/2015/03/25/on-the-joides-resolution-turbidite-transport-video/.)

ranging from a couple of centimeters in the shallow water column to a couple of meters in the deeper sea. These processes are involved in lithification and compaction of shallow sediments. Therefore, sediments constitute mineral grains and/or OM derived from the continental and ocean fossil remains of dead organisms, under prevailing climatic conditions.

3.3 GENESIS OF SEDIMENTARY ROCKS

The fabric and genesis of sedimentary is closely connected to minerology of host rock and the physical changes (texture) and chemical alteration (minerology) it under goes during erosion, transportation and deposition at final destination. During subsequent climatic changes these rocks are further altered. Study of sedimentary rock helps in revealing their mode of genesis and history of climate changes that occurred during ancient times. The constituent shape and size of the particles in the sedimentary rock reflect the extent of erosion and transportation the grains have undergone and draw inferences about the pattern and magnitude of velocity associated with shaping the sedimentary rock. The chemical elements that constitute the sedimentary rock provide information about its genesis and the absence of some elements in the host rock.

During transportation most of the suspended particles are sorted based on their size, shape, and specific gravity. This sorting operation has been categorized into two types: local sorting takes place locally or at the site of deposition. Particles are also progressively sorted along the direction of transportation. Both types of sorting depend on the size, shape, and specific gravity of the particles and the velocity, turbulence, viscosity, and specific gravity of the transporting agency (Russel, 1939). Based on the size of the particles, they are characterized into three distinct categories (Table 3.1). However further classification has also been used extensively in the literature.

Table 3.1 Wentworth grain-size scale. (Source: From C. K. Wentworth, 1922)

Wentworth grain-size scale		
Sediment	**Type**	**Diameter (mm)**
Gravel	Boulder	>256.0
	Cobble	65.5-256.0
	Pennle	4.0-64.0
	Granule	2.0-4.0
Sand	Very Coarse	1.0-2.0
	Coarse	0.50-1.0
	Medium	0.25-0.50
	Fine	0.125-0.25
	Very Fine	0.0625-0.125
Mud	Silt	0.0039-0.0625
	Clay	0.0002-0.0039
Colloid		<0.0002

Particle sizes greater than 2 mm are categorized as gravel and cover sizes ranging from 2 mm to >256 mm. The subdivision refers to granules, pebbles, cobbles, and boulders. Sand particles have grain sizes from 0.0625 to 2 mm. The entire range is further divided into very fine, fine, medium, coarse, and very coarse sands. The third category refers to mud, incorporating minute particles whose size can be distinguished by manual means either by smoothness of touch (silt) or smoothness of particles by crushing under teeth (clay). Colloid refers to particles whose sizes have to be determined under the microscope.

Based on the particle constituents and their mineralogical content, the sedimentary rocks have established nomenclature. Sedimentary rocks with predominantly higher particle size, such as gravels, are called conglomerate or breccia. The difference between these types of rocks is governed by the shape of the grains in the rock. Breccia have angular grains whereas conglomerate have gravels cast with rounded corners.

Sedimentary rocks mainly constituting sand particles are called sandstone. The sandstone comprises sand grains of the minerals quartz, feldspar, and other rock fragments. If the sandstone is mostly made up of quartz particles, it is called quartz sandstone. However, if the percentage of feldspar grains is appreciably higher along with quartz grains then it is called arkosic sandstone.

With a higher percentage of silt component the rocks are called siltstone and rocks having significant contributions from particles of clay are termed as claystone. For most rocks with a finer grain content, i.e., either silt or clay particles, the degree of bedding determines the nomenclature: rocks with a clear pattern of bedding are called shale and those without the bedding pattern are called mudstones.

Sedimentary rocks have been classified based on their mineral composition and this classification often refers to the genesis of the rocks. The data collected during the Deep-Sea Drilling Project (DSDP) expeditions resulted in sediment classification which is mostly descriptive and also includes some genetic relation to the source rocks (Kennet, 1982). In the descriptive component type of classification, sediments have gained their nomenclature based on the percentage of major mineral constituents (Hayes et al., 1975).

3.4 SEDIMENTATION IN OCEANS

Rocks of the continental and oceanic crusts are subjected to quite different physical and chemical conditions resulting in end products that differ significantly in their characteristics. In the early tectonic period during the rifting of continents, sedimentation was mainly due to the erosion of uplifted shoulders of rifts. The morphology of continental margins controls the processes of sedimentation and nature of sediment deposits and is closely connected to the evolution of the margins. Rupturing, rifting, breakup and erosion depositional history since seafloor spreading reflecting the complex sequences of geological events resulting in a great variety in the continental margin's dimension and style. Over the passive margins drifting of crust and mantle from the spreading ridge shoulders leads to their cooling, contraction and thickening and erosion of lifted shoulder at the ridge provides initial sediment deposits on the drifted part of the crust. On the contrary convergent margins the oceanic crust along with their sediments is stacked with continental, oceanic or island sediments over the trench axis resulting huge pile complex sediments on the convergent margins. Sediment on the seafloor originates from a variety of sources, including biota from the overlying ocean water, eroded material from land transported to the ocean by rivers or wind, ash from volcanoes, and chemical precipitates derived directly from seawater. A very small amount of sediment even originates from interstellar dust. In short, the particles found in sediment on the seafloor vary considerably in composition and record a complex interplay of processes that have acted to form, transport, and preserve them. Eroded continental mass and along with degraded organic matter is transported to oceans, where the fragmented rock particles are acted up on by oceanic currents to carry these particles to different regions of ocean for their deposition depending on the particle size of sediments and strength of the current prevailing that time in the oceans. Rate of sedimentation, (time duration of erosion) also controls the pattern and texture of sediments. During period lower rate of sedimentation the oceanic currents sufficient time to sort the sediment grains and leading uniform distribution of grain size in the sediment deposits. Faster rate of sedimentation sediment grains are not sorted well and sediments deposits have contribution from different grain sizes. Well sorted sediment grains may lead to uniform distribution of grains leading deposits such mud, sands, very fine sand and silts. High rate of sedimentation leads to sediments with mixed grain sizes such as gravel mixed with sand and sand mixed with mud (Pinet, 2009).

Marine deposits are broadly categorized into two distinct groups, terrigenous and pelagic. Most of the terrigenous sediments are deposited on the continental margins near the shore – river deltas, bays, estuaries, as well as continental rises. However, biogenic ooze and pelagic sediment is confined to the deep and quiet environment of the sea (see Fig. 3.2).

The pattern of worldwide distribution of sediment indicates latitude dependence, influence of large scale eddies, productivity of organic matter, nature of terrigenous inputs and climatic changes (Fig. 3.3). Biogenic sediments cover a large portion of continental shelves in the equatorial region and their areal

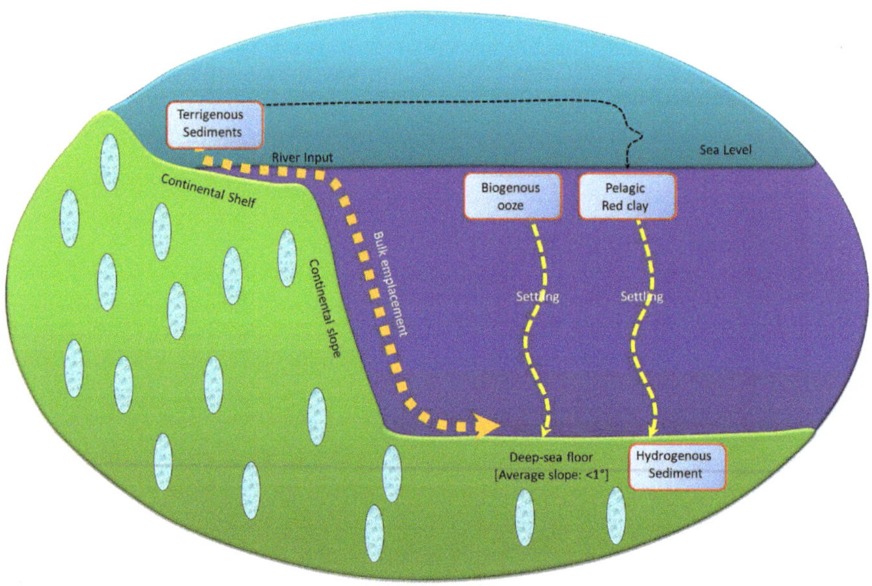

FIGURE 3.2

General seafloor morphological features. Different modes of transportation and types of sediment deposition into the sea.

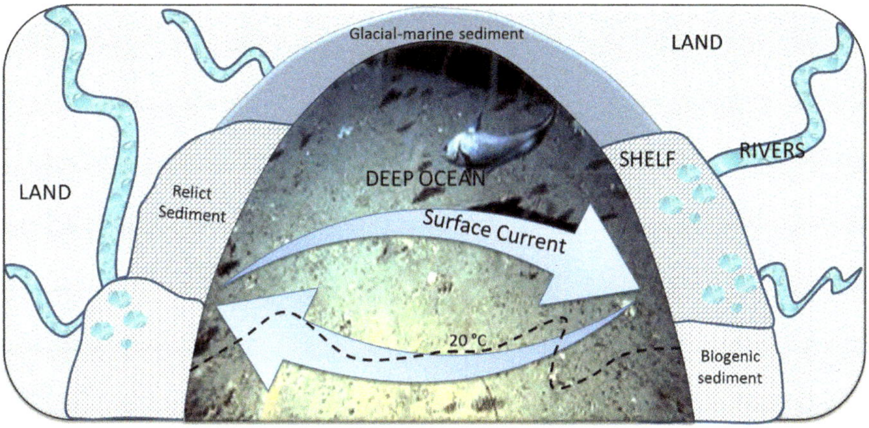

FIGURE 3.3

Distribution of shelf deposits. Sediments on the continental shelves of the world vary with latitude: biogenic sediments dominate the low latitudes, terrigenous sediments the middle latitude, and glacial-marine sediments the polar regions. (Modified from Davies and Gorsline, 1976; Emery, 1969.)

extent over western region extends to subtropics. On the contrary the eastern region the extent of these type sediments is rather confined to equatorial shelves. These sediments include coral reefs, remains hard parts of organism contributing the calcium carbonate rich sediments in these regions. The difference of their extent over the eastern-western regions may be attributed to difference temperature regime produced by large surface gyre in the ocean, which keeps the western region warmer and cold descending loop of the current from poles to equator making the western region little colder. Owing to temperature difference and the survival of coral and other calcareous matter to higher temperature makes the difference in pattern of distribution of these biogenic sediments over the equatorial region asymmetric. Higher regions of biogenic matter productivity are confined to eastern edges of oceans due to upwelling of nutrients and convergence of major oceanic gyres (Whitman and Davies, 1979). Continental shelves over the middle latitudes are mainly covered by terrigenous sediments brought in by mighty rivers that denude the land mass which primarily granitic in nature. Polar region a narrow region of the shelf is covered by sediments unsorted sediments like boulders, coarse grained gravel and mud dumped by glaciers, icebergs (Pinet, 2009, 2014). Higher regions of biogenic matter productivity are confined to eastern edges of the oceans due to the upwelling of nutrients and the convergence of major oceanic gyres (Whitman and Davies 1979).

Marine sedimentation primarily results from the combined contributions from external (allochthonous) terrestrial and internally produced matter (autochthonous) in the ocean water due to physical, chemical, and biological processes. Erosion of the continental landmass and its transportation by fluvial means, wind and glaciers, contributes significantly to shallow marine sedimentation. Terrigenous sediments owe their genesis to the breakdown of crystalline igneous rocks. The disintegration of rock mass provides particles of different grain size but still particles retain the mineralogical constituents of the host rock from which they are derived. Mighty rivers like Ganges erode the high mountains like Himalayas and dump the disintegrated fragments in to the nearby ocean to contributing terrigenous sediments in temperate latitude. Rivers not only carry terrigenous rock matter but also supply nutrients

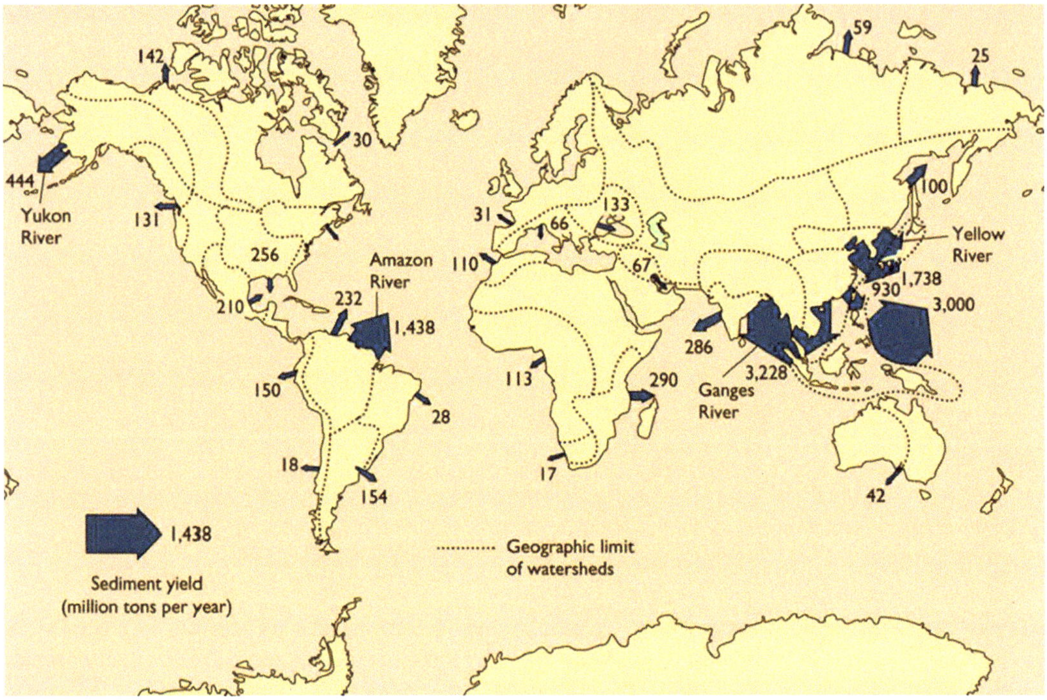

FIGURE 3.4

Worldwide annual discharges of suspended sediments from various drainage basins. Width of arrow indicates the magnitude in million tons. (From Milliman and Meade, 1983.)

(such as nitrogen, phosphorus) for supporting biological activity in the sea. Coarse-grained unsorted sediments, also transported by glaciers, are mainly confined to higher latitudes. Fine-grained sands from desert areas and arid regions are blown large distances over the sea to be deposited in the deeper part of the ocean. Terrigenous rocks are also transported by glaciers. Terrigenous sediments make up 45% of all the sediment that is deposited in the ocean. Their distribution over the globe is not uniform; the thickest accumulations (~10 km) occur in continental rises. Most of this type of sediment is confined to the mid-latitudes where mighty rivers like the Ganges transport a large quantum of eroded matter to the sea (Fig. 3.4).

Sediments in the ocean can be produced by either by authogenic, allochthogenic or cosmogenic means. In the authogenic process the sediments are formed in situ conditions, like precipitation of dissolved non-carbonate (silicates), and carbonates, hydrogenous Fe-Mn oxyhydroxides, organic matter through photosynthesis and decay of matter and its burial. In allochthogenic (terrigenous) process the sediment grains, degraded organic matter and soil is brought in oceanic regions via run-off rivers/streams and other modes of transportation. The volcanic ash and cosmogenic fine grained matter is mainly deposited in deep sea floor by wind driven processes. The contribution of external means comprising mainly allchthogenic processes is significantly higher in continental shelves and slope of continental margins when compared sediments formed due in situ processes.

Deep-sea sediments are those found at depths greater than 500 m. The basic sources of the sediments found in the deep-sea are erosion from land (terrigenous), ash from eruption of volcanoes (arial and submarine), precipitation (hydrogenous), production by pelagic organisms (biogenous), and cosmic fallout (cosmogenous). Fine- grained volcanic material also constitutes part of the deep-sea sedimentation. Volcanic material deposited on land may be transported to the sea by running water or blown long distances in the sea by the action of winds. Submarine volcanic material may also find its place in deep waters as unaltered lava fragments, volcanic glass, and mineral grains. The cosmogenic matter is either at the deepwater sites or brought to the site by wind and other transporting agencies. Pelagic sediments mostly contain fine-sized particles that settle through the water column in the deep-sea environment. The current velocities here are appreciably lower and the particles deposited are those transported in suspension, by wind action, and in situ production. Pelagic sediments that are mixed with terrigenous sediments are known as hemipelagic sediments. Terrigenous matter includes clays and silts, ice-rafted debris, and wind-blown pyroclastic matter. Turbidity currents and mass wasting due to gravity are the main mechanisms for transportation of terrigenous material to the deep sea. Ocean is covered by broken sediments which can be categorized into following

- Lithogeneous Sediments
- Biogenous Sediments
- Hydrogeneous sediments
- Cosmogeneous sediments
- Volcanogeneous Sediments

Sediments are produced by weathering of rocks and by transportation deposited at ocean floor. Sediments can also form by the accumulation of shells of dead organism and can consist of mineral and/or fossil. Deepest areas of ocean floor which is far from land are dominated by abyssal clay which is called "red clay" and consists of smaller sediments which are carried away by the currents and winds. These clay particles can be drifted beyond the continental shelves and out into the middle of the ocean because they sink very slowly. Some deep sea sediments are composed mainly of siliceous fossils including diatoms and radiolarians others are made of carbonate such as foraminifera and coccolithiphorids. In deep oceans, Ooze can be found as a deposit of soft mud. Ooze form on deepest areas of the sea floor where steady deposition of dead microorganism from overlying waters is not obscured by the sediments which are washed from the land.

Most of the deepwater sedimentation is the result of autochthonous biogenic activity. Most dead, decaying bodies and fecal matter are dissolved by ocean water, decomposed by bacteria, or consumed by animals as they slowly sink towards the bottom of the ocean. Only a small amount reaches the bottom (<= 1%). Organic matter in oceans exists in two forms as dissolved matter or particulate matter. Most of the OM gets dissolved in the sea and only a small percentage of it reaches the seafloor especially when the water column is appreciably thick. The biogenic sediment matter derived from dead or decaying organisms is also referred to as ooze. There are two types of oozes, deep-sea environment carbonate ooze and siliceous ooze. They primarily contain calcium carbonate and silica. The predominant deep sediment is carbonate ooze which covers nearly half the ocean floor (Fig. 3.5). Calcium carbonate is derived from the hard parts of shell or bones of organisms or grazing sea animals. Calcareous structures of animal origin are more abundant than those of plants. Two groups of plants contribute to calcareous material in sediments: single cell planktonic forms found in large quantities mainly in the open seas and the sessile calcareous algae which are abundant in the shallow depths of warm

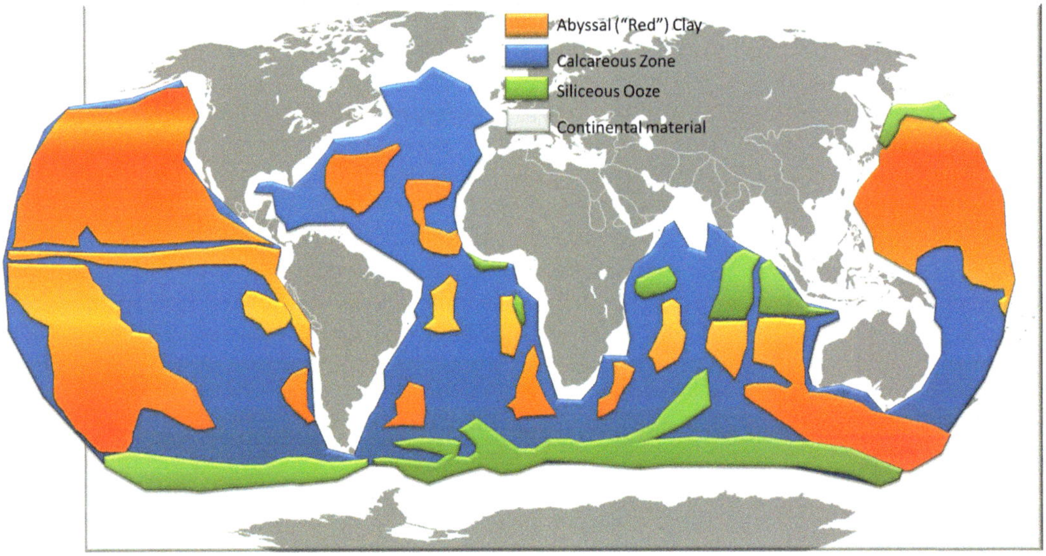

FIGURE 3.5

Worldwide distributions of deep-sea sediments. Calcareous ooze is confined to major parts of the oceans and in most parts of the Pacific the abyssal plane is covered with red clay.

waters in lower latitudes. Vast quantities of calcium carbonate are extracted from seawater by phytoplankton (e.g., coccolithophores) and zooplankton (e.g., foraminifers). The skeleton structures are subject to mechanical disintegration and chemical transformation which is generally related to solutions. Calcium carbonate easily dissolves in cold waters and hence most of the carbonate matter is confined to the shallow, elevated regions in the oceans such as ridges and shelves. The depth of the sea at which most of the shell matter gets completely dissolved is considered the calcium compensation depth. Calcium carbonate is not likely to occur below this depth level. When the supply of river sediment is low and the waters of the shelves are warm, there can be a build up of calcium carbonate material composed of organism shells. These shallow areas are called carbonate shelves or platforms.

Siliceous ooze is not as predominant as calcareous ooze and is confined mainly to the high latitudes and to some extent to the lower latitudes where conditions are favorable. The only plant types that secrete silica are the diatoms which grow only in the euphotic layer either as free-floating planktonic forms or in shallow water as benthic organisms. Diatoms are found in the regions of vertical mixing and upwelling. The excessive presence of these oozes in the polar regions is due to their affinity towards survival in cold oceanic conditions. Silica is also secreted by a group of protozoa (radiolarians) and is found to be most abundant in the low latitudes. It is also secreted to an extent by sponges. The most common siliceous rock is quartz or chert; other types include diatomite. Siliceous rocks in low latitude regions are confined to the western coasts of the continents (see Fig. 3.5). Hydrogenous sediments are formed by precipitation of minerals from the ocean's water or as a new mineral – the result of chemical reactions between ocean water and sediments that already exist on the ocean floor. Most of them either have high concentrations of iron, manganese, nickel, or cobalt.

3.5 RATE OF SEDIMENTATION

Fluvial agencies contribute significantly to the sedimentary budget of the oceans. Elevation of land, size of the particles, and type of weathering (disintegration and decomposition) primarily determine the extent of erosion and its subsequent deposition. Changes in the rate of erosion and transportation significantly alter the depositional pattern in the oceans. Terrigenous matter from land is mostly deposited nearer the coast. Only particles of smaller grain size (clay) are carried to the deep-sea environment. The rate of sedimentation of land-derived material deposited in the deep sea is of the order of a few millimeters ($<10/10^3$ year). The near-shore depositional rates are significantly higher, generally of the order of >10 mm/10^3 years. In the pelagic regions, the depositional rates are still lower (Chester, 1990). Sea level changes had a significant effect on altering the shape of elevation and extent of erosion on the ground. During low sea levels mechanical erosion and sediment load in the rivers doubled (Garrels and Mackenzie, 1971). Over the range of thousands of million of years, the effects of glaciation and deglaciation have left dominant effects on the erosion and transport of sea bottom sediments. Over the last two millions Earth experienced large climatic variations and recorded temperatures were appreciably lower resulting ice ages (Pleistocene age).The ice cap expanded and contracted over larger areal extent resulting in variation inputs terrigenous sediments to oceans. Sea level changes responded to the variation in extent ice caps growth by adjusting sea levels. During the low stand of sea levels, rivers dumped most of sediments in shelf breaks. These sediments mixed with water (slurries) with higher density, turbulently flew down the slope creating passage for themselves to deep seas by carving out the deep canyons and depositing unsorted sediments in deeper part of oceans. During the period of glaciations the ice caps grew and extended up to shelves in high and middle latitude, resulting in moving glaciers/icebergs erode the mountains hills, scoured the bottom of the shelves and dumped the sediments near the shelves. The erosion due the moving icebergs/ glacier not only provided sediments in narrow region on the shelves but even brought in morphological changes in land relief.

The average sedimentation rate for most regions of the world was determined during the DSDP expeditions, mainly over pelagic areas that exhibited several paleoceanographic lows and highs during the Cenozoic period (Fig. 3.6).

During the Paleocene to early Eocene, the rates of sedimentation were very low but were high during the middle Eocene. The rates were low during the late Eocene and middle Oligocene. The sedimentation rates during the late Oligocene to recent times vary with a marked high confined to the middle to late Oligocene and in the late Cenozoic (Davies and Worsley, 1981). Two general processes have been proposed that govern rates of regional erosion and associated rates of terrigenous sedimentation: (1) crustal deformation creates regionally elevated terrain that provides potential energy to the main agents of erosion, i.e., rivers and glaciers; and (2) climate, which has changed globally since 3 ± 4 Ma ago, controls the abundance and temporal distribution of these erosive agents, as well as the type, density, and distribution of vegetation that protects the earth's "nepheloid layers". Climate change offers the most plausible, globally synchronous process affecting erosion, especially given that both active and inactive mountain belts show accelerated erosion since 3 ± 4 Myr ago (Molnar and England, 1990). Rather, the change from a virtually unchanging climate to one that has been changing rapidly, as dictated by Milankovitch forcing (related to changes in the orbital parameters of the Earth) and amplified by a periodically icecover. Their study may unravel the genesis of the parent rock associated tectonics, origin of different ancient and present living species as well as past climatic conditions. Earth might

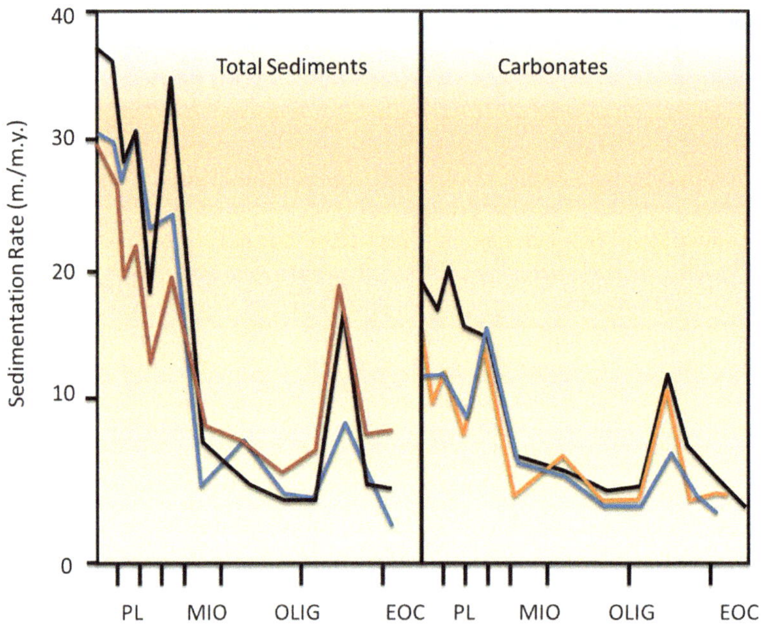

FIGURE 3.6

Sedimentation rates for different geological times during the Cenozoic Era. Appreciable enhancement rate of total sedimentation during Middle Eocene and Middle Miocene to recent times. Low rates of sedimentation have been observed during Paleocene to Early Eocene and Late Eocene to Early Miocene.

maintain a state in which erosive processes never reach equilibrium with the evolving landscape (Zhang et al., 2001). However, proxy records no evidence of increase in the past atmospheric carbon dioxide concentration (Pagani et al., 2010) and analysis of $^{10}Be/^{9}Be$ (von Balnckenburg and O'Nions, 1999) lead to the conclusion that processes different from an increase in denudation caused Cenozoic global cooling, and that the global cooling had no profound effect (Fig. 3.7) on spatially and temporally averaged weathering rates (Willenbring and Blanckenburg, 2010). It is estimated that about 18% of total carbon in sedimentary rocks is organic in nature and 82% is in the form of carbonate matter (Tissot and Welte, 1984). Organic matter mainly consists of carbon, oxygen, hydrogen, nitrogen, and to a certain extent phosphorus, sulfur, silicon, and other elements like potassium and iron. However, all these elements may be present in the sediments as inorganic constituents, such as nitrogen due to the activity of living organisms and ammonia and nitrate as minor components, which may be of organic origin (Trask, 1939).

Divergent living organisms differing in their sizes and chemical constituents are present in ocean water and near subsurface earth. Some organisms present in the water column are significantly smaller in size and free floating. They range from plankton such as bacteria (<1 μm) to crustaceans (>10 cm) or nektons that can swim across the circulating water and currents. The living organisms that survive near the seafloor environment are benthos. Those organisms that utilize inorganic substrate to produce OM are called autotrophs. Such organisms utilize light energy (photoautotrophs) and other inorganic chemicals (chemoautotrophs) to produce organic compounds. Organisms that utilize OM present in the

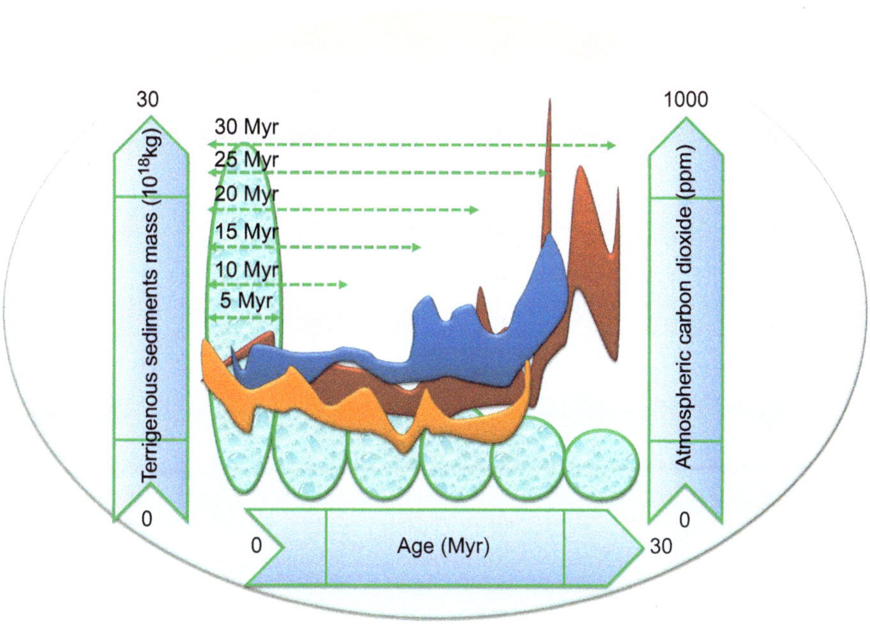

FIGURE 3.7

Plot of rates of terrigenous sediment deposition into the oceans and atmospheric carbon dioxide during the Cenozoic Era. Values are averaged 5 Myr bins. (Modified from Willenbring and von Blanckenburg, 2010.)

sea for their energy needs are called heterotrophs. In some cases nonphotochemical reactions with certain organic compounds in the sea produce or transform OM and in some cases cause removal of others (Duce and Duursma, 1977). Most of the OM is either produced in surface seawater or in the influx from terrestrial sources (fluvial and aeolian) and to some extent from other contributory (cosmic) sources. The OM in the sea is found to be in molecular continuum (Jiao et al., 2010). However, for the purposes of research convenience, OM is considered to exist as particulate organic matter (POM) or dissolved organic matter (DOM). The distinction is made mainly on the basis of size. Three sieve sizes, 0.1, 0.22, or 0.7 μm, are considered to separate these two classes. Matter passing through these sieve sizes is considered DOM and matter retained on the sieve is POM. Organic detritus and pelagic organisms are the main constituents of the POM. The contribution of POM to ocean is significantly low, approximately 10% (Legendre and Rassoulzadegan, 2008). Living organisms form only 25% of the POM and the rest is mainly from organic detritus. Phytoplanktons or marine algae (diatoms, coccolithophores, silicoflagellates) and zooplanktons (copepods, forminifera, pteropods, and radilarians) are the main contributors to POM in the ocean. The autotrophic algae are the main contributors to organic carbon in the ocean. Terrestrial OM containing living biomass, plant litter, and soil OM is brought to the marine realm through rivers, streams, and wind-driven processes. Terrigenous matter is contributed from recent primary production, diagenetically altered OM (aged soil), and ancient (fossil matter or kerogen) weathered rock material (Blair and Aller, 2012). Studies of carbon isotopes (Sackett et al., 1965) and the distribution of lignin in marine sediments (Hedges et al., 1981) suggest that the terrestrial supply of OM in the ocean is limited to a few tens of kilometers from the coast. The major part of the terrestrial OM is deposited in the deltaic environment and shelves and about 58 (±17) TgC yr^{-1} was buried in these regions. Most of the matter is confined to the shelf and the concentration decreases towards the deep sea. Only a small fraction (~20%) of POM of terrestrial origin is deposited in the pelagic regions of the ocean. This concentration can be as low as <10% over continental slopes of Washington (Hedges and Mann, 1979). Nearshore areas and deltas record appreciable organic inputs from land-derived organic materials like pollen, spores, and other plant debris.

Dissolved organic carbon in the oceans is the largest reservoir of reduced carbon, holding 200 times the carbon inventory of biomass. Oceanic DOM is mostly produced autochthonously by bacterial solubilization of POM in the surface waters and it serves as a substrate to a vast heterotrophic microbial population and a source of nitrogen and phosphorous to nutrient-starved autotrophs. The marine biotic carbon is generated either through exudation of organic substances by photosynthesis or excretion by zooplankton, and post death decay reactions also contribute to the oceanic DOC pool (Chester, 1990). The nonliving compounds come from the progressive dissolution of POM by bacterial release of hydrolytic enzymes (ecoenzymes) in seawater. Free-living bacteria use diversified hydrolytic ecoenzymes to progressively break down particles to polymers and then depolymerize the dimers and monomers. In addition to marine OM produced in situ, terrestrial and aeolian inputs also contribute to POM and DOM. The input of OM from the rivers and streams to the oceans is significantly high. Organic matter in the atmosphere exists as gas or particles and enters the ocean by direct impact, precipitation, and direct gas exchange. The influx of OM from the atmosphere is appreciably low.

The particulate organic carbon (POC) supplied by rivers to the oceans ranges from 135–231 × 10^{12} g/year of the total production to the order of 27 × 10^{15} g/year. Most of the OM in rivers is of humic and fluvic acid types that seldom reaches the open seas and decomposes or is deposited in the coastal areas. It was estimated that the dissolved OM in the major rivers of the world was of the order of 2–5 mg organic carbon/L (William, 1968; Menzel, 1974; Wangersky, 1978).

Organic matter preservation in sediments seems to result from complex interdependent parameters and may be controlled by the site-specific bio-physicochemical conditions. The environmental and geological conditions prevailing at the source and depositional sites play an important role in the in situ generation of marine biota, influx from the terrestrial and subsequent remineralization (respiration), burial, and ultimate preservation (Blair and Aller, 2012). Most of the OM produced in surface seawater or as influx from terrestrial sources (fluvial and aeolian) and to a lesser extent from other contributory sources (cosmic) undergoes remineralization due to biological processes (sustenance of various forms of organisms) or chemical reactions. The quantum of OM deposited on the seafloor at a site is mainly controlled by three factors: (1) productivity and the supply of OM in the overlying water (produced locally, autogenously) and produced elsewhere (allochthonously); (2) the rate of decomposition of OM while in water or accumulated at the site of deposition (by chemical actions in the water column or due to consumption by microorganisms; and (3) the movement of water in which the OM is deposited (Trask, 1939). Only a small fraction (0.5–1.0%) of total OM is generated either in situ (autochthonous) or produced elsewhere (allochthonous). This enters the marine system after remineralization and is buried in the surface sediments where it is subjected to further processes (diagenesis, catagenesis, and metagenesis) during its burial to greater depths, eventually leading to the formation of oil and gas. Only a small portion (<0.1%) of organic carbon from the photosynthesis cycle is preserved in the sedimentary rocks and an even smaller percentage of this is ultimately converted to oil and gas (Hunt, 1996).

3.6 PRODUCTIVITY OF ORGANIC MATTER

Over geological time, the productivity and preservation of OM on the continental margins is linked to prevailing physical and biogeological oceanographic conditions and processes which are responsible for its decomposition and burial. The processes of OM production and preservation are also controlled by the deposition sites for these processes. The principal mechanism for generation of OM in the sea is in situ photosynthesis by phytoplanktons, and in regions with sufficient light, the main source is marine phytobenthos. Marine OM is derived mainly from plankton within the upper few meters of the sea surface. The genesis of organic matter in the oceans is mainly from surface water plants (phytoplanktons), zooplanktons (microorganisms), and terrigenous, aeolian and cosmogenic sources. Free-floating OM at the sea surface produced by planktonic autotrophs (phytoplankton) has its genesis in photosynthetic reactions. Organic matter spreads over most parts of the surface of the oceans but its distribution differs from place to place depending on the local conditions, such as sunlight and mineral content of saltwater. The primary productivity of OM is mainly governed by the availability of sunlight and nutrients such as nitrogen and phosphorus, and is the consequence of the combined operating processes of photosynthesis, chemosynthesis, and biosynthesis. Proper illumination of the water column by sunlight and the presence of dissolved nutrients such as nitrogen and phosphorus favor the growth of phytoplankton. Its rapid growth provides a sufficient supply of nutrients in the shallower parts of the ocean.

The position of the sun in the sky according to latitude and its seasonal variation, and the stratification in the water column affect the quantum of OM produced through photosynthesis, which shows distinct patterns across the globe (Calvert and Pederson, 1992). The productivity of algae is constrained due to the limited availability of nutrients. Higher order productivity requires nutrients to be brought to the region of productivity, a task mainly done by ocean currents. The high productivity rate of OM is confined to areas of coastal and equatorial upwelling and regions of high fluvial inputs where the influx of nutrients is comparatively higher (Stein, 1990). In order to maintain the growth of phytoplankton the nutrients have to be replenished by vertical mixing of seawaters and by the influx from fluvial resources such as streams. In the northern latitudes, during winter the water is not properly illuminated and even in tropical regions with strong sunlight, the absence of the proper quantum of nutrients curtails plant growth. However, the twilight regions of the Arctic are exposed to light for longer periods of time during the day and in Antarctica additional vertical mixing provides an abundance of nutrients leading to higher organic productivity in both these regions. Oceanic sedimentation is controlled by the interaction between the surface and deep ocean currents. The thermocline boundary separates these two current configurations. The regions of upwelling bring nutrients to shallow parts of the ocean thus helping the productivity of OM. Conversely, downwelling cellscarry OM from surface water to the deeper parts of the ocean for its further decomposition and deposition on the ocean floor (Hay and Southam, 1977). The areas of highest production of OM are confined to regions around landmasses where upwelling is a predominant factor for preservation and concentration of the OM (Tissot and Welte, 1984). Most of the upwelling is confined to coastal regions, where the shallow shelf and sharp gradients in the seafloor relief lead to turbulence and mixing makes the nutrients flow in vertical and horizontal directions leading to a supply of nutrients in the region of productivity. Coastal regions, such as the west coast of America and equatorial western Africa characterized by proper circulation of nutrients due to mixing, have higher productivity of phytoplankton (Fig. 3.8). Stagnant waters restrict productivity despite the presence of proper sunlight owing to the limited circulation of ocean waters.

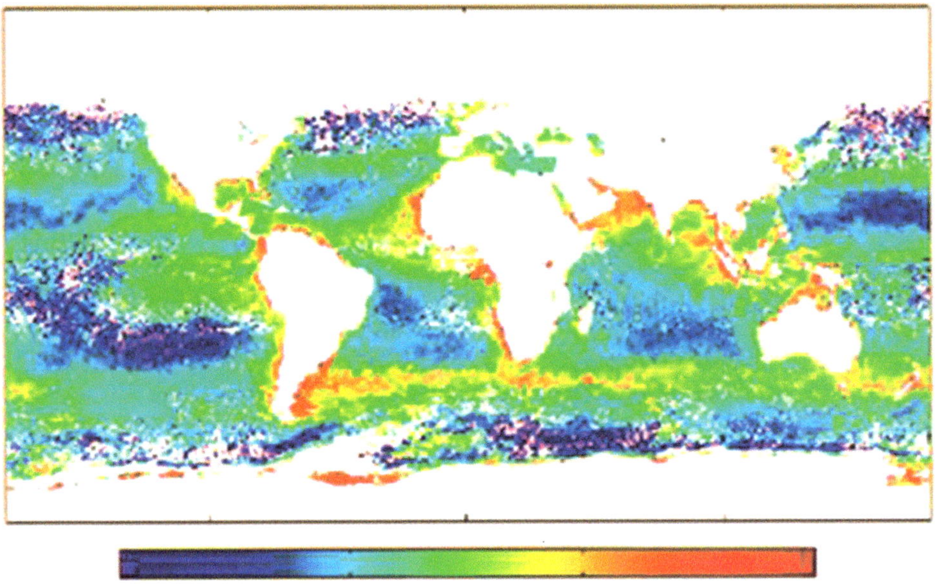

FIGURE 3.8

Average of all models for global primary organic productivity of the world for January 1998. Units are in $gC.m^{-2}.d^{-1}$. (From Carr et al., 2006.)

Rivers too bring in numerous nutrients to the nearshore areas thereby increasing the productivity of OM. Owing to higher productivity, the microorganisms and zooplankton thrive in this region leading to a further increase in the OM due to the decay and excretion of living organisms. Nearshore sediments contain relatively higher amounts of OM than those far from the shore. Nearshore sediment in general has about 1–8% OM whereas sediments in the offshore regions have organic content in the range of 0.3–1.5% (Trask, 1939). The influx of terrestrial OM, which is mostly POC in nature, is appreciably higher in the nearshore and deltaic regions. The global OM distribution shown in Figure 3.8 indicates that most of the high productivity is confined to coastal regions where conditions are favorable for growth of organisms.

Higher order productivity leads to an increase in the population of heterotrophs (bacteria, fungi, yeasts, and zooplankton such as protozoa and metazoa). These living organisms consume OM by osmosis and ingestion. Heterotrophs utilize OM for their growth and energy needs. Microorganisms excrete undigested OM which may add to already decomposed matter. Carnivorous zooplankton eat the herbivore organisms. Zooplankton consume phytoplankton to produce their own secondary and tertiary type of OM. A substantial part of the OM produced is also consumed by microorganisms (filter feeders) at shallow levels of the oceans and only a small portion of OM settles downwards and enters the nutrient-deficient oligotrophic region with a lower content of OM (10%) and a nutrient-rich eutrophic region (30%) (Eppley and Peterson, 1979). However, the mortal remains of all the organisms and their excretions add very little OM to the oceans. It is estimated that modern oceanic productivity is about 50×10^{15} g Cyr^{-1} and the total organic content in the sediments ranges from <25 mg C × gdw^{-1} to about

200 mg C × gdw^{-1} in coastal and continental margins under oxygen-deficient conditions and intense upwelling (Burdige, 2007). It has been estimated that on average the contribution from marine biota by the action of photosynthesis is the highest (~200 × 10^{14} g/yr) and by rivers is <1 × 10^{14}; the contribution from the atmosphere is relatively low (Duce and Duursma, 1977).

3.7 REMINERALIZATION (DECOMPOSITION) PRESERVATION OF ORGANIC MATTER

The preservation of OM in sediment is due to the complex interplay of many physical, chemical, and biological factors prevailing in the oceans in different depositional locations operating over geological timescales. Marine OM is preserved in ocean sediments after degradation, respiration, and/or remineralization due to several physical, chemical, and biological factors operating in the ocean water column and near the seafloor environment. It seems that the OM which does not suffer remineralization is preserved under normal circumstances. Is the efficiency of preservation directly connected to the effective remineralization processes? The factors responsible for remineralization must have a major bearing on the preservation of OM in sediments (Burdige, 2007). Preservation to an extent can be seen in the context that those factors or processes which preclude or circumvent degradation, decomposion, and remineralization may favor preservation. Factors that influence preservation include the productivity of OM, sedimentation rate, bottom water oxygen minima, sediment particle size (surface area), the exposure of matter to different biological activity, and the percentage of land-derived terrestrial and marine OM in total organic content. The amount of OM produced in and supplied to the upper ocean waters, together with the amount and type of mineral ballast, determines the rain rate and the composition of OM delivered to the seafloor (Zonneveld et al., 2010). This OM input is either generated by primary production in the photic zone or through transport of terrestrial OM to the marine realm. Organic matter from both sources may contain compounds that are intrinsically labile or refractory purely on the basis of molecular structure, presence of functional groups, and physical form in which they exist in nature (Hedges and Oades, 1997). Many microorganisms act on the OM for their sustenance under different oceanic conditions and in the process most OM is decomposed and altered before it is buried in the ocean bottom sediments. The terrigenous OM supply is already in altered form and the processes controlling the degradation of soil alter the molecular structure before their entry to the marine environment. The entire process of OM from productivity to burial becomes a sink for carbon dioxide (CO_2) and in a way is an integral part of CO_2 circulation in the atmosphere.

Preservation of OM results from the interplay of many processes; isolating the effect of each of them is rather difficult and may lead to entirely different conclusions. A comparison of organic carbon accumulation with the primary productivity for different rates of sedimentation has been studied to evaluate the pattern of deposition under different rates of sedimentation. Sediment deposition rate is the most important factor influencing the preservation of OM in the sediments (Stein, 1986). As an approximation it can be surmised that a tenfold increase in the sedimentation rate may lead to the doubling of organic carbon content in sediments under favorable oceanic conditions (Müller and Suess, 1979). Rapid sedimentation precludes the OM from destruction and consumption and its exposure to the bottom oxygen-bearing layer, eventually enhancing its preservation. In regions of high sedimentation rates, the decomposition of OM takes place through sulfate reduction (anaerobic), which is not an efficient process for decomposition (Bastviken et al., 2004). Where sediment deposition takes place at

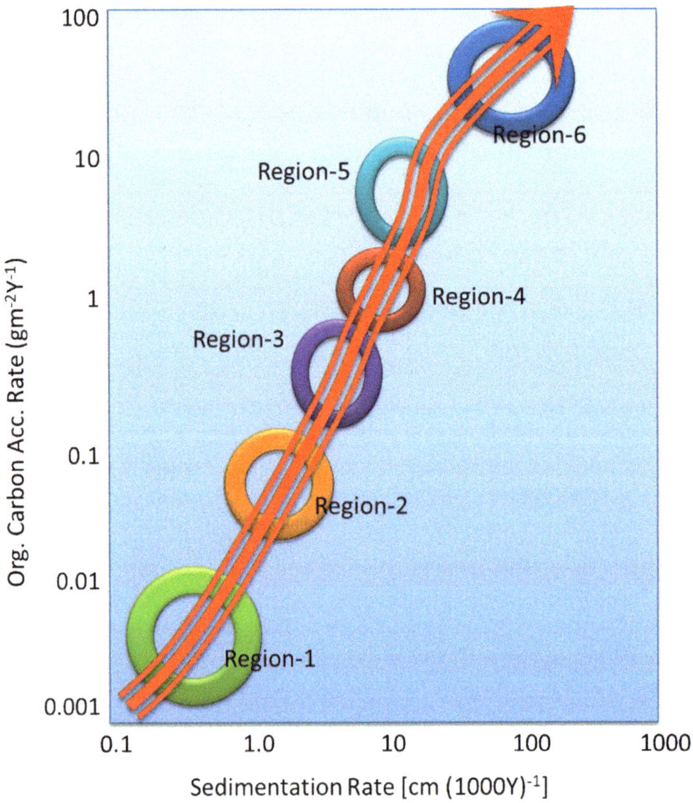

FIGURE 3.9

Organic carbon accumulation rates as a function of sedimentation rates for different margins. Line indicates the linear fit for the observed data sets. (Modified from Müller and Suess, 1979.)

a lower rate, the OM is exposed to oxidation and less matter enters the anaerobic zone leading to a lower order of OM preservation in the sediments (Canfield, 1993). Sediment data from different continental margins has indicated that OM content from the last glacial period is substantially higher than that from the interglacial period (Müller, 1975; Stevenson and Cheng, 1972; Hartmann et al., 1976; Müller and Suess, 1979). These studies suggest that the increase in organic carbon correlates well with increase in sedimentation rate (Fig. 3.9). Experiments have shown that in anoxic conditions, higher rates of sedimentation may lead to preservation of labile OM (Berner, 1978) and very high sedimentation rates may result in good preservation of deposited OM with an initial carbon phosphorous ratio close to the Redfield value (Ingall and Cappellen, 1990).

However, a direct relationship between sediment burial rates of 0.1–10 mg/cm^{-2} yr^{-1} (Müller and Suess, 1979) and preservation of OM is not seen universally. It is observed that in coastal and deltaic depositional sites where most of the OM is stored, the carbon concentration rapidly decreases for rates of sedimentation exceeding 10 mg/cm^{-2} yr^{-1} (Doyle and Garrels, 1985). The examination of a correlation between the rate of sediment deposition and burial efficiency has shown that a linear relation holds

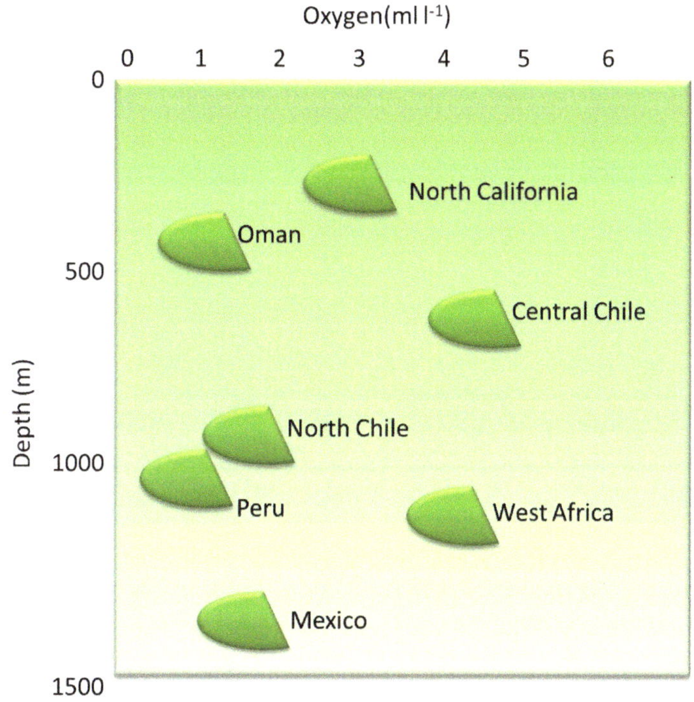

FIGURE 3.10

Pattern of dissolved oxygen profile for different continental margins of the world. Oxygen minimum in most oceans is between depth range of 100 and 1500 m. (Modified from Wyrtki, 1962.)

good for a rate of sedimentation of the order of 15 mg/cm^{-2} yr^{-1} and above this the burial efficiency remains almost constant at 100%, suggesting that control of the rate of sedimentation over the preservation of OM is not valid for very high rates of sedimentation (Betts and Holland, 1991).

Dissolved oxygen in the oceans is mainly from diffusion through the atmosphere and the near surface photosynthetic activity of plankton. Dissolved oxygen levels in the marine environment can range from zero to over 10 ml/L in temperate waters. The oxygen content in the upper layers of ocean waters varies from place to place. The concentration of dissolved oxygen in most oceans decreases from the surface downwards, reaching relatively low values in a zone at intermediate depths. The oxygen minimum zone (OMZ) is defined as regions where the concentration of oxygen is <0.5 ml/L (saturation <22 μM). Oxygen minimum zones are created in the ocean due to biochemical oxygen consumption with circulation affecting its distribution and position in the ocean (Wyrtki, 1962). Oxygen minimum zones in most oceans are found to occur between depths of 100 and 1500 m (Fig. 3.10). The OMZ occurs in the regions of high nutrient content due to upwelling leading to higher productivity, which sinks and degrades, depleting oxygen in the water column (Levin, 2003). Minimum oxygen conditions also prevail in stagnant waters with depleted source waters (Sarmiento et al., 1988). The OMZ represents a major oceanographic boundary. It expands and contracts over geological time and may influence the genetic diversity and play a key role in the evolution of species at bathyal depths (Levin, 2003). The pattern of the oxygen profile determines to some extent the decomposition of organic material when it passes through this region.

Organic carbon burial efficiency shows a strong correlation with the length of time accumulating particles are exposed to molecular oxygen present in the sediment pores (Harnett et al., 1998). It is suggested that the pattern of oxygen profile mainly determines the degradation or preservation of OM (Pratt, 1984; Canfield, 1993). The OM that escapes or is subjected to a smaller time exposure to the oxygen zone suffers little or no decomposition and thereby leads to preservation of OM in the sediments (Canfield, 1994). Under reduced vertical circulation (density stratification), the deeper parts of the ocean are not exposed to light leading to oxygen-deficient layers. In these regions, the preservation of matter over longer periods of time is controlled by anoxic conditions (Keil et al., 1994).

Regions of higher productivity and higher rates of sedimentation are often accompanied by bottom waters with low oxygen content (Reichart et al., 1998). Results from North California, a region of upwelling with higher productivity, have indicated high preservation of OM in sediments of the Pleistocene age, where the oxygen minimum zone has suboxic and anoxic conditions (Dean et al., 1994). Analysis of the elemental composition of OM from ancient sediments has revealed that under oxygen-free conditions, matter is not degraded as much and is better preserved compared to normal marine sediments (Pratt, 1984; Ingall et al., 1993). Two cases were studied in the North Pacific Ocean; where one site exhibits a pattern of OMZ (Mexican margin) in the bottom waters and the other site shows the normal ocean profile of oxygen distribution (Off-Washington State) (Fig. 3.11a). The Mexican region is characterized by very low concentrations of oxygen with high order productivity in the near surface waters. The oxygen content off the Washington State coast is of the order of 20 μmol/L. The pattern of oxygen content and the extent of burial in both places show changes from the shelf towards the slope. Sediment samples collected from the two sites indicate that the Mexican region has a very high order of preservation of OM (12.0 wt%) and at the Washington region the concentration of OM is less than 2.0 wt% (Fig. 3.11b) (Harnett et al., 1998).

The correlation between preservation of OM and the percentage of oxygen suggests that the sedimentary OM is of recent origin. However, there seems to be little uncertainty about the mode of decomposition in the presence or absence of oxygen (Pedersen and Calvert, 1990; Calvert and Pedersen, 1992). The settling OM is subject to different environments in the sub-bottom water column and consequently the decomposition and preservation patterns are significantly different. In regions of high preservation in an oxygen-free (euxinic) regime, comparatively less efficient anaerobic processes operate the decomposition. The large amount of preserved OM in some regions of the world may be due to incomplete degradation of matter under anoxic conditions and retentive carbon in sediments over longer periods of time. Locations of higher order concentrations of OM in sediments are not always associated with regions of oxygen-free environment and processes other than anoxia are responsible for such preservation. It has been observed that the composition of newly deposited OM both under high and low saturation of oxygen is similar, thereby inferring that concentration of oxygen distribution is not a significant factor in the preservation of OM in sediments (Calvert and Pedersen, 1992). Laboratory experiments have demonstrated the lack of evidence to support the influence of oxygen in the preservation of OM (Westrich and Berner, 1984; Kristensen and Balckburn, 1987). In normal circumstances if OM exposed to oxygen is the main control for decomposition of OM then seafloor sediments should be more degraded in general. However, sediment samples of highly concentrated OM from Cape Lookout Bight field of North Carolina suggest that such a relationship is not always valid (Hedges et al., 1995).

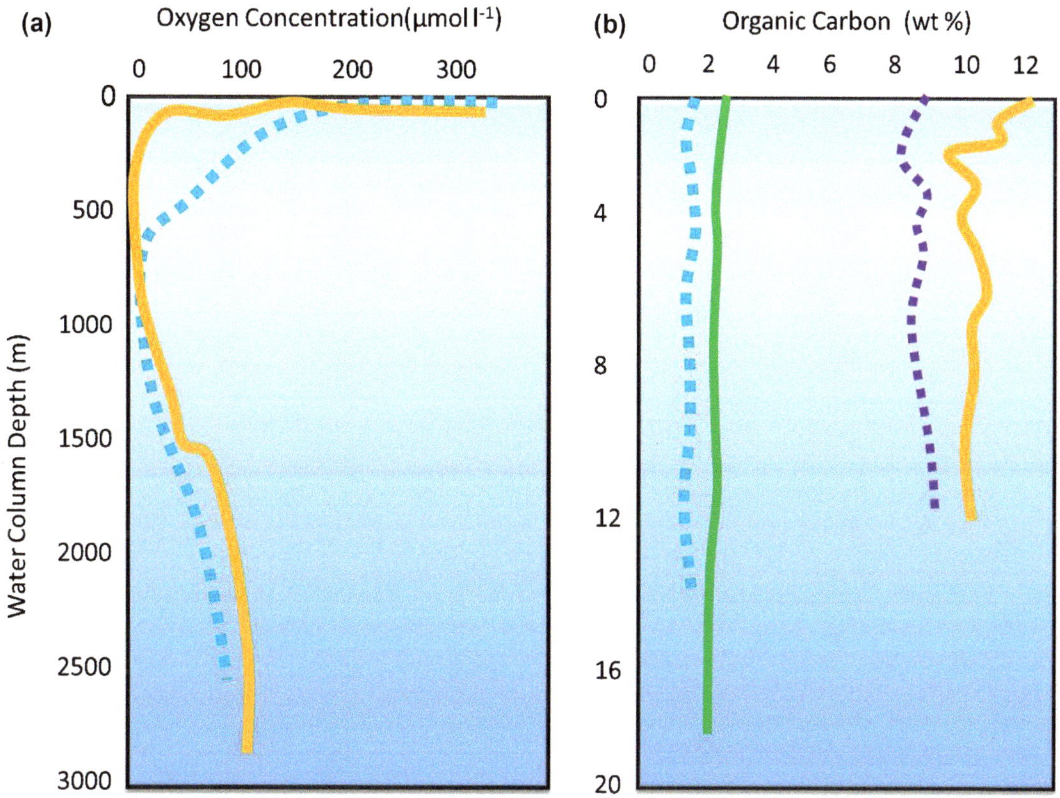

FIGURE 3.11

(a) Effect of oxygen content on the accumulation of organic carbon in the sea as evidenced from two different locations of the USA's continental margin: Gulf of Mexico with presence of oxygen zone (solid line) and Washington Coast with normal oxygen distribution (squares) and associated carbon content. (b) The effect of degradation of organic matter with respect to oxygen concentrations. (Source: Modified from Harnett et al., 1998.)

Burial and preservation of OM in sediments is the combined effect of interplay between different parameters such as productivity, oxygen minimum, sedimentation rate, and the environment prevailing in source and deposition regions. Already degraded terrigenous matter, irrespective of prevailing conditions, does not undergo substantial decomposition, and resists easier remineralization; on the contrary marine biogenic polymers are much more susceptible to remineralization. Stein (1990) has presented a composite scenario by incorporating most of the relevant parameters to ascertain the importance of each of these parameters for the preservation of matter in different marine environments. In an open sea environment, productivity is appreciably lower and the accumulation of sediments is mainly governed by particles settling under oxic conditions where a change in the sedimentation rate hardly has an effect (Field A, Fig. 3.12).

Regions of upwelling and high productivity result in higher rain of OM at the seafloor and thereby provide proper avenues for preservation. With an increased rate of production, the population of

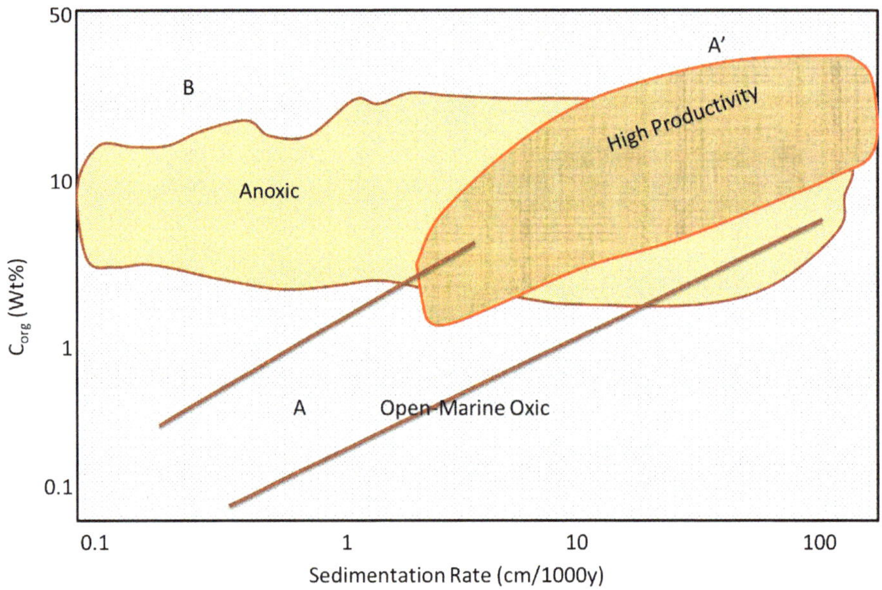

FIGURE 3.12

Correlation between (marine) organic matter and sedimentation rates (OCSR). Recent to Miocene sediments deposited in open sea environment (Field A), upwelling and higher productivity regions (Field A'), and anoxic areas (Field B) for varying rates of sediment accumulations. (From Stein, 1990.)

aerobically respiring organisms increases, leading to the depletion of oxygen and with continuous influx of OM and a lesser extent of oxidation, the OM has a better chance of preservation (see Fig. 3.12). Higher productivity and a higher rate of sedimentation mainly in the regions of upwelling provide the best opportunity for OM burial as the exposure time for OM in the OMZ is rather small before OM enters the less efficient remineralization anaerobic zone, thereby leading to enhanced preservation of the sediment.

Over the coastal margins an inverse correlation has been noticed between the grain size and organic content (Premuzic et al., 1982; Keil et al., 1994). This correlation is frequently ascribed pertaining to soil as stabilization of OM by adsorption (Anderson, 1988). More than 90% of the total sedimentary matter from a wide variety of marine depositional sites cannot be physically separated from its mineral grain (Hedges and Keil, 1995). Natural minerals have rough surfaces and the topography of the surface can be expressed in terms of pore size distribution. Mineral surface area is primarily controlled by the smaller pores (Mayer et al., 1988). Roughness of the surface area of a mineral depends on the species as well as exposure to chemical and mechanical processes. Intense leaching, chemical weathering, precipitation of clay minerals, oxy-hydroxides, and poorly ordered siliceous coating results in rough surfaces (Ransom et al., 1998). In a quiet, hydrodynamic ocean environment the correlation of grain size and organic content may be the result of impedance of fine-grained sediments to a diffusive exchange between sediment grains and organic particles or sorption of OM to higher surface areas of sediments (Keil et al., 1998). The strongly associated organic component varies directly in concentration with sediment surface area. The OM is supposed to get adsorbed on mineral grains and reside in fractures and crevasses in the grains. Correlation between the total organic content and mineral surface area has been

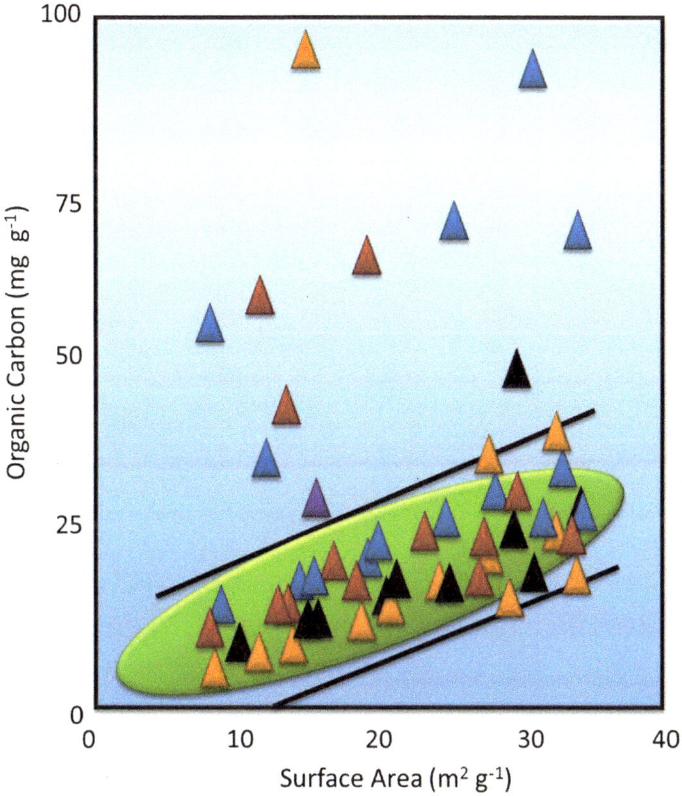

FIGURE 3.13

Percentage of organic carbon versus mineral surface area for multiple samples in the north-east Arabian Sea. Lines denote the range of organic carbon typically observed in the continental margin sediments. (Modified form Keil and Cowie, 1999.)

shown (Fig. 3.13) irrespective of mineral species (Mayer et al., 1988). With time and deeper burial into the sediment (more than a couple of centimeters), where the slow process of anaerobic respiration by bacteria operates, the fraction of the degrading material that is adsorbed to mineral surfaces within the mesopores increases. During these later stages the organic material is old and normally defined as refractory (recalcitrant) and structurally complex. When this material is adsorbed within mesopores it is protected from hydrolytic degradation by anaerobic bacteria because of size exclusion by bacterial eco-enzymes. In other words, the poor ability of anoxic bacteria to carry out the first degradation step (enzymatic hydrolysis) on old refractory, structurally complex material may be due to adsorption to mineral surfaces of the fraction of decomposing organic material during the later stages of diagenesis. The observation of close association between OM and low permeability, clay-rich sediments may arise from the interaction of organic clay minerals and other biologic components in the sediments. Such characteristics of the host medium provide suitable shielding from the attack of digestive enzymes of mesofauna and other bacterial communities, thus preserving the OM in the sediments (Ransom et al., 1998).

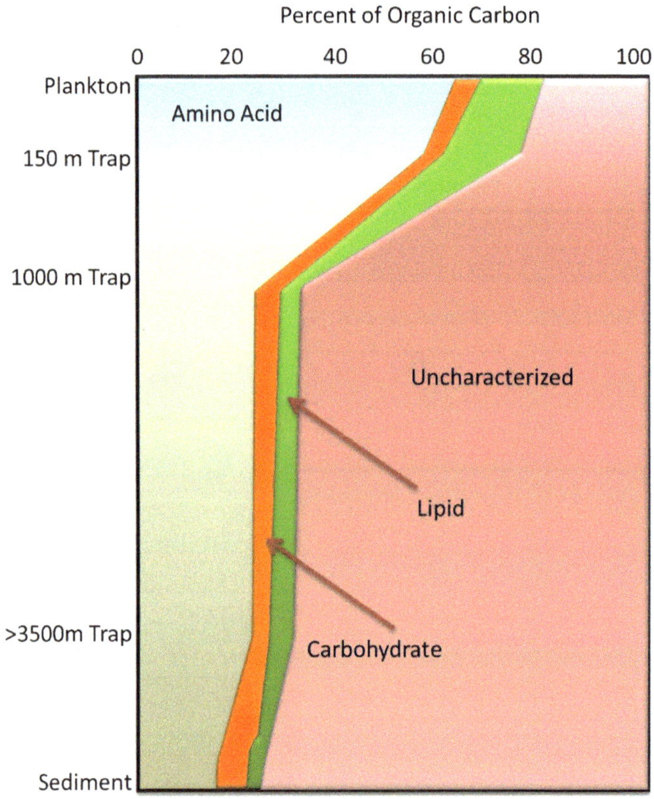

FIGURE 3.14

Percentage of amino acids, carbohydrate, lipid, and molecularly uncharacteristic origin matter in the particulate planktonic organic matter. The concentration of molecularly uncharacteristic origin matter increases with depth. (Modified from Wakeham et al., 1997.)

The environmental and geological conditions prevailing at the source and depositional sites play an important role in situ generating marine biota, influx from the land and subsequent remineralization, burial, bioturbation, bioirrigation, and its ultimate preservation (Blair and Aller, 2012). In the context of preservation, it becomes important to understand the formation process and molecular structure of the OM existing in the marine environment in different forms. Only a small percentage of total organic carbon (TOC) buried in the marine realm can be isolated and identified by biochemical and molecular signatures of marine sediments and endmembers. The geochemical signature of POM decreases rapidly with depth and becomes a minor component (<25%) with attenuation (Wakeham et al., 1997; Hedges et al., 2000) (Fig. 3.14). Owing to the complexity associated with the formation and status of technological development, the identification of the remaining organic compounds still remains unrealistic and speculative and such matter is considered to be preserved and inhibited remineralization.

OM generated in situ is altered, due to the activity of functional groups or chemical reactions to different substrates, whose characteristics may be more recalcitrant for further remineralization.

Molecular and chemical characteristics of OM not identified by conventional means and matter that escapes detection during its burial is grouped in the category of molecularly uncharacterized OM. This matter is generated either by biological or chemical processes operating in different environments acted upon by different agencies over a period of time. Such matter either flows into the marine realm as altered matter from the terrigenous source or is generated in situ during the process of remineralization of OM. Most of the terrigenous altered OM is confined to soils, which may constitute black carbon due to incomplete burning of the biomass. Black carbon is the product of incomplete combustion, forming under conditions of insufficient oxygen to fuel ratios. Black carbon includes graphite, and soot which are formed in the gas phase, from the recondensation of elemental carbon atoms produced by the pyrolytic cracking of the carbon-carbon bonds in organic matter (Goldberg, 1985). Terrestrial OM (TOM) soils may also contain kerogen (higher molecular weight and insoluble in organic solvents and water), which may resist further remineralization and may constitute important elements for preservation.

Organic matter produced in the surface waters, either due to in situ processes such as photosynthesis or chemosynthesis or due to the influx of terrigenous sediments, is acted upon by living organisms (bacteria or benthic animals) and also by physical and chemical processes. Most OM is degraded or decomposed within the oceanic water column and near the seafloor where different living microorganisms and animals consume or use diversified hydrolytic enzymes (exoenzymes) to progressively break down the OM and in the process lead to the synthesis of new substrates with altered chemical characteristics. At each stage of its transit and temporary storage, OM undergoes alteration and decomposition. The behavior of sedimentary OM following isolation from the new source involves both degradative and synthetic processes (Blair and Aller, 2012). The cycle of alteration and synthesizing mostly to lower molecular weight OM continues over a long period of time where the OM is degraded, remineralized (liberating CO_2, ammonia, and water), and generates lower molecular order organic substrates of either bacterial or faunal biomass. Different types of biologically produced matter have different remineralization potential (reactivities) and selective preservation (Burdige, 2007). The process of disintegration of OM depends on the type of OM as well as the rate constant (period required for disintegration) of degradation (Canfield 1994). The remineralization process also depends on the chemical status, agencies (bacteria and/or animals), and biological functional groups. Preservation of OM in sediments is estimated by the efficiency of organic carbon burial. Burial efficiency is mainly controlled by the rate of sedimentation. As the rate of sedimentation is substantially higher in the deltaic and near shelf regions compared to deep oceans the burial efficiency is higher in such regions. Preservation efficiency is the ratio of OM burial sediments divided by the rain rate of its input in the oceans, i.e., riverine TOM and primary productivity for the marine organic matter (MOM) (Burdige, 2007). Preservation of OM implies that it inhibits remineralization, becomes more recalcitrant, and protects itself from further degradation or decomposition either by encapsulating within more refractory chemical compounds or adsorption to mineral surfaces. OM produced in the marine surface mainly contains proteins, carbohydrates, and lipids, whereas terrigenous OM is primarily derived from the decay of plant materials and soils rich in lignin, resin, and waxes. MOM is mostly of recent genesis, has high reactivity (labile), and very easily undergoes degradation. The reactivity of such OM decreases with age and with the availability of electron acceptors. Terrigenous matter on land exposed to altering climatic conditions and on its journey to depositional locations suffers degradation. The TOM consists of organic elements which span in age from recent to quite ancient (fossil) and hence have different reactive capabilities. Organic matter in soils is a complex mixture of substances formed largely by microbial degradation of plant tissues, incompletely burnt biomass, and kerogen. Altered TOM is more resistive (refractory) to remineralization compared to more reactive or labile MOM. Therefore it is assumed that MOM is less efficient and TOM is more efficient for preservation. An examination of the carbon budget for the oceans suggests that roughly two to three as much organic carbon is transported to the oceans from land by rivers than is buried in marine sediments (Hedges and Keil, 1995). The estimated preservation efficiency of TOM is of the order of ~30–50% or less depending on its proportionate contribution with respect to MOM. The percentage of TOM preservation in the sediment suggests that the marine environments are fairly efficient media for its remineralization (Burdige, 2007).

Conventional models proposed disintegration of OM leading to resistant geopolymers, a precursor to kerogen. Such models postulate stepwise biological and enzymatic breakdown of biomacromolecules to bimolecular levels leading to the destruction of proteins and carbohydrates and the formation of water-soluble intermediates such as low molecular weight amino acids and sugars or perhaps higher molecular weight peptides. Most of the hydrolysable compounds are destroyed by microbial activity during sedimentation and early digenesis at shallow depths or recombine due to abiotic chemical reactions progressively leading to the formation of polycondensate structures, "geopolymers" (Tissot and Welte, 1984; Chester, 1990). The abiotic condensate with high concentration and temperature produces molecules that exhibit brown colored aromatics (melanoidins) due to the reaction between carbohydrates and amino acids and/or proteins (Hedges et al., 2000). This chemically complex material is generally referred to as humic which is amorphous, hydrolytic, and is recalcitrant to both chemical and biological degradation. The most important hydrocarbon formed during diagenesis is methane. Heavy hydrocarbon compounds along with CO_2 are generated during later stages of OM diagenesis. These condensate reactions transform and degrade OM to the point where it is biologically unavailable and goes undetected by conventional analytical techniques. However, in recent times the process of formation of geopolymers has been debated extensively (Hedges et al., 2000; Burdige, 2007; Zonneveld et al., 2010). One contention is that conventionally solution-based fractions of OM, such as fulvic acid, humic acid, and residual humin, cannot be related directly to the microbial turnover rate as the fraction is physically isolated based on size and density characteristics. The physical forms are not as important as the chemical states (Hedges and Oades, 1997; Hedges et al., 2000). Organic matter degradation appears to be predominantly a process of attrition, during which relatively resistant biochemicals are selectively concentrated. An analysis of neutron magnetic resonance (NMR) (^{15}N) observations from organic soils and modern sediments has indicated that constituent organic nitrogen in samples is in amide form (Knicker, 1995; Knicker et al., 1996), whereas abiotic condensation products, melanoidins, result from reaction of amino acids and carbohydrates (Benzing-Prudie et al., 1983).

In recent times condensation reactions (polymerization) of different organic chemical compounds have been suggested for the generation of refractory OM. Such phenomena could either be aggregation or cross-linking between different biochemical compounds (Burdige, 2007; Zonneveld et al., 2010). Protein in sediment may be preserved and may become hydrolysis-resistant through the process of aggregation and cross-linking both between proteins and perhaps carbohydrates too (Nguyen and Harvey, 2001). Functional lipids can be polymerized under oxic conditions by oxygen cross-linking (Versteegh et al., 2004). In similar fashion, under anoxic conditions, lipids and even sugars can react with hydrogen sulfide or polysulfide, resulting in the formation of polymer substances cross-linked via (poly) sulfur bridges (Zonneveld et al., 2010).

During the earlier stages, the metabolic activity of bacteria distorts the molecular structure of the biogenic macromolecules by bacterial (Ogawa et al., 2001) and physiochemical reactions (Harvey et al., 1983). During the selective utilization of OM pooled organisms consume more reactive components leaving behind more resistive refractory macromolecules. The newly formed polymer (macromolecule) during degradation or aging can isolate OM from the degrading agent, and as such protect against further degradation, leading to the formation of uncharacteristic OM (Zonneveld et al., 2010). During remineralization, some fraction of the sediment OM undergoes mineralization and is re-assimilated at the monomer or oligomer level and repacked as new in situ bacterial mass. This derived bacterial mass in the sediments includes cell exudates, cell lysis products, or remains of cell walls (necromass) (Parkes et al., 1993; McCarthy et al., 1998). A large number of organisms produce highly aliphatic macromolecules that are insoluble, nonhydrolysable, and resistant to biological degradation. Such molecular refractory matter is produced by vascular plants, cuticles of higher plants, and algal cell walls (Hartmann et al., 1976). Refractory biomacromolecules are likely to represent a very small fraction of initial biomass produced by marine and terrestrial organisms; however, during diagenetic maturity they may constitute substantially more to preserve OM (Burdige, 2007).

In the uppermost sedimentary layer, microorganism respiration takes place in the presence of oxygen. Most living organisms utilize the dissolved inorganic oxygen (aerobic) for respiration and metabolic activities for their growth and sustenance. Organic matter entering this part of the layer undergoes oxidative processes and the degradation is relatively fast, decreasing in an exponential fashion as the residual OM becomes more recalcitrant. Much of hydrolysable compounds are destroyed or altered at shallow depths especially in the oxic environment and nondegradable residues. With continued sedimentation or bioturbation owing to benthic animals, the labile OM is transported to deeper parts where it may stimulate degradation of refractory OM and partially degraded material may be transported upwards to other redox zones (Zonnevelde et al., 2010). In the anaerobic region, the degradation is not so efficient and residual matter from this region is mostly preserved. Modeling suggests that depending on the degradation of refractory and labile components, bioturbation may enhance preservation of carbon either in euxinic or normal marine sediments independent of efficiencies of the aerobic and anaerobic oxidation processes (Fig. 3.15) (Canfield, 1994; Zonneveld et al., 2010).

A lower rate of sedimentation in the abyssal planes or pelagic regions provides considerable time for remineralization of OM due to aerobic respiration (Archer et al., 2002) so the quantum of total organic content is appreciably low (~0.1–0.2%, Premuzic et al., 1982), and the OM that does get preserved is of low reactivity. Over the coastal region, two factors dominate the processes of preservation, i.e., higher rate of productivity and the rate of sediment burial. The sedimentation in these regions occurs under suboxic or unoxic conditions leading to the preservation of more reactive OM (Burdige, 2006). In some coastal regions the OM is remineralized in alternating oxic and unoxic conditions due to the bioturbation and vertical mixing of sediments (Aller and Blair, 2004). Redox oscillations may enhance degradation by promoting the symbiosis of aerobes and anaerobes (Aller, 2001; Aller and Blair, 2006; Zonneveld et al., 2010).

The dynamics of OM and more generally life in the oceans is largely determined by interaction between POM and DOM. Only a very small percentage of organic detritus and pelagic organisms after consumption or decomposition are retained in the form of POM or DOM along with minerals and deposited at the seafloor. Heterotrophic organisms utilize OM either by osmosis or through ingestion. To this end, the OM has to be hydrolyzed before its utilization by the microorganisms. Most microorganisms are unable to transport molecules greater than 600 amu. Larger molecules have to be dismantled outside the cell by exoenzymes. At moderate seafloor depths, the bacteria accomplish decomposition of OM by enzymatic processes for energy and other needs. During this process, proteins and carbohydrates are hydrolyzed providing amino acids and sugars. POM represents only 10% of DOM. However,

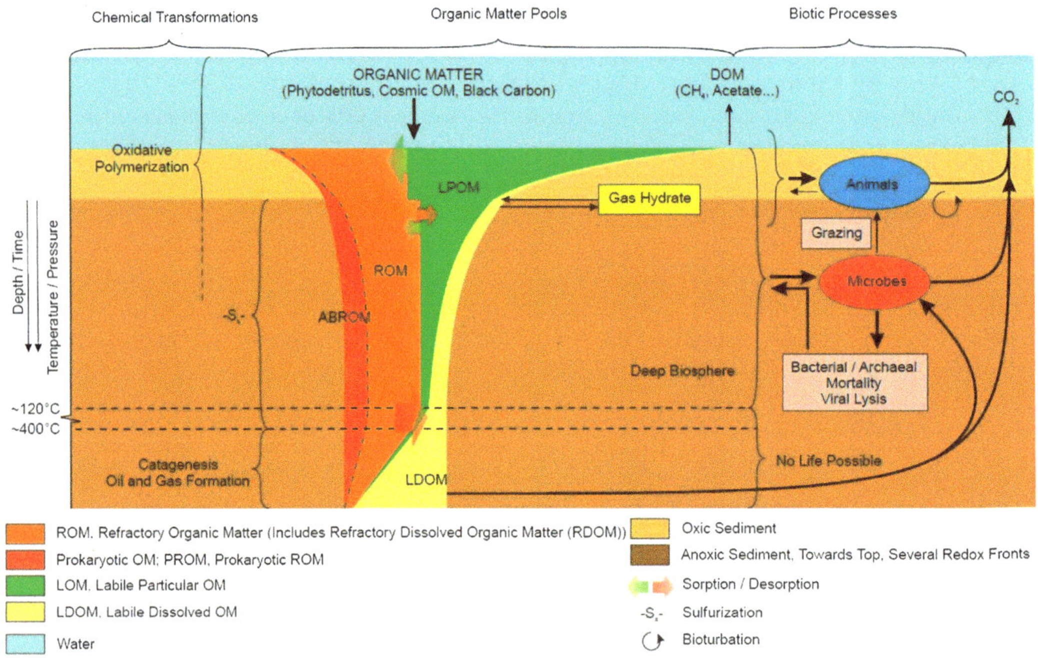

FIGURE 3.15

Different modes of degradation and transformation of organic matter in sediments in marine environment. Left panel: chemical transformation; center panel: organic matter in pools; right panel: biotic processes. (From Zonneveld et al., 2010.)

in reality, most of the DOM is derived from POM, which itself has autotrophic genesis. The nonliving compounds come from progressive dissolution of the POM by bacterial release of exoenzymes. POM consists of many organic components that differ in reactivity potential ranging from labile (LPOM) to very high refractive matter such as black carbon (RPOM). RPOM also includes labile OM which is intrinsically biodegradable but physically protected from biodegradation by means of adsorption to inorganic mineral or organic surfaces or due to encapsulation. In the process of encapsulation, the labile matter is shielded by degraded algal cells (Knicker and Hatcher, 2001). Similar classification has also been attributed to DOM. It contains both the refractive (RDOM) and labile (LDOM) matter. However, for simplicity RPOM is also included in RDOM (see Fig. 3.15) (Zonneweld et al., 2010). Bulk analysis of DOM components that can be isolated from seawater by ultrafiltration have chemical characteristics consistent with marine rather than riverine origin (Hedges et al., 1997). The DOM also consists mostly of nonliving compounds and may include living viruses. Phytoplankton spontaneously release DOM to surrounding waters under various circumstances, including when essential nutrients are scarce (Legendre and Rassoulzadegan, 2008). Based on the consideration of rate constant, semilabile matter is considered as SDOM. Accordingly, the LDOM may be consumed by heterotrophic microorganisms within days whereas SDOM can persist for months to years and accounts for most of the DOM exported from the euphotic zone to deeper parts of the ocean. Microbial species and functional groups consume part of the LDOM in the euphotic regions. In contrast, microorganisms in the bathypelagic depths (1000 to

4000 m) have developed metabolic tendencies to utilize the low reactivity of deep-sea DOM (Jiao et al., 2010). Determination of sources of exact mechanisms and rates of RDOM generation is intriguing and complex. More than 75% of DOM is concentrated with uncharacteristic matter at the molecular level (Benner, 1998). Some portion of DOM survived multiple mixing during its passage through the worldwide oceanic circulation system with no shielding from mineral surfaces, the presence of complexing ions making this matter extremely resistant to photolytic attacking and its cycling through submarine hydrothermal vents (Hedges et al., 2000).

In the upper sedimentary layers, the organisms responsible for the degradation of OM include bacteria, Archaea, and unicellular and multicellular eukaryotes. The bacteria and Archaea are grazed (consumed) by unicellular protists and in turn the unicellular and multicellular eukaryotes are consumed by Metazoa. Increasing evidence indicates that viral infection may be responsible for the high mortality of autotrophic and heterotrophic organisms over the sea surface (Procter and Fuhrman, 1990; Suttle et al., 1990). This produces cascading effects on carbon cycling and nutrient regeneration. Viral lysis of infected microbes transforms their cell contents and biomass into organic detritus (as POM and DOM), which is subsequently consumed by noninfected prokaryotes (i.e., viral shunt; Danovaro et al., 2008; Jiao et al., 2010). Such environments promote prokaryotic heterotrophic production but it also restricts the efficiency of carbon transfer to higher trophic levels. A large portion of DOM produced is mostly consumed by microbial activity but a small portion of OM derived from bacteria and Archaea likely contributes to refractory OM preserved in the sediments as archaeal and bacterial OM produced in the sediments (ABROM). However, all OM that is encapsulated, adsorbed on mineral or organic grains, and has escaped remineralization undergoes degradation or decomposition when exposed to oxygen over a long period of time (Zonneveld et al., 2010).

To a large extent, the factors operating on the decomposition of OM have been explored and reliable information obtained. However, a major portion of OM still remains uncharacteristic and associated complexities forbid tracing its genesis. Only a small part of the OM produced either in situ or brought into the marine realms after its degradation or decomposition by biological activity, chemical and physical processes reaches the upper layers of sediments to be acted upon by different agencies. The preserved OM at shallow subsurface levels is subjected to further alteration under predominantly high temperatures and pressures (catagenesis, metagenesis) leading to the formation of oil and gas. Isolation of each operating mechanism for characterizing such matter becomes an enormous and difficult task. Integrated efforts from all disciplines of science, field data, and laboratory synthesis may provide some valuable information in unraveling the enigma associated with uncharacteristic OM.

REFERENCES

Aller, R.C., Blair, N.E., 2004. Early diagenetic remineralization of sedimentary organic C in the Gulf of Papua deltaic complex (Papua New Guinea): net loss of terrestrial C and diagenetic fragmentation of C isotopes. Geochem. Coschem. Acta 68, 1815–1825.

Aller, R.C., Blair, N.E., 2006. Carbon remineralization in the Amazon-Guiana's tropical mobile mud belt: a sedimentary incinerator. Cont. Shelf Res. 26, 2241–2259.

Aller, R.C., 2001. Transport and reactions in the bioirrigated zone. In: Boudreau, B.P., Jorgensen, B.B. (Eds.), The Benthic Boundary Layer. Oxford University Press, Oxford, pp. 260–301.

Anderson, D.W., 1988. The effect of parent material and soil development on nutrient cycling in temperate ecosystems. Biogeochemistry 5, 71–97.

Archer, D.E., Morford, J.L., Emerson, S.R., 2002. A model for suboxic sedimentary diagenesis suitable for automatic tuning and gridded global domains. Glob. Biogeochem. Cycles 16 (1). http://dx.doi.org/10.1029/2000GB001288.

Bastviken, D., Persson, L., Odham, G., Tranvin, L., 2004. Degradation of dissolved organic matter in oxic and anoxic lake waters. Limnol. Oceanogr. 49, 109–116.

Benner, R., 1998. Cycling of dissolved organic matter in the oceans. In: Hessen, D.O., Tranvik, L.J. (Eds.), Aquatic Humic Substances: Ecology and Biogeochemistry. Springer, New York, pp. 317–331.

Benzing-Prudie, L., Ripmeester, J., Preston, C., 1983. Elucidation of nitrogen forms in the melanoidins and humic acid by nitrogen-15 cross polarization-majic angle spinning nuclear magnetic resonance spectroscopy. J. Agri. Food Chem. 31, 913–915.

Berner, R.A., 1978. Sulfate reduction and rate of deposition of marine sediments. Earth Planet. Sci. Lett. 37, 492–498.

Betts, J.N., Holland, N.D., 1991. The oxygen content of ocean bottom waters, the burial efficiency of organic carbon and regulation of atmospheric oxygen. Paleogeogr. Paleoclim. Paleoecol. 97, 5–18.

Biscay, P.E., Eittreim, S.L., 1977. Suspended particle loads and transport in the Nepheloid layer of the abyssal Atlantic Ocean. Mar. Geo. 23, 155–172.

Blair, N.E., Aller, R.C., 2012. The fate of terrestrial organic carbon in the marine environment. Annu. Rev. Mar. Sci. 4, 401–423. http://dx.doi.org/10.1146/annurev-marine-120709-142717.

Burdige, D.J., 2006. Geochemistry of Marine Sediments PhD thesis. Princeton University Press, Princeton.

Burdige, D.J., 2007. Preservation of organic matter in marine sediments: controls mechanisms, and an imbalance in sediment organic carbon budget? Chem. Rev. 107, 467–485.

Calvert, S.E., Pedersen, T.F., 1992. Organic carbon accumulation and preservation in marine sediments: how important is anoxia? In: Whelan, J.K., Farrington, J.W. (Eds.), Productivity, Accumulation, Preservation of Organic Matter in Recent and Ancient Sediments. Columbia University Press, New York, pp. 231–263.

Canfield, D.E., 1993. Organic matter oxidation in marine sediments. NATO ASI Ser. 14, 333–363.

Canfield, D.E., 1994. Factors influencing carbon preservation in marine sediments. Chem. Geol. 114, 315–319.

Carr, M.E., Friedrichs, M.A.M., Schmeltz, M., et al., 2006. A comparison of global estimates of marine primary productivity from sea floor color. Deep Sea Res. 53, 741–770.

Chester, R., 1990. Marine Geochemistry. Unwin Hyman Ltd, London.

Danovaro, R., Dell'Anno, A., Corinaldesi, C., Magagnini, M., Noble, R., et al., 2008. Major viral impact on the functioning of benthic deep-sea ecosystems. Nature 454, 1084–1087.

Davies, T.A., Gorsline, D.S., 1976. Oceanic sediments and sedimentary processes. In: Riley, J.P., Chester, R. (Eds.), Chemical Oceanography. Academic Press, New York, pp. 1–80.

Davies, T.A., Worsley, T.R., 1981. Paleoenvironmental implication of ocean carbonate sedimentation rates. In: Warme, J.E., Douglas, R.G., Winterer, E.L. (Eds.), The Deep Sea Drilling Project: A Decade of Progress (Society of Economic Paleontologists and Mineralogists, Special Publication No. 32). Society of Economic Paleontologists and Mineralogists. AAPG, USA.

Dean, W.E., Gardner, J.V., Anderson, R.Y., 1994. Geochemical evidence of enhanced preservation of organic material in the oxygen minimum zone of the continental margin of North California during the late Pleistocene. Paleoceanography 9, 47–61.

Doyle, L.J., Garrels, R.M., 1985. What does percent organic carbon in sediments measure? Geo. Mar. Lett. 5, 51–53.

Duce, R.E., Duursma, E.K., 1977. Inputs of organic matter to the ocean. Mar. Geochem. 5, 319–339.

Emery, O.K., 1969. The continental shelves. Sci. Am. 221, 106–122.

Eppley, R.W., Peterson, B.J., 1979. Particulate organic matter flux and planktonic new production in the deep ocean. Nature 282, 677–680.

Flood, R., 1978. Studies of Deep Sedimentary Microtopography in the North Atlantic Ocean PhD Dissertation. Wood Hole Oceanographic Institute and Massachusetts Institute of Technology.

Folger, D.W., 1970. Wind transport of land derived mineral, biogenic and industrial matter over the North Atlantic. Deep Sea Res. 17, 337–352.

Gardner, W.D., 1978. Fluxes, Dynamics, and Chemistry of Particulates in the Ocean Ph.D. Thesis. Wood Hole Oceanographic Institute and Massachusetts Institute of Technology.

Garrels, R.M., Mackenzie, F.T., 1971. Evolution of Sedimentary Rocks. Norton, New York.

Goldberg, E.D., 1985. Black Carbon in the Environment. Properties and Distribution. Wiley-Interscience, New York.

Harnett, H.E., Keil, R.G., Hedges, J.H., Devol, A.H., 1998. Oxygen exposure as a controlling factor in preservation of organic carbon in continental margin sediments. Nature 39, 572–574.

Hartmann, M., Müller, P.J., Suess, E., van der Weijden, C.H., 1976. Chemistry of Late Quaternary sediments and their interstitial waters from the NW African continental margin. Meteor. Forsch. C 24, 1–67.

Harvey, G.R., Boram, D.A., Chesal, L.A., Tokar, J.M., 1983. The structure marine fluvic and humic acid. Mar. Chem. 12, 119–132.

Hasse, L., 1983. Introductory meteorology and fluid mechanics. In: Liss, P.S., Slin, W.G.N. (Eds.), Air–sea Exchange of Gases and Particles. Reidel, Dordrecht, pp. 1–51.

Hay, W., Southam, J., 1977. Modulation of marine sedimentation by continental shelf. In: Anderson, N.R., Malahoff, A. (Eds.), The Fate of Fossil Fuel CO_2 in the Ocean. Plenum, New York, pp. 569–604.

Hayes, D.E., et al., 1975. Initial Reports of the Deep Sea Drilling Project Leg. 28, Vol. 28. Government Printing Office, Washington, DC, US.

Hedges, J.I., et al., 2000. The molecularly uncharacterized component of nonliving organic matter in natural environments. Org. Geochem. 31, 945–958.

Hedges, J.I., Kiel, R.G., 1995. Sedimentary organic matter preservation: an assessment and speculative synthesis. Mar. Chem. 49, 81–115.

Hedges, J.I., Mann, D.C., 1979. The characterization of plant tissues by their lignin oxidation products. Geochim. Cosmochim. Acta 43, 1803–1807.

Hedges, J.I., Oades, J.M., 1997. Comparative organic geochemistries of soils, and marine sediments. Org. Geochem. 27, 319–361.

Hedges, J.I., Keil, R.G., Benner, R., 1981. What happens to land-derived organic matter in the ocean? Org. Geochem. 27, 195–212.

Hedges, J.I., Keil, R.J., Benner, R., 1997. What happens to land-derived organic matter in ocean? Org. Geochem. 27, 195–212.

Hjulström, F., 1939. Transportation of debris by moving water. In: Trask, P.D. (Ed.), Recent Marine Sediments; A Symposium. American Association of Petroleum Geologists, Tulsa, Oklahoma, pp. 5–31.

Hjulström, F., 1955. Transportation of debris by moving water. In: Trask, P.D. (Ed.), Recent Marine Sediments. 4, SEPM Spec. Publ, pp. 5–31.

Holeman, J.N., 1968. The sediment yield of major rivers. Water Resour. Res. 6, 737.

Hunt, J.M., 1996. Petroleum Geochemistry and Geology. W.H. Freeman, New York.

Ingall, E.D., Cappellen, P.V., 1990. Relation between sedimentation rate burial of organic phosphorus and organic carbon in marine sediments. Geochim. Cosmochim. Acta 54, 373–386.

Ingall, R.D., Bustin, R.M., Van Cappallen, P., 1993. Influence of water-column anoxia on the burial and preservation of carbon and phosphorus in marine shales. Geochim. Cosmochim. Acta 57, 303–316.

Jiao, N., et al., 2010. Microbial production of recalcitrant dissolved organic matter: long-term global carbon storage in the global ocean. Nat. Rev. Microbiol. http://dx.doi.org/10:1038/nrmicro2386.

Jones Peter, A., 1995. Influence of mass wasting on sediment yield and soil producitivity in the little appleagate watershed. http://soda.sou.edu/Data/Library1/021120a4.pdf.

Keil, R.G., Hu, F.H., Tsamaki, E.C., Hedges, R.I., 1994. Pollen in marine sediments as indicator of oxidation of organic matter. Nature 369, 639–641.

Keil, R.G., Cowie, G.L., 1999. Organic matter preservation through oxygen deficient zone of NE Arabian Sea as discerned by organic carbon: mineral surface area ratio. Mar. Geol. 161, 13–22.

Keil, R.G., Tsamakis, E., Giddings, J.C., Hedges, J.I., 1998. Biochemical distributions (amino acids, neutral sugars and lignin phenols) among size-classes of modern marine sediments from the Washington Coast. Geochimica et Cosmochimica Acta 62, 1347–1364.

Kennet, J., 1982. Marine Geology. Prentice-Hall, Inc, Eaglewood Cliff, NJ.

Knicker, H., Hatcher, P.G., 2001. Sequestration of organic nitrogen in the sapropel from Mangrove Lakes Bermuda. Org. Geochem. 32, 733–744.

Knicker, H., Lüdemann, H.D., 1995. N-15 and C-13 CPMAS and solution NMR studies of N-15 enriched plant material during days of microbial 381 degradation. Org. Geochem. 23, 329–341.

Knicker, H., Scaroni, A.W., Hatcher, P.G., 1996. ^{13}C and ^{15}N NMR spectroscopy investigation on the formation of fossil algal residual. Org. Geochem. 24, 661–669.

Kristensen, E., Balckburn, T.H., 1987. The fate of carbon and nitrogen in experimental marine system: influence of bioturbation and anoxia. J. Mar. Res. 45, 231–257.

Legendre, L., Rassoulzadegan, F., 2008. Biological Oceanography. In: Nihoul, J.C.J., Chen, C.T.A. (Eds.), Oceanography: Encyclopedia of Life Support Systems (EOLSS), Developed under the Auspices of the UNESCO. EOLSS Publishers, Oxford, UK, pp. 1–35.

Levin, L.A., 2003. Oxygen minimum benthos: adaptation and community response to hypoxia. In: Gibson, R.N., Atkinson, R.J.A. (Eds.), Oceanography and Marine Biology. An Annual Review, vol. 41, pp. 1–45. Taylor & Francis Group, Abingdon.

Mayer, L.M., Macko, S.A., Cammen, L., 1988. Provenance concentration and nature of sedimentary organic nitrogen in Gulf of Maine. Mar. Chem. 25, 491–504.

McCarthy, M., Hedges, J.I., Benner, R., 1998. Major bacterial contribution to marine dissolved organic nitrogen. Science 281, 231–234.

McCave, I.N., 1975. Vertical flux of particles in the ocean. Deep Sea Res. 22, 297–310.

Menzel, D.W., 1974. Primary productivity, dissolved and particulate organic matter and the sites of oxidation of organic matter. In: Goldberg, E.D. (Ed.), The Sea. Wiley, New York, pp. 654–678.

Milliman, J.D., Meade, R.H., 1983. World-wide delivery of river sediments to the oceans. J. Geol. 91, 1–21.

Molnar, P., England, P., 1990. Late Cenozoic uplift of mountains ranges and glacial climate changes, chicken or egg. Nature 346, 29–34.

Müller, P.J., Suess, E., 1979. Productivity, sedimentation rate and sedimentary organic matter in the oceans-I. Organic Carbon Preservation. Deep Sea Res. 26, 1347–1362.

Müller, P.J., 1975. Zur Diagenese stickstoffhaltiger Substazen in marine sedimenten unter oxydierenden und reduzierenden Bedingungen Dissertation. Unversitat Kiel, p. 179.

Nguyen, R.T., Harvey, H.R., 2001. Preservation of protein in marine systems: hydrophobic and other noncovalent association of major stabilizing forces. Geochim. Cosmochim. Acta 69, 167–180.

Ogawa, H., Amagai, Y., Koke, T., Kaiser, K., Benner, R., 2001. Production of refractory dissolved organic matter by bacteria. Science 29, 217–220.

Pagani, et al., 2010. High earth-system climate sensitivity determined from Pliocene carbon dioxide concentration. Nat. Geosci. 3, 27–30.

Pak, H., Zaneveld, J., Kitchen, J., 1980. Intermediate nepheloid layers observed off Oregon and Washington. J. Geophys. Res. 85, 6697–6708.

Paphitis, et al., 2002. A quantitative study of microbial decomposition of biopolymers in recent sediments from the Peru Margin. Mar. Geol. 113, 55–66.

Parkes, et al., 1993. A quantitative study of microbial decomposition of biopolymers in Recent sediments from Peru Margin. Mar. Geol. 113, 55–66.

Pedersen, T.F., Calvert, S.E., 1990. Anoxia vs productivity: what controls the formation of organic carbon-rich sediments and sedimentary rocks? Bull. Am. Assoc. Petrol. Geol. 74, 454–466.

Pinet Paul, R., 2009. Invitation to Oceanography, Jones & Bartlett Learning ASIN: B00N4EI542. http://www.amazon.com/Paul-Pinet-Invitation-Oceanography-Edition/dp/B00N4EI542.

Pinet Paul, R., 2014. Invitation to Oceanography, Jones & Bartlett Learning ISBN-13: 9781284057072. http://www.jblearning.com/catalog/9781284057072/.

Pratt, L.M., 1984. Influence of paleoenvironmental factor in the preservation of organic matter in Middle Cretaceous Greenhorn formation Pueblo Colorado. Bull. Am. Assoc. Petrol. Geol. 68, 1148–1159.

Premuzic, E.T., Benovitz, C.M., Gaffney, J.S., Walsh, J.J., 1982. The nature and distribution of organic matter in near surface sediments of world oceans and Seas. Org. Geochem. 4, 63–77.

Procter, L.M., Fuhrman, J.A., 1990. Viral mortality of marine bacteria and cynobacteria. Nature 343, 60–62.

Ransom, B., Kim, D., Kastner, M., Wainwright, S., 1998. Organic matter preservation on continental slopes: importance of mineralogy and surface area. Geochim. Cosmochim. Acta 62, 1329–1345.

Reichart, G.L., Lourens, L.J., Zachariasse, W.J., 1998. Temporal variability in the northern Arabian Sea oxygen minimum zone (OMZ) during the last 225,000 years. Paleoceanography 13, 607–621.

Russel, R.D., 1939. Effects of transportation on sedimentary particles. In: Trask, P.D. (Ed.), Recent Marine Sediments. American Association of Petroleum Geologists, Tusla, Oklahoma, pp. 32–47.

Sackett, W.M., Eckelmann, W.R., Bender, M.L., Be, A.W.H., 1965. Temperature dependence of carbon composition in marine planktons and sediments. Science 148, 235–237.

Sarmiento, J.L., Herbert, T.D., Toggweiler, J.R., 1988. Causes of anoxia in the world ocean. Glob. Biogeochem. Cycles 2, 115–128.

Schumm, S.A., 1963. The disparity between present rates of denudation and orogeny. US Geol. Soc. Prof. 13S Paper 454-H.

Shaw, J., 1977. Till body morphology and structure related to glacier flow. Boreas 6, 189–201.

Stein, R., 1986. Surface water paleo-productivity as inferred from sediments deposited in oxic and anoxic deep waters. SCOPE/UNEP Sonderband 60, 55–70.

Stein, R., 1990. Organic carbon content/sedimentation rate relationship and its paleoenvironmetal significance for marine sediments. Geo. Mar. Lett. 10, 37–44.

Stevenson, F.J., Cheng, C.N., 1972. Organic geochemistry in Argentine Basin sediments: carbon-nitrogen relationship and Quaternary correlations. Geochim. Cosmochim. Acta 36, 653–671.

Sundborg, A., 1956. The river Klaralven: a study of fluvial processes (Geopgrafiska Annaler). Esselte Aktiebolag 125–316. Stockholm, Esselte Aktiebola.

Sundborg, A., 1956. The River Klarålven: Chapter 2. The morphological activity of flowing water—erosion of the stream bed: Geografiska Annaler, Vol. 38, pp. 165–221. Stockholm, Esselte Aktiebolag.

Suttle, C.A., Chan, A.M., Cotrell, M.T., 1990. Infection of phytoplankton by viruses and reduction of primary productivity. Nat. Rev. Microbiol. 5, 801–812.

Tissot, B.P., Welte, D.H., 1984. Petroleum Formation and Occurrence. Springer-Verlag, New York.

Trask, P.D., 1939. Organic content of recent marine sediments. In: Recent Marine Sediments. American Association of Petroleum Geologists, Tusla, Oklahoma, pp. 428–453.

Twenhofel, W.H., 1939. Principles of Sedimentation. MaGraw-Hill Book Co, New York.

Versteegh, G.J.M., Blokker, P., Wood, G., et al., 2004. Oxidative polymerization of unsaturated fatty acids as a preservation pathway for microalgal organic matter. Org. Geochem. 35, 1129–1139.

von Balnckenburg, F., O'Nions, R.K., 1999. Response of beryllium and radio generic isotope ratios in North Atlantic Deep Water to the onset of northern hemispheric glaciations. Earth Planet. Sci. Lett. 167, 175–182.

Wakeham, S.G., Lee, C., Hedges, J.I., Hernes, P.J., Peterson, M.L., 1997. Molecular indicators of diagenetic status in marine organic matter. Geochim. Cosmochim. Acta 61, 5363–5369.

Wangersky, P.J., 1978. Production of dissolved organic matter. In: Kinne, O. (Ed.), Marine Ecology, Vol. IV. Wiley and Sons, New York, pp. 115–220.

Weertman, J., 1961. Mechanism for the formation of inner moraines found near the edge of cold ice caps and ice sheets. J. Glaciol. 3, 965–978.

Wentworth, C.K., 1922. A scale of grade and class terms for classic sediments. J. Geol. 30, 377–392.
Westrich, J.T., Berner, R.A., 1984. The role of sedimentary organic matter in bacterial sulfate reduction: the G model tested. Limnol. Oceanogr. 29, 236–249.
Whitman, J.M., Davies, T.A., 1979. Cenozoic oceanic sedimentation rates: how good are the data? Mar. Geol. 30, 269–284.
Willenbring, J.N., von Blanckenburg, F., 2010. Long-term stability of glacial erosion rates and weathering during Late Cenozoic cooling. Nature. http://dx.doi.org/10:1038/nature09044.
William, P.M., 1968. Organic and inorganic constituents of the Amazon river. Nature 218, 937–938.
Wyrtki, K., 1962. The oxygen minimum in relation to ocean circulation. Deep Sea. Res. 9, 11–23.
Zhang, P., Molnar, P., Downs, W.R., 2001. Increased sedimentation rate and grain sizes 2–4 Myr ago due to the influence of climate change and erosion rates. Nature 410, 891–897.
Zonneveld, K.A.F., Versteegh, G.J.M., Kasten, S., et al., 2010. Selective preservation of organic matter in marine environments; processes and impact on the sedimentary record. Biogeosciences 7, 483–511.

CHAPTER

TECTONICS AND GAS HYDRATES

4

"The only true wisdom is in knowing you know nothing."

Socrates

CHAPTER OUTLINE

4.1 INTRODUCTION

Gas hydrates are found in passive and active continental margins, inland lakes (such as Lake Baikal), and permafrost regions. In most deep marine margins of the world, the thermobaric conditions present offer the correct environment for natural gas hydrate formation as measured in model systems consisting of various proportions of commercial paraffinic-type asphaltene -resin -paraffin deposits (ARPDs); the shapes of crystallization are affected by isotherm signals. Various other physicochemical and biological conditions found in specific locations around the world are favorable for gas hydrate formation. Methane, the main constituent of gas hydrates, is generated in subsurface earth by complex biological, chemical, and physical processes. The accumulation of methane sources in continental margin sediments can be explained by the higher productivity of organic matter and faster deposition of organic-rich sediments in the sea in these areas. All these conditions cause gas hydrate formation in specific locations at the continental margins of the world. These margins show great diversity in their physical, chemical, morphological, and geological properties, and these characteristic differences in conditions cause gas hydrates to be formed and stabilized in varying subsurface depths and environments.

Geological configuration and sedimentation in continental margins of the world show distinct patterns. Continental margins have been categorized into two distinct domains based on the morphological features, i.e., passive and active margins (see Chapter 1). Active and passive margins differ in structural style and nature with respect to morphology, tectonics, and geological factors such as porosity, permeability, and type of organic content. Furthermore, passive margins may evolve with marked volcanism or in its absence.

Geological Controls for Gas Hydrates and Unconventionals. http://dx.doi.org/10.1016/B978-0-12-802020-3.00004-7

The initiation of igneous processes associated with volcanic passive margins occurs before and/or during the rifting process, depending on the cause of the rift (as detailed in Chapter 1). The nonvolcanic and volcanic margins differ appreciably in morphological and geological characteristics and tectonic style.

Passive margins are typified by wide coastal plains and continental shelves that separate deepwater far from the eroding land surfaces, which are the source of sediment. Smooth shelf and slope relief of such margins provide a suitable environment for the accumulation of a thick pile of sediments. Over the nonvolcanic passive margins with low geomorphic relief, sediment particles are well sorted and show a gradational pattern from near shore to deep-sea environment with particle size reducing towards the deep sea. The relief of margins and the hydrodynamic conditions, i.e., water velocity and its interaction with sediments, primarily determines the characteristics of particles, their size, and their onwards transport into the sea. The result of this physical separation is the efficient sorting of sediment. While sands are trapped in rivers and along the shore, the only grains capable of staying suspended in the water long enough to reach the deep ocean are very fine-grained silts and clays. As a result, the sediment in deepwater passive margin settings is commonly homogeneous mud. The hydrodynamic sorting of fine-grained (clays) and/or low-density (small plant) material from coarse-grained or denser sediments will be more pronounced in either low-gradient systems or during low-discharge events. On the contrary during high precipitations in the high gradients, the uplands will mobilize a large aggregate of sediments that will include large biogenic debris (Blair and Aller, 2012). The distance between upland precipitation and the sea provides space (watershed) and time for the proper filtering of sediments before they are deposited on the margins. In this context, export of particulate organic matter (POM) from large rivers would have a greater proportion of highly degraded, complex fine-grained material as compared to steep-gradient rivers with a smaller watershed (Singh et al., 2007; Syvitski and Saito, 2007).

Along passive margins where thick piles of sediment may develop over considerable periods of time, sediments that are unaffected by tectonic forces get altered due to compaction and thermal and chemical changes that continuously occur during depositional history. The principal diagenetic alteration of sediment is attributed to porosity, permeability chemical, and biological changes in the organic content. Continuous sedimentation leads to the expulsion of fluids from the pore spaces, and reduction of porosity and permeability takes place in a broadly similar, depth-dependent fashion, mainly through compaction.

Passive margins are petroleum storehouses because they are associated with favorable conditions for accumulation and maturation of organic matter. Early continental rifting conditions led to the development of anoxic basins, large sediment and organic flux, and the preservation of organic matter that led to oil and gas deposits; crude oil will form from these deposits. Mann (2004) examined 592 giant oil fields (which contain proven reserves >500 million barrels of oil or >3 trillion cubic feet of natural gas) estimated to contain ~65% of all oil reserves. These oil fields are confined to six basin and tectonic-setting categories, and continental passive margins hold 31% of the observed fields. Continental rifts which are likely to evolve into passive margins with time contain another 30% of the world's giant oil fields.

Faster rates of sedimentation resulting in the accumulation of a thick pile of sediments with high organic content derived from terrestrial and marine productivity and suitable pressure and temperature conditions make the passive continental margins suitable avenues for the formation of gas hydrates in deep-sea environments. Blake Ridge in the Atlantic Ocean, off the eastern coast of the United States has been most extensively studied and is considered to be one of the most formidable storehouses of gas hydrates over the passive continental margins (Borowski, 2004).

Active (convergent) margin regions occur along plate boundaries in which one plate composed of oceanic crust is subducted beneath another continental or old oceanic plate along a deep ocean trench. A variety of crustal interactions can occur at the convergent margin; this includes the oceanic crust underthrusting the continental crust (west coast of South America) and the oceanic crust underthrusting the modified ocean crust. An example of an ocean–continent subduction zone is the Middle America Trench offshore Guatemala and that of an ocean–ocean subduction zone is the Nankai Trough off Japan.

The interaction of two separate blocks of continent–ocean and ocean-modified ocean block at the subduction zone results in complex geological configurations, including the trench, accretionary prism, forearc basin, frontal arc, volcanic arc and backarc basin, and remnant arc; these also include strike-slip and oblique-slip segments forming convergent transform margins. The sediment input is quite variable, depending on the distance of the trench or backarc basin from major river systems.

Compaction of the sedimentary section and its included organic detritus entering the subduction zone is determined by the age and motion of the plate upon which it resides, coupled with the magnitude of the near trench and trench floor sediment supply. The interaction between two sedimentary regimes (open ocean, pelagic, and near-shore terrigenous) commonly results in a sedimentary section entering the subduction environment, which changes in origin and character from the base to the top of the prism. At the base, the section is dominated by biogenic debris derived from plankton, airborne dust, and chemical precipitates of seawater. The up-section sediment grades clays and fine silts of terrestrial origin, derived from subaerial weathering processes, mixed with pelagic material (hemipelagic). The upper portion is often dominated by coarse-grained silts and sands of continental origin (terrigenous) (Fig. 4.1). Tides, waves, and currents strongly control continental shelf sedimentation. Ocean basins adjacent to poorly drained landmasses have sediment accumulation of the order of 200–500 m. Shoreline turbulence from waves is one of the most notable influences because it keeps small particles from settling. Surf and waves carry the smallest particles out to sea, which is why most beaches and immediate shorelines are sandy rather than muddy. Convergent margins drained by large rivers deposit a large quantity of sediments in the adjoining ocean basins. Variations in the sediment input, thermal structure, fluid flow regime, and structural style of accretionary prisms have a substantial effect on sediment that enters the gas generation window and the amount and type of hydrocarbon generation. Organic detritus in the accretionary system is primarily terrestrial in origin and gas prone. Factors favorable for maximum evolution of gas include a large, thick accretionary prism, a thick incoming sedimentary section, substantial axial trench sedimentation fed with continental detritus, development of decollement near the top of the incoming section, substantial underplating, a young subducting plate, and slow to moderate convergence rates (Stevenson, 1996).

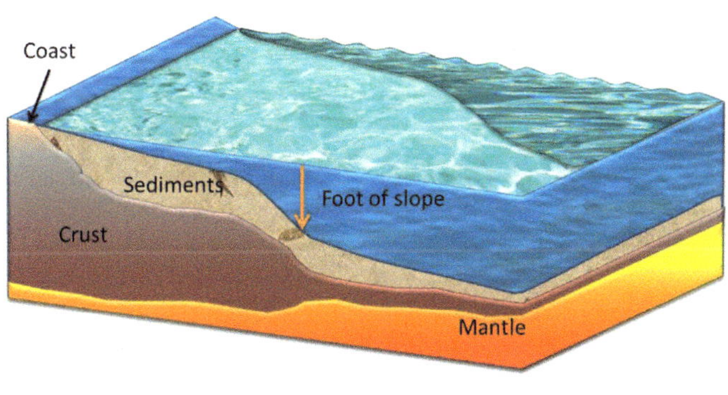

FIGURE 4.1

Generalized distribution of sediments over continental margins.

Convergent margins provide the most favorable environment for gas hydrate accumulation because of their characteristically high sedimentation rates and intense tectonic activity, which provide vertical and horizontal conduits by which gases and fluids can migrate, and average overall higher organic matter concentrations because of upwelling initiated by distinctively narrow shelves and steeper slopes. An accretionary wedge sediment complex is an example of a "structural" hydrate deposit, in which tectonics and gas flow play obvious roles in controlling the abundance and distribution of methane hydrate, as opposed to the "stratigraphic" deposits (Milkov, 2004), where the sediment accumulates into depositional layers.

Studies from different gas hydrate-prone regions have broadly distinguished formation of gas hydrates as primarily controlled by: (a) availability of organic matter, either particulate or dissolved, derived from primary production in the marine regions; (b) kinetics of microbial degradation and generation of methane; (c) rate of sedimentation; (d) thickness of the gas hydrate stability zone; (e) methane-rich fluid flows either from a biogenic or deeper thermogenic source (compression driven, faults, fractures); and (f) methane solubility profile. All these aspects are related to the biological, geological, and tectonic framework of the continental margins.

4.2 ORGANIC CONTENT

Gas hydrates tend to concentrate in regions where the primary productivity is appreciably higher and this is of great importance for the sustenance of marine species. The productivity shows a high order of variability over the continents and oceanic regions (details given in Chapter 2). Primary productivity is dependent on sunlight and availability of water and nutrients. The high productivity rate of organic matter is confined to areas of coastal and equatorial upwelling and regions of high fluvial inputs, where the influx of nutrients is comparatively higher (Stein, 1990). Phytoplankton produces most of the organic matter and is mainly produced in the open ocean. The food chain is very short in coastal areas, which implies that the recycling of nutrients is limited and a large fraction of primary production might be exported either to sediments or to the slope. In the open ocean, phytoplankton is dominated by organisms of small size and the food chain is rather long, consequently most of the organic matter is remineralized in the euphotic zone and little is exported to deep waters (Wollast, 2003). In order to maintain the growth of phytoplankton, nutrients have to be replenished either by vertical mixing of sea waters or by influx from fluvial resources such as streams. The regions of upwelling bring nutrients to the shallow parts of the ocean and thus help in the productivity of organic matter. Conversely, downwelling cells carry organic matter from the surface water to deeper parts of the ocean for decomposition and deposition on the ocean floor (Hay and Southam, 1977). About 90% of organic matter is preserved on the continental margins and only 10% reaches the seafloor; only a small proportion of this (0.5%) is retained in the sedimentary rocks.

Most of the organic matter mainly produced by the photosynthesis processes and is confined to sea surface. There is significant decay of the organic matter with depth in ocean as sunlight hardly penetrates in the deeper parts of the sea. Not only production of organic matter, even its preservation at different depth in water column is significantly important factor for the formation of petroleum products. Pattern of distribution of nutrients in the water column play vital role in the /production/preservation of organic matter. Surface wind pattern, temperature distribution with depth, Coriolis effect and the configuration of coastal region, produce currents which the make water to sink (down-welling) or push upwards (upwelling) regions over the entire continental margins and deep marine environments (Fig. 4.2). Such current pattern alter the distribution of dissolved nutrients such as oxygen, nitrogen and phosphate compounds and help sustenance of organic matter in the sea. In the regions of upwelling the nutrients from

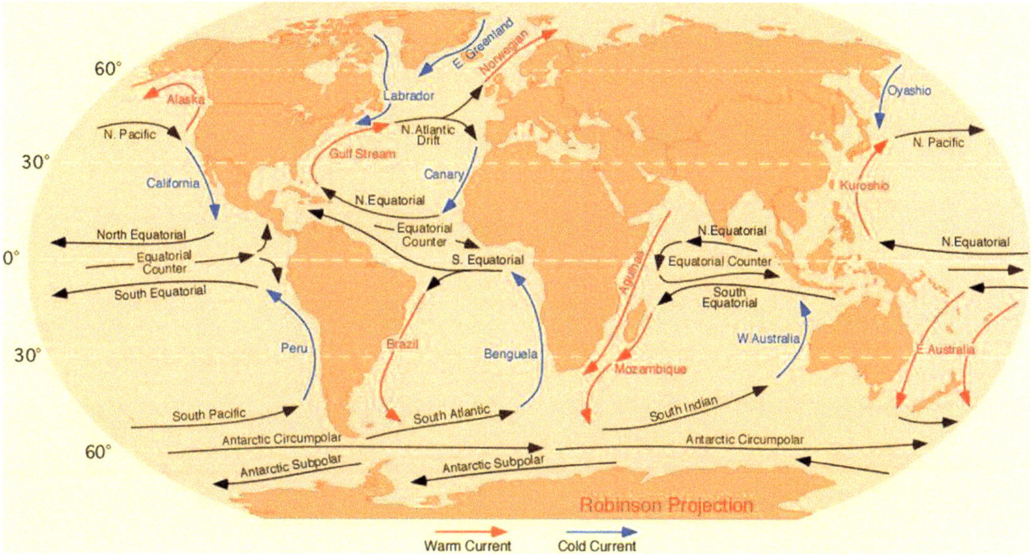

FIGURE 4.2

Major ocean currents of the world. On this illustration red arrows indicate warm currents, while cold currents are displayed in blue. [Source Pidwirny, (2006)]

the deeper parts are brought to sea surface increasing the productivity of organic matter. On the other hand over the regions of down -welling dissolved oxygen from the sea surface is taken to deeper part of the ocean for the sustenance of organic matter at these depths. In absence proper circulation of these nutrients the organic matter may decay very fast over short period of time (Hays and Southam, 1977).

Passive margins mainly surround large continental plains and have wider shelves compared to the narrow shelves of the active margins; these wider shelves are more favorable places for primary productivity. Higher productivity and higher rates of sedimentation make the passive margins suitable regions for gas hydrate formation.

Total organic carbon (TOC) is comprised of material derived from both marine and terrestrial sources. The total quantum of organic matter present in the marine environment is the sum of what is produced in situ and what is produced on land, brought to the marine environment by rivers, wind, and volcanic activity. Continental organic matter is more resistive to decomposition. River-laden terrestrial particulate organic matter (Corg) is complex as it has contributions from: (a) organic matter from recent primary production; (b) aged and diagenetically altered material (aged soil); and (c) ancient kerogens derived from sedimentary rocks on land. The terrestrial component of TOC decreases with the distance from the shore; it changes from being the dominant fraction in shelf sediments to a minor but nonetheless quantitatively significant fraction in slope and basin sediments. The terrestrial component in slope and basin sediments is more biodegraded than that found in shelf sediments and looks almost like bulk river sediments (Prahl et al., 1994). The residence time of organic matter in active margins, such as mountainous, rapidly eroding source regions, is less than that of passive margins, such as lower gradient river systems, thereby resulting in different distribution patterns of organic matter components in continental margins (Fig. 4.3a; Blair and Aller, 2012). Organic matter in ocean basins in the vicinity of

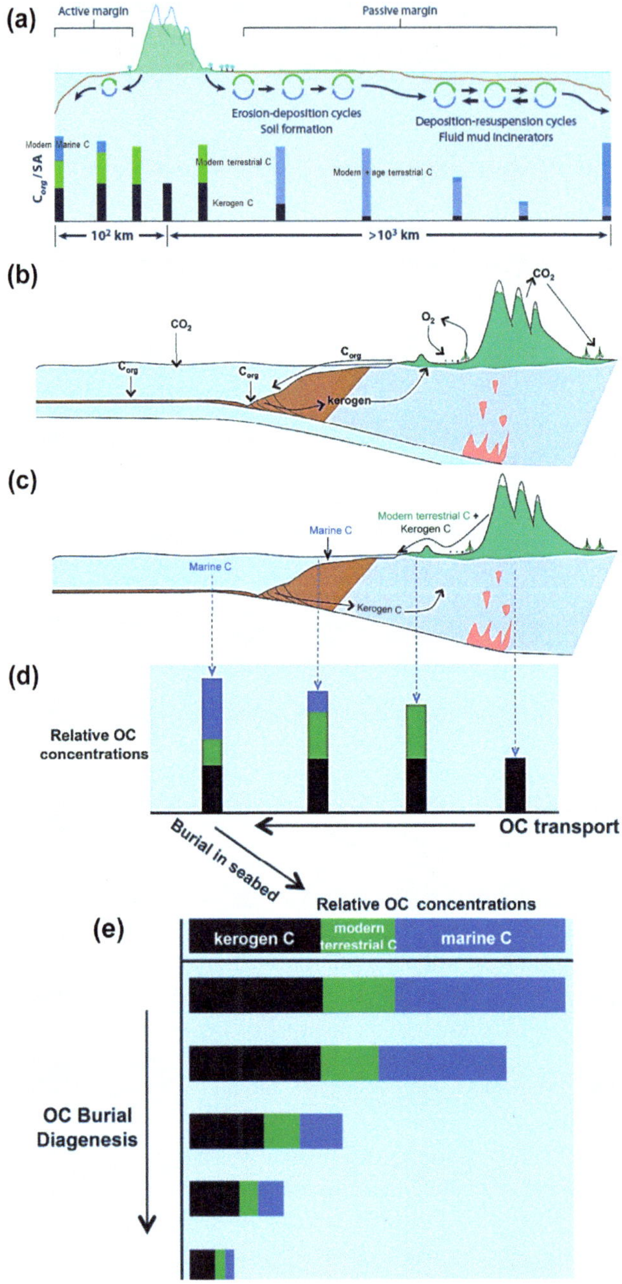

FIGURE 4.3

Distribution of organic matter in the continental margins. (a) The contrasting active and passive margin, serial transport- reaction system. (From Blair and Aller, 2011.) (b) The active margin carbon cycle. Accretion, uplift, and erosion of sedimentary rock on the continent bring previously buried organic carbon (OC) to the surface. If mass wasting is sufficiently rapid, as is the norm on these margins, the exposed fossil carbon (C) is recycled into the sedimentary system thereby avoiding oxidation in subaerial outcrops. The recycled fossil carbon is blended with younger material as sediments move across the surface. (c) The active margin and accretionary wedge carbon cycle. (d) Kerogen formation begins with the diagenesis of organic matter. Tectonically uplifted kerogen will combine with modern terrestrial sources and the mixed pool will be transported by rivers to the marine realm, where the marine pool of organic carbon will be added prior to burial, while terrestrial carbon is concurrently lost. (e) During burial and diagenesis, marine and modern terrestrial carbon will be lost, while kerogen will be preferentially preserved. (parts b–e: From Childress and Blair, 2013.)

high-gradient convergent landmasses is mainly terrigenous in nature, and in sediment accumulations the presence of resistant organic material such as kerogen is significant. The amount of organic matter in convergent margins decreases rapidly towards the deep sea. Sediments in ocean basins adjoining low-relief landmasses (divergent passive margins) have significant contributions from recently produced labile organic matter (see Fig. 4.3a). In an active margin system, the recycled pool represents a significant portion of the organic carbon buried in the marine environment (Blair et al., 2004) (see Fig. 4.3b). Small mountainous rivers along active margins export particulate organic carbon that is 7–75% fossil carbon (kerogen) in content, with the remainder derived from modern vegetation and millennial-age soil sources (Leithold et al., 2006; Drenzek et al., 2009; Blair et al., 2010). Recycled carbon is more inert than younger forms derived from plants and soils, and this inherent recalcitrance should lead to persistence, transit to the deep marine environment, and incorporation into the subduction zone.

4.3 GAS HYDRATE STABILITY ZONE

The presence of gas hydrates is controlled by several factors, among which temperature, pressure, ionic strength of the water, gas composition, and abundance are key parameters (Sloan, 1998). Pressure and temperature are the main factors that govern the thickness of the gas hydrate stability zone. However, the presence of a higher quantity of methane more than the solubility of methane in water is a prerequisite for the formation of gas hydrates in a region. The pressure and temperature conditions required for pure methane hydrate stability are illustrated in Fig. 4.4. In this case, methane hydrate is stable at temperatures higher than 15°C only at high pressures (>10 MPa). At lower pressures, the stability of methane

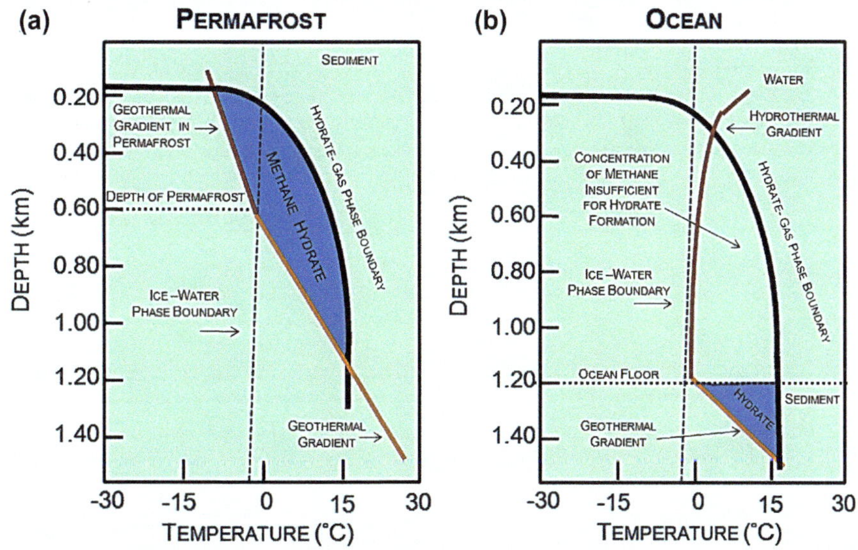

FIGURE 4.4

Methane hydrate stability zones (blue) for (a) permafrost and (b) oceanic environment. (From http://large.stanford.edu/courses/2010/ph240/harrison1/)

hydrate requires cooler temperatures (e.g., for pressures <6 MPa; temperatures <10°C). In the pressure and temperature field, the phase boundary is determined by the gas composition and also by the ionic strength of water. Gas hydrates occur naturally in the pore spaces of sediments, which have the appropriate high pressure and temperature conditions. Over the continental margins, the seawater attains temperatures of the order of 2–3°C at seafloor depths of 2–3 km. The areas colored blue in the generic curves shown in Fig. 4.4 illustrate regions in permafrost and oceanic sediments where the pressure and temperature conditions and the concentration of methane gas are within the hydrate formation and stability zone. These curves are based on pressure–temperature phase equilibrium data and correspond with reflection seismic data collected in these environments. Gas hydrates occur under specific conditions of pressure and temperature, where the supply of methane is sufficient to stabilize the hydrate structure. Since these hydrates form in relatively low temperature environments, the geothermal gradient limits hydrate occurrence to shallow (<1000 m) regions of the sedimentary section. Furthermore, composition of the hydrocarbon gases and pore-water salinity are also important controlling factors in the gas hydrate stability zone (Sloan, 1990, 1998). For the given normal geothermal gradient of 3°C/100 m below the seafloor, it is expected that gas hydrates will be stable up to 300 m below the seafloor. These conditions make gas hydrates stable in the upper few hundred meters of sediments: this is called the gas hydrate stability zone. Using global estimates of pressure–temperature distribution, the gas hydrate stability over continental margins and in the polar region can be calculated from the different procedures available.

The global pattern of hydrate stability zone thickness based on global pressure–temperature with constant geothermal gradient for the entire region is given in Fig. 4.5 It can be visualized that in most of the polar regions and at the continental margins the necessary conditions of stability exist, and these regions can hold gas hydrates provided the availability of excess methane gas far exceeds the solubility

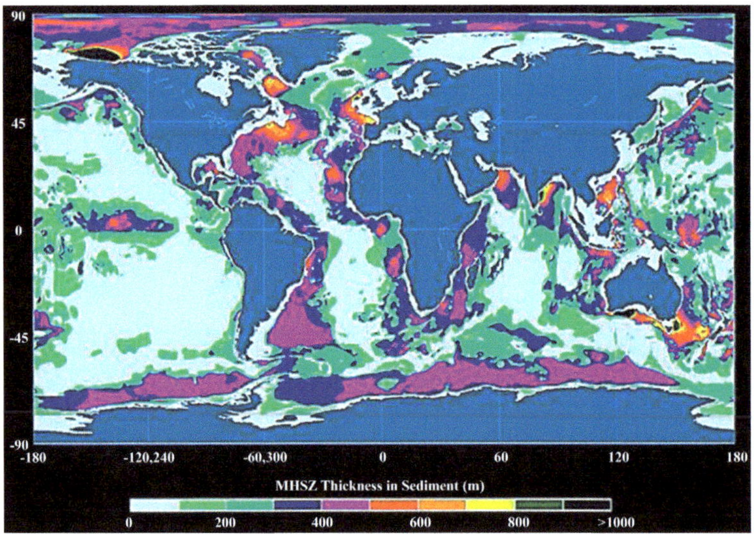

FIGURE 4.5

Global pattern of hydrate stability zone thickness based on global pressure–temperature with constant geothermal gradient for the entire region.

of methane in water. The presence of carbon dioxide, hydrogen sulfide, ethane, and/or propane will have the effect of shifting the stability curve to a higher temperature at a given pressure, increasing the stability of the methane hydrate. Conversely, the presence of dissolved ions in the pore fluids inhibits the stability of the hydrate. With increase in temperature depth hydrates become unstable and entrapment of methane gas in the sediment's pore spaces is expected; such gas is considered free gas.

The presence of gas hydrate and free gas in the region alter the physical properties of the host rock and thereby bring in anomalous variations in the geophysical observations. Such anomalous behavior is clearly observed in seismic surveys owing to changes in velocity and density due to the presence of hydrate and free gas in the regions under investigation. Owing to significant change in impedance contrast (product of velocity and density) across the base of the gas hydrate zone and free gas interface, a significant change in reflection characteristic, namely the bottom-simulating reflector (BSR), is observed in seismic reflection surveys. The observation of BSR in a region generally infers the presence of gas hydrates. As BSR is mainly controlled by the pressure–temperature regime and has no relevance to local geological subsurface layers, it has a tendency to cut across the inclined or folded subsurface layers and run parallel to the seafloor relief (Fig. 4.6). Cutting across the inclined bedding of the BSR in a seismic section can easily be distinguished and the identification of BSR over many active margins has been used for the detection of gas hydrates.

The areal extent of BSR on the passive margin is comparatively small compared to an area covered by BSRs on convergent margins (Haacke et al., 2007). Different mechanisms or models have been proposed for drawing inferences of distribution of hydrates and free gas based on the characteristics of BSR (Hyndman and Davis, 1992; Ginsburg and Soloviev, 1997; Davie and Buffett, 2003; Davie et al.,

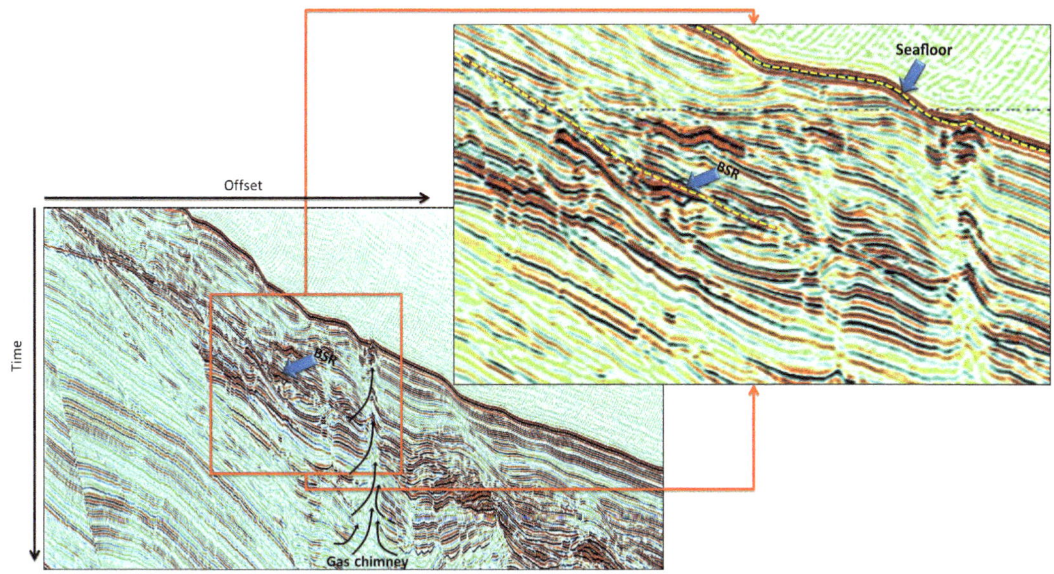

FIGURE 4.6

An example of the extent of bottom-simulating reflector (BSR) from southeast Asian continental margins. Gas is migrating in an upwards direction and a gas chimney is clearly marked on the seismic section.

2004; Haacke et al., 2007). Modeling of BSRs from active and passive margins has been attempted considering the aspects of the quantum availability of methane, its solubility (saturation), and fluid flow patterns to account for the distribution of hydrate and free gas for different margin environments. On the passive margins, such as the Atlantic off the US coast, it has been suggested that in situ generation of methane with a high rate of sedimentation accounts for the occurrence of hydrates and association with BSRs. Methane-rich fluid expelled during the interaction of two boundaries at the subduction zone and a faster rate of sedimentation are hypothesized for the occurrence of gas hydrates and observed BSRs on the active margins. Apart from this, regions of coarse, clastic sediments with higher permeability favor formation of gas hydrates over accretionary regions (Hyndman et al., 1993).

In addition to pressure and geothermal gradients, other variables that affect the stability of gas hydrates are the hydrothermal gradient, the sediment's thermal conductivity, heat flow, and geological heterogeneity (Macleod, 1982). Geological processes can shape a potential gas hydrate reservoir; this is evident from the explanation about the reasons for gas hydrates occupying a position well above their regional phase boundary. The geothermal gradient is also affected by the heterogeneities in the local geological configuration and associated thermal conductivity, as observed in the Gulf of Mexico (Ruppel et al., 2005) and the Barents Sea (Bugge et al., 2002). The proximity-specific geological features, such as mud diapirs, shale diapirs and salt domes, shallow basement, and the presence of faults cause focused fluid flow (Chand et al., 2008). It has been observed globally that shallow salt structures can enhance the temperature of the nearby sediments up to 30°C due to the high conductivity of salt (Thomsen and Lerche, 1991). Vertical fluid migration due to the mud diapirism brings in gases along with heat to shallower levels thereby disturbing the local stability conditions (Chow et al., 2000). The thermal conductivity of salt domes is high thereby lowering geothermal gradients in the dome. This results in compressed isotherms above the salt dome, which may produce thinning of the stability zone above the salt dome in comparison to its flanks (Dillon et al., 1980). A thinning hydrate stability zone over the salt dome has been observed in the southeast offshore areas of the USA (Dillon and Max, 2003). The presence of salt ions in the interstitial fluids act as inhibitors resulting in higher temperatures and thereby preventing the formation of gas hydrates (Taylor et al., 2000). The thermal conductivity of shale is lower in general. In case of shale diapirism, the geothermal gradient inside the diapir is high making the isotherm deepen in the diaper, which results in the deepening of the gas hydrate stability zone over the diaper (Macleod, 1982) (Fig. 4.7).

Hydrate occurrences are controlled by temperature–pressure conditions that in turn define the solubility profile in the presence of adequate methane. Persistence of hydrates in sediments requires a continual supply of methane to maintain a sufficient concentration of dissolved methane against the losses due to diffusion and advection (Davie and Buffett, 2003). Methane distribution with depth differs from region to region. The methane concentration profile changes with temperature, pressure, and the presence or absence of hydrates (see Chapter 3 for details). Studies have shown that the TOC in the sediments is not the only factor that controls the gas hydrate formation but methane-rich fluid migration also plays an important role. The prevailing pressure–temperature conditions in Blake Ridge and the Peruvian margins provide a thicker hydrate stability zone (450–400 m). In the tectonically active margins of the Cascadian region and the Chile Triple Junction, the gas hydrate stability zone is thinner, and the TOC in these two zones of differing hydrate thickness also differs significantly. The thicker hydrate zones are characterized by high TOC content whereas tectonically active regions have lower concentrations of TOC. Owing to this, the solubility profile in these regions differs significantly to maintain the supply of methane required for persistence of hydrate in these regions (Fig. 4.8). The higher gradient

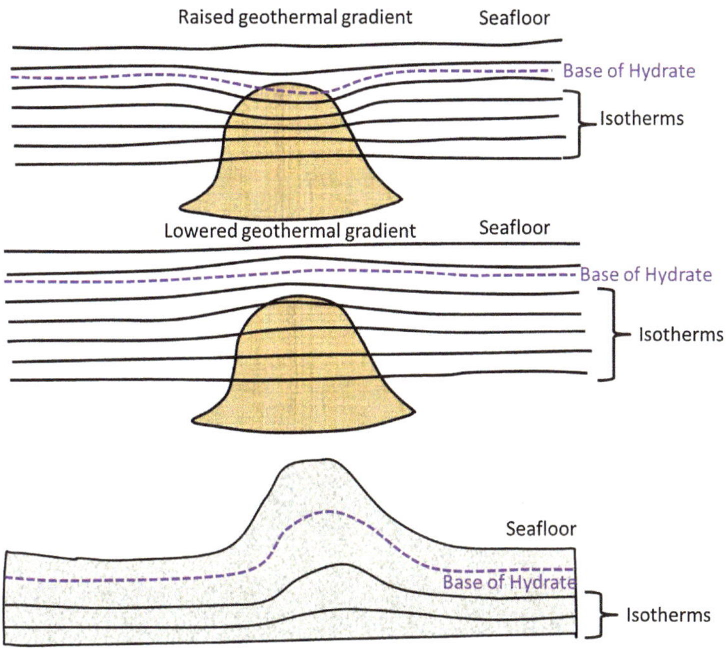

FIGURE 4.7

Schematic illustration of shale diapirism. The geothermal gradient inside the diapir is high making the isotherm deepen in the diaper, which results in the deepening of the gas hydrate stability zone over the diaper. (Modified from Macleod, 1982.)

solubility profile in active regions suggests that fluid migration accounts for persistence of hydrate in these regions (Davie et al., 2004).

Over the passive continental margins with the accumulation of thick pile sediments barring the tectonic activity, the bedding planes are generally horizontal and the isothermal lines more likely to run parallel to the seafloor. It is suggested that hydrate may not form in well-bedded sediments due to reduced permeability that may impede free flow of methane and fluids (Zuehlsdorff et al., 2000). In the regions of well-bedded slope basins and areas of thick sedimentations, the identification of BSR is rather difficult. In such scenarios, the cross-cutting characteristic of BSR is usually absent. In the absence of cross-cutting, the identification of BSR on seismic reflection becomes speculative and it is difficult to draw inferences on the presence of gas hydrates in a region. In such regions, other geological proxies, such as pockmarks, transparent sediments, gas venting, and chimneys, can be used. An excellent example of pockmarks and associated seismic signatures for gas hydrates is described by Jiangong et al. (2015) and is shown in Fig. 4.9. Jiangong and colleagues also concluded that the gas hydrate distribution and thermal regime in the pockmark are largely controlled by the intensity, periodicity, and direction of fluid flow. The joint interaction between fluid flow, gas hydrate formation and dissolution, and the thermal regime governs pockmark formation and evolution on the Nigerian continental margin. It has been suggested that the presence of transparent strata (blanking) in seismic data has been observed over Blake Ridge (Lee et al., 1993, 1996; Dillon and Max, 2003), Lake Baikal (Vanneste et al., 2001), and many

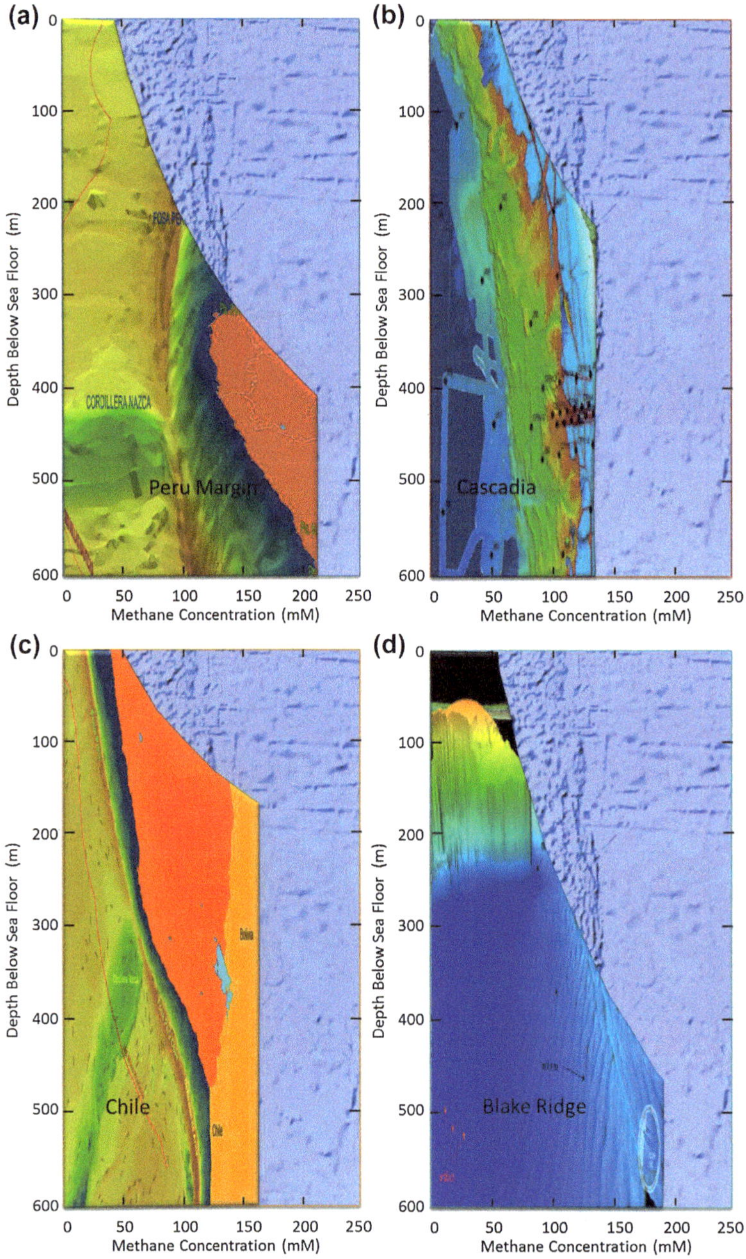

FIGURE 4.8

Methane concentration plots for Peru, Cascadia, Chile, and Blake Ridge with respect to the depth below seafloor. Background shows seafloor bathometry. (a) Peru Margin; 3D representation shows methane concentration with respect to the depth below seafloor. At ~400 m depth below seafloor methane concentration is >200 mM. (b) Cascadia, at ~225 m depth below seafloor methane concentration is ~175 mM. (c) Chile, at ~175 m depth below seafloor methane concentration is close to 160 mM. (d) Blake Ridge, at >450 m depth below seafloor the concentration of methane is ~175 mM. (Modified and reinterpreted from Davie et al., 2004.)

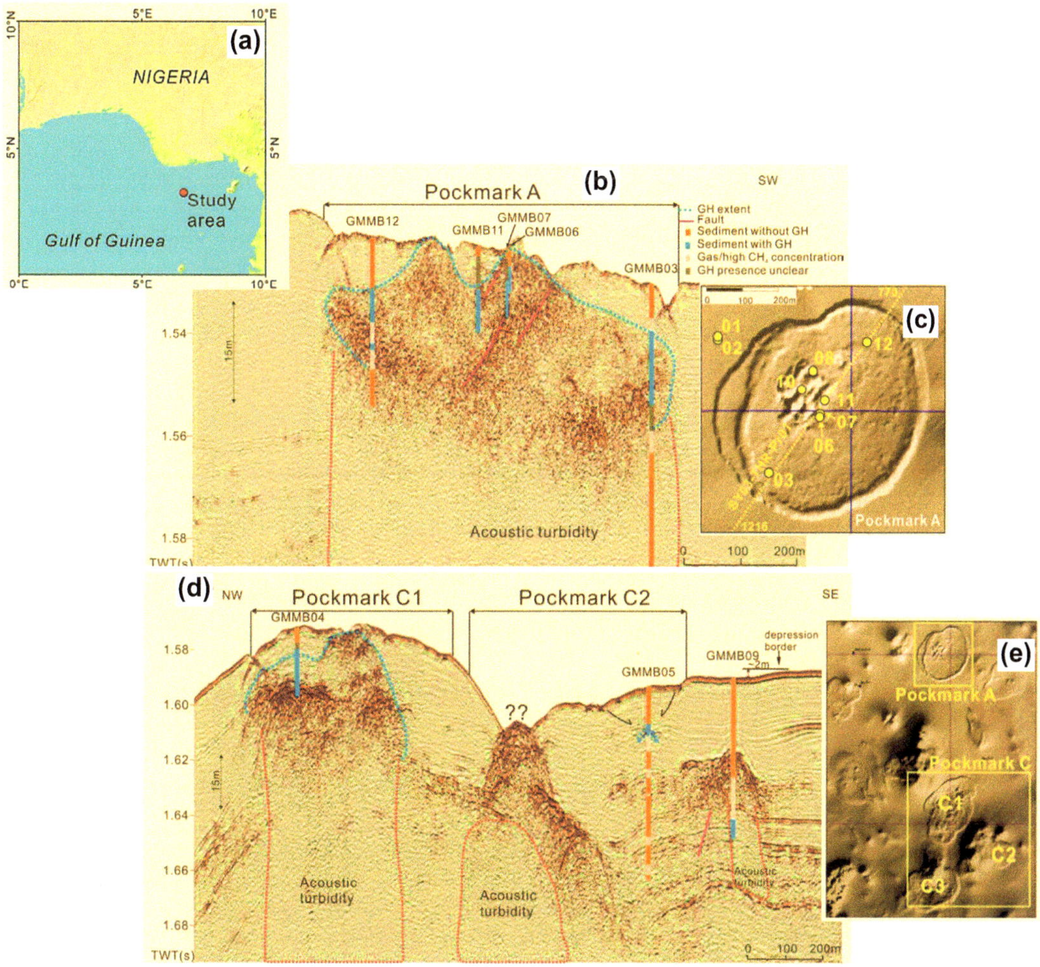

FIGURE 4.9

Examples of pockmark and associated seismic signatures over Nigerian continental margin. (a) Location of the pockmark field at the Nigerian continental margin. (b) Seismic section crossing pockmark A. Interpretations from cores are projected on the seismic lines. High-amplitude reflectors are widespread in the seismic profile. (c) Pockmark overview. (d) Seismic section crossing pockmark C. Interpretations from cores are projected on the seismic lines. High-amplitude reflectors are widespread in the seismic profile. (d) Pockmark overview. (Modified from Jiangong et al., 2015.)

other regions. It is argued that cementation of sediment grains with gas hydrates may lead to uniformity within the sediment strata, thereby showing no reflection in this zone. The observed blanking in many parts of the world has been attributed to the presence of gas hydrates in the region. At low saturations, hydrates may reduce the impedance contrast between regions of low and high porosity and thereby cause blanking (Holbrook et al., 2002; Max and Dillon, 2003). A "blank zone" in seismic data may imply a paucity of horizontal reflections rather than a lack of impedance contrast. Modeling results have indicated enhancement in the blanking pattern with increase in saturation of gas hydrates in sediments.

Gas or gas-charged fluids trapped below the hydrate stability zone move upwards through the faults or fractures and thereby disturb the characteristic signature of gas hydrate-bearing sediments above and change the reflectivity of the medium. This predominantly vertical pathway provides the gas or gas-charged fluids with an escape route through the gas hydrate-bearing layers. If the venting of gas does not get confined to the sedimentary layer, the gas and fluids escape in overlying ocean waters and in the process destabilize the seafloor relief with the creation of small-scale depressions (pockmarks), generally subcircular features of several meters across occurring at shallow depths. Pockmarks are craters in the seabed caused by fluids (gas and liquids) erupting and streaming through the sediments. Pock marks are considered as a geological indicator of gas hydrates as venting of gas through the sedimentary strata suggests the possible existence of hydrates and free gas molecules.

4.4 SEDIMENTATION

Sedimentation is one of the main factors that control gas hydrate accumulation on continental margins. A high rate of sedimentation favors the formation of gas hydrates on the continental margins and the sedimentation rate generally decreases with increasing distance from the coastline and increasing water depth (e.g., Burwicz et al., 2011). Sedimentation rates for major ocean basins during the Cenozoic Era are given in Fig. 4.10 (Whitman and Davies, 1979). Rapid sedimentation at rates >1 mm/year generate overpressure in many sedimentary basins around the world (Rubey and Hubbert, 1959; Fertl, 1976).

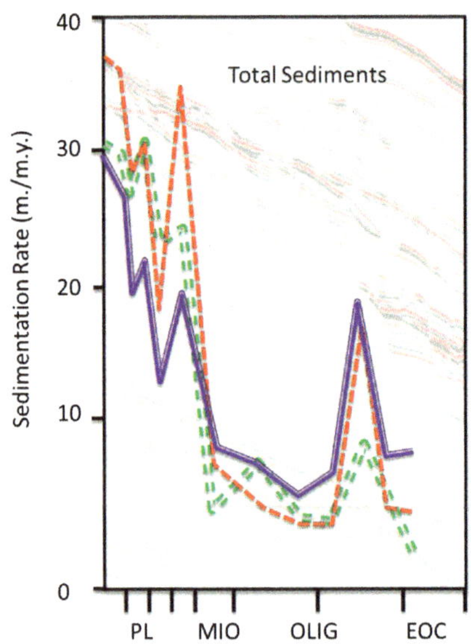

FIGURE 4.10

Sedimentation rates for major ocean basins during the Cenozoic Era. A representative seismic section is shown in the background. (Modified from Whitman and Davies, 1979.)

Regions of higher sedimentation bring in a higher quantum of organic matter. Sedimentary organic carbon content (% Org-C) doubles with each 10-fold increase in sedimentation rate (S), assuming that other factors remain constant, i.e., primary production (R), porosity (φ), and sediment density (ps) (Müller and Suess, 1979) When low-permeability sediments are rapidly loaded, pore fluids cannot escape and the fluids bear some of the overlying sediment load. In this situation, there is development of pore pressure that exceeds the hydrostatic pressure (overpressure, P*). Numerous processes generate overpressure but in many passive margins rapid deposition is thought to be a dominant mechanism, producing a two-way coupling between sedimentary processes and fluid flow. The process of flow expelled fluid is dependent on the rate of sedimentation, lithology, and stratigraphy. Fluids are often trapped in pore spaces between sediments and rocks, which will affect the fluids when sedimentation and compaction occurs. In the subsurface the compaction and sedimentation will increase with depth which would influence the liquids to be displaced upward because the density is lower than the surrounding solids. The fluid flow at the subsurface occurs mainly from complex fault systems, salt diapirs and will mainly be trapped in the highest point of a permeable rock unit (Selley, 1998). Fluid migration is defined when petroleum migrates from the source rock to the reservoir rock. This process is defined when petroleum is transported as a separated phase and mainly driven by buoyancy of petroleum relative to water (Bjørlykke, 2010). Geological structures such as faults, diapirs and landward dipping bedding planes are favorable zones for rising fluid flows (Ginsburg and Soloviev, 1997).

4.5 FAULTING

Evolution of divergent (passive) continental margins initiates with thinning, rifting, and breaking of the lithosphere along preexisting weak zones. In major detachment areas, low-angle normal faults play a key role in the lithospheric extension process and displacement may be of the order of tens of kilometers (Whitmarsh et al., 2001; Thinon et al., 2003; Reston, 2005; Tucholke et al., 2007). Architecture of the basement created during the rifting, faulting, changes in the stress pattern, and dewatering during the earthquake activity may have an important bearing on fluid flow. Faults get rejuvenated during subsequent tectonic activity and deposition of sediments. These fault zones act as conduits or pathways for fluids that are caused by sedimentation. There is a clear association between gas hydrate occurrence and fault zones, as well as other tectonically active related features, as observed by Hyndman and Davis (1992), and Soloviev and Ginsburg (1994). All but one of 15 drill hole sites in which hydrates are present lie within tectonically active continental margin environments. Blake Ridge off the southeastern coast of the USA is the only site in a passive margin but nonetheless is characterized by the presence of extensive fault zones. Fig. 4.11 shows faulting and gas venting structures in the Kerala-Konkan and Krishna-Godavari gas hydrate-bearing basins in the Indian subcontinent.

Fluid flow in accretionary prisms occurs due to spacing of pores and their interconnection through permeability typically along faulting and fractures in the sediments. During the sediment accretion sufficient fluid pressure built occurs making the fluids to rise and expelled across the medium. Fault pattern over the convergent margins determine the flow pattern of the methane rich fluids. Thrust faults are more common in accretionary regimes. In such type faulting the overburden of sedimentary strata decrease the pore space with depth thereby reduces free flow methane rich fluids from down below. On the contrary normal and strike- slip faults provide good avenues to flow fluids. Due to lateral movement of different blocks provide enough space/direction for the fluids to flow freely. However, during a long

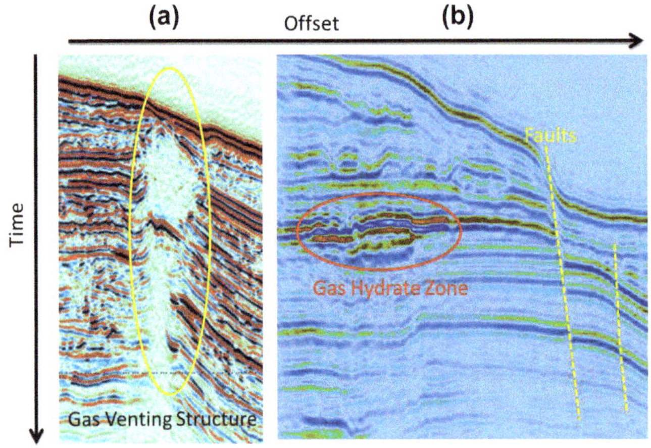

FIGURE 4.11

Examples of gas venting structures and gas hydrate structures from the Indian subcontinent. (a) Seismic section showing gas venting structures from Krishna-Godavari Basin. (b) Seismic section showing faulting and gas hydrate signatures over Kerala-Konkan Basin in the Indian subcontinent

period evolution of accretion prism most of the distributed flows are restricted due to compaction of sediments and cementation of sediments grains.

As permeability tends to be anisotropic (more permeable normal to the direction of compaction) in the consolidating sediments, porous beds may also favor hydrate accumulation through pathways that trend towards horizontal (Rajput et al., 2010). Structural and metamorphic features are affected at all scales by fluid pressures or fluid flow in accretionary prisms. High porosity and permeability favor the accumulation of hydrate. Mature hydrocarbons expulsed along prism faults indicated seep sources and many tens of kilometers of lateral transport of fluids.

The occurrence of numerous well-connected shallow fault systems in the North Sea Basin was termed "polygonal fault" (Cartwright, 1994). Polygonal faults occur in many sedimentary basins all over the world and have been attributed to the result of sediment contraction and fluid expulsion (Cartwright and Dewhurst, 1998; Lonergan et al., 1998; Gay et al., 2004). The polygonal faults are thought to play a major role connecting deeper and shallow reservoirs (Berndt et al., 2003; Gay et al., 2007). Three processes are believed to cause polygonal fault systems: (i) syneresis of clay minerals, (ii) density inversions, and (iii) gravitational collapse. However, it is not entirely clear how they interact and which is the most important (Cartwright et al., 2003). The polygonal and linear distribution of the anomalies following the main orientation trend of deep faults but not matching the exact fault extension or location suggests that faults drive fluids towards the free gas zone through an inhomogeneous mechanism where only segments of faults are active players. Similarly, the distribution of fluid escape chimneys is controlled by local excess pore pressure compared to lithostatic pressure at exclusive zones within the contacts between the sediments and upward migrating gas. Faults are therefore indirect controllers of chimney distributions in Nyegga in the Norwegian region (Plaza-Faverola et al., 2010).

Sandy sediments generally have high permeability and hydraulic conductivity, which allows pore fluids to dissipate quickly even if loading and associated fluid expulsion from minerals or advection rates

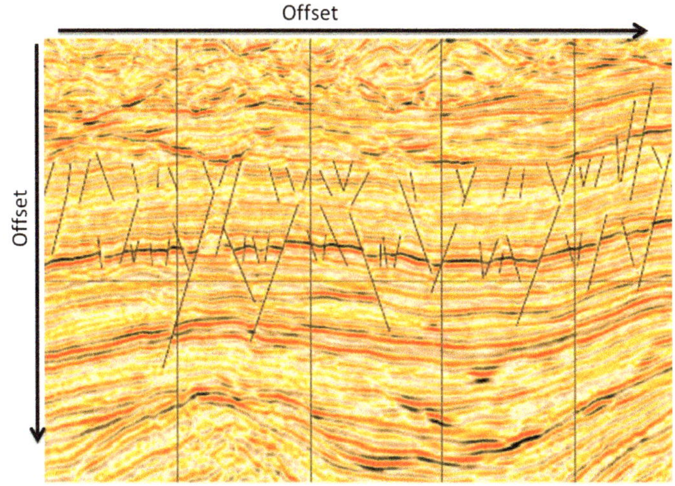

FIGURE 4.12

Occurrence of gas hydrates in deep waters of the South China Sea; heavy faulting is shown. (Modified from Chen et al., 2011.)

from deeper strata are high (Bredehoeft and Hanshaw, 1968). However, some contracting and water-expelling rocks are under high excess pore pressure. This can be caused by high sedimentation rates preventing pore fluids from dissipating at the same rate as the load increases during sediment compaction (Dugan and Flemings, 2000; Flemings et al., 2003). It can also be caused by the presence of impermeable rocks above. Such a situation can arise as a result of deposition, i.e., a stratigraphy in which water-rich sediments are overlain by impermeable sediments. In some cases, the overpressure changes the fluid flow pattern from homogeneous dissipation to focused flow by self-enhanced permeability.

Presence of polygonal faults in many oceanic sedimentary basins have lead to suggestion that these faults are mainly signature of extensive fluid expulsion with spontaneous contraction of fine grained sediments They are indicative of connection between deeper weaker zone (faulting) to shallower multidimensional faulting (Cartwright and Dewhurst, 1998; Lonergan et al., 1998; Berndt et al., 2003; Gay et al., 2004; Gay and Berndt, 2007). These shallow faulting may not run to the full extent of deep fault, suggesting fluid driven control for their formation and their chimney like vertical extent is indicative of excess pore fluid pressure as evident in Nyegga in the Norwegian region. (Plaza-Faverola 2010). Polygonal faulting in the Norwegian region is mostly associated with gas hydrates and indicate such type of faulting may provide a reliable signature of occurrence of gas hydrate in a region of investigation (Fig. 4.12). These faults probably connect deep thermogenic gases to shallow biogenic gases in the hydrate stability zone (Chen et al., 2011).

4.6 FLUID FLOW

Subsurface fluid flow plays a significant role in many geological processes and is increasingly being studied in the time scale of sedimentary basin evolution and relevance to hydrocarbon generation. Many economic resources such as petroleum and mineral deposits are products of basin-scale fluid flow operating over large periods of time. Fluid flow occurs both in passive and active continental margin settings. However, this activity is found to be quite prevalent in the active regions. On passive continental margins, fluid flow is most common in areas with rapid sedimentation and undercompaction, which prevents fluids from being expelled during sedimentation. On active continental margins, fluid flow relates mainly to compressional geological processes (Judd and Hovland, 2007). Fluids are also able to break through the capillary pressure of overlying sediments while oil and water generally use permeable formations, faults, and fractures to migrate (Ligtenberg, 2005). The assumption of basic proportionality between porosity and permeability suggests that higher porosity may favor greater capability to transport fluids. Accretionary prisms are composed of initially saturated sediments caught in subduction zone tectonism. Fluid flow from the accretionary prism feeds surface biological cases, precipitates and dissolves minerals, and causes temperature and geochemical anomalies. Structural and metamorphic features are affected at all scales by fluid pressures or fluid flow in accretionary prisms (Moore and Vrolijk, 1992). In nonaccretionary margins where sediments are subducted, a large volume of fluids are expelled in the narrow zone of the continental slope (Hyndman et al., 1993). As the incoming sedimentary sections are weak, porous sediment consolidation must be responsible for a large proportion of cumulative fluid production. Commonly seismic wipe outs, which are indicative of fluid escape, occur above the fault systems. As these wipe outs frequently do not reach the surface, they allow dating of compaction-related faulting that gives rise to fluid expulsion. Outcropping of faults and evidence for fluid expulsion in the overburden imply that faulting and compaction start shortly after deposition and continue after the faulted sediments are buried by subsequent sedimentation.

Variations in the sediment input, thermal structure, fluid flow regime, and structural style of accretionary prisms have a substantial effect on the sediment that enters the gas generation window and the amount and type of hydrocarbon generation. Organic detritus in the accretionary system is primarily terrestrial in origin and gas prone (Stevenson, 1993). Factors favorable for maximum evolution of gas include a large, thick acccretionary prism, a thick incoming sedimentary section, substantial axial trench sedimentation fed with continental detritus, development of decollement near the top of the incoming section, substantial underplating, a young subducting plate, and slow to moderate convergence rates. The thick pile of sediments from the accretionary prism is the major source for fluids expelled at shallower depths. Apart from fluids expelled due to compression of the accretionary column, the sediment veneer over the incoming plate subducting below the decollement also expels fluids (Hyndman and Davis, 1992) (Fig. 4.13).

In most cases, dissolved methane is the source of gas hydrate formation, where the methane-rich fluids migrate to hydrate stability zones. Fluid flow and gas hydrate formation or dissociation may be strongly linked phenomena. The quantum of biogenetically produced methane in sediments is insufficient for the formation of gas hydrates. In this regard, migrating methane-rich fluids complement locally generated biogenic methane for the formation of gas hydrates in a region (Fig. 4.14). Apart from the large quantum of methane that is present in hydrates, geophysical studies in most of the cases have inferred the presence of free gas below the gas hydrate zone to account for the observed BSRs. The geological setting controls the rates of fluid flow and the amount of methane available for gas hydrate formation. Rich, shallow gas hydrate deposits form near the seafloor only where geological structure focuses on methane. Lithology, fracture distribution, and pore-space geometry also impact fluid flow and gas hydrate nucleation and growth, and thus control the fine-scale gas hydrate distribution.

Gas hydrates are meta-stable state. Any alteration in physical or chemical condition will alter their stability and in the process the encaged gases or fluids (water) are released in the surrounding medium.

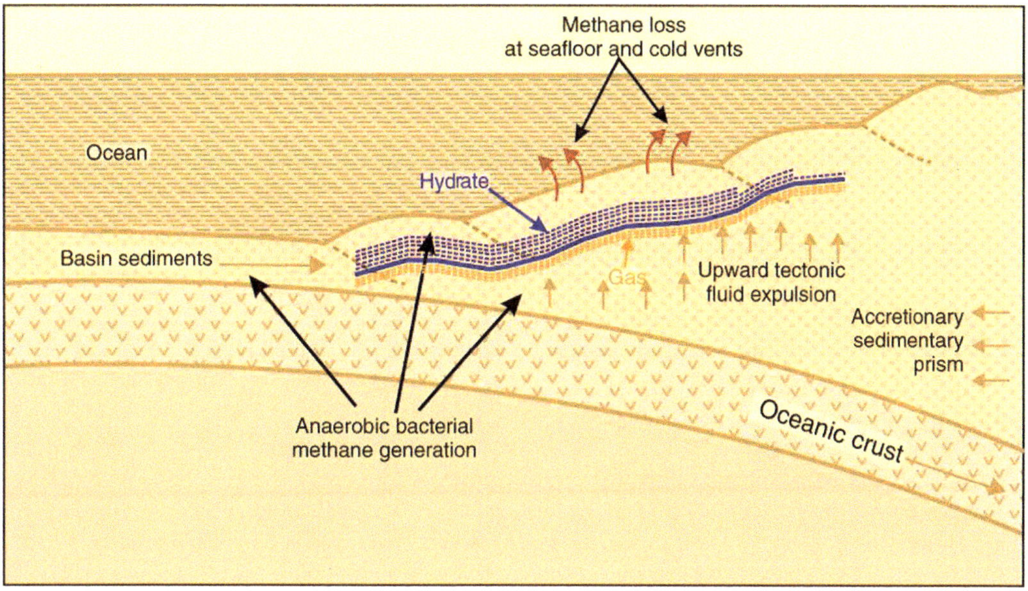

FIGURE 4.13

Schematic illustration of the first-order processes thought to be important in the formation of gas hydrates and associated bottom-simulating reflectors in subduction zone accretionary prisms. (After Hyndman and Davis, 1992; Rajput, 2010; Riedel et al., 2010.)

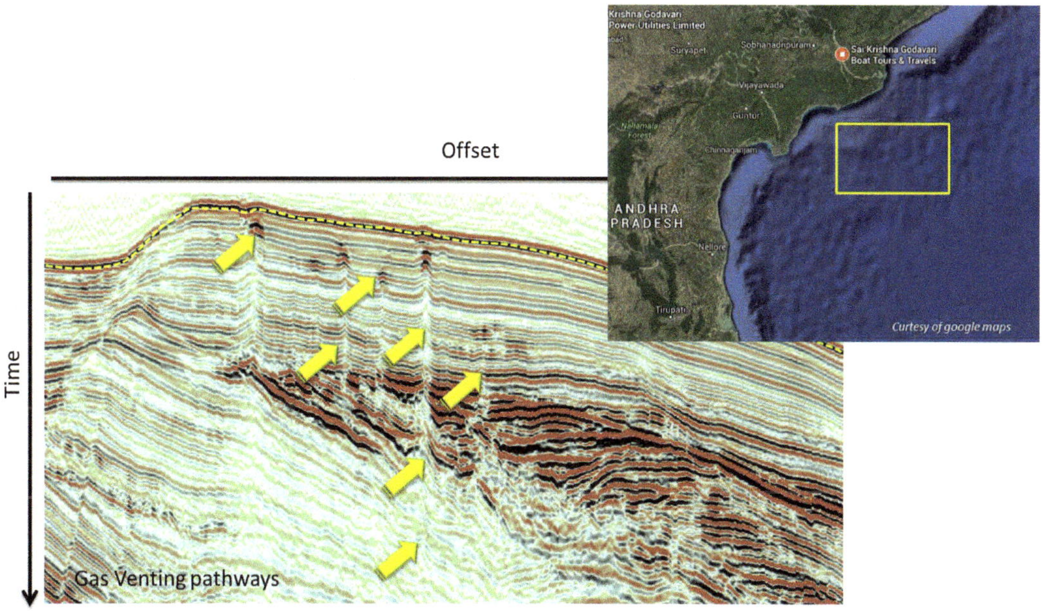

FIGURE 4.14

Example of gas venting structures from Krishna-Godavari Basin in the Indian subcontinent. (Courtesy of National Geophysical Research Institute (NGRI).)

Owing to their stability in high pressure regime, their episodic dissociation would damage the structural configuration of medium under which they found to occur and surrounding regions also. Any change in climatic conditions would lead to their instability, thereby immense release gases escaping through the sea floor and causing imbalance to atmosphere. Favourable climatic changes may also thicken the gas hydrate stability zone thereby trapping more gases emanating below and make work out a prospect for exploration of natural gases.

4.7 DIAPIRISM

Diapirism underlying less dense rock rises through buoyant forces. Diapirs (Fig. 4.15) are most commonly composed of evaporatic salt deposits (salt domes) or gas-charged muds, but may be igneous and found commonly associated with trapped salts during different episodes of sediment deposition and evolution of continental margins. Diapirism is also associated with shale and mud as the lesser density and lower viscosity enable upward flow or creep. Diapirism also seems to be associated with extension on divergent continental margins or in intracontinental rifts (Jackson et al., 1994). Mud diapirs are found generally associated with volcanism, producing near surface mud volcanoes. Even in the nonpermeable regions, gas-related flows overcome the capillary pressures whereas oil and water may use permeable and fractured media as pathways. In regions of known diapirism, when overpressure builds up due to gravitational or fluid loading, gas migration can occur via faults and fractures. Here a combination of overpressure and the buoyancy of expanding gas drive the flow. These characteristics are typical in regions such as the Gulf of Mexico (Ruppel et al., 2005), where overpressured fracture zones that surround moving salt diapirs provide active conduits for vertical migration from deep reservoirs to shallow subsurface (Sassen and Macdonald, 1994).

Hydrate formation may plug the migration conduits, and as high gas pressure builds up, the gas may be rerouted into soft sediment layers. The dynamic processes that interact with a complicated plumbing system may be responsible for the large variety of gas hydrate and sediment fabrics observed. Hydrate

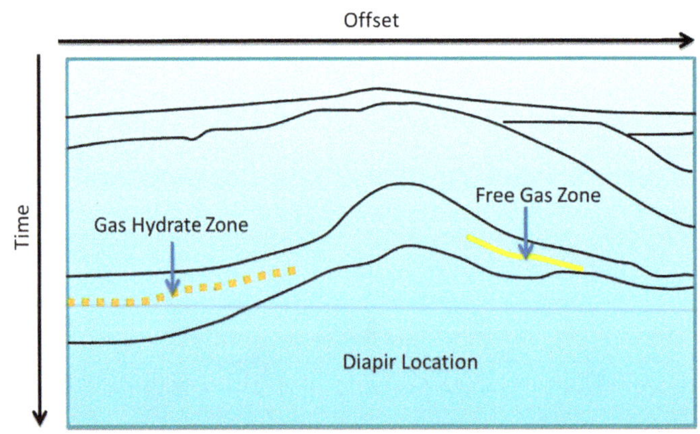

FIGURE 4.15

Example of a diaper structure resulting from seismic interpretation. A Gas hydrate zone and free gas zone are also shown.

formation may plug the migration conduits and as high gas pressure builds up, the gas may be rerouted into soft sediment layers. Further penetration of the diapir causes extensional faulting above the diapir and the base of the gas hydrate to recede locally above the diapir. Gas collects beneath a hydrate seal. Gas hydrates are concentrated around the diapir and its structure becomes very complex in the vicinity of the diapirs. Migration of a diapir through the subsurface forms a gas trap which could be derived from the breakdown of hydrate above the diapir and accumulation of gas from the surrounding area. This process causes the development of an extension fault above the diapir which could be extended through the base of the hydrate stability field.

Analysis of seismic reflection data over the diapirs in the gas hydrate-rich province of Blake Ridge in the Atlantic Ocean have indicated a significant shallowing of the base of gas hydrate stability (BGHS) over the center of the diapir. The seafloor is warped by about 100 m above the diapir from about 2300 m to about 2200 m. The BGHS, as indicated by a BSR, is about 4.5 s off the flanks of the diapir, rising to about 4.15 s at the center. Above the diapir, a fault system appears to rise vertically from the BGHS to about 0.05 s below the seafloor (40–50 m); it then diverges into several steeply dipping faults that breach the seafloor and cover an area ~700 m in diameter (Taylor et al., 2000).

REFERENCES

Berndt, C., Bünz, S., Mienert, J., 2003. Polygonal fault systems on the mid-Norwegian margin: a long term source for fluid flow. In: van Rensbergen, P., Hillis, R.R., Maltman, A.J., Morley, C.K. (Eds.), Subsurface Sediment Mobilization. Special Publication 216. Geological Society of London, London, pp. 283–290.

Blair, N., Aller, R., 2011. The fate of terrestrial organic carbon in the marine environment. Annu. Rev. Mar. Sci. 4, 401–423.

Blair, N., Leithold, E., Aller, R., 2004. From bedrock to burial: the evolution of particulate organic carbon across couple watershed-continental margin systems. Mar. Chem. 92, 141–156.

Blair, N., Leithold, E., Brackley, H., et al., 2010. Terrestrial sources and export of particulate organic carbon in the Waipaoa sedimentary system: Problems, progress and processes. Mar. Geol. 270, 108–118.

Blair, N.E., Aller, R.C., 2012. The fate of terrestrial organic matter in the marine environment. Annu. Rev. Mar. Sci. 4, 171–193.

Borowski, W.S., 2004. A review of methane and gas hydrates in the dynamic stratified system of Blake Ridge region offshore southeastern North America. Chem. Geol. 205, 311–346.

Bredehoeft, J.D., Hanshaw, B.B., 1968. On the maintenance of anomalous fluid pressures. 1. Thick sedimentary sequences. Geol. Soc. Am. Bull. 79 (9), 85–94. http://dx.doi.org/10.1130/0016-7606(1968)79[1097:OTMOAF]2.0.CO;2.

Bjørlykke, K., 2010. Petroleum Geoscience: From Sedimentary Environments to Rock Physics. Springer-Verlag Berlin Heidelberg, Berlin, Heidelberg.

Bugge, T., Elvebakk, G., Fanavoll, S., et al., 2002. Shallow stratigraphic drilling applied in hydrocarbon exploration of the Nordkapp Basin. Barents Sea. Mar. Petrol. Geol. 19, 13–37.

Burwicz, E.B., Rüpke, L.H., Wallmann, K., 2011. Estimation of the global amount of submarine gas hydrates formed via microbial methane formation based on numerical reaction-transport modeling and a novel parameterization of Holocene sedimentation. Geochim. Cosmochim. Acta 75, 4562–4576.

Cartwright, J.A., 1994. Episodic basin-wide hydrofracturing of overpressured Early Cenozoic mudrock sequences in the North Sea Basin. Mar. Petrol. Geol. 11, 587–607.

Cartwright, J.A., Dewhurst, D.N., 1998. Layer-bound compaction faults in fine-grained sediments. Geol. Soc. Am. Bull. 110, 1242–1257.

Cartwright, J.A., James, D., Bolton, A., 2003. The genesis of polygonal fault systems: a review. In: van Rensbergen, P., Hillis, R.R., Maltman, A.J., Morley, C.K. (Eds.), Subsurface Sediment Mobilization. Special Publication 216. Geological Society of London, London, pp. 223–243.

Chand, S., Mienert, J., Andreassen, K., et al., 2008. Gas hydrate stability zone modelling in areas of salt tectonics and pockmarks of the Barents Sea suggests an active hydrocarbon venting system. Mar. Petrol. Geol. 25, 625–636.

Chen, D., Wu, S., Wang, X., Lv, F., 2011. Seismic expression of polygonal faults and its impact on fluid flow migration for gas hydrates formation in deep water of the South China Sea. J. Geol. Res. http://dx.doi.org/10.1155/2011/384785.

Childress, L.B., Blair, N.E., 2013. The Active Margin Carbon Cycle. White Paper, GeoPRISMS Planning Workshop for the New Zealand Primary Site. April 15–17, 2013. Te Papa Museum, Wellington, New Zealand. http://geoprisms.org/meetings/past-meetings/newzealand-apr2013/.

Chow, J., Lee, J.S., Sun, R., Liu, C.S., Lundberg, N., 2000. Characteristics of the bottom simulating reflectors near mud diapers: offshore southwestern Taiwan. Geo. Mar. Lett. 2000 (20), 3–9.

Davie, M.K., Buffett, B.A., 2003. A steady state model for marine hydrate formation: constraints on methane supply from pore water sulfate profiles. J. Geophy. Res. 10/2003; 108(B10). http://dx.doi.org/10.1029/2002JB002300.

Davie, M.K., Zatsepina, O.Y., Buffet, B.A., 2004. Methane solubility in marine environment. Mar. Geol. 203, 177–184.

Dillon, W.P., Max, M.D., 2003. Oceanic hydrates. In: Max, M.D. (Ed.), Coastal Systems and Continental Margins, Natural Gas Hydrate in Oceanic and Permafrost Environment, 2003. Kluwer Academic Press, London, pp. 61–76.

Dillon, W.P., Grow, J.A., Paull, C.K., 1980. Unconventional gas hydrate seals may trap gas off southeast US. Oil Gas J. 78, 124–130.

Drenzek, N., Hughen, K., Montlucon, D., et al., 2009. A new look at old carbon in active margin sediments. Geology 37 (3), 239–242.

Dugan, B., Flemings, P., 2000. Overpressure and fluid flow in the New Jersey continental slope: implications for slope failure and cold seeps. Science 289, 288–291.

Fertl, W.H., 1976. Abnormal Formation Pressures. Elsevier, Amsterdam.

Flemings, P.B., Liu, X., Winters, W.J., 2003. Critical pressure and multiphase flow in Blake Ridge gas hydrates. Geology 31, 1057–1060.

Gay, A., Lopez, M., Cochonat, P., Sermondadaz, G., 2004. Polygonal faults-furrows system related to early stages of compaction: upper Miocene to Recent sediments of the Lower Congo Basin. Basin Res. 16, 101–111.

Gay, A., Lopez, M., Berndt, C., Seranne, M., 2007. Geological controls on focused fluid flow associated with seafloor seeps in lower Congo basin. Mar. Geo. 244, 68–92.

Ginsburg, G.D., Soloviev, V.A., 1997. Methane migration within the submarine gas-hydrate stability zone under deep-water conditions. Mar. Geol. 127, 49–57.

Haacke, R.R., Westbrook, G.K., Hyndman, R.D., 2007. Gas hydrates, fluid flow free-gas formation of bottom simulating reflector. Earth Planet. Sci. Lett. 261, 407–420.

Hay, W.W., Southam, J.R., 1977. Modulation of marine sedimentation by the continental shelves. In: Malahoff, A., Andersen, N.R. (Eds.), The Fate of Fossil Fuel CO2 in the Oceans. Plenum Press, New York. pp. 564–604.

Holbrook, et al., 2002. Seismic gas hydrate-bearing sediments in Lake Baikal Siberia. Mar. Geol. 172, 1–21.

Hyndman, R., Davis, E., 1992. A mechanism for formation of methane hydrate and sea floor bottom simulating reflectors by vertical fluid expulsion. J. Geophys. Res. 97, 7025–7041.

Hyndman, R.D., Wang, K., Yuan, T., Spence, D., 1993. Tectonic sediment thickening, fluid expulsion, and the thermal regime of subduction zone accretionary prisms: the Cascadia Margin of Vancouver Island. J. Geophys. Res. 98, 21865–21876.

Jackson, M.P.A., Vendeville, B.C., Schultz-Ela, D.D., 1994. Structural dynamics of salt systems. Ann. Rev. Earth Planet. Sci. 22, 93–117.

Jiangong, W., Thomas, P., Nabil, S., et al., 2015. Gas hydrate distributions in sediments of pockmarks from the Nigerian Margin: results and interpretation from shallow drilling. Mar. Petrol. Geol. 59, 359–370. Publisher's official version: http://dx.doi.org/10.1016/j.marpetgeo.2014.09.013.

Judd, A.G., Hovland, M., 2007. Seabed Fluid Flow: The Impact of Geology, Biology and the Marine Environment. Cambridge University Press, Cambridge.

Lee, M.W., Hutchinson, D.R., Dillon, W.P., et al., 1993. Method of estimating the amount of in situ gas hydrate in deep marine sediments. Mar. Petrol. Geol. 10, 493–506.

Lee, M.W., Hutchinson, D.R., Collet, T.S., Dillon, W.P., 1996. Seismic velocities in hydrate bearing sediments using weighted equation. J. Geophys. Res. 101, 20347–20358.

Leithold, E., Blair, N., Perkey, D., 2006. Geomorphic controls on the age of particulate organic carbon from small mountainous and upland rivers. Glob. Biogeochem. Cycles 20 GB3022.

Ligtenberg, H., 2005. Detection of fluid migration pathways in seismic data: implications for fault seal analysis. Basin Res. 17, 141–153.

Lonergan, L., Cartwright, J.A., Jolly, R.J.H., 1998. The geometry of polygonal fault systems in tertiary mudrocks of the North Sea. J. Struct. Geol. 20, 529–548.

Macleod, M.K., 1982. Gas hydrates in ocean bottom sediments. AAPG Bull. 1982 (66), 2649–2662.

Mann, P., 2004. Tectonics Setting of World's Giant Oil and Gas Fields. HGS Northsiders Committee – invited talk, 19 October, 2004. http://www.hgs.org/articleshttp://www.hgs.org/articles.

Milkov, A.V., 2004. Global estimates of hydrate-bound gas in marine sediments: how much is really out there? Earth Sci. Rev. 66, 183–197.

Moore, J.C., Vrolijk, P., 1992. Fluids in accretionary prisms. Rev. Geophys. 3, 113–135.

Müller, P.J., Suess, E., 1979. Productivity, sedimentation rate and sedimentary organic matter in the oceans-I. Organic carbon preservation. Deep Sea Res. 26, 1347–1362.

Pidwirny, M., 2006. Surface and Subsurface Ocean Currents: Ocean Current Map. In: Fundamentals of Physical Geography, second ed. http://www.physicalgeography.net/fundamentals/8q_1.html (accessed on September–December 2015).

Plaza-Faverola, A., Bünz, S., Mienert, J., 2010. Fluid distributions inferred from P-wave velocity and reflection seismic amplitude anomalies beneath the Nyegga pockmark field of the mid-Norwegian margin. Mar. Petrol. Geol. 27, 46–60.

Prahl, F.G., Ertel, J.R., Goni, M.A., Sparrow, M.A., Eversmeyer, B., 1994. Terrestrial organic carbon contributions to sediments on the Washington margin. Geochim. Cosmochim. Acta 58, 3035–3048.

Rajput, S., Thakur, N.K., Rao, P.P., Joshi, A., 2010. AVO response for a complex double bottom simulating reflectors model. Curr. Sci. 98, 1354–1358.

Reston, T.J., 2005. Polyphase faulting during the development of the west Galicia rifted margin. Earth Planet. Sci. Lett. 237, 561–576.

Riedel, M., Collett, T.S., Kumar, P., Sathe, A.V., Cook, A., 2010. Seismic imaging of a fractured gas hydrate system in the Krishna–Godavari Basin offshore India. Mar. and Petrol. Geol. 27, 1476–1493.

Rubey, W.R., Hubbert, M.K., 1959. Role of fluid pressure on the mechanics of over-thrusting faulting. Geol. Soc. Am. Bull. 70, 167–206.

Ruppel, C., Dickens, G.R., Castellini, D.G., Gilhooly, W., Lizarralde, D., 2005. Heat and salt inhibition of gas hydrate formation in the northern Gulf of Mexico. Geophys. Res. Lett. 32 L04605.

Sassen, R., Macdonald, I.R., 1994. Evidence of structure H hydrates. Gulf Mexico slope. Org. Geochem. 22, 1029–1039.

Selley, R.C., 1998. Elements of Petroleum Geology. Elsevier Science: Academic.

Singh, M., Singh, I.B., Muller, G., 2007. Sediment characteristics and transportation dynamics of the Ganga River. Geomorphology 86, 144–175.

Sloan Jr., E.D., 1990. Clathrate hydrates of natural gases. Dekker, New York and Basel, p. 641.
Sloan Jr., E.D., 1998. Clathrate hydrates of natural gases. Marcel Dekker, New York, p. 705.
Soloviev, V., Ginsburg, G.D., 1994. Formation of submarine gas hydrates. Bull. Geol. Soc. Den. 41, 86–94 Copenhagen.
Stein, R., 1990. Organic carbon content/sedimentation rate relationship and its paleoenvironmetal significance for marine sediments. Geol. Mar. Lett. 10, 37–44.
Stevenson, A.J., 1993. Generation, migration, and resource potential for hydrocarbons in accretionary subduction systems - a large, unconventional hydrocarbon resource. United States Geological Survey, Professional Paper; (United States) 1570.
Stevenson, A.J., 1996. Generation, migration and resource potential for hydrocarbons in accretionary subduction systems: a large unconventional hydrocarbon resource? In: Howell, D.G. (Ed.), The Future of Energy Gases, pp. 353–363. USGS, Numbered Series 1590, Professional paper, Washington United States Government Printing Office.
Syvitski, J.P.M., Saito, Y., 2007. Morphodynamics of deltas under the influence of humans. Glob. Planet. Change. 57, 261–282.
Taylor, M.H., Dillom, W.P., Pecher, I.A., 2000. Trapping and migration of methane associated with gas hydrate stability zone at the Blake Ridge Diapir: new insights from seismic data. Mar. Geol. 164, 79–89.
Thinon, I., Matias, L., Réhault, J.P., et al., 2003. Deep Structure of the Armorican Basin (Bay of Biscay): a review of Norgasis seismic reflection and refraction data. J. Geol. Soc. Lond. 160, 99–116.
Thomsen, R.O., Lerche, I., 1991. Salt diapir velocity assessment from temperature and thermal indicator anomalies: application to Lulu-1 Danish North Sea. Terra Nova 3, 500–509.
Tucholke, B.E., Sawyer, D.S., Sibuet, J.C., 2007. Breakup of the Newfoundland-Iberia rift. In: Karner, G., Manatschal, G., Pinheiro, L. (Eds.), Imaging, Mapping and Modelling Continental Lithosphere Extension and Breakup. Geological Society of London Special Publications, No. 282. Geological Society of London, London, pp. 9–46.
Vanneste, M., De Batist, M., Versteeg, W., Golmshtok, A., Kremlev, A., 2001. Multi-frequency seismic study of the gas hydrate-bearing sediments in Lake Baikal, Siberia. Mar. Geol. 171 (1/2), 1–21.
Whitman, J.M., Davies, T.A., 1979. Cenozoic oceanic sedimentation rates: how good are the data? Mar. Geol. 30, 269–284.
Whitmarsh, R.B., Manatschal, G., Minshull, T.A., 2001. Evolution of magma-poor continental margin from rifting to seafloor spreading. Nature 413, 150–154.
Wollast, R., 2003. Continental margins: review of geochemical settings. In: Wefer, G., et al. (Ed.), Ocean Margin Systems, pp. 15–31. ISBN 978-3-662-05127-6, Springer Berlin Heidelberg.
Zuehlsdorff, L., Spiess, V., Huebscher, C., Villnger, H., Rosenberger, A., 2000. BSR occurrence near surface reflectivity anomalies and small scale tectonism imaged in multi-frequency seismic data set from the Canadian accretionary prism. Geol. Rundsch. 88, 655–667.

CHAPTER

ROCK PROPERTIES

5

"If we knew what it was we were doing, it would not be called research, would it?"

Albert Einstein

CHAPTER OUTLINE

5.1 INTRODUCTION

The current generation is blessed with technology that has been developed over the twentieth and twenty-first centuries. Technological advances have led to tremendous increases in oil and natural gas production, creating jobs and spurring economic development. The source of oil and gas technologies has changed over time. It is estimated that industrial production increased more than 50-fold during the twentieth century and that four-fifths of this growth happened in the second half of the century.

Geological Controls for Gas Hydrates and Unconventionals. http://dx.doi.org/10.1016/B978-0-12-802020-3.00005-9

Hydrocarbon fuel contributes ~60% of the world energy demand (International Energy Agency, http://www.iea.org). Our thirst for energy is increasing and more advanced technologies are needed to supply that energy. The global energy system is under pressure and falls short of our hopes and expectations. Technological advances and efficiency are the hope for future sustainability. Global energy demand is growing tremendously and it is expected to increase by ~40% by 2040. By that time, the world's energy supply will be dependent on oil, gas, and coal, and low-carbon fuel. These hydrocarbon energy sources face a distinct set of challenges. The most significant challenge is to understand the description of reservoir heterogeneity. It is very well understood by now that reservoir characteristics such as porous media properties, permeability, porosity, heterogeneity, and spatial distribution of oil and water predominantly control the flow field, reservoir performance, development strategies, and hence the economic returns of investments, which are of great concern to major oil producers. Industry is working on constructing better models for achieving better results with reduced uncertainties in exploration and production programs. Modern reservoir models provide the fundamental framework for reservoir management and flexible logistic applications.

One needs to understand the quote by George Edward Pelham Box, "All models are wrong, but some models are useful," and humility about the limitations of our knowledge. Because all models are wrong, models cannot be validated and truthfulness cannot be established. The focus should be on creating models that are useful, the process of testing, the ongoing comparison of the model against all the data, and the continuous iteration of experiments within the virtual world of the model and experiments in the real world. In oil and gas exploration and development programs, these models are constructed by integrating geology, geophysics, petrophysics, and reservoir engineering information.

Improved reservoir characterization integrates all available data to define the reservoir geometry, distribution of physical parameters including rock properties, and flow properties of a petroleum reservoir and requires useful models for a successful field development. Therefore reservoir characterization begins with developing an understanding of seismic data that offers an image of the earth below the surface. Seismic data shows different rock formations as a layer of reflectors. Rock types and fluids cause seismic reflection events (Fig. 5.1). Mathematically, seismic traces are the result of convolving (summing) the wavelet with the reflection coefficient series.

Seismic surveys are routinely performed to gather information about the geology and fluid fill of hydrocarbon fields. Seismic data, even when it has relatively low resolution, can still deliver a structural model for the reservoir. The availability and coverage area of the data provides valuable information to constrain the reservoir model. Geophysicists are now able to extract quantitative information about reservoir rocks and fluid parameters from seismic data by working backwards using seismic inversion methods. Described as the flip-side of forward modeling, *seismic inversion* is the process of characterizing rocks and fluids that have produced the seismic data. In a sense, the process involves trying to determine the input (rock properties) by looking at the output (seismic data). It starts with actual measured data and removes the wavelet and derives acoustic impedance from the reflection coefficient (Fig. 5.2) with a low-frequency model to infer *rock properties* and deliver an interpreted earth model. Fig. 5.2d shows an example of an inverted acoustic impedance section of seismic data from Fig. 5.1.

Typically, however, seismic inversion results have limited vertical resolution compared to the scale of the well logs. The output of the inversion process is limited to the input seismic resolution capturing only a bulk elastic response and with all layering details smeared over the seismic wavelength. Filters can be applied to reduce the noise contribution to seismic data. Highly detailed seismic analysis e.g. interpretation of stochastic inversion derivatives shows that there can be multiple combinations of layer stacks that generate the same seismic response above the same threshold limit. Therefore it can be said that the

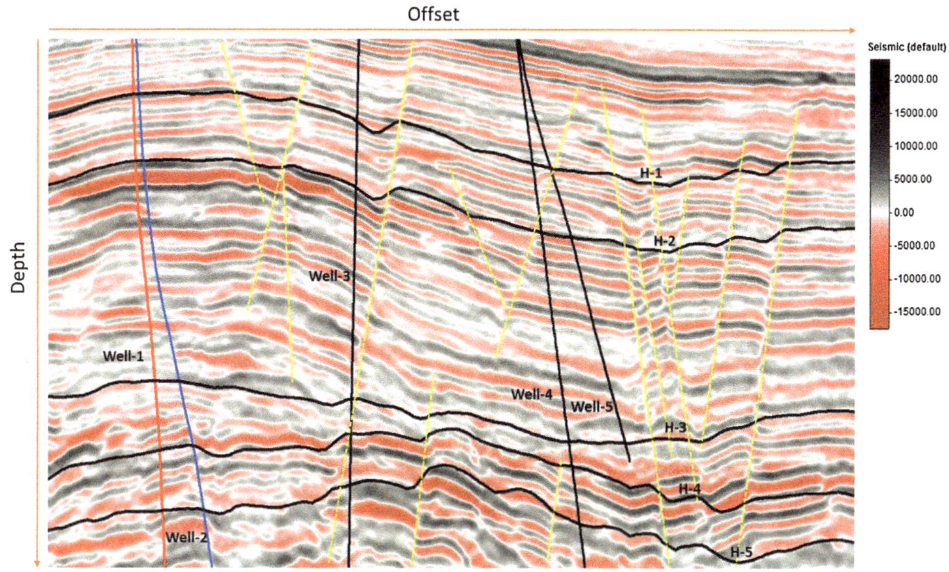

FIGURE 5.1

An example of seismic reflection events. Dotted yellow lines represent faults. Well and seismic horizons are marked.

seismic inversion process boosts the resolution of conventional seismic data and sets the study of reservoir characteristics at a different level. It has a variety of applications and varying levels of complexity. Owing to the increasing need for reservoir characterization, several researchers have put their best efforts towards understanding and developing the inversion methods (Aki and Richards, 1980; Stolt and Weglein, 1985; Menke, 1984; Tarantola, 1987; Sen and Stoffa, 1995; Sen, 2006). There are various case studies available that utilize seismic inversion for reservoir characterization and well planning (Connolly, 1999; Goodway et al., 1997; Latimer et al., 2000; Baman, et al., 2004; Thakur and Rajput, 2010; Torres-Verdin et al., 1999; Russell et al., 2003; Veeken et al., 2002; Tarantola, 1984, 1986). In addition, the process of seismic inversion is inherently non-unique; therefore, it can be said that there could be an infinite number of elastic property models that represent the best fit of the seismic data. There are several industry-scale applications of seismic inversion including the extraction of layer lithologies and fluid saturation, estimation of porosity and permeability, compressional and shear impedances, density, net to gross, and many others. The main benefit of seismic inversion is minimizing the tuning effect and maximizing the vertical resolution by broader bandwidth of inverted impedance volume. Researchers use different types of inversion, for example, tomographic inversion for velocity and amplitude inversion. The inversion methods are either deterministic or probabilistic and the approach can be post- and/or pre-stack (Veeken and Silva, 2004) but most of them are model driven. All of the inversion-based algorithms predict rock properties by using seismic and well data. Inverted rock properties provide a basic understanding of reservoir description and heterogeneity, which are the fundamental problems that need to be addressed in exploration, appraisal, and field development programs. Elastic properties include Young's modulus, shear modulus, bulk modulus, and Poisson's ratio, and inelastic properties include fracture gradient and formation strength. Basic definitions of rock properties that are useful for hydrocarbon field exploration, appraisal, and development programs are defined below.

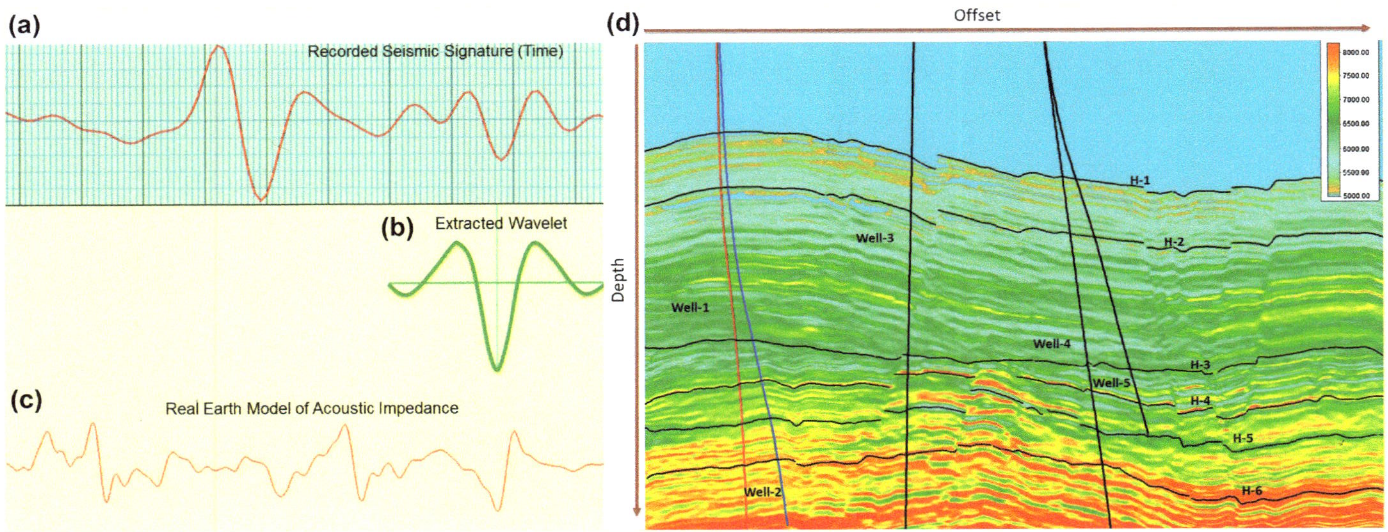

FIGURE 5.2

Recorded seismic signature removes the effect of an extracted wavelet and estimates the real earth model of acoustic impedance at every time sample. (a) Recorded seismic data in time; (b) extracted wavelet; (c) real earth model of acoustic impedance; (d) inverted acoustic impedance section of seismic data shown in Fig. 5.1.

5.1.1 ACOUSTIC IMPEDANCE (*Z* OR AI)

This is the product of density and seismic velocity, which varies among different rock layers, commonly symbolized by Z. The difference in acoustic impedance between rock layers affects the reflection coefficient.

$$\text{Acoustic impedance } (Z \text{ or AI}) = \text{velocity } (V) \times \text{density } (\rho)$$

5.1.2 SHEAR IMPEDANCE (Z_S OR SI)

Shear wave impedance is a product of bulk density and S-wave velocity (V_s) and can be expressed as:

$$Z_s = \rho_b V_s$$

5.1.3 DENSITY (ρ)

The general definition of the bulk density of any material is given as:

$$\rho = \frac{m}{V}$$

where m is the mass of the object and V is the volume it occupies.

For a two-phase system, e.g., a saturated clastic rock, the following densities can be given: the dry ρ_{dry} and the wet ρ_{wet}. The dry density uses the mass of the matrix in the expression given above. The wet density takes into account the weight of the fluid (often being water, oil, or brine).

5.1.4 POROSITY (*Ø*)

The definition of bulk porosity is given as:

$$\varphi = \frac{V_v}{V_b}$$

where V_v is the void occupied volume and V_b is the bulk volume (typically measured by a caliper). Porosity may be inter- or intragranular. In clastic sedimentary rocks, "intergranular" means that the grains are still in place, while the voids are around the grains. In clastic sedimentary rocks, the two types of porosity cannot be in coexistence whereas in carbonate rock, they can be, due to their depositional patterns. Shale is said to have microporosity. Shale acts as an impermeable trap in an oil reservoir.

5.1.5 YOUNG'S MODULUS (*E*) (FIG. 5.3)

Young's modulus (E) reflects the stiffness of earth material and is expressed as the ratio of stress to strain in a uniaxial setup. Stress is the force per area in the direction of the force that causes the deformation. Strain is the relative decrease in length Δl caused by deformation to the original value of the length:

$$\Delta l = l - l_x \text{ to the original length } l$$

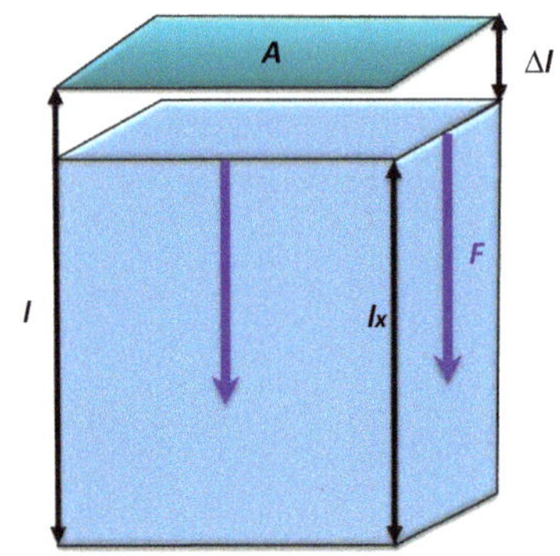

FIGURE 5.3

Young's modulus. F is the force, A is the area, Δl is the displacement, and l_x is the initial length.

$$E = \frac{\text{Stress}}{\text{Strain}}$$

The greater the value of E, the larger the stress that is needed to achieve the deformation. Strain is thus, a measure of the deformation of the material and is a non-dimensional quantity i.e. it has no units. It is simply a ratio of two quantities with the same unit. It is often referred to simply as the elastic modulus. "E" can be expressed in terms of compressional and shear wave velocities and densities as:

$$E = \rho V_s^2 \frac{\left(3V_p^2 - 4V_s^2\right)}{V_p^2 - V_s^2}$$

Young's modulus, E, is the ratio of uniaxial compressive (tensile) stress to the resultant strain. For unconventional hydrocarbon related studies this property influence the closure stress and therefore the hydrocarbon fracture height. If large difference is observed between the vertically and horizontally measures Young's moduli, the closure stress will be higher than in isotropic rocks. The anisotropic intervals are normally associated with rocks that have higher clay volume as well as with expandable clays. These rocks are poor considered as poor candidates for both horizontal well placement and hydraulic fracture stimulation due to higher stress in these intervals, it will be more difficult to retain fracture conductivity during production because the proppant is more likely to embed into the ductile formation.

5.1.6 SHEAR MODULUS (μ) (FIG. 5.4)

The shear modulus is the earth's material response to the shear deformation. It is defined as the ratio of shear stress and shear strain. This valuable property tells us in advance how resistant a material is to shearing deformation. If a material is very resistant to attempted shearing, then it will transmit the shear energy very quickly.

$$\mu = \frac{\text{Shear Stress}}{\text{Shear Strain}}$$

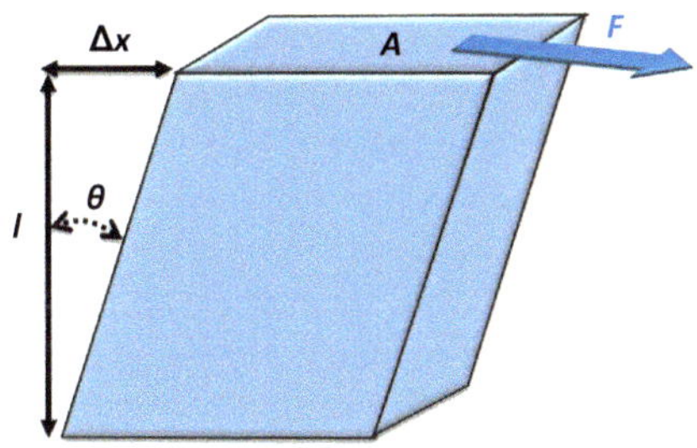

FIGURE 5.4

Shear modulus. F is the force, A is the area, Δx is the displacement, l is the initial length, and θ is the displacement angle.

$$\mu = \frac{F}{A \times \theta}$$

This is also known as the modulus of rigidity. The shear modulus' derived SI unit is the pascal (Pa), although it is usually expressed in gigapascals (GPa) or in thousands of pounds per square inch (ksi). Shear modulus is always positive. This can be expressed in terms of shear wave velocity and density as:

$$\mu = \rho V_s^2$$

This is also known as the modulus of rigidity. Seismically derived shear moduli are typically greater than those measured by in-lab strength tests because seismic measurements are made under elastic, low-strain conditions.

5.1.7 BULK MODULUS (*K*)

This elastic constant reflects the resistance of the earth's material to an overall gain or loss of volume in conditions of hydrostatic stress. It is defined as the ratio of the infinitesimal pressure increase to the resulting relative decrease of the volume. Its SI unit is the pascal. A material that has a large bulk modulus could be imagined to consist of very tightly bound spring and ball materials which will quickly transmit a wave. "*K*" can be expressed in terms of compressional and shear wave velocity and density as:

$$K = \left(V_p^2 - \frac{4}{3} V_s^2 \right)$$

5.1.8 POISSON'S RATIO (σ)

This elastic constant is named after Simeon Poisson's, a French mathematician. This is the measure of the compressibility of material perpendicular to the applied stress. This could also be represented as the ratio of latitudinal or longitudinal strain.

$$\sigma = \frac{1}{2} \left(\frac{V_p^2 - 2V_s^2}{V_p^2 - V_s^2} \right)$$

Poisson's ratio is often related to the V_p/V_s ratio.

5.2 ESTIMATION METHODS OF ROCK PROPERTIES

Rock and fluid prediction between the seismic lines which are away from well bores requires understanding of how rock's bulk and seismic properties are linked to each other and how they vary with geologic age, depth, and location. The ultimate goal of reservoir and elastic property prediction away from well control (a distance away from the well location) is to quantify lithology, fluid, and porosity from seismic data. This goal can be achieved by integration of a variety of information from multidisciplinary techniques. These techniques include forward and inverse modeling approaches as described in the following.

5.2.1 AMPLITUDE VERSUS OFFSET/AMPLITUDE VERSUS ANGLE ANALYSIS

A lack of sufficient data and an excessive initial investment have hindered the exploration of challenging hydrocarbon prospects. Since 1984, when Ostrander showed an amplitude variation with offset for gas sand capped by shale in prestack seismic data, amplitude versus offset (AVO) has become a commercial tool for hydrocarbon prediction. Forward modeling and rock physics models together with amplitude versus angle (AVA) responses play an important role in preparing the model for a prospect. AVA analysis is preferable to AVO analysis when comparing a deeper target with a shallower one. However, AVA analysis requires information about velocities and raypaths that is not necessary for AVO analysis. In practice, both offset and angle displays may be helpful to the interpreter because each enables a different data perspective. The concept of incident P-wave, reflected P-wave and S-wave, and transmitted P-wave and S-wave is illustrated in Fig. 5.5a.

In AVO or AVA studies, the synthetics (see Fig. 5.5c–e) are usually calculated from various approximations of the Zoeppritz equations. Classical approximations include those of Bortfeld (1961), Aki and Richards (1980), and Shuey (1985). In Shuey's two-term approximation of the Zoeppritz equations, the P-wave reflection coefficient can be written as a function with two parameters: AVO intercept (*I*) and AVO gradient (*G*) (Shuey, 1985) (see Fig. 5.5g & h). In deep-water environments, the sediments follow normal compaction processes which define the background trends. If the two parameters (A and B) are placed on a crossplot, AVA anomalies can be identified because they deviate from the background trend. To apply the principle, it is necessary to establish AVA trends for rock properties and to understand the difference between conventional petrophysics and seismic petrophysics, which can be achieved with A and B crossplot analysis and seismic inversion.

AVO/AVA attributes intercept (A), and gradient (B) can be explained by Shuey's two-term approximation (Shuey, 1985) to the Zoeppritz equations (Zoeppritz, 1919). Zoeppritz equations were derived by Karl Bernhard Zoeppritz in 1919. These are a set of four equations that describe the partitioning of seismic wave energy at an interface, typically a boundary between two different layers of rock.

Shuey's two-term approximation:

$$R(\theta) = A + B\ \sin^2(\theta) + \cdots$$

where R = reflection coefficient, θ = angle of incidence, A = AVO intercept, and B = AVO gradient.

The two-term Shuey approximation to the Zoeppritz equations represents the angular dependence of P-wave reflection coefficients with two parameters: the AVO intercept (A) and the AVO gradient (B). In practice, the AVO intercept is a band-limited measure of the normal incidence amplitude, while the AVO gradient is a measure of amplitude variation with offset. Assuming appropriate amplitude calibration, A

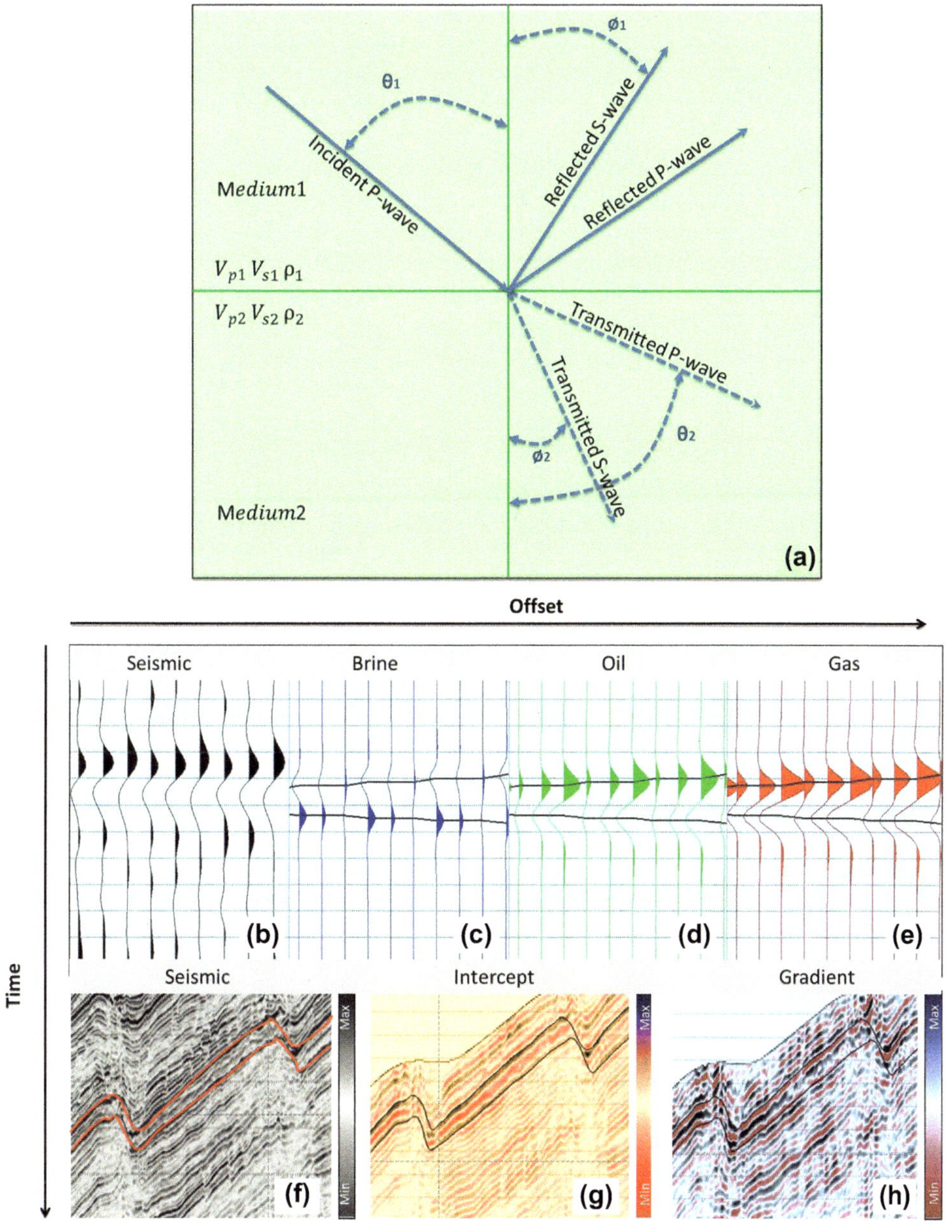

FIGURE 5.5

Figure shows wave propagation in an isotropic medium. Compressional wave velocity (V_p), shear wave velocity (V_S), and density (ρ) are different for two mediums: seismic and synthetic examples for brine oil and gas scenarios are highlighted. AVO attributes intercept and gradient are also shown. (a) Incident P- wave, reflected P-wave and S-wave, transmitted P-wave and S-wave propagation in an isotropic medium; (b) real seismic data (angle gathers) example from an oil field located offshore Australia; (c) synthetic generated for brine case; (d) synthetic generated for oil case; (e) synthetic generated for gas case; (f) full stack seismic; (g) AVO attribute intercept; (h) AVO attribute gradient.

is the normal incidence reflection coefficient and B is a measure of offset-dependent reflectivity. An example of calculated intercept and gradient form seismic data is shown in Fig. 5.5g & h.

The petroleum industry employed seismic inversion and AVA/AVO modeling more than three decades ago. The fundamentals of seismic character start with the seismic data because of its nonunique nature. Therefore, whatever is used or developed for a better understanding of the seismic behavior of earth comprises a nonunique solution. There will always be uncertainties in seismically driven derivatives. Seismic inversion helps to interpret and understand the subsurface using direct measurement obtained from wellbore and seismic data and is tied to the rocks. Seismic petrophysics and rock physics modeling help in understanding the relationship between reservoir properties and elastic properties. "Seismic Petrophysics" refers to "the careful and purposeful use of rock physics data and theory in the interpretation of seismic observations" (Pennington, 1997). Conventional reservoir petrophysics mainly focuses on the reservoir rocks to determine the saturation, porosity, lithology, and permeability, whereas seismic petrophysics determines the rock properties (i.e., saturation, porosity, and lithology) over the zone of interest for seismic inversion; the goal is to understand the reflections caused by the contrast in acoustic impedance between two rocks. Seismic petrophysics provides the input for rock physics modeling (i.e., porosity, situation, and V-clay) to relate seismic observations to rock properties that are important in the reservoir. To develop a predictive rock physics model for a field, it is important to ensure that the petrophysical parameters used are of high quality and are consistent between wells and with the process applied in the rock physics modeling.

The goal of AVO analysis is to investigate the amplitude response from various sand models with respect to oil and gas saturation. A logical workflow for conducting AVA modeling includes the quality control and preconditioning of seismic data and seismic petrophysics-conditioned wells logs. The data should be taken through an iterative process of log calibration and wavelet extraction until an optimum wavelet is achieved. This wavelet should then be used for the calculation of AVA responses from various models generated using fluid-substituted logs. Fluid type affects the AVA characteristics. AVA attributes (intercept and gradient) must be calculated to establish the relationship between lithology and rock properties. Fig. 5.6 shows integrated AVO/AVA modeling workflow.

5.2.1.1 AVA Modeling Example

To study the seismic response for a hydrocarbon prospect, a wedge model was constructed using available well log information. Seismic to well ties were performed for fluid-substituted logs, and synthetics seismic responses were calculated as illustrated in Fig. 5.7. Seismic AVA and bandpass-filtered property responses for brine, oil, and gas show various characteristics. The oil and gas cases show increasing amplitudes with angles and a polarity reversal when compared to the brine case. Correlating these cases with real seismic data helps to improve the understanding of seismic response. Fig. 5.7 shows the comparison of AVA response with real seismic angle gathers.

At this stage, prestack seismic inversion was performed on a wedge model for various fluid scenarios and on real seismic data. A comparison of inverted wedge-model rock properties and real seismic data suggests that seismic anomalies can be mapped. Fig. 5.8 shows the inverted bandpass P-impedance and Vp/Vs ratio for real seismic data and the wedge model.

To analyze the extended effect of AVO/AVA analysis, a mudrock line-based AVO index can be calculated. An AVO index highlights fluid effects. The AVO index is calculated as a distance of separation of samples from the mudrock line. Positive values of the AVO index highlight a potential reservoir, which should be further calibrated with seismic inversion and geological models. An example of derisking a prospect based on the AVO index is shown in Fig. 5.9.

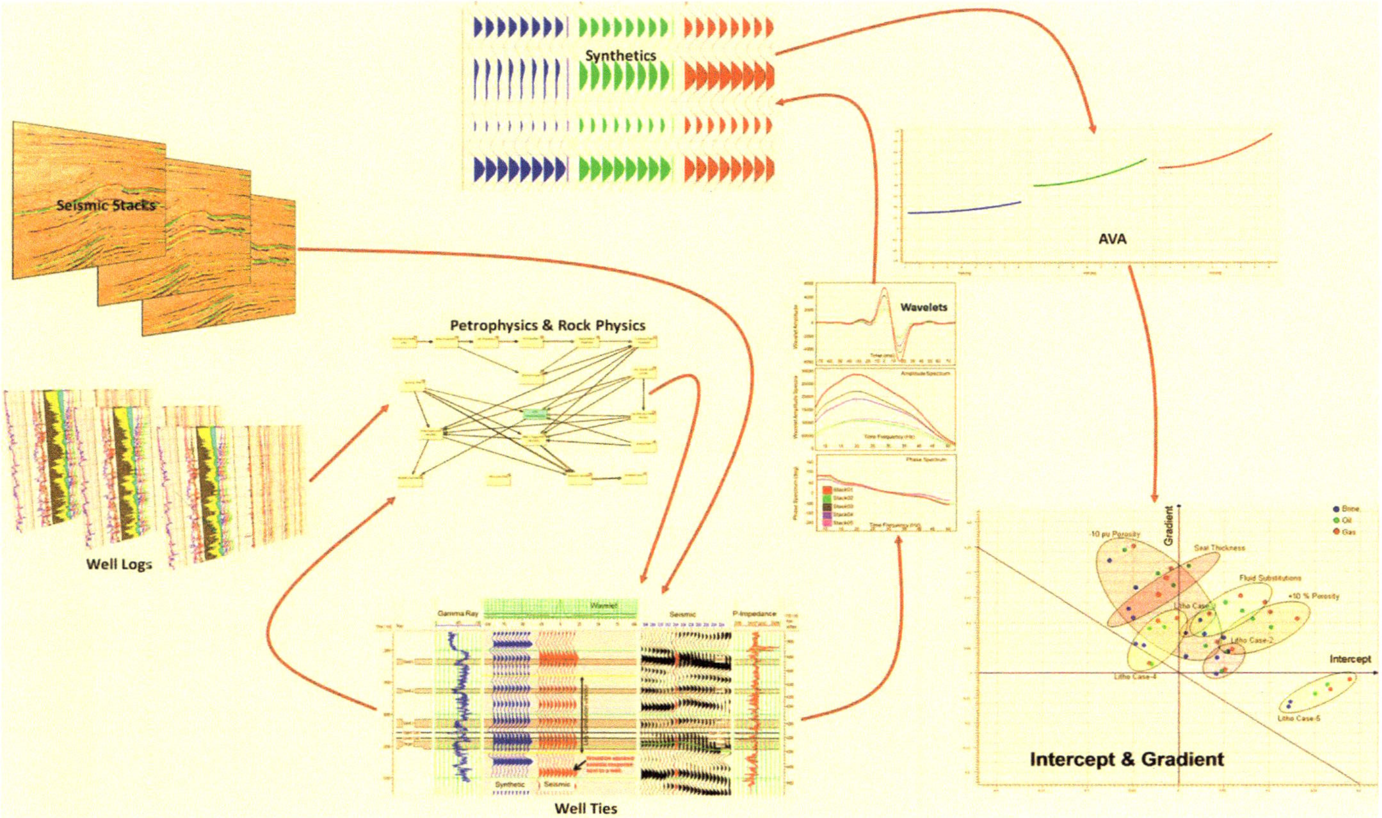

FIGURE 5.6

Integrated AVO/AVA modeling and analysis workflow. Main elements include seismic data, well logs, petrophysics and rock physics modeling, synthetic seismograms highlighting brine (blue), oil (green) and gas (red), AVA curves for synthetics, wavelet extraction, seismic to well ties and AVO/AVA attributes (intercept and gradient) crossplot. Fluid substitution is a part of petrophysics and rock physics modeling.

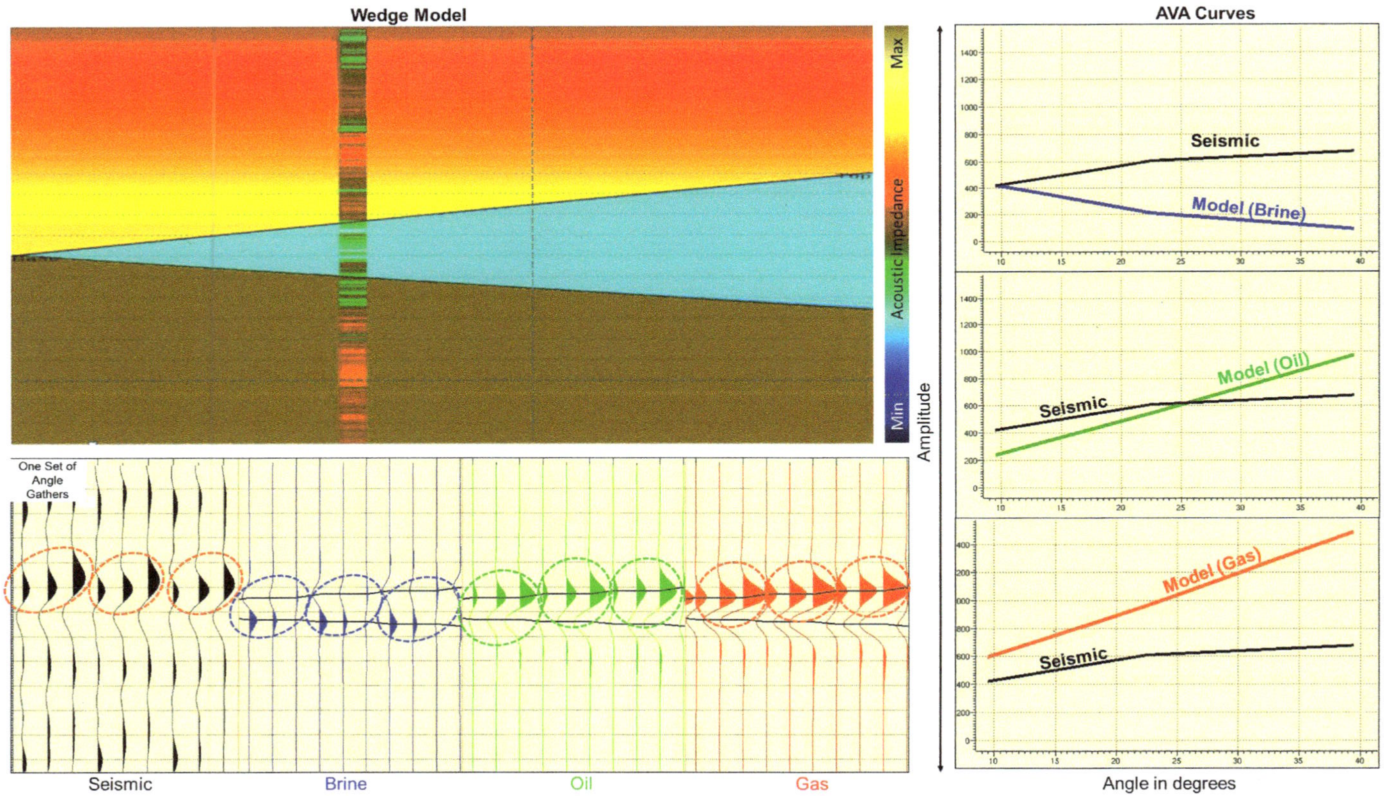

FIGURE 5.7

The AVA response for fluid-substituted logs over the wedge model (top) for brine (blue), oil (green), and gas (red) scenarios. A set of three traces represents angle gathers created from angle stacks. An individual trace represents one angle stack. AVA synthetics are calculated for various fluid types and show variation in amplitude response.

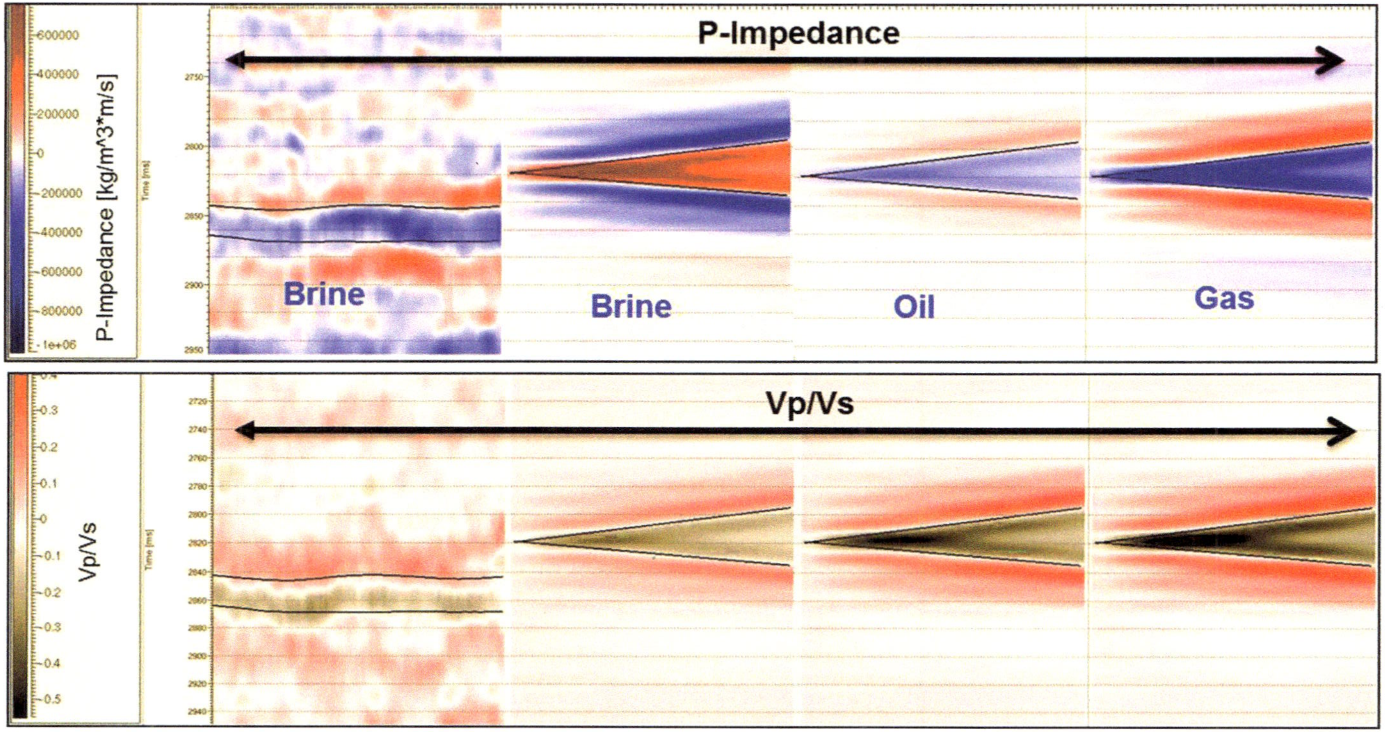

FIGURE 5.8

Comparison between real seismic and inverted seismic data together with synthetic seismic data simulated for a wedge model and inverted synthetic seismic for brine, oil, and gas scenarios. Bandpass P-impedance (top) and V_pV_s (bottom) captured the anomaly from hydrocarbons.

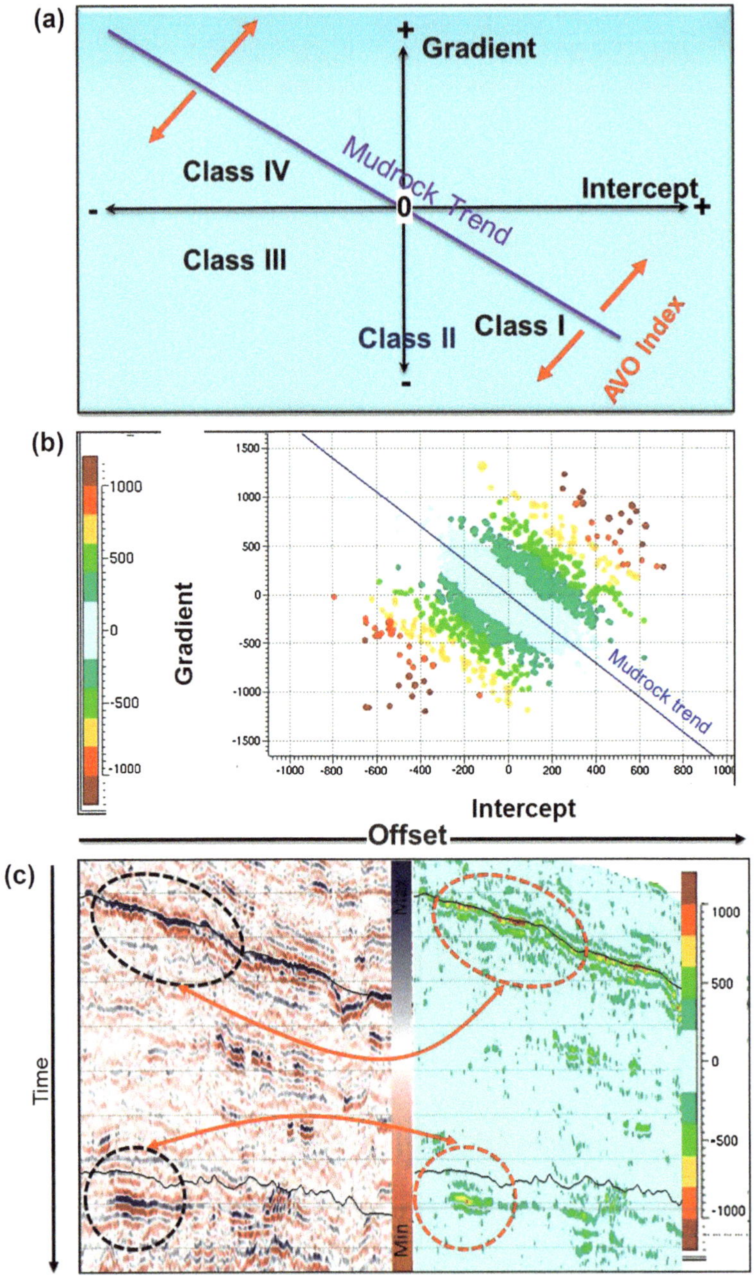

FIGURE 5.9

This figure highlights AVO attributes, intercept and gradient, and AVO index: (a) Schematic shows the crossplot of intercept and gradient together with AVO classes and defined mudrock line; (b) crossplot of AVO attributes (intercept and gradient) at 20 selected seismic lines; (c) post-stack seismic section with prospect highlighted by dashed black ovals and calculated AVO index on right; positive AVO index highlights AVO anomaly which could be a reservoir.

5.2.2 AVO FOR UNCONVENTIONAL RESOURCE PLAYS

Unconventional reservoirs are composed of complex structural and geological systems, fluids and total organic carbons embedded in heterogeneous seismic facies; they are subject to various pressure and temperature regimes and commercial pressures. These reservoirs have always been challenging when employing quantitative interpretation methods for well placement or optimal fracturing. Owing to the low permeability of unconventional plays, hydraulic fracture stimulation is required to achieve oil or gas production and effective stimulation requires either connection to existing natural fractures or the presence of geomechanical brittleness capable of supporting extensive induced fractures. The application of AVO methods is simplified to primarily extracting the mechanical properties and lithofacies using seismic data. The use of 3D seismic AVO and AVO variation with azimuth (AVAZ) helps to detect anisotropy due to fractures, stress, or over-pressure, which are essential parameters for developing a plan to drill and produce. An example of AVO analysis from a shale play in China is shown in Fig. 5.10, but the methods are equally applicable to Coalbed methane (CBM or coal-bed methane) or gas sands. Poisson ratio decreases with increasing Young's modulus for a given rock type. The analysis is similar to the Barnett Shale trend equation from Grigg (2004). The inverse linear relationship between Young's modulus and Poisson's ratio is simply the result of an increase in rigidity.

For gas hydrate studies, the bottom-simulating reflector (BSR) marks the base of gas hydrate stability (BGHS), above which conditions are generally favorable for gas hydrate formation, and below which they are not (Pecher et al., 2004). The BSR is the response to a decrease in seismic velocity and/or density, which is usually attributed to free gas underlying gas hydrate (Bangs et al., 1993). Free gas associated with the gas hydrate system can potentially intersect the seafloor at BSR pinch-outs, where

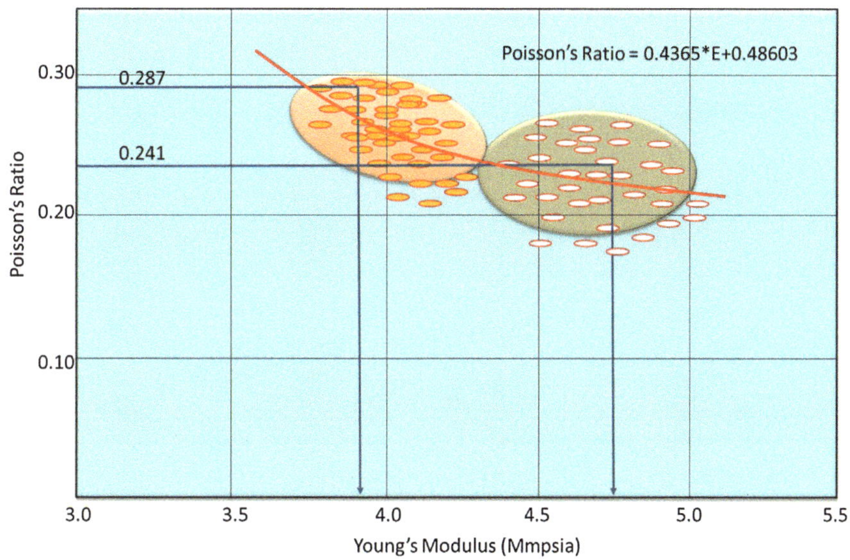

FIGURE 5.10

Poisson's ratio vs. Young's modulus relationship for a shale play from China. Poisson's ratio decreases with increases in Young's modulus.

the BGHS coincides with the seafloor (e.g., Crutchley et al., 1999). BSR also can form where there is only a small volume of the hydrates and only a small quantity of free gas in the underlying layer because even a small amount of gas reduces velocity significantlyHowever, as we shall see presently, in some cases, BSRs are not visible even where there are proven gas hydrate deposits. In addition to this, the presence of BSR-like features does not confirm that there are definitely gas hydrates. The computation of reflection coefficient for primary and secondary BSR configurations, using the Haskell–Thomson method (Haskell, 1953, 1960, 1962; Thomson, 1950; Aki and Richards, 1980), can be utilized to demonstrate the AVO effect in porous gas-saturated sandstones. This computation should be carried out in two steps: (1) determination of scattering matrices for a set of plane waves with different frequencies; and (2) selected reflection coefficients from the scattering matrices should be convolved with a Ricker wavelet of a chosen frequency. This approach allows the reflection and transmission coefficient vs. plane wave dip angle dependency to be stabilized. This fact is well illustrated with the linearized formula for P-wave reflection coefficient (Shuey, 1985). This formula also permits estimation of the magnitude of change in the effective Poisson ratio ($\Delta\sigma$), using the least-squares method. The character of reflection coefficient of BSR's on a global scale is observed as a "negative trend". The data from offshore India, New Zealand, Japan, and USA show a similar trend; a generalized schematic is shown in Fig. 5.11.

Dealing with more than one layer makes the AVO behavior complex. The seismic energy gets attenuated in the presence of low-velocity layers and the reflection coefficients in general exhibit frequency-dependent phase shift. However, there is not much significant change in reflective characteristics for low and high frequencies for a small angle of incidence. Secondary BSR or double BSR can be generated if the sediments and hydrate matrix contain low-velocity fluids in the hydrate stability zone even without incorporating higher order hydrocarbons.

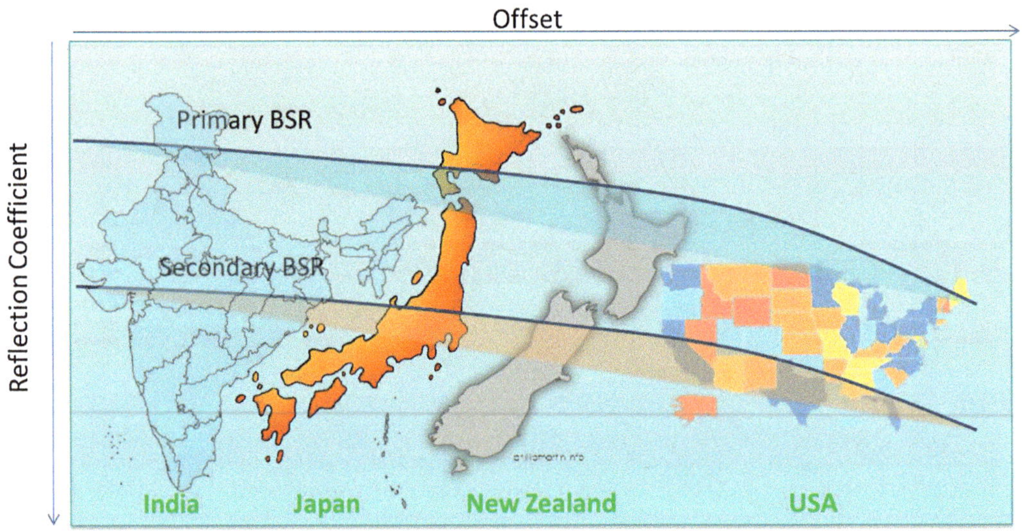

FIGURE 5.11

Schematic of reflection coefficient for primary BSR and secondary BSR (double BSR) from offshore India, Japan, New Zealand, and USA. A similar pattern of "negative trend of reflection coefficient" for primary and secondary BSR is interpreted.

5.2.3 SEISMIC INVERSION

Seismic inversion is the process of estimating what physical characteristics of rocks and fluids could have produced the seismic records we have, in other words seismic inversion converts seismic data from interface properties to layer properties. In the seismic inversion process, we try to determine the input by looking at the output. It is the reverse side of forward seismic modeling and can be explained with the help of Fig. 5.1. The basic outline of the methodology in seismic inversion starts with establishing rock physics trends that relate the reservoir properties to the elastic properties of the rock. Rock physics plays an important role in linking elastic parameters such as impedances and velocities to reservoir properties of interest such as lithology, porosity, and pore fluids. Once this relationship is established based on rock physics models, one can determine whether the input data is suitable for inversion or not. Provided the inversion can be performed on the data, it is recommended to start working on the main elements of a logical inversion workflow (Fig. 5.12). The main elements that need to be considered for a good inversion workflow are as follows:

- Quality control of seismic data
- Seismic data conditioning
- Quality control of interpreted time horizons and faults
- Seismic interpretation data conditioning
- Well log conditioning
- Seismic petrophysics and rock physics modeling for wells
- Well log calibration and wavelet extraction
- Relative impedance inversion
- Revision of seismic interpretation
- Low-frequency modeling
- Simultaneous inversion

5.2.3.1 Quality Control of Seismic Data and Conditioning

This is one of the most important steps in the inversion process. It is well known that the seismic data is imperfect due to errors that occur during acquisition and processing. Relative amplitude preserved processing, rigorous noise analysis, stringent quality control (QC), and velocity must be optimized. Quality control procedures for seismic data involve the following steps:

- Display seismic data headers information for inline, crossline, coordinates, and information on processing sequence applied to the data
- Study the amplitude distribution of full stack for poststack inversion and partial stacks for prestack inversion
- Display time horizons on seismic data and identification of inversion window
- Perform seismic amplitude spectra together with frequency QC and seismic detectability
- Display of mean central frequency, which can provide information about frequency variation from near to far stack

After performing proper migration of the seismic data, post-stack/pre-stack time alignment should be carried out. There are various flattening (time alignment) techniques available such as calculating the displacement for each volume element and matching it with the seismic cube by applying the estimated displacement to each volume element (Nickel and Soenneland, 1999). This procedure is based on the nonrigid

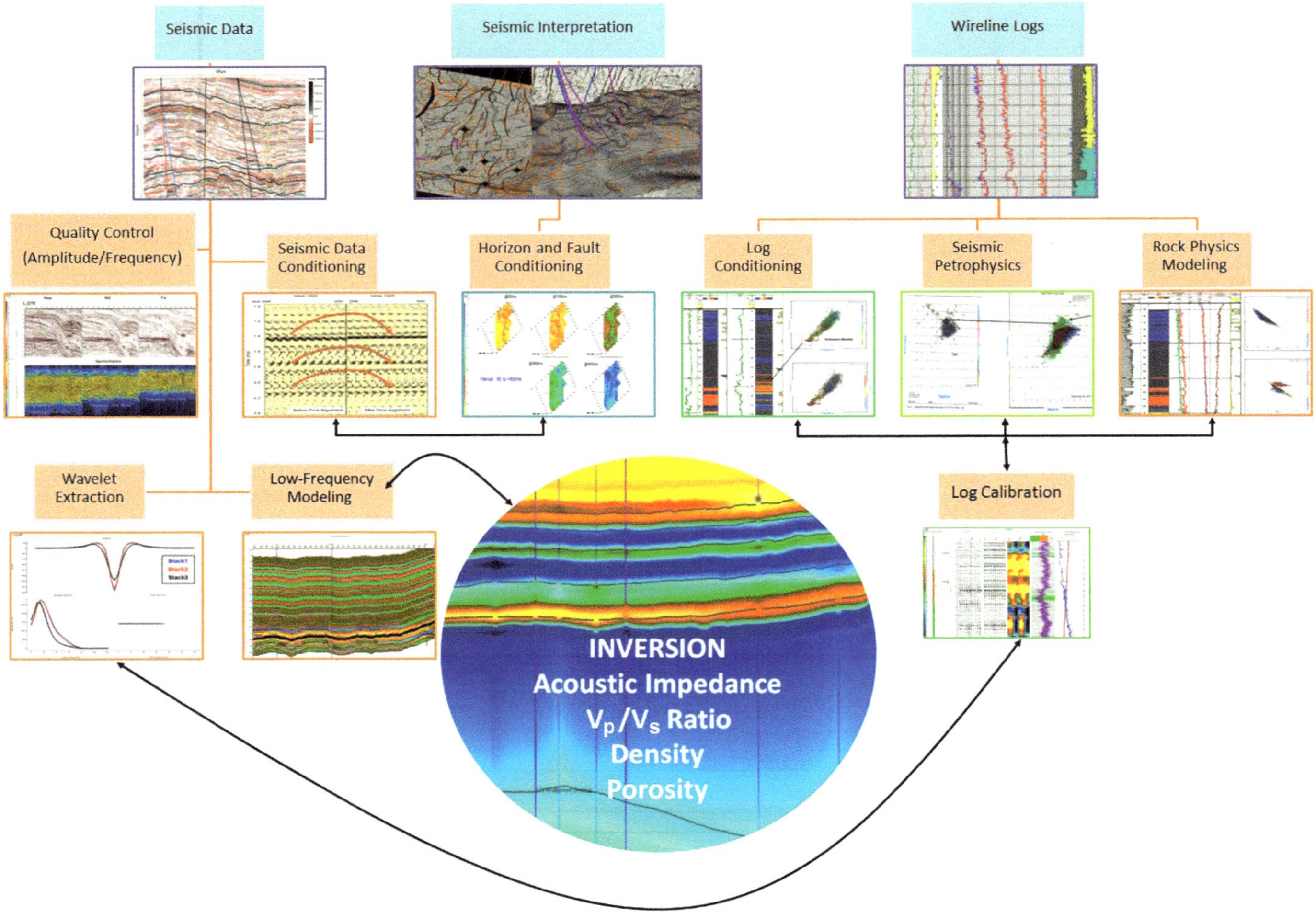

FIGURE 5.12

Seismic inversion workflow for reservoir characterization.

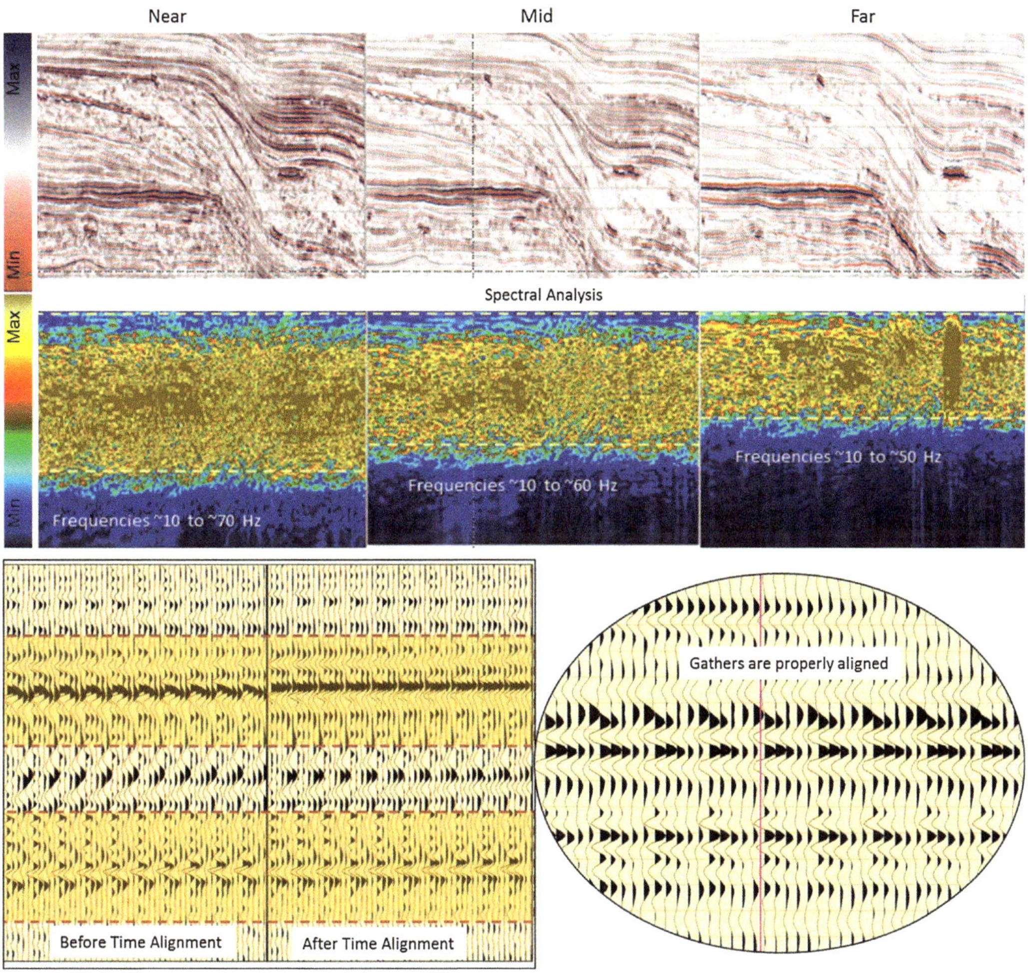

FIGURE 5.13

Seismic data quality control and conditioning. Top part shows near mid and far seismic amplitude stacks; middle figure highlights the spectral analysis for near mid and far angle stacks. Frequency contents are decreasing from near to far angle stacks. Bottom figure shows angle gathers before and after time alignment. Bottom right figure highlights properly aligned seismic angle gathers.

matching method. An example from offshore Australia for time alignment of seismic data is shown in Fig. 5.13. Significantly improved results of a conventional matching procedure were obtained, which could provide useful information to analysts for further interpretation of the reservoir characteristics.

5.2.3.2 Seismic Petrophysics

Hydrocarbon reservoirs are characterized by seismic, well log, and petrophysical information, which is dissimilar in spatial distribution, scale, and relationship to reservoir properties. Well log measurements are subject to irregularities, which should be addressed while processing the data to obtain consistent and accurate well logs for all the available wells. The logs are required for petrophysical interpretation and subsequent rock physics modeling for calibration of elastic logs (density, p-sonic, and often, s-sonic logs). In order to perform a high-quality QC procedure, the following questions need to be answered efficiently:

- Are the provided logs consistent with other wells?
- Are the provided logs covering the vertical zones of interest?
- Are the provided logs covering reservoir and nonreservoir rocks?
- Are the provided logs representing the true, undistributed rock properties?

There could be various reasons for not satisfying the above-mentioned conditions such as tools used in measurements, borehole fluids, logging conditions, missing data, etc. Therefore, well log data need to be conditioned, corrected, and synthesized to provide complete and reliable input for seismic reservoir characterization. Well log conditioning primarily consists of three steps: editing, petrophysical well log analysis, and geophysical well log analysis. While performing the log editing one should consider editing the shallow reading devices, washouts, cycle-skipping, and shale alteration. Petrophysical log analysis starts with applying corrections for mud-filtrate invasion to lithology determination. It is very important that well logs are analyzed from a geophysical prospective involving porosity and pore fluid determination, rock physics modeling, fluid substitution and porosity modeling, shear prediction, and calculations of compressional and shear impedances, Poisson's ratio and production of outputs for acoustic modeling. Additionally, well logs must represent the rock as observed in the seismic domain. Some type of dispersion correction should be applied to eradicate the differences in frequencies between logging tools and surface seismic data.

Seismic petrophysics provides the input for quantitative seismic interpretation and is mainly used by geophysicists; conventional reservoir petrophysics provides input data for reservoir engineers. The goals are different but they both use the same inputs. Seismic petrophysical analysis can differ from standard petrophysical evaluation in several ways. One of the biggest differences is the emphasis placed on the evaluation of nonreservoir rocks. Since the well logs are used in their entirety for processes such as wavelet estimation and low-frequency modeling, ensuring high-quality log data in nonreservoir rocks is just as important as in the reservoir rocks. Another significant difference is that the rock physics modeling requires mineral volumes such as volume of clay rather than lithology volumes such as volume of shale. These differences may not necessarily change the interpretation of the reservoir but can have a significant effect for the nonreservoir rocks. In seismic petrophysics, shales and wet sands are as important seismically and analytically (and sometimes more so) as pay intervals.

Conventional reservoir petrophysics focuses on reservoir intervals. In most cases, the well logs acquired over the reservoir section are of good quality, and so they are not the focus of the well log data conditioning effort. Seismic petrophysics includes rock physics, but also includes the proper conditioning and interpretation of log data that must be undertaken prior to the application of rock physics and seismic modeling. The well log data conditioning includes the whole zone of interest, comprising reservoir and nonreservoir rocks. Therefore, the primary goal in processing well log data is to rid the data of measurement-related errors and to obtain consistent and accurate logs from well to well. These errors might arise from borehole irregularities (rugosity), casing points, and missing data over bit change intervals or cycle-skipping. The well log data conditioning plays an important role in producing high-quality and consistent petrophysical results throughout the reservoir and nonreservoir intervals that have a key role in the development of the predictive rock physics model for the field. The main differences between reservoir petrophysics and seismic petrophysics are highlighted in Table 5.1. Rock physics modeling can help scientists understand the behavior of the reservoir and nonreservoir zones and make corrections for some of the problems encountered in well log data (Avseth, et al., 2001a,b). Fig. 5.14 shows the well data preparation workflow.

Mud-filtrate invasion modifies the connate fluid distribution in rock formations penetrated by a well. This, in turn, perturbs the density and P- and S-wave velocities in the near well-bore region. One of the common goals for seismic petrophysics and rock physics modeling is to correct these logs for the mud-filtrate invasion effects, which is a common problem that occurs in porous and permeable rocks. Mud-filtrate invasion occurs during drilling with over-balanced mud weight conditions. The implications for reservoir geophysics are primarily related to the density and sonic logs. These logs, having shallow depths of investigation, are likely samples of the invaded zone. For example, if a gas sand reservoir is drilled with

Table 5.1 Main Differences Between Reservoir Petrophysics and Seismic Petrophysics

Reservoir Petrophysics	Seismic Petrophysics
Focus mainly on reservoir intervals	Focus on reservoir and nonreservoir intervals
Determined saturation, porosity, lithology, and permeability	Determined saturation, porosity, and lithology
V-Clay or VSH	Require mineral volume such as V-Clay; not VSH
Goal: f*(1 – Sw), k*height	Goal: reflection characterization

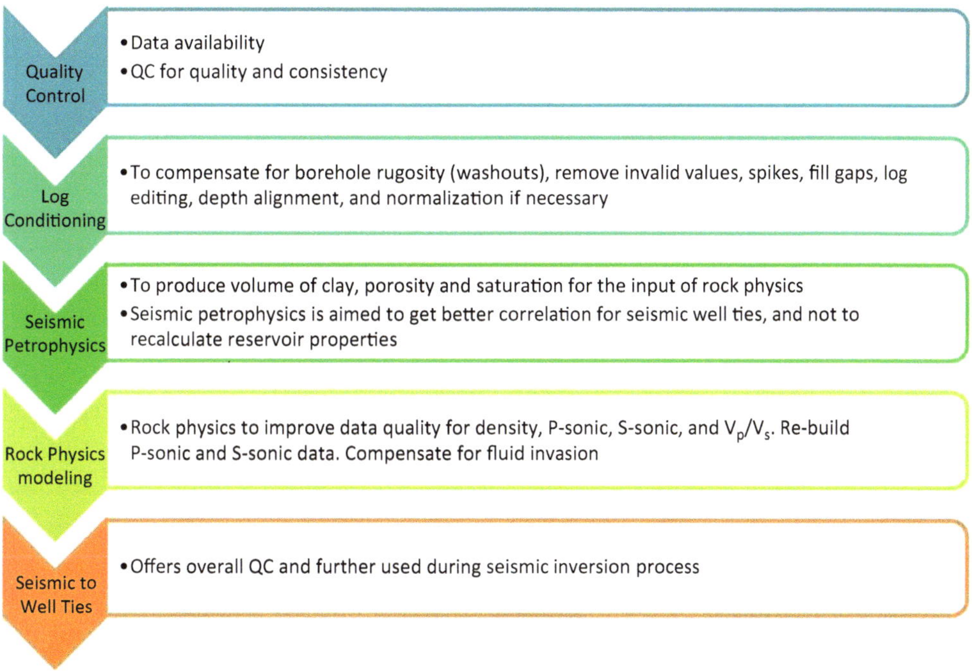

FIGURE 5.14

Seismic petrophysics workflow for data preparation.

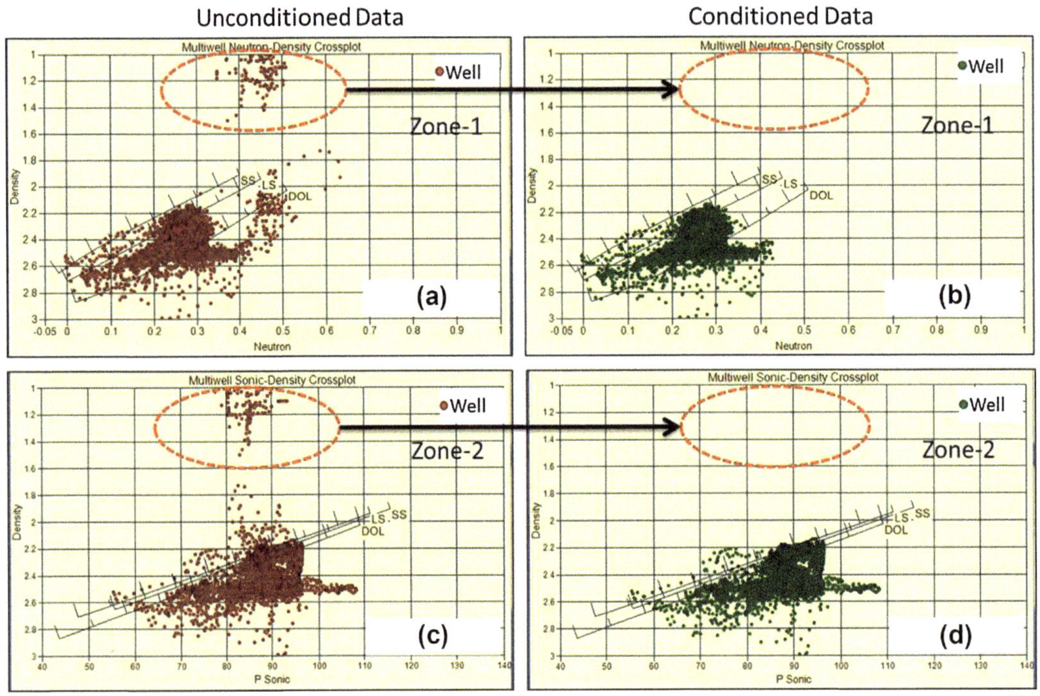

FIGURE 5.15

Neutron-density crossplots of real well data from offshore Australia before and after well log conditioning workflow and seismic petrophysics. (a) Before conditioning and seismic petrophysics for zone-1; (b) after conditioning and seismic petrophysics for zone-1; (c) before conditioning and seismic petrophysics for zone-2; (d) after conditioning and seismic petrophysics for zone-2.

water-based mud, the invasion has higher water saturation than the noninvaded zone. If synthetic seismograms are made from the uncorrected elastic logs (i.e., sonic and density), the results are poor well-to-seismic ties, and must be managed in the seismic petrophysics workflow. Fig. 5.15 shows an example of a neutron-density crossplot before and after seismic petrophysics in addition to log conditioning workflow.

5.2.3.3 Rock Physics Modeling

Rock physics modeling has proven very useful during the seismic inversion process. One should use rock physics modeling for predicting elastic properties from the geology and/or predicting geology from seismic data. Before starting any inversion process, the available well log data is crossplotted to get a basic understanding about the area and to develop some lithology and hydrocarbon trends. Rock physics modeling helps to better understand the lithologies and rock physics behavior of the area. The objectives of the rock physics modeling are to provide a consistent set of elastic logs at each well for wavelet estimation, to provide a link between the petrophysical properties and the elastic properties of the rocks [P-velocity (V_p), S-velocity (V_s), and density (ρ_b)]. This link will allow the elastic properties of the rocks determined through seismic inversion to be interpreted in terms of reservoir properties and can only be achieved accurately for full bandwidth inversion results. The methodology to construct the rock physics models should be consistent with the petrophysical analysis and the elastic properties of the rocks as given by the conditioned well log data. The parameters of the rock physics model are determined where the measured data is complete and considered to be of very good quality. The developed rock physics models are used to generate the synthetic elastic logs such as shear sonic. If necessary the synthesized data can be used to replace the poor quality data and to supply data where the measurements are not available. An example of the rock physics modeled log is shown in Fig. 5.16.

5.2.3.4 Log Calibration and Wavelet Extraction

Seismic data can provide a very good image of the subsurface. However, without knowing the seismic wavelet there can be many equally valid surface geologic interpretations of the actual subsurface geology. The wavelet connects seismic and well data. Before estimation of the wavelets, the well logs must be calibrated to align with the seismic time data. To calibrate seismic data the log's log set

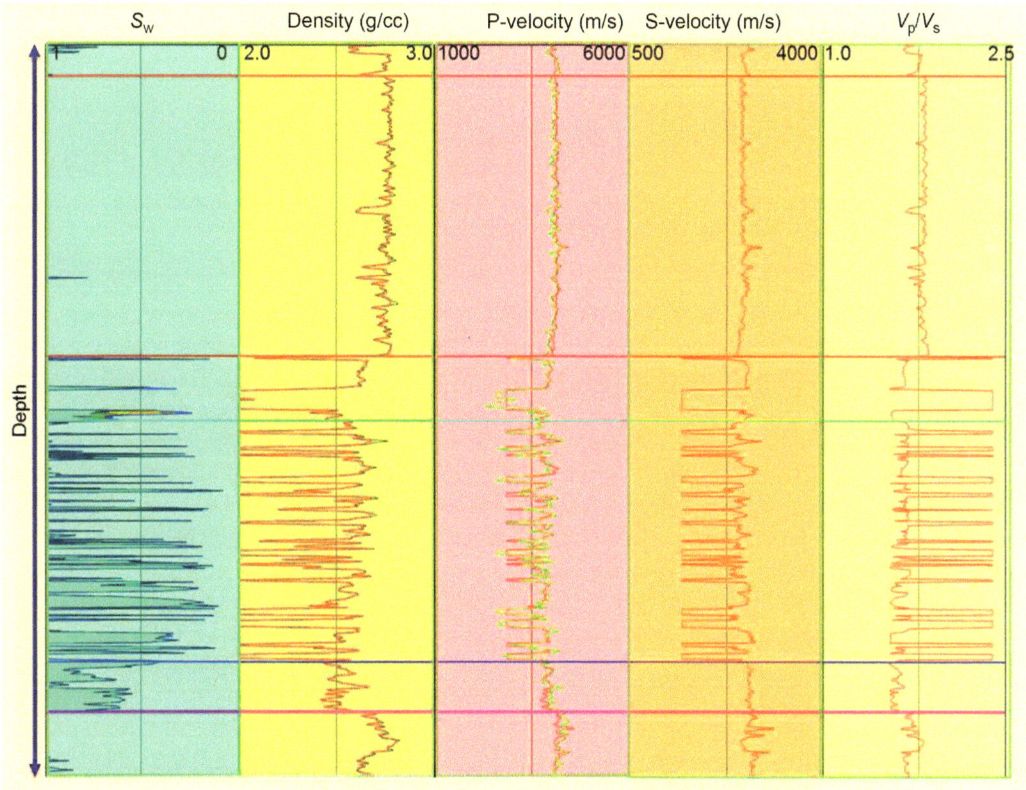

FIGURE 5.16

Rock physics modeling results for one of the wells (Well-A) which is used for parameter optimization. Measured logs are plotted in black and rock physics modeled logs are in red. P-velocity corrected for invasion effect is represented by green color.

(e.g., compressional/shear sonic log) and a check-shot are used. The purpose of the log calibration is usually to create:

- A reflectivity series with the same sampling rate as the seismic data for use in the wavelet estimation
- An acoustic impedance series for use in construction of the low-frequency model

Using available check-shots, the sonic and density logs are converted from the depth domain to the two-way travel-time domain. The acoustic impedance series is computed by multiplication of the calibrated density log and the velocity log derived from the calibrated sonic log. The reflectivity series is then derived by differentiating the acoustic impedance series. If necessary, visual ties are added to shift, stretch, or squeeze the logs. Visual ties are determined by comparing the calibrated logs directly with the seismic data, or by comparing the seismic trace at the well location with the synthetic seismic trace derived by convolving a wavelet with the calibrated reflectivity log. Visual ties may be required for the following reasons:

- The well logs are measured at much higher frequencies than the seismic data. Consequently, the well log velocities can be up to 10% higher than the reflection seismic velocities. This error is largely compensated for by using check-shots.
- The filters used in the seismic processing may introduce a static time shift into the seismic data. It is therefore sometimes necessary to apply a compensatory static shift using the visual ties.
- The seismic migration does not perfectly move the reflectors back to their correct position as the migration velocities and the migration algorithms are imperfect. This will introduce a varying time shift dependent on depth and dip of the reflectors.
- The normal move-out correction for the check-shots is imperfect, particularly for deep and dipping reflectors.

Once the optimum wavelet for the inversion has been chosen, synthetic seismic traces are calculated by convolving the wavelet with the reflectivity series derived from the well logs. The resulting synthetic traces are compared with the seismic traces at the well locations. The well logs are then finally calibrated (stretched and squeezed) for optimum match. The estimation of wavelets at the well location is derived by matching the reflectivity at the well with seismic data at the well location. If some wells exhibit poor synthetic-to-seismic matches due to washout in the borehole, these should not be used in the wavelet extraction process. The ultimate aim is getting the low residual value between seismic data used for the wavelet and the resultant synthetic. Seismic wavelet can also be estimated by model based phase and amplitude spectrum estimation from the seismic data only or with well control. In the presence of multiple partial stacks, the wavelet extraction process can be carried out for each stack individually. There is a clear consistency between these wavelets with frequency decreasing with angle/offset and the phase slightly rotating with frequency. Once the optimum wavelet for the inversion has been chosen, synthetic seismic traces are calculated by convolving the wavelet with the reflectivity series derived from the well logs. The resulting synthetic traces are then compared with the seismic traces at the well locations. The well logs are then finally calibrated (stretched and squeezed) for optimum match. Fig. 5.17 shows extracted wavelets by using five offset stacks and AVO well ties for the three wells (Well-A, Well-B, and Well-C) that are used to optimize the inversion workflow across the main zone of interest. Good quality of well ties is obtained. The correlation between seismic and synthetic is reasonable.

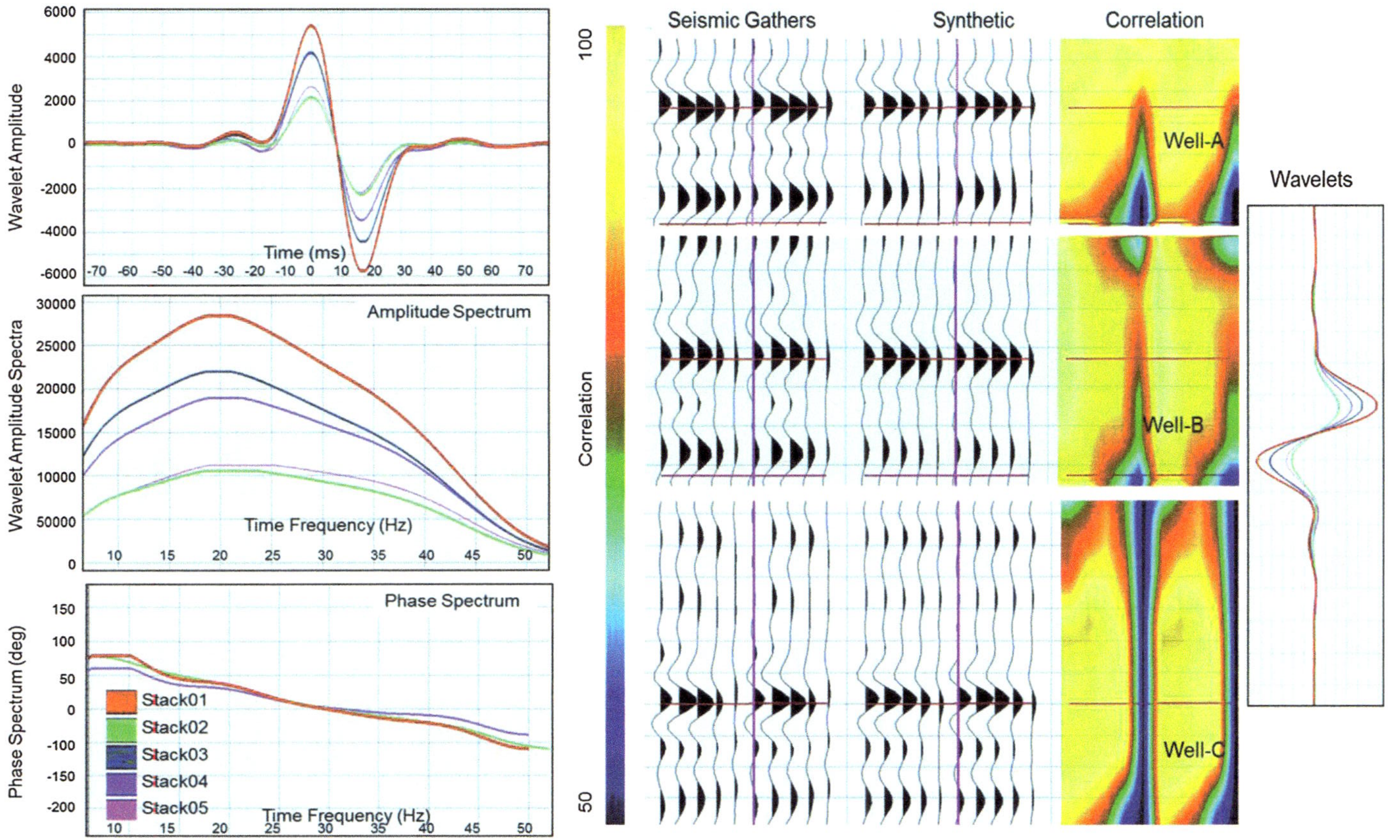

FIGURE 5.17

An example of extracted wavelets for partial stacks. Wavelet (top left), amplitude spectrum of extracted wavelets (middle left), and phase spectrum (middle bottom) are displayed. AVO well ties (Well-A, Well-B, and Well-C) for three different wells in the inversion zone. Seismic gathers are calculated from the five stacks. Each trace represents one stack and the purple straight line shows the well location. Extracted AVO wavelets from five stacks are displayed in the rightmost figure.

5.2.3.5 Low-Frequency Modeling

Acquiring low-frequency information is very important in calibrating the inversion results of processed seismic data. Low frequencies refer to the frequencies below the seismic band, i.e., the frequency gap from zero to the low end of the seismic bandwidth. It is normally assumed that the seismic data has been processed to zero phase and that some additional low-frequency information (from an impedance model) is available to constrain the inversion. The low-frequency model is constructed by using a combination of all the available data including seismic stacking velocities as a guide, elastic well data, and time horizons/faults from seismic interpretation. One should also consider the compaction trends and attenuation effect while constricting the low-frequency model. A correct low-frequency model will remove side lobes and provide absolute values for the elastic properties that are required for quantitative interpretation. The low-frequency model must be consistent with the geology and therefore often requires interpretation and iteration to achieve the desired results. For simultaneous inversion as carried out in this study, a low-frequency model is required for P-impedance, V_p/V_s, and density. Horizon interpretations and wells were tied into a strata model built in time. The strata model comprises layers bounded by horizons. Each layer contains a micro-layering scheme representative of the seismic stratigraphy. Well logs are interpolated through the strata model along micro-layers. An example of a low-frequency model for acoustic impedance at two well locations (Well-C and Well-D) that are used in this study is shown in Fig. 5.18.

5.2.3.6 Simultaneous Inversion

Simultaneous offset-dependent seismic inversion converts the seismic reflectivity of multiple partial offset stacks or angle stacks to elastic property contrasts and simultaneously integrates low-frequency information to produce a broadband model of the elastic properties of the subsurface. The outputs of simultaneous inversion are those of rock properties, e.g., acoustic impedance (AI), shear impedance (SI), and density. Once the low-frequency model is produced, all the data is incorporated into an inversion engine to generate the elastic properties cubes. To deal with the nonunique inversion results, the relative inversion is performed first, which is not influenced by the effect of low-frequency information and filters out all of the nonunique and potentially misleading information. Alternatively, the best available way of dealing with these nonunique solutions is stochastic or probabilistic inversion, which is performed at a later stage. In model-based inversion, the synthetic data is generated for an assumed model and compared with the observed data to get the best match. The best match is then accepted as the solution otherwise the inversion process needs to be repeated until the best match is obtained. Fig. 5.19a shows how successive iterations through inversion workflow update the impedance model in a manner that converges on a solution which shows good match between P-impedance logs and inverted seismic data. At this point, the inversion algorithm moves onto the next input trace to be inverted and this process is continued until the entire seismic volume is inverted. Seismic inversion is a very powerful technique for extracting reservoir rock and fluid properties from seismic data. Table 5.2 summarizes the applications of seismic inversion where the end products can be used at different stages of reservoir characterization purposes.

An example of bandpass (relative) and full bandwidth P-impedance is illustrated in Fig. 5.19b & c. The sections vary smoothly and there is a good agreement between the logs and adjacent traces. As the bandpass (relative) impedance inversion results are not influenced by the background trend (low-frequency model), the seismic horizons and faults interpretation could be improved. The bandpass inversion results are used to constrain the structural elements (horizon/faults) and the low-frequency model is updated with the new inversion results. The final low-frequency model is robust and quality controlled

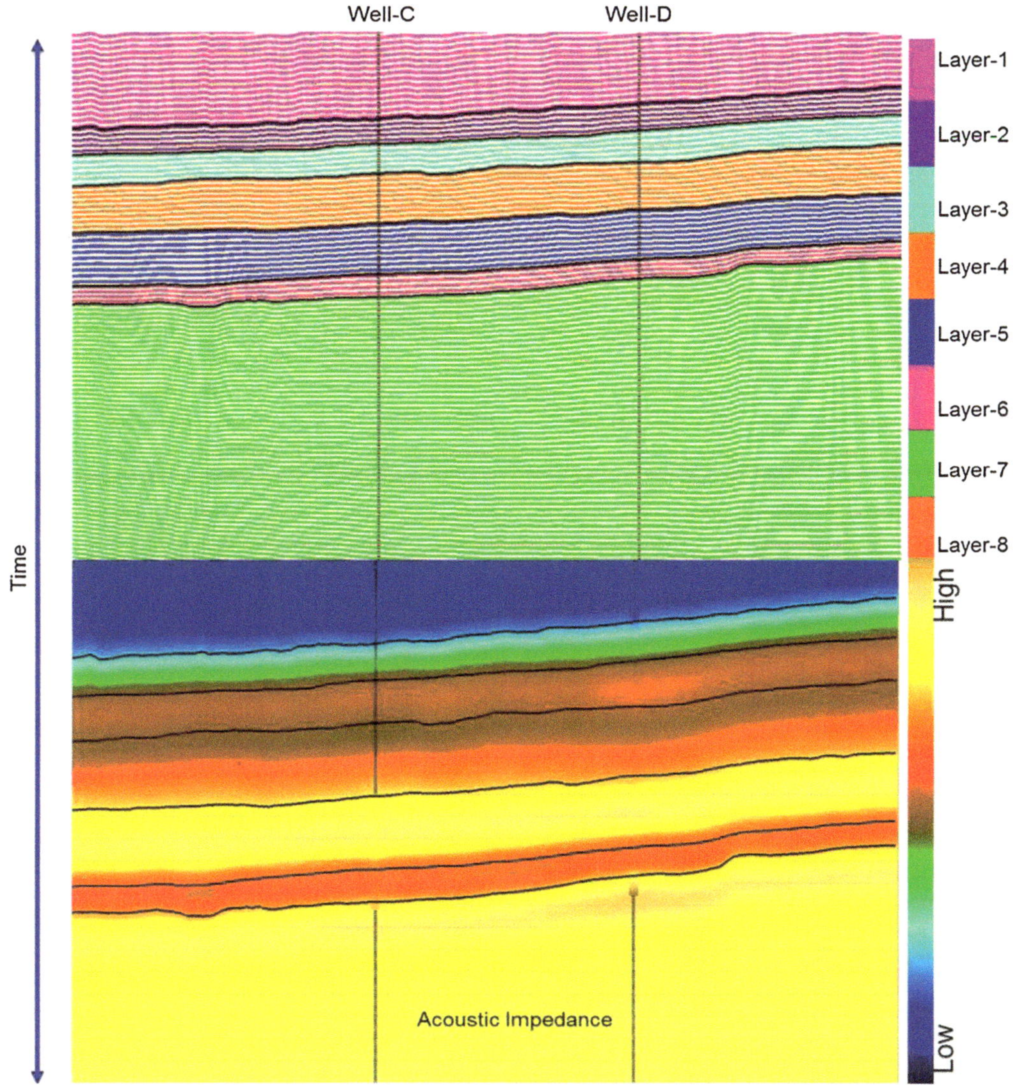

FIGURE 5.18

Display shows strata model (upper figure) constructed by horizon interpolation together with well calibration. The layers are divided into microlayers that represent the seismic stratigraphy. Lower figure shows the low-frequency model (0–8 Hz) for acoustic impedance. Well logs are interpolated through the strata model microlayers. Bandpass-filtered (0–8 Hz) well logs are superimposed on acoustic impedance (0–8 Hz) section.

in terms of the stratigraphic relationship and identification of bull's eyes (unusual features) around wells. An example for inverted bandpass V_p/V_s and full bandwidth V_p/V_s is illustrated in Fig. 5.19d & e.

The hydrate-bearing sediment has an anomalously high compressional velocity, and gas hydrate saturation can be estimated when a reference velocity curve is known (Tinivella, 1999). However, a

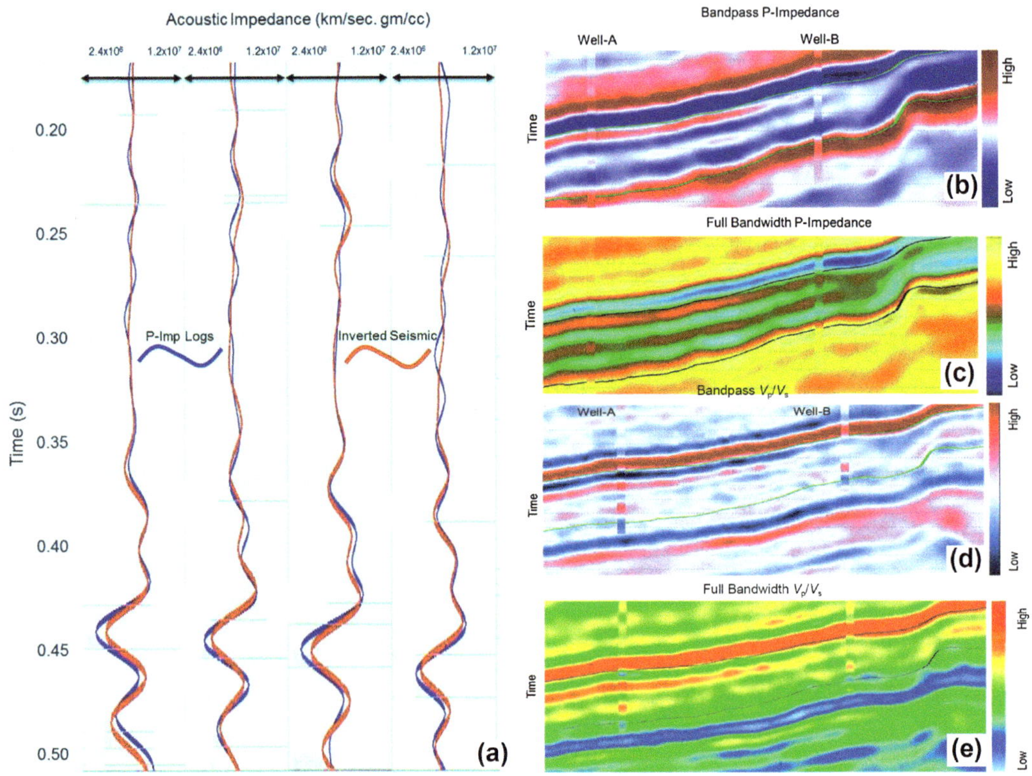

FIGURE 5.19

Impedance models and simultaneous inversion results. (a) Convergence of impedance models with successive iterations in model-based inversion. User supplied initial estimate (in light green) and successive iterations converging on the final answer (pink) models are shown in blue and red (solid line). They differ only in their low-frequency content (lower than the seismic bandpass), which causes different impedance value. (b & c) Example of inverted bandpass impedance (above) and full bandwidth P-impedance (below). Bandpass and full bandwidth (inversion bandwidth) impedance logs are superimposed on the sections. The sections are smoothly varying and there is a good agreement between the logs and adjacent traces. (d & e) Example of inverted bandpass V_p/V_s (above) and full bandwidth V_p/V_s (below). Bandpass and full bandwidth (inversion bandwidth) V_p/V_s logs are superimposed on the sections. The sections are smoothly varying and there is a good agreement between the logs and adjacent traces.

Table 5.2 Applications of Seismic Inversion Schemes

Type	Application
Full-stack (post-stack) inversion	PI (P-wave impedance), porosity
Pre-stack (partial-stack) AVO inversion	PI (P-wave impedance), SI (shear impedance), density, V_p/V_s ratio, porosity
Time lapse (full stack)	Simultaneous time-lapse inversion for baseline PI and change in PI for each time interval
Time lapse (partial stack)	Simultaneous time-lapse inversion for base properties (PI, SI, density) and changes for each time interval
Stochastic seismic inversion	Lithology models, highly detailed P-impedance, V_p/V_s models, porosity models

low compressional velocity zone may indicate the presence of free gas in the pore space. The strong acoustic impedance contrast between sediment containing gas hydrate and sediments with free gas is evident on marine seismic reflection profiles as BSRs (e.g., Shipley et al., 1979; Hyndman and Spence, 1992). The elastic properties of pure gas hydrate are well known, but little is known about the elastic properties of hydrate-bearing sediments. An example of inverted acoustic impedance for gas hydrate estimation is illustrated in Fig. 5.20. A low impedance anomaly at the base of gas hydrate is observed where free gas is present. Acoustic impedance profiling derived from simultaneous inversion helps in differentiating sand from shale. The impedance anomalies permitted the estimatation of gas hydrate and free gas saturation based on well log data. The presence of gas hydrates in sediments dramatically alters some of the normal physical properties of the sediment. These changes can be detected by field measurements and by down-hole logs. An understanding of the physical properties of hydrate-bearing sediments is necessary for interpretation of geophysical data collected in field settings, borehole, and slope stability analyses; reservoir simulation; and production models. The presence of gas hydrate in sediments can result in anomalies in resistivity, chloride concentration, P-wave velocity, acoustic impedance, and water-filled porosity. All of these anomalies can be used to indicate the presence of gas hydrate above the BSR. Acoustic impedance can be predicted by seismic inversion and is considered one of the main rock properties for locating the presence of gas hydrate. The estimation of the hydrate content using down-hole electrical measurements based on Archie's law requires the knowledge of the saturation exponent. The saturation exponent is an empirical parameter that includes influences from the internal rock structure such as pore shape, size, connectivity of the pore network, and the distribution of the conducting phase. An example of a gas hydrate saturation section is shown in Fig. 5.21.

Another example of an unconventional reservoir (shale gas) from an Australian basin is shown in Fig. 5.22. Seismic and well data was utilized to perform seismic inversion and acoustic impedance, porosity, and total organic carbon (TOC) were predicted. Porosity prediction from acoustic impedance is not always reliable so it should be used with caution. Rock properties should be calibrated with geological models before launching any drilling campaign. Fig. 5.22 illustrates seismic data, seismic to well ties, predicted bandpass and full bandwidth acoustic impedance, TOC, and impedance-derived porosity.

5.3 SOME THOUGHTS

Most of the unconventional reservoirs are complex, heterogeneous, and/or anisotropic in nature, and their geoprediction implies a thorough integrated assessment of facies quality on a project by project basis. Several workflows are available but an optimization is always preferred. In order to detect unconventional "sweet spots" with the highest quality of reservoirs and optimal fracture placement, the following should be considered:

- Getting the maximum information from seismic with reasonable correlation of other available data
- Compute compressional and shear moduli and pore content
- Use core and log data from wells for TOC estimation at log scale
- Perform seismic inversion to predict acoustic impedance and ratio of compressional to shear velocity

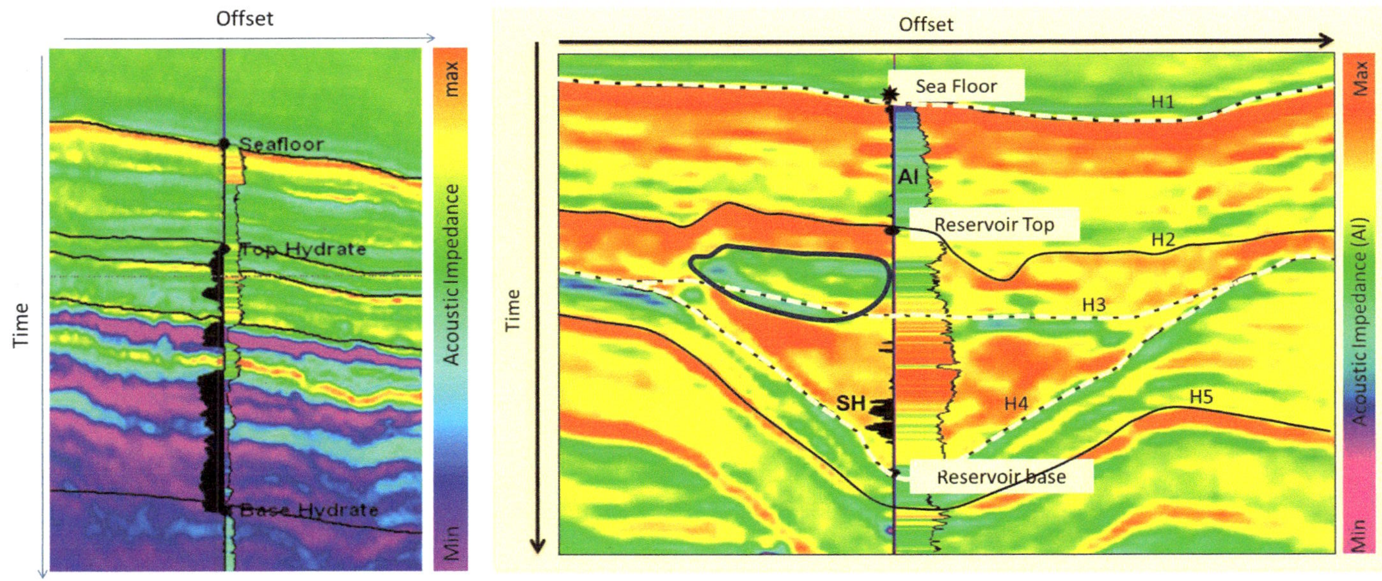

FIGURE 5.20

Inverted acoustic impedance section from South-East Asian region is shown. Left figure illustrates log impedance superimposed on inverted impedance. Hydrate saturation log is highlighted in black. Low impedance anomaly at the base of hydrate is marked (gas hydrate and free gas boundary). Similarly on the right figure, inverted acoustic impedance section showing the top and base of the hydrate reservoir. H1 to H5 are the seismic horizons. Acoustic impedance (AI) and saturation logs are superimposed on the impedance section.

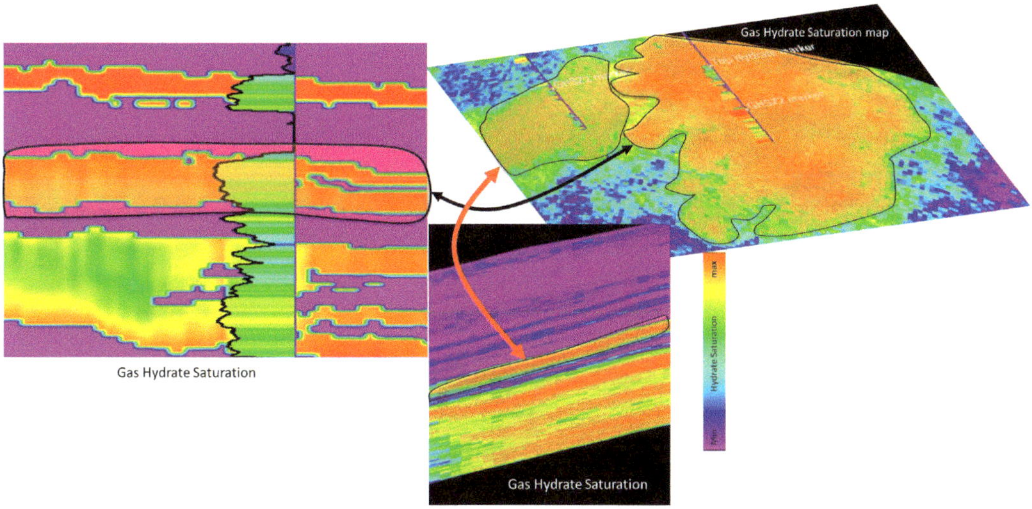

FIGURE 5.21

Gas hydrate saturation section and map view. On left, hydrate saturation estimated from seismic inversion is shown and gas hydrate saturation log is superimposed. A good correlation is highlighted. A corresponding map view is shown on right. Highlighted area shows the highest saturation section (shown on left). The middle section highlights saturation display. Highest saturation area is highlighted in map view.

- Use inversion derivatives to predict TOC away from wells
- Determine the relationship between elastic, geomechanical, and geological properties by facies
- Correlate seismic inversion-derived elastic properties to facies under the presence of fracture network
- Develop geocellular models by utilizing pressure regime, shale diagenesis, and their effect on porosity preservation
- Integrate stratigraphic interpretation to predicted elastic properties
- Use geostatistics to derive seismic facies
- Correlate seismic facies to geological facies
- Quantitatively interpret most likely indicators of reservoir distribution and quality at sweet spot locations
- Use geomechanical properties in a static mechanical earth model, which is required for drilling and stimulation
- Perform correlation of elastic properties with image logs and perform integrated interpretation

Integrated workflows for unconventional reservoirs provide a crucial quantitative link between well and seismic data by utilizing rock physics models and mechanical earth models. Rock properties prediction workflow should be utilized to characterize unconventional facies for organic-rich shale plays, coal bed methane plays, shale oil and gas plays, and most of the other unconventional plays. AVO inversion attributes calibrated with direct information including rock physics models add value in exploration and development programs for unconventional hydrocarbon reservoirs.

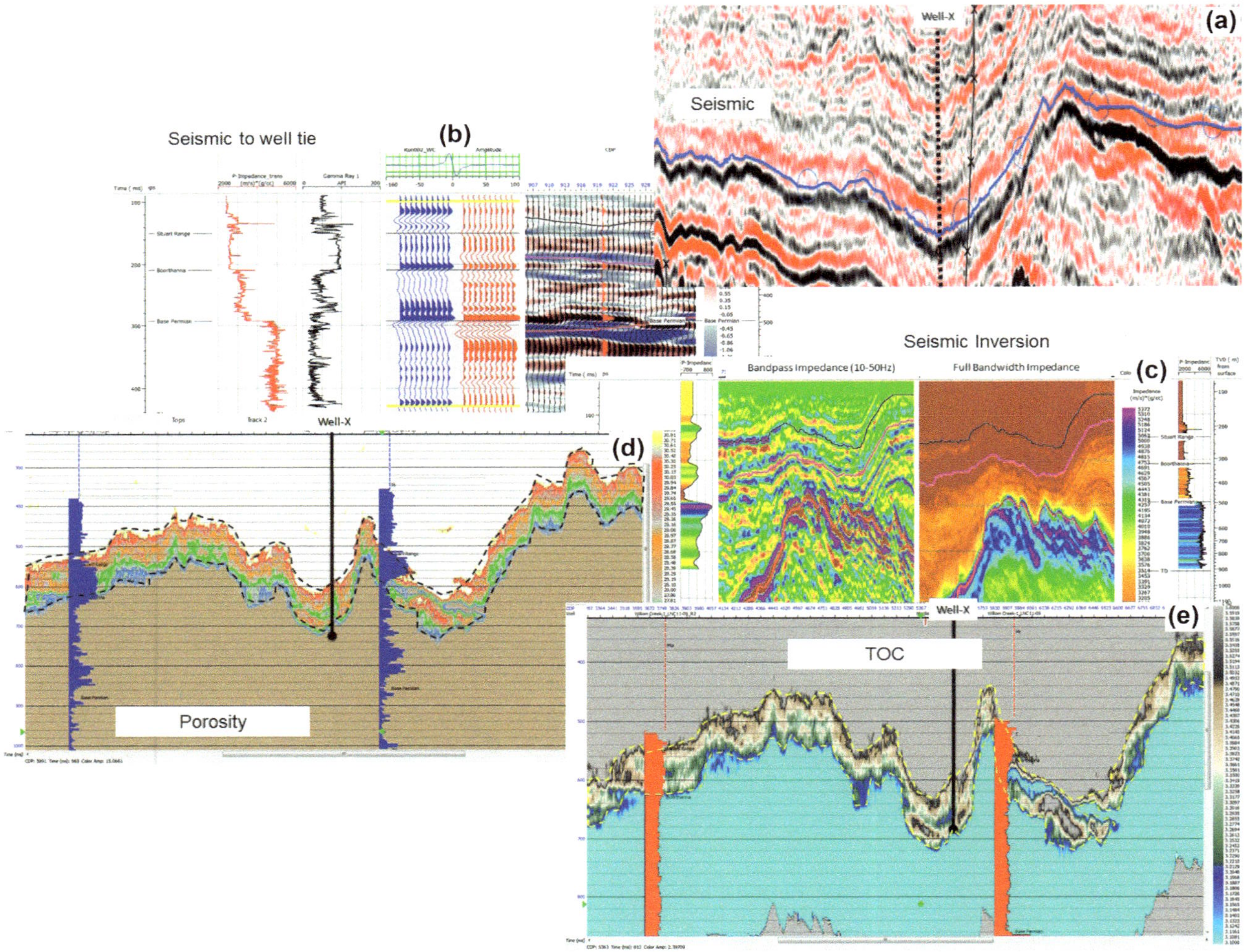

FIGURE 5.22

Derived porosity and TOC from acoustic impedance. (a) Seismic amplitude section. (b) Wavelet estimation and seismic to well tie. (c) Inversion-derived acoustic impedance. (d) Acoustic impedance-derived porosity with target location. Blue logs are the well logs of porosity at two different locations. (d) Acoustic impedance-derived TOC with target location. Red logs are well logs of TOC at two different locations. (e) Acoustic Impedance and TOC relationship from well log is used to predict TOC content over the seismic line. The logs are superimposed over the zone of interest.

REFERENCES

Aki, K., Richards, P., 1980. Quantitative Seismology. W. H. Freeman, San Francisco, California.

Avseth, P., Mukerji, T., Jorstad, A., Mavko, G., Veggeland, T., 2001a. Seismic reservoir mapping from 3-D AVO in a North Sea turbidite system. Geophysics, Soc. Expl. Geophys. 66, 1157–1176.

Avseth, P., Mavko, G., Dvorkin, J., Mukerji, T., 2001b. Rock physics and seismic properties of sands and shales as a function of burial depth. 71st Ann. Internat. Mtg: Soc. Expl. Geophys. 1780–1783.

Baman, F., Xia, H., Yun, L., Lideeng, G., 2004. On seismic inversion. Appl. Geophys. 1 (1), 63–64.

Bangs, N.L.B., Sawyer, D.S., Golovchenko, X., 1993. Free gas at the base of the gas hydrate zone in the vicinity of the Chile triple junction. Geology 21, 905–908.

Bortfeld, R., 1961. Approximations to the reflection and transmission coefficients of plane longitudinal and transverse waves. Geophys. Prospect 9 (4), 485–503.

Connolly, P., 1999. Elastic inversion. Lead. Edge 18 (4), 440.

Crutchley, G.J., Pecher, I.A., Gorman, A.R., Henrys, S., Greinert, J., 1999. Submitted. Seismic imaging of gas conduits beneath seafloor vent sites in a shallow marine gas hydrate province. Hikurangi Margin, New Zealand. Mar. Geol. Dickens GR 1999. The blast in the past. Nature 401, 752–753.

Goodway, B., Chen, J., Downton, J., 1997. Avo and prestack inversion. CSEG Ann. Mtg. Abs. 148.

Grigg, M., 2004. Emphasis on mineralogy and basin stress for shale gas exploration. In: SPE meeting on Gas Shale Technology Exchange.

Haskell, N.A., 1953. The dispersion of surface waves in multilayered media. Bull. Seismol. Soc. Am. 53, 17–34.

Haskell, N.A., 1960. Crustal reflection of plane $H waves. J. Geophys. Res. 65, 4147–4150.

Haskell, N.A., 1962. Crustal reflection of P and SV waves. J. Geophys. Res. 67, 4751–4767.

Hyndman, R.D., Spence, G.D., 1992. A seismic study of methane hydrate marine bottom simulating reflectors. J. Geophys. Res. 97, 6683–6698.

Latimer, R.B., Davison, R., van Riel, P., 2000. An interpreter's guide to understanding and working with seismic derived acoustic impedance data. Lead. Edge 242–256.

Menke, W., 1984. Geophysical Data Analysis: Discrete Inverse Theory. Academic Press. ISBN: 9780123971609.

Nickel, M., Soenneland, L., 1999. Non-rigid matching of migrated time-lapse seismic. In: SEG Annual Meeting, Expanded Abstracts. SEG Technical Program Expanded Abstracts 1999: 872–875.

Pecher, I.A., Henrys, S.A., Zhu, H., 2004. Seismic images of gas conduits beneath vents and gas hydrates on Ritchie Ridge, Hikurangi margin, New Zealand. N. Z. J. Geol. Geophys. 47, 275–279.

Pennington, W.D., 1997. Seismic petrophysics: an applied science for reservoir geophysics. Lead. Edge. 16 (3), 241. http://dx.doi.org/10.1190/1.1437608.

Russell, B.R., Hedlin, K., Hilterman, F.J., Lines, L.R., 2003. Fluid-property discrimination with AVO: a Biot-Gassmann perspective. Geophysics 68, 29–39.

Sen, M.K., Stoffa, P.L., 1995. In: Global Optimization Methods in Geophysical Inversion: Advances in Exploration Geophysics, vol. 4. Elsevier, Amsterdam.

Sen, M.K., 2006. Seismic Inversion. Society of Petroleum Engineers. ISBN: 978-1-55563-110-9. USA.

Shipley, T.H., Houston, M.H., Buffler, R.T., et al., 1979. Seismic evidence for widespread possible gas hydrate horizon on continental slopes and rises. AAPG Bull. 63, 2204–2213.

Shuey, R.T., 1985. A simplification of the Zeoppritz equations. Geophysics 50, 609–614.

Stolt, R.H., Weglein, A.B., December, 1985. Migration and Inversion of Seismic Data. Invited paper for Golden Anniversary Issue of Geophysics, 50 (12), 2458–2472. Reprinted in "Inversion of Geophysical Data." Geophysics reprint series; no. 3, L. R. Lines, Editor, 543 pp. (1989) Society of Exploration Geophysicists.

Tarantola, A., 1984. Inversion of seismic reflection data in the acoustic approximation. Geophysics 49, 1259–1266.

Tarantola, A., 1986. A strategy for nonlinear elastic inversion of seismic reflections data. Geophysics 51, 1893–1903.

Tarantola, A., 1987. Inverse Problem Theory: Methods for Data Fitting and Model Parameter Estimation. Elsevier Science Publ. Co., Inc. Springer-Verlag Berlin Heidelberg.

Thakur, N.K., Rajput, S., 2010. Exploration of Gas Hydrates: Geophysical Techniques. Springer Publications. Springer-Verlag Berlin Heidelberg.

Thomson, W.T., 1950. Transmission of elastic waves through a stratified media. J. Appl. Phys. 21, 89–93.

Tinivella, U., 1999. A method for estimating gas hydrate and free gas concentrations in marine sediments. Boll. Geofis. Teor. Appl. 40, 19–30.

Torres-Verdin, C., Victoria, M., Merletti, G., Pendrel, J., 1999. Trace-based and geostatistical inversion of 3-d seismic data for thin sand delineation: an application to San Jorge Basin, Argentina. Lead. Edge 18 (9), 107.

Veeken, P.C.H., Da Silva, M.D., 2004. Seismic inversion methods and some of their constraints. First Break 22, 47–70.

Veeken, P.C.H., Rauch, M., Gallardo, R., Guzman, E., Vila Villasenor, R., 2002. Seismic inversion of the Fortuna National 3D survey, Tabasco, Mexico. First Break 20, 287–294.

Zoeppritz, K., 1919. Erdbebenwellen VII. VIIb. Über Reflexion und Durchgang seismischer Wellen durch Unstetigkeitsflächen. Nachrichten von der Königlichen Gesellschaft der Wissenschaften zu Göttingen, Mathematisch-physikalische Klasse 66–84.

CHAPTER

FLUID FLOWS

6

"Curving back within myself I create again and again."

The Bhagavad Gita

CHAPTER OUTLINE

6.1 INTRODUCTION

Unconventional reservoirs such as gas hydrates, coal seam gas (CSG), shale gas, tight oil and gas, and geothermal reservoirs are characterized by extremely low porosity and permeability. Porosity is a measure of how much of a rock is open space, which can be between grains or within cracks or cavities of the rock. Or in other words, porosity is the proportion of a reservoir rock that is not rock. Porosity is normally expressed as a percentage (%) of the total rock which is taken up by pore space. Porosity does not give any information about pore sizes, pore distribution, or their degree of connectivity. It depends on stress conditions. Rocks of the same porosity can have widely different physical properties. It can be measured by core data, geophysical well log data, and seismic data. Permeability is a measure of the ease with which a fluid (water in this case) can move through a porous rock. Permeability in petroleum-producing rocks is usually expressed in units called millidarcys (one millidarcy is 1/1000th of a darcy). Shale has a lot of porosity (much more than sandstone), but extremely low permeability. Permeability depends on direction and fluid type (wettability and viscosity). Permeability can be estimated from core data, pressure transient tests (PTT), well logs, and well log to core correlation.

The term fluid flow includes both liquid and gas. Fluid migration is the movement of fluids through formations with sufficient permeability. Transportation of the liquid and gaseous form of hydrocarbons through fluid flow and venting structure systems is a widespread process in continental margins and sedimentary basins. Understanding fluid flow mechanisms helps in the assessment of geohazards, geological phenomena, including submarine slope stability and dynamics of unconventional resources, environment

Geological Controls for Gas Hydrates and Unconventionals. http://dx.doi.org/10.1016/B978-0-12-802020-3.00006-0

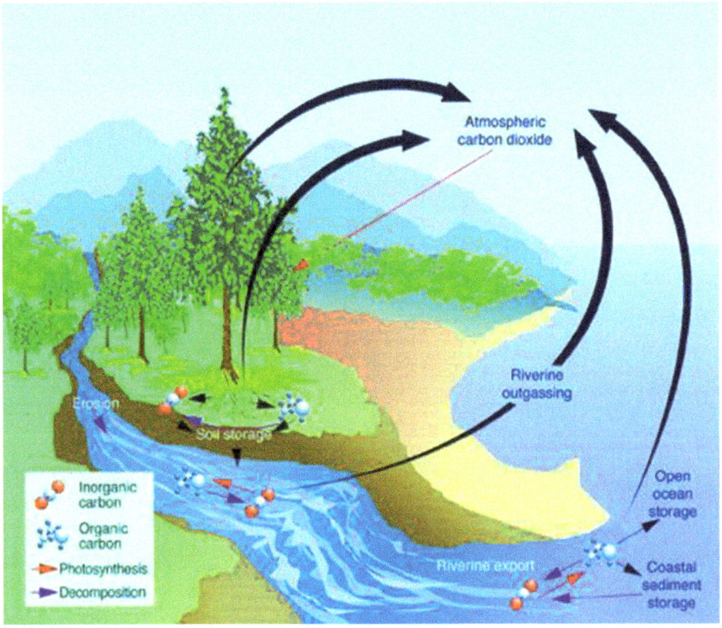

FIGURE 6.1

In the earth's carbon cycle, carbon is exchanged between the planet's land, atmosphere, and oceans. In the process, it is transformed from inorganic carbon to organic carbon and back again. (Source: https://str.llnl.gov/str/March06/Brown.html.)

conservation, and securing fossil energy resources. Therefore studying the abundance, distribution, and drivers for fluid flow characteristics is very important. The fluid flow mechanism has a significant role to play in order to understand atmospheric carbon budget (Fig. 6.1), earth's climate, dynamics of unconventional, e.g., gas hydrate, accumulations, and destabilization. Moreover, since natural gas seepage is a significant source of methane to the ocean and atmosphere (Archer, 2007), it may impact ocean chemistry and might have played a role in earth's past maximum-thermal events (Kvenvolden and Rogers, 2005). In particular, structural surfaces along bedrock (Shaw et al., 1997), salt diapirs (Taylor et al., 2000; Satyavani et al., 2005), and faults and faulted anticlines (Boe et al., 1998; Soter, 1999; Vogt et al., 1999; Eichhubl et al., 2000; Dimitrov and Woodside, 2003) create pathways for fluid migration. However, neither the timing of fluid expulsion (episodic, cyclic, catastrophic) nor the driving processes (underlying overpressure, overlying sediment loading), or their mutual feed-back have been clearly defined yet. The main driving force of fluid is explained using Darcy's law, where the fluid flow rate is dependent on the permeability of the fluid flow medium, the hydraulic potential, and the fluid viscosity (Freeze, 1979).

Methane hydrate is a class of clathrate, composed of water and low molecular weight gases, mainly methane, which forms under low temperature, high pressure, and appropriate methane concentrations. Hydrates with free gas beneath them form a strong acoustic interface, which is often conspicuous in seismic sections as bright reflections known as bottom-simulating reflectors (BSRs). These are interpreted as the primary geophysical indicator for inferring the presence of gas hydrate (Andreassen et al., 1990; Shipley et al., 1979). The BSR corresponds to a thermodynamic phase boundary at the base of the hydrate stability zone (HSZ), a region that includes the uppermost several hundred meters of sediment where low temperatures and high pressures force the excess methane dissolved in pore water to form hydrates (Gorman et al., 2002). A BSR is identified on multichannel or multicomponent (4C) seismic sections as a high-amplitude seismic reflection that mimics the seafloor reflection but with inverse polarity. As the BSR is a thermodynamic phase boundary it may cross-cut the bedding plane of sedimentary layers (Hyndman and Spence, 1992; Thakur and Rajput, 2010). The ice-like structure of gas hydrate consists of light hydrocarbons (mostly methane) trapped by a rigid cage of water molecules.

Gas hydrates occur naturally in the pore space of sediments when appropriate high-pressure and low-temperature conditions exist (Sloan, 1998). These conditions confine gas hydrates to the upper few hundred meters of sediments, the gas hydrate stability zone (GHSZ). Pure gas hydrate has a high P-wave velocity (3000 m/s), and therefore sediments partially saturated with hydrate are characterized by higher velocities than brine-saturated sediments (Ecker et al., 1998). The presence of hydrates in the pore space reduces porosity and permeability of the sediment, and they typically form a partial hydrological seal for upward migrating fluid and gas (Nimblett and Ruppel, 2003). The BSR is characterized by increase in negative amplitude with increasing offset, i.e., the compressional velocity above the BSR is larger than that below BSR (Ecker et al., 1998). Hence, if only a small percentage of free gas exists below the GHSZ, the P-wave velocity there may sharply decrease to below the acoustic velocity of water (i.e., 1475 m/s; Ecker et al., 1998). The velocity reversal occurs at the BSR which delineates the base of the GHSZ, and allows the determination of gas hydrate without drilling. Below the BSR, the hydrate destabilizes, so that in methane-saturated pore fluids, additional methane is in gas form and is considered as free gas (Buffett and Zatsepina, 1999).

In recent times, the occurrence of double bottom-simulating reflector (DBSRs) has been identified on seismic reflection sections (Posevang and Mienert, 1999; Bangs et al., 2005; Rajput et al., 2010). Tréhu et al. (1999) suggested that the DBSR at Hydrate Ridge, offshore USA, is caused by tectonic uplifting, growth of an anticline structure, sea level change, or a change in the water bottom temperature. Bangs et al. (2005) made a case for interpreting the DBSR as a relic of the glacial GHSZ. Below a certain depth, depending on the local temperature conditions, hydrates cannot form, and only free methane exists. Between this depth, known as the base of the gas hydrate stability zone (BGHSZ), and the top of the GHSZ is where the hydrates are. Musgrave et al. (2006) suggested that slow migration of the BGHSZ in the closing stages of the last glacial period allowed accumulation of free methane below the hydrate zone. This established a rock magnetic signal that has survived the subsequent rapid upward movement of the BGHSZ as the pulse of heat from water bottom warming reset the geothermal profile over Hydrate Ridge. Recently, the thermal nature of the DBSR has been studied (Golmshtok and Soloviev, 2006). With little information available about the DBSRs and lack of suitable models for the interpretation, their formation mechanism is still subject to speculation.

The characteristics of BSRs and DBSRs may vary within a study area and even on a single seismic profile (Rajput et al., 2010). During the past two decades, AVO techniques have been used for direct detection of gas reservoirs (Ostrander, 1984; Castagna and Backus, 1993; Santoso et al., 1995). For complex geological structures, AVO responses can considerably reduce uncertainty when predicting the hydrocarbon reserves. Seismic reflection amplitudes that change with offset or incidence angle can be analyzed in terms of the reflection coefficient characteristics. For isotropic media, Zoeppritz (1919) produced a set of equations to compute the particle displacement amplitude of reflected and transmitted waves. In the presence of anisotropy, the behavior of elastic waves becomes more complicated and requires sophisticated analysis techniques (e.g., Rueger, 1996; Tsvankin, 2005; Behura and Tsvankin, 2009). The formulation proposed by Rueger (1996) to derive the reflection coefficient of the BSR was used for both synthetic and real data.

Gas hydrate provinces are associated with methane gas seepage in both passive and active margin settings (e.g., Holbrook et al., 2002). Structures for hydrocarbon accumulation and structural pathways (faults and fractures) for fluid migration are important factors contributing to the presence of methane seepage. In low permeability environments, vertically focused fluid flow is generally initiated through the process of natural fracturing (e.g., Morley, 2003; Tingay et al., 2003; Zuhlsdorff and Spiess, 2004). Conventional seismic analyzes typically reveal these venting structures as near-vertical, well-defined distorted columns of seismic "wipe-out" (loss of coherency), and are termed seismic chimneys or pipes (Heggland, 1997; Løseth et al., 2001).

To establish a linkage of methane seepage to fluid flow mechanism an example of a naturally occurring BSR over Kerala-Konkan basin (KK), offshore Indian margin and that of a double bottom-simulating reflector (DBSR) from Green Canyon (GC) Gulf of Mexico (GOM), offshore USA are discussed in the following sections and different models are compared. Conventional seismic data analyzes together with amplitude versus offset (AVO) characteristics were evaluated with real anisotropy parameters from other comparable regions (Blake Ridge, Pecher et al., 2003). Reflection coefficient characteristics for real and modeled BSR's were studied. At this stage the whispers from seismic derivatives are compared from two different regions (GC and KK). This analysis indicates that the variability in AVO signatures for gas hydrate-saturated sediments is potentially linked to discrete zones of steeply inclined fractures that are responsible for the migration of deep gas and its escape through the seabed.

6.2 SEISMIC DATA ANALYSIS AND INTERPRETATION

Methane hydrates can be found in the shallow sediments of many deep ocean areas. The clathrate/hydrate itself forms an impermeable, thin, high-velocity layer, and below this, an accumulation of free gas can be seen, giving rise to a low-velocity zone immediately below the high-velocity clathrate layer. These form zones which strongly attenuate seismic amplitudes and are characterized by strong reflectivity due to the higher-velocity hydrates overlying the lower-velocity free gas interval. Dim zones occur beneath these gas hydrate pockets. The size and shape of the dim zones vary over different offsets. The thickness of the GHSZ depends on the seafloor depth, geothermal gradient, and composition of gas, bottom water temperature, lithostatic and hydrostatic pressure, pore water salinity, as well as the physical properties of the host sediments (e.g., Clennell et al., 1999; Henry et al., 1999). In the Arctic, where cold air temperatures create thick zones of permanently frozen soils (permafrost), the top of the GHSZ typically lies about 300 to 400 m below the land surface, often in the midst of the permafrost. In regions of relatively thick permafrost, the GHSZ often extends 500 m or more below the base of the permafrost. In oceans or deep inland lakes, where high pressures are generated by 300 to 500 m or more of overlying water, the top of the GHSZ (Fig. 6.2) occurs within the water column, and the base is some distance below the seafloor. Moreover these physical conditions determine whether free gas and water, expelled from reservoirs, will migrate through the GHSZ or if they will form gas hydrates (Tréhu et al., 2005). A phase diagrams of stability conditions of methane in the marine environment is shown in Fig. 6.2.

The sedimentary basins adjoining the west coast of India include, from north to south, Kutch Offshore basin, Cambay basin, Saurashtra basin, Surat basin, Bombay Offshore basin (collectively termed Northern basins), the Konkan basin, and the Kerala basin. The Kerala and Konkan basins are particularly useful in that they provide a sedimentary record since ca. 80–90 Ma. The KK basins constitute a major depocenter for sediment from the onshore hinterland of Western India and as such provide

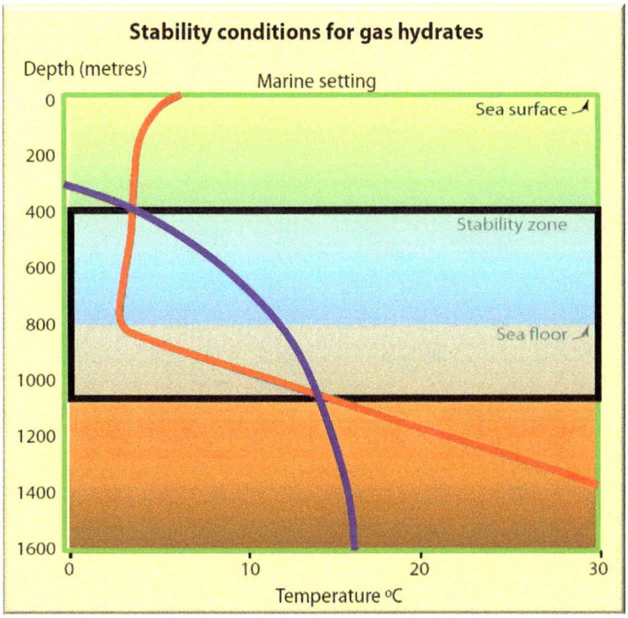

FIGURE 6.2

Phase diagrams illustrating where methane hydrate is stable in marine. Hydrate can exist at depths where the temperature as highlighted by red curve is less than the maximum stability temperature for gas hydrate as shown by violet curve. Pressure and temperature both increase with depth in the Earth, and though hydrates can exist at warmer temperatures when the pressure is high as highlighted by violet curve, the temperature in the Earth (red curve) gets too hot for hydrate to be stable, limiting hydrate stability to the upper ~1km or less of sediment.

a valuable record of the timing and magnitude of Cenozoic denudation along the continental margin. Sedimentation in the Kerala and Konkan basins is within coast parallel N–S trending grabens separated by local basement highs. The analyzes of seismic data from KK basin suggest that the dimming anomalies are present (Fig. 6.3) in the region and represent permanent deformation independently of whether or not these dimming zones contain free gas. Fig. 6.3 shows that the BSR observed on the seismic section from the KK basin exhibits the reverse polarity and stronger seismic amplitude for hydrate-saturated sediments than for equivalent water-saturated sediments. Hydrate recycling and vertical migration of fluids from deep sources are processes evoked as controllers of the formation and stability of the free gas zone (FGZ) beneath the BGHSZ. Gas hydrate formation appears to be closely linked to fluid flow because, in most settings, it is predicted to require a supply of gas from below into the GHSZ (Ruppel and Kinoshita, 2000; Rajan, 2012). Models for gas hydrate formation based on single-phase fluid flow (Zatsepina and Buffet, 1997; Xu and Ruppel, 1999; Ruppel and Kinoshita, 2000; Xu, 2004) assume that gas is transported into the hydrate stability zone in solution in upward migrating pore water.

The seismic data from the GC area, as shown in Fig. 6.4, show a "leaky" gas hydrate province (labelled as "A", "B", and "C" in Fig. 6.4) in which tectonically controlled structural elements promote the rapid migration of thermogenic gas from sub-seabed reservoirs to the seafloor. Localized

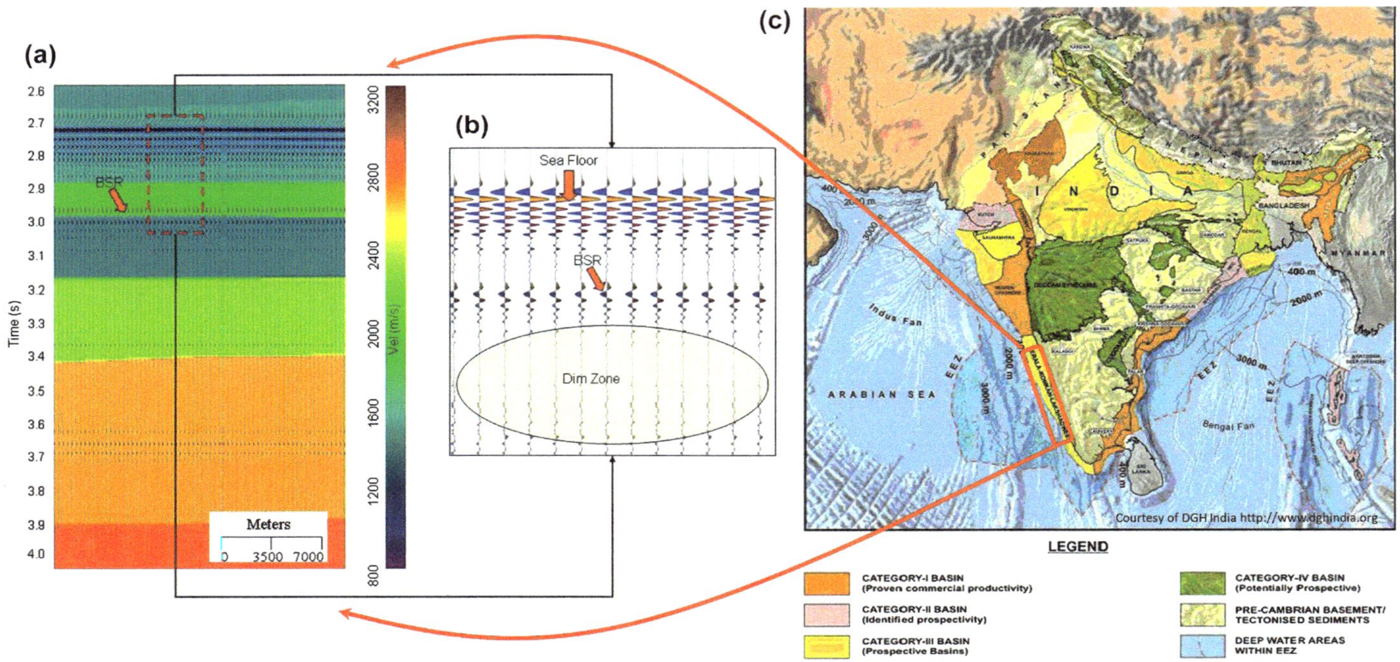

FIGURE 6.3

Example from KK Basin. (a) - Migrated seismic image of a part of the KK basin is superimposed on the seismic stacking velocities. Different colours represent variations in velocities. At 3.0 s, a BSR is identified as a reverse polarity event and marked. (b) Zoomed in part of the red dotted box of (a). The sea floor and BSR show reverse polarity. Weak amplitudes are seen in the zone referred to as dim zone. (c) Location of Kerala Konkan Basin, offshore India (Source: courtesy of DGH India, http://www.dghindia.org/)

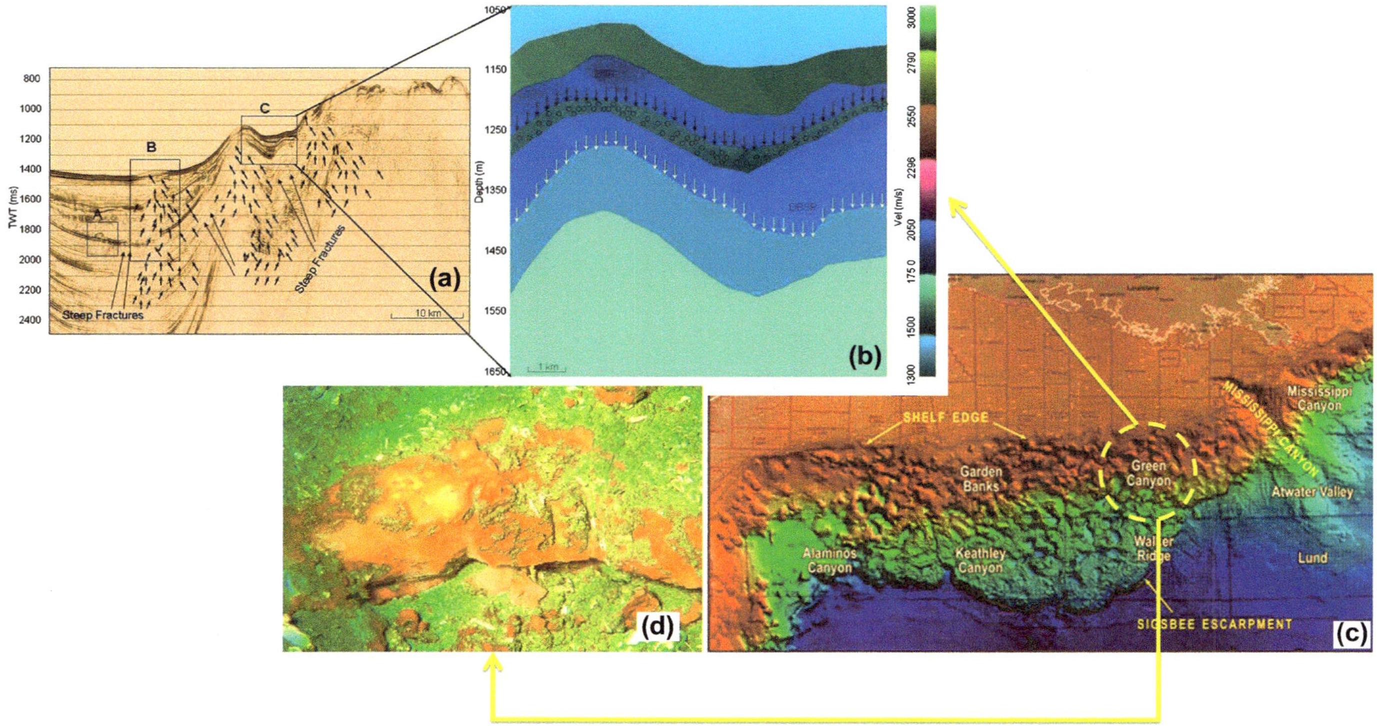

FIGURE 6.4

Time migrated section from the the Green Canyon region, offshore USA and velocity model for around the DBSR and gas hydrate on seafloor of Green Canyon (a). Three structures (labeled "A", "B", and "C") are discussed in the text. Long arrows indicate near-vertical fractures. Small arrows show the migration of gas from deeper in the section to the seabed. (b) Seismic velocity model of the feature "C" in Fig. 6.4a. Black arrows represent the location of the BSR and white arrows show the location of the DBSR. Black circles show the accumulation of fluids and gas within the GHSZ. (c) Location of Green Canyon in the Gulf of Mexico. (d) Gas Hydrates on the seafloor in Green Canyon. (Photograph courtesy of GERG/Texas A & M University.) TWT; Two way Time.

amplitude dimming with seismic pull-up and mound-like features are observed. In typical gas hydrates, mound-like structures form when the proper temperature, pressure, and chemical composition for the gas are achieved (feature A in Fig. 6.4). The apparent gas migration pathways (seismic chimney) and funneling of deep gases into the base of vent feature B suggest a high gas flux into this vent. A high upward flux of gas would also account for the large velocity anomaly due to the hydrate accumulation in the vent.

The DBSR observed at feature C (Fig. 6.4) is caused by a low-velocity layer (LVL) lying within the GHSZ as shown in Fig. 6.4b. This LVL is probably caused by the structural elements (gas vent structures, faults, folds, mounds, fractures, etc.) present in the GC region (e.g., Gorman et al., 2002). The seepage of methane could be from destabilization of gas hydrates or free gas through structural elements such as faults and fractures or could be from a gas-bearing source rock by filtration and diffusion processes (e.g., Zuhlsdorff and Spiess, 2004). Within feature C (Fig. 6.4), the permeable pathways locally connect the FGZ to the seafloor. This occurs where the impermeable cap of young, unfaulted and unfractured sediments has been removed by erosion or breached by gently dipping sediments, where onlapping sediments will guide fluid flow. Seismically similar features have been investigated from a variety of geographical areas, including the Niger Delta (Hovland et al., 1997), Norwegian margin (Hovland and Svensen, 2006), Blake Ridge (Paull et al., 1995), the Cascadia margin (Suess et al., 1999; Riedel et al., 2006), and offshore Korea (Haacke et al., 2009).

Close inspection of the migrated seismic images (Fig. 6.5) shows that the vertical vent (seismic chimney) is composed of steep amplitude striations that are probably swarms of cracks and fractures, which are described by Haacke et al. (2009) and Zuhlsdorff and Spiess (2004). The seismic pull-up of BSR around and beneath the hydrate mound (feature A in Fig. 6.4 that corresponds to Fig. 6.5a) could indicate some degree of upward bending of the GHSZ due to warm upwelling fluids, as suggested by Wood et al. (2002). The gas hydrates at hydrate mounds are not dispersed in sediments as nodules or thin seams, but instead occur as continuous masses. Velocity analysis suggests that in this structure there is already a high concentration of hydrate and rapid migration of gas through these sediments resulted in the disruption of the structural relationship that gives the surrounding rock its reflective character. The incoherent reflections at feature "B" (Fig. 6.4 and Fig. 6.5b) indicate the focusing of lower concentrations of the gas towards the base of the seismic chimney, which carries the gas through the seafloor and into the overlying ocean. Fluid flow is very important in the formation of gas hydrates as it provides a pathway for the movement of fluid from the hydrate. In hydrodynamic environments, although gas usually moves vertically upward due to buoyancy, water may move in any direction (Zhang et al., 2011).

Thus, fluid migration (including both water and gas) can be vertical along faults, which act as pathways for fluid transport. It also can be horizontal within high-permeability sandy sediments. These sediments may act as either pathways or intermediate reservoirs to concentrate fluids (Gay et al., 2007). The apparent gas migration pathways (seismic chimney) and funneling of deep gases into the base of vent feature B (Fig. 6.4) suggest a high gas flux into this vent. A high upward flux of gas would also account for the large velocity anomaly due to the hydrate accumulation in the vent. In the sediment-filled plain of the GC, a number of buried faults produce a series of surface spots or diffuse gas vents (e.g., Gorman et al., 2002; Haacke et al., 2007, 2009). Gas flow there was found to be spatially discontinuous and variable in nature. Gas migration pathways ranged from relatively small, high-flux points (due to fault intersections), to more diffuse, low-flux areas (represent large brittle fracture zones).

Seismic data analysis suggests that the possibility of rapid formation of hydrate in the GC region from migrating gas decoupled from upwelling liquid that consumes water while excluding salts exists.

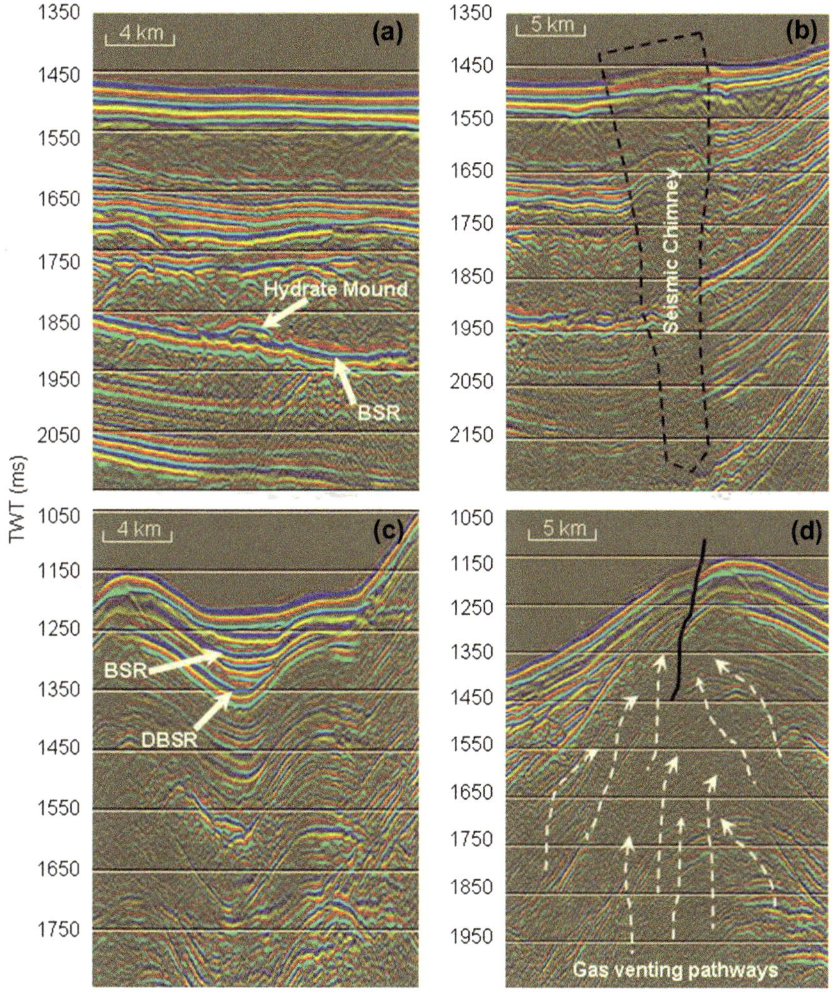

FIGURE 6.5

Detailed seismic reflection sections from three boxed areas of Fig. 6.4 and an observed steep fault. Steep amplitude striations in and beneath vent structure "B" are fractures referred to in the text. Seismic pull-up at feature "A" indicates the accumulation of hydrate as a hydrate mound. At structure "C", a DBSR is observed. (a) Corresponds to feature "A" in Fig. 6.2. (b) Corresponds to feature "B" in Fig. 6.4, representing a seismic chimney. (c) Corresponds to feature "C" in Fig. 6.4. (d) Shows the vertical gas venting pathways though a steep fault from the Green Canyon in the Gulf of Mexico, offshore USA. White arrows represent the migration of free gas through this structure as independent feature. TWT, Two way Time

This could change the local stability conditions until hydrate coexists with liquid and gas and free gas can escape through the GHSZ to the ocean. Near the seabed, because of the large salinity gradient, the anomaly could be reduced by upward diffusion, which could enable further formation of hydrate at or near the seafloor. The interpretation shows that the physical characteristics of various gas venting structures control their source and distribution in the subsurface and their eventual linking into the seabed.

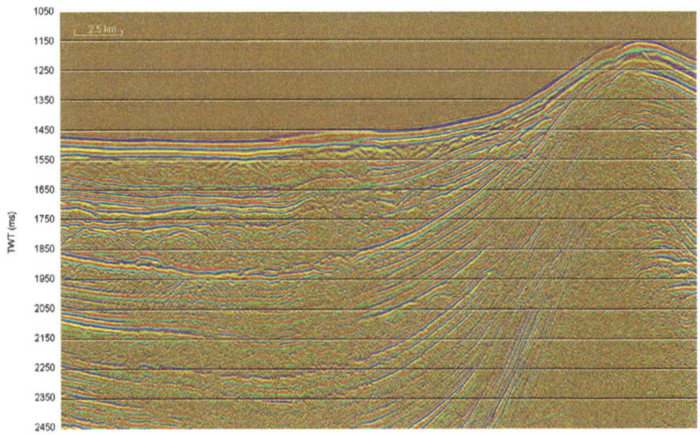

FIGURE 6.6

Uninterpreted time migrated seismic image from a part of the Green Canyon region, offshore USA.

Gas hydrate in the GC area occurs near the seafloor, indicating that gassy fluids move into the GHSZ. Large faults are clearly observed in the study area and one example is shown in Fig. 6.5d. The presence of the fault is very important for the vertical migration of free gas. Vertical gas venting pathways though faults are inferred (Fig. 6.5d). Seismic interpretation indicates that the fault serves as a conduit for gas migration in an upward direction.

The uninterpreted (Fig. 6.6) and interpreted seismic sections (Fig. 6.7) indicate that chimneys, consisting of vertical zones of disrupted stratal reflections, appear to be rooted within or below highly reflective strata underneath the BSR. This section from the GC shows that the upper internal amplitude anomalies have apparently positive polarity compared to the seafloor reflection, whereas the lower anomaly shows reversed polarity (indication of a BSR). This may indicate the presence of locally high concentrations of gas hydrate near the seafloor. The internal reflection of the seismic chimney appears highly disrupted. The focused gas flow within the conduits could occur through the cracks initiated by natural hydraulic fracturing (e.g., Morley, 2003). Hydrofracturing may occur when pore pressure exceeds the minimum horizontal stress and the tensile strength of the host sediment (Hubbert and Willis, 1957). Seismic data typically reveal these venting sites as near-vertical distorted zones with low reflectivity, and are termed acoustic chimneys or pipes. The seismic chimney observed over the GC area comprises dimmed reflection with variable continuity. The transition from the dimmed reflections and the outside stratigraphic reflections becomes more diffuse with depth. Therefore it can be said that seismic chimneys, once created, represent long-term permeability structure.

6.3 UNCERTAINTY ANALYSIS FOR BSR

To investigate the detailed variation in BSR's seismic reflection characteristics, seismic data from the KK basin, India and GC basin, Gulf of Mexico, offshore USA were studied. For anisotropic media, the reflection coefficients of the BSR have been estimated by the formulation for P-waves and SV-waves in vertically transverse isotropic (VTI) medium given by Rueger (1996). These equations are good

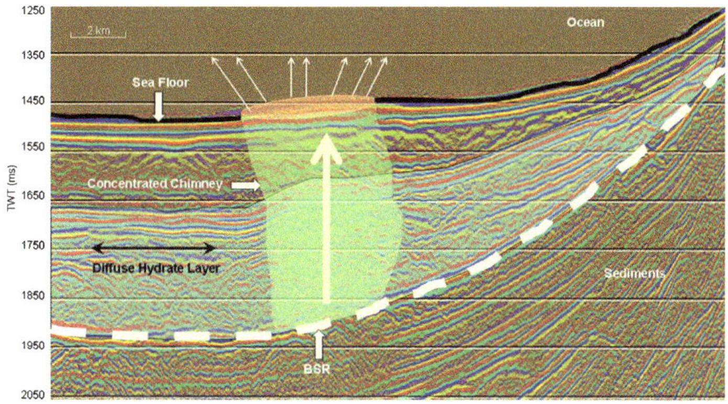

FIGURE 6.7

Interpreted time migrated seismic image from a part of the Green Canyon region, offshore USA. This represents hydrate formation and gas venting through vertical structures termed seismic chimneys. The thick white dashed line marks the BSR, which varies from ~ 1500 to 1950 ms two way time (TWT).

approximations to the exact plane wave reflection coefficients for precritical angles of incidence in the context of weak anisotropy. The gas hydrate seismic response was modeled using velocity estimations from the KK basin (Fig. 6.2) and GC basin (Fig. 6.4). For realistic calculations, a reasonable amount of noise is included and compared with the real data. The main focus of this analysis was on gas hydrate reservoirs with relatively high concentrations of hydrates.

6.3.1 MODELLING

Before describing the detailed modeling approach, a brief description of anisotropy is presented:

- The words "isotropic" and "anisotropic" describe the properties of materials and minerals and contain the element of direction in their description.
- Isotropic is independent of direction while anisotropic refers to the properties of a material that is dependent on the direction.
- Anisotropy is responsible for the largest variation in seismic velocities; changes in the preferred orientation of mantle minerals, or in the direction of seismic wave, cause changes in the velocity that can be accounted for by change in temperature, composition, or mineralogy.
- Almost 30 years ago in 1986, Leon Thomsen's classic paper (Thomsen, 1986), Weak elastic anisotropy was published in "Geophysics". This paper introduced three parameters that are needed alongside the compressional velocity (Vp), Shear wave velocity (Vs) and Density (ρ) to explain anisotropy. They are δ (delta), ε (epsilon), and γ (gamma), collectively referred to as Thomsen's parameters.
  - **δ** (Delta) (Thomsen, 1986),
    - It captures the relationship between the velocity required to flatten gathers (the normal move out (NMO) velocity) and the zero-offset average velocity as recorded by checkshots. It's easy to measure, but perhaps hard to understand in physical terms. This is the short offset effect.

- **ε (epsilon)** (Thomsen, 1986),
  - According to Thomsen it is "the fractional difference between vertical and horizontal P velocities; i.e., it is the parameter usually referred to as 'the' anisotropy of a rock". Unfortunately, the horizontal velocity is rather hard to measure. This is the long offset effect.
- **γ (gamma)** (Thomsen, 1986),
  - This relates a horizontal shear wave with horizontal polarization to a vertical shear wave. Colin Sayers describes "γ can be determined in a single well using sonic. So the correlation with ε and δ is of great interest."

In this example the velocities of water and sedimentary layers vary from 1500 m/s to 3200 m/s for the KK basin and the GC region. The velocities of gas hydrates and free gas vary from 1950 to 2060 m/s and 1400 to 1600 m/s, respectively. Single BSR cases are analyzed from the KK basin and GC region seismic data and for the double BSR (DBSR) only the GC seismic data are used. The model parameters correspond to the estimated velocities estimated for the KK (Fig. 6.2) and GC regions (Fig. 6.4). In modeling the VTI case, the velocities of vertically travelling waves differ from horizontal velocities. In vertical transverse isotropy, the horizontal velocity is different from the vertical velocity. In this example γ values are not considered as this affects only horizontally polarized shear waves. To calculate the reflection coefficient of the BSR in a VTI medium the values of anisotropic parameters ($\varepsilon = 0.06$ and $\delta = 0.05$) for the gas hydrate layer, and ($\varepsilon = 0.17$ and $\delta = 0.22$) for the free gas layer are considered. These are the typical values obtained for gas hydrates and free gas respectively from a walkaway vertical seismic profile experiment for the Blake Ridge region (Pecher et al., 2003). First, the reflection coefficient is calculated for real data (GC region from USA and KK basin from offshore India); second, the reflection coefficients for modeled data for the GC and KK basins have been estimated. To this end the reflection coefficients are plotted for real and modeled data for correlation (Figs. 6.8, 6.9, and 6.10). Uncertainties (mismatch) in the real and modeled reflection coefficient have been estimated by probability distribution and found under admissible limit.

To assess the reliability of AVO analysis in marine gas hydrate studies, the AVO response of BSRs is modeled. The scattering in the amplitude observed for both real and modeled BSRs (single and double BSR) appears to be caused by the vertical venting structures, which are directly related to the "plumbing mechanism" and may help explain the seepage of marine gases into the atmosphere. The uncertainty estimation in the AVO response is in an acceptable range. The maximum uncertainty regions correspond to a large number of the gas vent structures whereas minimum uncertainty is observed when fewer venting structures are seen.

At this stage, to study the relationship between amplitude scattering and venting structures, the reflection coefficient of BSR was calculated at a location further south from the GC study area; a continuous BSR and no gas venting structure (Fig. 6.11) were identified on seismic data. The reflection coefficient showed a negative trend and a good fit for the data points was obtained.

This exercise was repeated at a location where gas venting structure was observed together with relatively low-amplitude strength BSR (Fig. 6.12). The reflection coefficient of a continuous BSR (Fig. 6.11b) does not show amplitude scattering data whereas the reflection coefficient of BSR where gas venting structure is evident (Fig. 6.12b) represents amplitude scattering. Studying the behavior of the AVO trend at these two locations (Figs. 6.11 and 6.12), it can be said that the amplitude scattering corresponds to the gas venting structure. The underlying zones of reduced reflection amplitude indicate the presence of gas, which migrates upward through near-vertical conduits to feed the vent structure. The local geology in the GC area and underlying plumbing system indicate a high flux of gases migrating through the region.

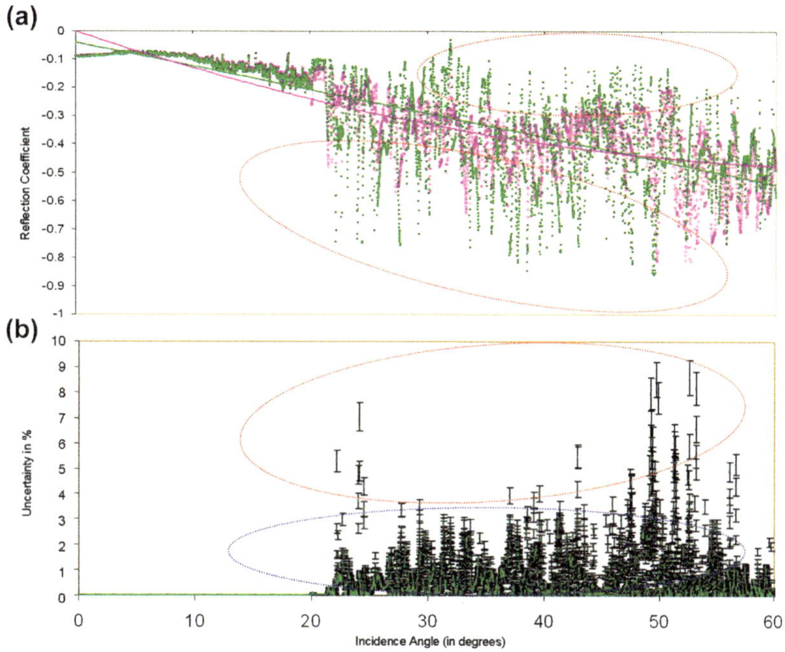

FIGURE 6.8

(a) Reflection coefficient vs. offset of the modeled BSR using the estimated seismic velocities (pink dots) and the real BSR identified in the Kerala-Konkan basin, offshore India (green dots). The reflection coefficients of the real and modeled BSR show a negative trend. The least square polynomial fit for real (green line) and modeled (pink line) reflection coefficients of the BSR correlate well. Dotted ovals show the scattered data points that are caused by structural elements discussed in the text. (b) Estimated uncertainty by probability distribution in the real and modeled values of the reflection coefficients of the BSR. The maximum uncertainty is 5%. The blue dotted oval shows that the majority of data points have less than 2% uncertainty. The red dotted oval shows those data points with uncertainty between 2% and 5%.

6.3.2 SOME THOUGHTS

The interpretation of the seismic reflection data from the KK and GC basins indicates that dim zones up to >5 km wide above the GHSZ demonstrate relatively high velocities (compared to the free gas) produced by the higher concentration of gas hydrates. Owing to active vertical migration of gas to the seafloor under the stability conditions, thermogenic gas hydrates occur within the GC region. Huge amounts of thermogenic gas hydrates occur in association with the outlets of hydrocarbon vents. The second example from the GC region (Fig. 6.4) shows high velocities attributed to gas hydrate just beneath the seabed and it appears that the gases vent directly into the ocean. The example shows a high influx of upwelling gases connected to the deeper source zone of highly concentrated gas through a discrete network of steep fractures and faults. Gas hydrate is also rapidly precipitated in seafloor experiments using natural vent gas as the starting material. One of the experiments of this type is conducted by using thermogenic hydrocarbon gases as the starting material that vent to the water column in association with the seafloor mounds of natural gas hydrate (Sassen and Macdonald, 1997). The seismic image from the hydrate mound and DBSR regions (Fig. 6.4 and 6.5a) indicates that these structures are

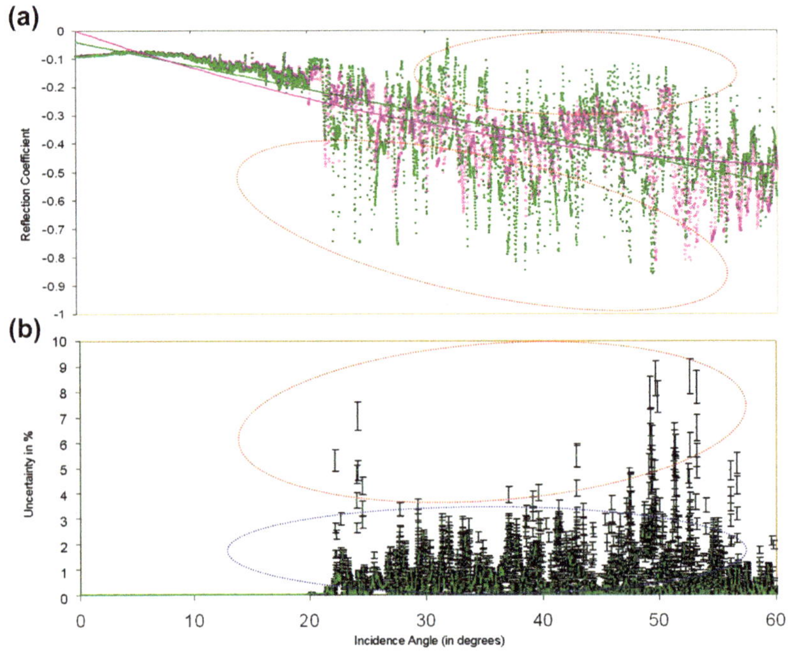

FIGURE 6.9

(a) Reflection coefficient vs. offset of the modeled single BSR (pink dots) using the estimated seismic velocities from Fig. 6.3 and the real BSR (green dots) identified in the Green Canyon region offshore USA, showing a negative trend. The least square polynomial fit for real (green line) and modeled (pink line) reflection coefficients of the BSR correlate well. Dotted ovals show the scattered data points that are caused by structural elements discussed in the text. (b) Estimated uncertainty in the real and modeled values of reflection coefficients of the BSR. The maximum uncertainty is 9%. The blue dotted oval shows that the majority of data points have less than 3% uncertainty. The red dotted oval shows those data points with uncertainty between 3% and 9%.

the result of a low flux of gas input from a diffuse region of the underlying structural elements. In both of the examples, the free gases migrated from deeper structures to feed the hydrate vents. There is a strong possibility of a salinity effect driven by the migrating gases as one of the principal mechanisms, allowing free gases to escape through the GHSZ in the high flux vent. In the mound and DBSR structures of low flux, however, the gases escape through the seabed as a dissolved phase. The results of seismic data analysis and modeling reveal high and low flux styles of gas venting from adjacent, local systems several kilometers apart that appear to be good examples of "seepage-to-leakage" and could help in explaining the complex plumbing system.

AVO analysis shows that the seepage-to-leakage mechanism appears to be linked to the amplitude variations of BSRs. The local scattering of the BSR seismic amplitudes is probably due to structural elements (vertical discrete fracture network, faults, vertical venting structures, hydrate mounds, etc.) that are causing the seepage of underlying gases. Modeled BSR reflection coefficients in a realistic anisotropic medium can explain the variations observed in real data. This suggests that anisotropy is also contributing to the complexity of the seepage-to-leakage plumbing

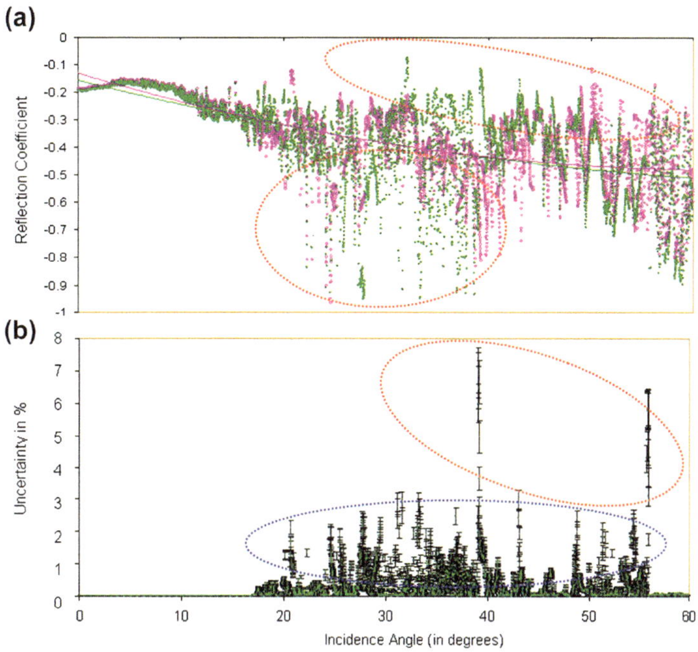

FIGURE 6.10

(a) Reflection coefficient vs. offset of the modeled DBSR (pink dots) using the estimated seismic velocities from Fig. 6.3 and the real DBSR (green dots) identified in the Green Canyon region offshore USA, showing a negative trend. The least square polynomial fit for real (green line) and modelled (pink line) reflection coefficients of the DBSR correlate well. Dotted ovals show the scattered data points that are caused by structural elements discussed in the text. (b) Estimated uncertainty in the real and modeled values of reflection coefficients of the DBSR. The maximum uncertainty is 8%. The blue dotted oval shows that the majority of data points have less than 3% uncertainty. The red dotted oval shows those data points with uncertainty between 3% and 8%.

system. The lateral variability in the reflection strength may be due to the variation in hydrate concentration or due to the variation in the amount of free gas beneath the hydrate. The hydrate may be distributed laterally in a discontinuous or disseminated layer. The uncertainty estimates are quite low. More sophisticated experiments from 4C data, permanent monitoring systems, and research submarine platforms could significantly enhance our understanding of thermogenic gas hydrate formation in the deep sea.

6.4 COLD SEEP PLUMBING SYSTEM

More than 30 years after the discovery of cold seeps, the complex plumbing system that feeds deep-water seeps remains poorly explained (Hovland and Judd, 1988; King and MacLean, 1970; Levin, 2005; Paull et al., 1984; Van Dover et al., 2003). Methane seeps liberate carbon from methane hydrate

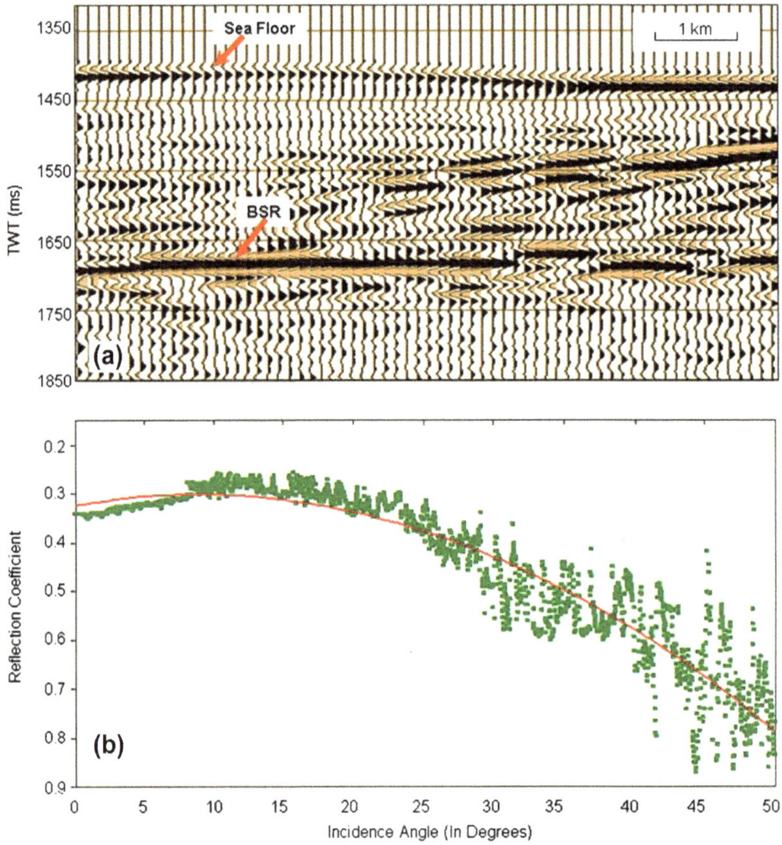

FIGURE 6.11

(a) Time migrated seismic section from the southern part of the Green Canyon region where we observed a continuous BSR, which is showing the negative polarity with respect to seafloor. (b) Reflection coefficient of BSR showing a sharp negative trend without much amplitude scattering.

reservoirs and often sustain exotic seafloor communities (Hornbach et al., 2007). Pockmarks, craters, seeps, and mounds are not the only features that are associated with fluid expulsion. Gas hydrates, mud volcanoes, and gas chimneys are also commonly documented at seafloor fluid escape sites (Virs, 2015). Submarine cold seeps refer to the seepage and venting of fluids involving gas, water, and sediments. Seafloor sites of methane-rich fluid expulsion may be associated with hydroacoustic flares (Greinert et al., 2006; Solomon et al., 2009; Law et al., 2010), pockmark depressions (Pilcher and Argent, 2007; Chand et al., 2009), mud volcanoes (Kopf, 2002; Van Rensbergen et al., 2005; Jerosch et al., 2007; Savini et al., 2009), hydrate mounds (Van Dover et al., 2003; Haeckel et al., 2004), methane-derived authigenic carbonate deposits producing mounds, slabs and/or chimneys (Ritger et al., 1987) and chemosynthetic biological communities (Kulm et al., 1986). Cold seeps are found in sediments of both passive continental margins and subduction zones, in association with

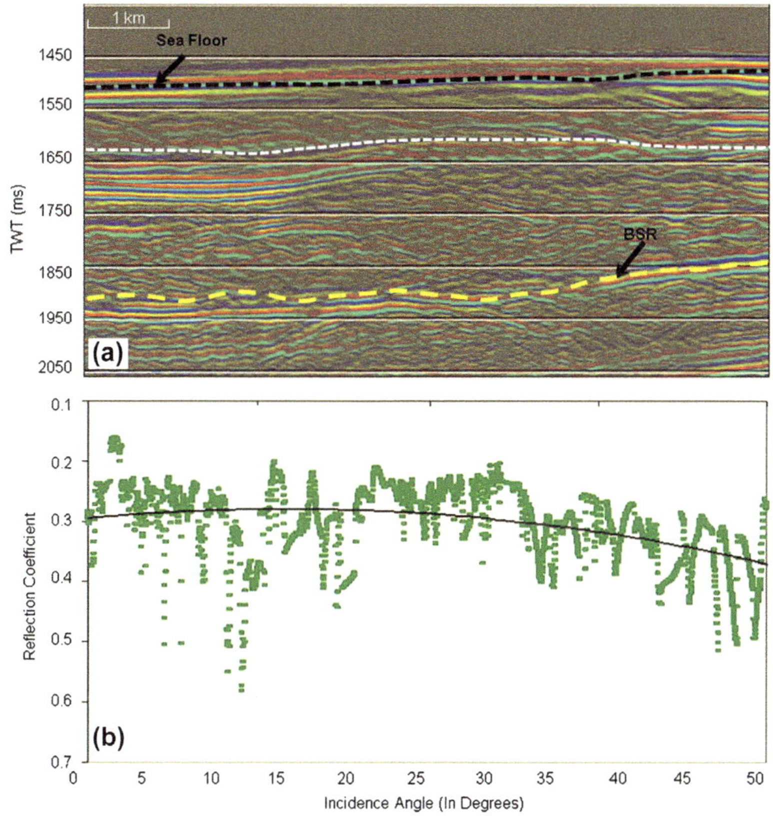

FIGURE 6.12

(a) Time migrated seismic section over a gas venting structure from the Green Canyon region. (b) Reflection coefficient of BSR showing a negative trend with amplitude scattering.

subsurface hydrocarbon reservoirs. The migration of hydrocarbon-rich seep fluids is driven by a variety of geophysical processes, such as plate subduction, salt diapirism, gravity compression, or the dissociation of methane hydrates. Furthermore, the complex "plumbing system" covers the area from the source to the seabed in the water column, where fluid expulsion, venting, and seeping are present (Talukder, 2012). Cold seeps are shallow areas on the ocean floor where gases percolate through underlying rock and sediment layers and emerge on the ocean bottom. The gases found in the seep are methane and sulfur-rich gases and sediments releasing petroleum. They are places where hydrocarbons – mostly methane but also ethane, propane, or even oil – seep from the sediment. Unlike at hydrothermal vents, the emanating fluids (gases and liquids) are no hotter than the surrounding seawater, and they are not necessarily under high pressure. Cold seeps are rich in methane and support large and diverse communities sustained by the primary productivity of chemosynthetic bacteria; methane is taken up by the gills of the dominant mussels (*Bathymodiolus*), where it is oxidized by intracellular bacteria to release energy. The term "cold seep" often refers only to the seeping or venting structures on the seabed. These systems are complex in nature and comprise the following three structural elements (Talukder, 2012):

- Source of the fluid
- Plumbing system
- Venting structures or seeping features

To understand the cold seep plumbing system in light of the above elements, the following points need to be considered:

- Mechanical properties of the sediments
- High-pressure source
- Upward migration through venting structures:
  - Tectonic elements
    - Faults and fractures
  - Overburden erosion
  - Seismological activities
  - Subduction zones

In the past plumbing system have typically been studied by two-dimensional seismic data. The advancement of three-dimensional seismic acquisition and improved processing algorithms have helped in the discovery of new class of fluid conduits known as gas venting structures or pipes, which are very narrow vertical fluid conduits. The pipe structures, chimneys and polygonal faults are features that allow fluids to migrate upwards. The examples of gas pipes and gas could identified on seismic section from eastern Indian offshore is shown in Figure 6.13. Pipes were commonly ignored in two-dimensional seismic because the vertical to sub vertical pipe geometry was easily confused with seismic artifacts such as migration anomalies, scattering artifacts, lateral velocity anomalies and attenuation variations related to shallow diffractions (Talukder, 2012).

Gas clouds and gas chimneys/pipes have often been considered as a source of seismic noise that degrades the quality of seismic reflection events. Much effort has been devoted to filter out the impact of gas clouds and provide interpretable sections by imaging through them. The geological features associated with cold seep plumbing system can be structural (faults and fractures) or stratigraphical which include bedding planes along the crest of folds and sedimentary ridges, erosional surfaces at the base, flanks and margins of canyons, paleo canyons and near side scraps, entrapped high permeable layers such as buried channels (Talukder, 2012). The anaerobic oxidation of methane is one of the post-genetic alteration processes which can affect systems of the different fluid venting structures that occur worldwide. Furthermore, most of the produced methane in marine sediments is consumed by the anaerobic oxidation of methane (AOM) (Knittel and Boetius, 2009). Aerobic methanotrophy is conducted by groups of methanotrophic bacteria, instead of methanotrophic archaea and sulfate reducing bacteria described for anoxic conditions (AOM).

6.4.1 PLUMBING MECHANISM

Fluid migration in sedimentary basins is an important process as the input of greenhouse gases into the atmosphere and ocean system is an important element of the atmospheric carbon budget and the fluid expulsion at the seafloor has a significant role to play in seafloor stability. There may be several

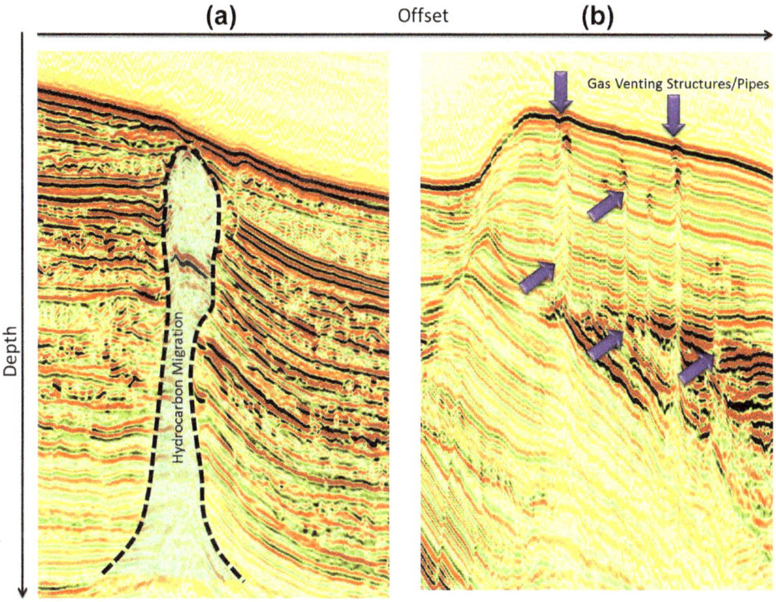

FIGURE 6.13

Example of gas venting structures. (a) Gas cloud; (b) gas is migrating through pipes/chimneys, as highlighted by solid violet arrows.

pathways for fluid flow and that can include pockmarks on the seafloor, faults and fractures, mud and diapirs, and fluid injection. Pockmarks, or seafloor craters, occur worldwide in a variety of geological settings and are often associated with fluid discharge and are among the most enigmatic mud landforms in the world. Pockmarks are identified during offshore hydrocarbon exploration and scientific surveys in various depositional systems at water depth ranging from 25m to >3000m. King and MacLean (1970) considered that "the crater-like nature of the pockmarks strongly suggests that they are erosional features". After discussing various possible mechanisms, they concluded that the association with the underlying coastal plain sediments suggested a link and surmised that water or gas rising from these sediments (or underlying coal-bearing Upper Carboniferous strata) to the seabed was the most likely cause or agent. Pockmarks generally appear in unconsolidated, fine grained sediments as cone-shaped circular or elliptical depressions, ranging from a few meters to 300 m or more in diameter and from 1 m to 80 m in depth, and they concentrate in fields extending over several square kilometers. Sea floor venting in active fluid-escape zones including the areas of hydrocarbon production, hydrate ridges, deltas, and tectonically active areas. There are evidences of pockmark occurrences in estuaries (Wildish et al., 2008), fjords (Webb et al., 2009) and lakes (Manley et al., 2004) exhibiting minimal fluid venting. Pockmarks generally interact with flow. One of the world's biggest pockmark is detected offshore New Zealand in 2013 (Figure 6.14). These pockmarks are formed by seafloor eruptions of gas and fluid. The most common pockmarks probably are standard circular and elliptical. For example, on the slope of the Norwegian Trench the axes are arranged parallel to the slope, signifying that there is a relationship between pockmark shape and slope. Asymmetric pockmarks on side-scan sonar records look as if they

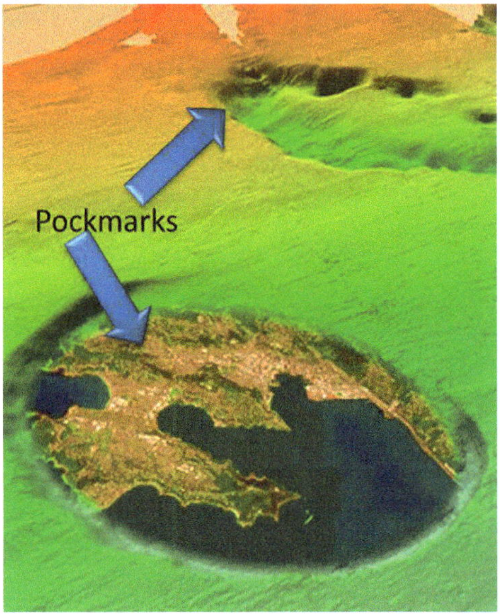

FIGURE 6.14

Example of Pockmark that was discovered offshore New Zealand. (Source: http://strangesounds.org/2013/04/words- biggest- craters- formed –by seafloor-eruptions -pockmarks-found-in-waters-off-new-zealand.html)

have a diverse and very long 'tail' and on one side they show strong back wall reflection. Elongated pockmarks remind of gullies or troughs. Unit pockmarks are very small seabed depressions found in isolation, in groups and in association with larger pockmarks. Composite pockmarks occur in places where normal pockmarks fuse in one another, later some of them can form into complex shapes. The pockmarks are situated at a depth of about 1,000 meters on the seafloor of the Chatham Rise, about 500 km east of Christchurch. The three giant pockmarks, the largest measuring 11 km by 6 km in diameter and 100 meters deep, are possibly twice the size of the largest pockmarks recorded in scientific literature. As you can see in the first image, the largest pockmark is about the size of Wellington, New Zealand. Pockmarks are formed by the expulsion of fluids from the seabed. The escaping fluid may be liquid (e.g. groundwater, as in Eckernförde Bucht, Germany), but in the majority of cases it is gas. The gas may originate from either the microbial decomposition of organic matter within the near-seabed sediments ('microbial', 'bacterial' or 'biogenic' gas), or from the thermocatalytic destruction of kerogens deep within the sediments ('thermogenic' gas). Typically the gas is primarily composed of methane (CH_4) which is commonly present in concentrations of >95%. The higher hydrocarbon gases (ethane, propane, butane and pentane) may also be present if the source is thermogenic (Judd, 2001).

The formation of this fluid venting feature is dependent on fluid escape capacity of sediments. These are formed in suitable environment including soft, silty clays; if there is no suitable environment is present there may be no morphological feature as evidence of fluid escape. In erosional environment the evidence of pockmark activity may be vanished. The nature of the leakage to seepage mechanism or plumbing mechanism is dependent on the stress environment within the sediments. Plumbing systems and chimney structures are different in their characteristics and in their sediments.

The style of fluid migration and seabed leakage is determined by the integration of different information. Key structures that act as a gateway for gas and gaseous fluids are called pockmarks. Faults, fractures, sedimentation rate, permeability, compaction, overpressure generation, gas hydrates, free gas and source are features that are relevant to the plumbing system and the functioning mode of a chimney. Structural changes could open new migration pathways within the seep plumbing system. Seismic activity would exert an important influence in opening new meshes of faults and fractures which would enhance fluid flow. Changes in seep plumbing system resulting from blockage of seepage pathways, seismically induced structural changes, exposure by erosion or burial by deposition would tend to irreversibly alter seep distribution. Plumbing system can be controlled by two fluid flow mechanisms:

1. *Structural fluid flow*
   a. Structural fluid flow can be described through faults and fractures. Faults are classified according to the kind of motion that occurs on them e.g. in Joints there is no movement, in Strike-slip horizontal Motion and Dip-Slip has vertical motion. Due to tectonic stress including subduction or diapirism faults are formed, which induce the formation of physically week pathways where fluid may arise. Fault zones and fault systems have a key role in the development of the Earth's crust. They control the mechanics and fluid flow properties of the crust, and the architecture of sedimentary deposits in basins (Faulkner et al., 2010). In the faulting process, rock is comminuted and clay is mixed in with sand, producing various types of fault rock with different magnitudes of permeability reduction (Talukder, 2012). As with top seals, the capillary capacity of a fault seal relates to pore throat size rather than to permeability. Faults are important controls on hydrocarbon migration and ore mineralization and, in areas of active deformation, are the most important source of seismic hazard (Kurz et al., 2008). However, faults are rarely discrete surfaces and the internal structure of fault zones (e.g., the thickness, nature and continuity of the fault rocks, the distribution and segmentation of slip surfaces, and the orientation, distribution and connectivity of subsidiary faults and fractures) is a key control on their bulk fluid flow and mechanical properties. The fluid flow in faults is controlled by dialatancy compaction, mechanical properties, segmentation roughness, structural consideration and permeability. All these processes as highlighted in figure 6.15 are important to understand the fluid flow mechanism in faults.
      The question of how faults affect the fluid flow and migration in hydrocarbon reservoir is challenging as many faults contribute significantly to formation permeability and allow the fluids migration between different reservoir units. Some faults act as effective barriers separating distinct reservoir compartments. In many cases critically stressed faults are responsible for promoting hydrocarbon leakage and migration. Stress, pore pressure, and fault orientation appear to be important factors in controlling hydrocarbon leakage and migration. Some important highlights for fault related fluid flow are given here under:
      i. Fault zones are ubiquitous structures throughout the Earth's crust and as a fault evolves it can significantly influence the rheological and hydrological properties of the crust. These are important geological structures for fluid, energy, and solute transport.
      ii. Structure, mechanics and fluid flow properties of faults and faults system are inextricably coupled as highlighted in Figure 6.14. Understanding the heterogeneity and along fault variability is till challenging.

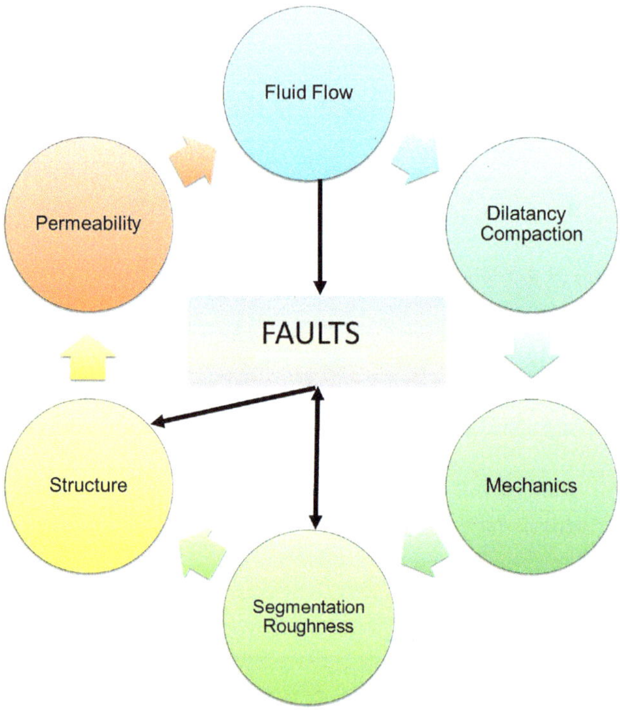

FIGURE 6.15

Fluid flow mechanism in faults is controlled by dialatancy compaction, mechanical properties, segmentation roughness, structural consideration and permeability.

iii. The impact of new 3D seismic datasets has the potential to greatly enhance our understanding of fault growth by providing a detailed view of growth strata that are intimately associated with the growth of faults and development of fault systems.
iv. Structural faults can act as seal to migration as well as conduits
v. Fault zone have varying permeability. High permeability faults, acting as preferential pathways for fluid migration.
vi. Fluid flow through fault zones is typically associated with fault-fluid-rock interactions and these interactions can affect the mineralogy, strength and evolution of a fault zone. Fluid flow around faults is dictated by 3D fault zone structure
vii. The changing fluid pressure determines whether a fault remains a conduit for fluid or becomes barrier to it, hence determining the temporal evolution of the seeps.
viii. There could be several heat and fluid flow regime can be recognized at fault plane including conductive, advective, steady convective and unsteady convective.
ix. Abundant vein material and associated alteration in fault zones gives clear evidence of enhanced fluid flow along faults. Estimates of fluid-rock ratio can yield minimum fluxes.
x. Up-fault fluid migration can be locally enhanced at fault irregularities

xi. In structural fluid flow regime, primary control on flow rates is controlled by fault zone geometry and connectivity
xii. Geochemical reactions that occur within a fault are controlled in part by the evolutionary stage of the fault, the fault rock permeability and the volume of fluids that pass through the fault
xiii. Factors affecting the fluid and heat flow regimes at fault zones include heat flux, gradients, length, width and depth of the faults, thermal conductivity and areal extent.
xiv. Faults are the most important geological stricture that controls the fluid flow mechanism. Fluid flow in faults is controlled by permeability, fault gouge ratio, clay smearing, length, width and depth of faults, stress field, mechanical properties, reactivation capability, degree of lithification and the evoluation of fluid pressure.

2. *Sedimentary fluid flow*
   a. Through salt diaper, mud diaper, mud volcanoes
      i. A mud diapir is an intrusive structure characterized by a slowly upward migrating mass of clay-rich sediment and fluid discharge (Kopf, 2002). Mud diapirs can be recognized on seismic profile in terms of acoustically amorphous piercement structures. The sedimentary strata on both sides of a mud diapir exhibit onlapping structures due to the uplifting of the diaper
      ii. Some mud diapirs and mud volcanoes have been linked to both extensive and compressive contexts, such as in the Western Alboran Sea area (Talukder et al., 2003)
      iii. High pressure in sedimentary layers due to rapid sedimentation and gas generation and tectonic processes are considered to be the main driving mechanism for the development of mud diapirism.
      iv. Discharge of hydrocarbon bearing fluids can increase the bud buoyancy to facilitate diapiric intrusion.
      v. Mud diapir is an efficient pathway for the vertical migration of thermogenic gas fluid from the deeper part of the sedimentary layers
      vi. Diapirs generally do not reach the seabed but it is possible. If diapirs display evidence of extruding sediment or fluid, the would be termed as "mud volcano"
      vii. A mud volcano usually occurs above the diapir, as a result of fluid migration directly along the body of the mud diapir or through faults (fractures) connected to the mud diapirs (Milkov, 2000; Kopf, 2002). Mudvolcanos are the most important pathways for methane emission from deep marine sediments into atmosphere (Dimitrov, 2002, 2003). Guliev (1992) defined mud volcanos as "periodic expulsion from deep parts of the sediment cover of mixtures of water, various gases and solid material". Mud volcanoes, similarly to gas chimneys, occur in environments with pressurized fluids beneath the surface.
      viii. Formation of fluid venting structure and mud volcanoes in passive and active continental marging are due to the high sediment accumulation rate and the lateral tectonic compression. Talukder et al. (2007) suggested that mud diapirs are driven by buoyancy forces due to bulk density contrast between an over-pressured muddy mass and an overburden of greater density.
      ix. The common presence of methane in mud volcanoes provides a driving force for hydrogeologic and intrusive systems because of voluminous quantities of fluid, decreasing

mud densities and increasing buoyancy forces in diapirs. The shape of a mud volcano is closely related to the viscosity of the flows. The high viscosity of flows causes a dome-like cone with a steep slope

x. Methane is a greenhouse gas and considered to play a significant role in the mechanism of mud dispirsm and volcanism as above the mud volcanoes the concentration of methane is observed very high.

xxv. The mechanisms contributing to the formation of the mud diapirs and mud volcanoes are not only due to the existence of overpressure layers and tectonic compressional forces but also due to the gas-bearing fluids in many areas ie.g. Offshore SW Taiwan (Chen et al., 2014)

3. The plumbing system is complex with changing conduit systems, which make the fluid flow highly transient and spatially and temporally variable (Talukder, 2012). Geological structures that are associated with seeps include faults and fractures, bedding planes, diapirs and sand intrusion. Pockmarks, mud diapirs, mud volcanoes, pipes, chimneys, are the indicators of gas escape events but not all the pockmarks are actively seeping hydrocarbons. Seepage to leakage can happen with or without pockmarks. Fault reactivation and hydrocarbon leakage can be associated with locally activated pore pressure due to byoyant hydrocarbon abutting faults, orientation of faults that are nearly optimally oriented for frictional slip in the current stress field and recent perturbation of the compressional stress. The combination of these elements can induce slippage and hydrocarbon leakage.

6.5 INJECTED SANDS AND FLUID FLOW

Fluid follows different pathways and generates diverse geological structures as highlighted in Fig. 6.16. The occurrence of fluid seepage on the sea floor is manifested by the presence of hydrothermal vents (metal-rich deposits) and cold seeps (typically carbonate deposits). Cold seeps are associated with carbonate mounds, coral reefs, pockmarks, mud volcanoes and seamounts (Mazzini et al., 2007). Zoned cement in seeps can offer the understanding of geochemical properties of the cementing fluid. Sometime it is difficult to identify the cross-cutting features between cemented fractures and sand injectities which suggest rapid cementation following fracturing and sand injection.

Following sand injection, hydrocarbons continued to migrate into and saturate the injected sandstones, as indicated by the oil inclusions in their carbonate and quartz cements and within the deformation bands. The relationship between hydrocarbons and injected sandstones assigns a key role in soft-sediment deformation. Fluid overpressure in the reservoir causes rapid fluid escape through hydraulic fracturing and sand injection. The newly created fluid flow pathways are quickly cemented. Once cemented, fluid escape can continue to take place through seepage at the top of the structures (Figure 6.16).

Fluid escape from reservoirs can occur significantly in a shorter time span through hydraulic fracturing and sand injection but continued fluid escape can take place over a prolonged time through seepage as supported by fluid inclusion. The relationship between hydrocarbons and injected sandstones assigns a key role in soft-sediment deformation. This relation has also great economic significance, because injected units may comprise considerable reserves and become production targets during development phase.

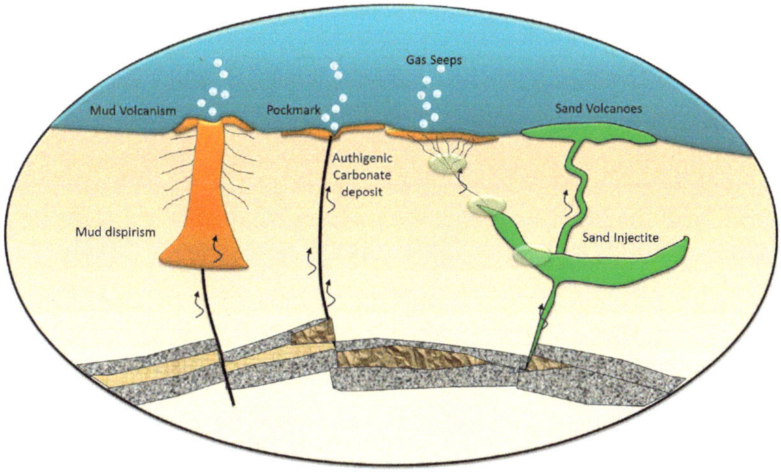

FIGURE 6.16

Schematic cross section showing a range of features through which fluids can escape from a reservoir: mud diapir structures eventually reaching the sea floor as mud volcanoes where chemosynthetic communities can occur; dense networks of faults and fractures act as conduits for seepage (conventional cold seeps) or can be used by sand injection and pockmarks. Injected sandstones can reach the sea floor or can terminate at various depths eventually allowing fluids to reach the sea floor through fractures and seepage. Light green ovals correspond to carbonate cemented features (Source: modified from Mazzini et al., 2007).

REFERENCES

Andreassen, K., Hogstead, K., Berteussen, K.A., 1990. Gas hydrate in the southern Barents Sea, indicated by a shallow anomaly. First Break 8 (6), 235–245.

Archer, D., 2007. Methane hydrate stability and anthropogenic climate change. Biogeosciences 4, 993–1057. http://dx.doi.org/10.5194/bgd-4-993-2007.

Bangs, N.L., Musgrave, R.J., Tréhu, A.M., 2005. Upward shifts in the south Hydrate Ridge gas hydrate stability zone following post-glacial warming offshore Oregon. J. Geophys. Res. 110, B03102. http://dx.doi.org/10.1029/2004JB003293.

Behura, J., Tsvankin, I., 2009. Reflection coefficients in attenuative anisotropic media. Geophysics 74 (5), WB193–WB202.

Boe, R., Rise, L., Ottesen, D., 1998. Elongate depressions on the southern slope of the Norwegian trench (Skagerrak): morphology and evolution. Mar. Geol. 146, 191–203.

Buffett, B.A., Zatsepina, O.Y., 1999. Metastability of gas hydrate. Geophys. Res. Lett. 26, 2981–2984.

Castagna, J.P., Backus, M.M., 1993. Offset-dependent Reflectivity - Theory and Practice of AVO Analysis. Society of Exploration Geophysicists, Tulsa, OK.

Chand, S., Rise, L., Ottesen, D., et al., 2009. Pockmark-like depressions near the Goliat hydrocarbon field, Barents Sea: morphology and genesis. Mar. Petrol. Geol. 26, 1035–1042.

Chen, S.C., Hsu, S.K., Wang, Y., et al., 2014. Distribution and characters of the mud diapirs and mud volcanoes off southwest Taiwan. J. Asian Earth Sci. 92, 201–214.

Clennell, M.B., Hovland, M., Booth, J.S., Henry, P., Winters, W.J., 1999. Formation of natural gas hydrates in marine sediments 1. Conceptual model of gas hydrate growth conditioned by host sediment properties. J. Geophy. Res. 104. http://dx.doi.org/10.1029/1999JB900175 issn: 0148–0227.

Dimitrov, L., Woodside, J., 2003. Deep sea pockmark environments in the eastern Mediterranean. Mar. Geol. 195 (1–4), 263–276.
Dimitrov, L.I., 2002. Mud volcanoes: the most important pathway for degassing deeply buried sediments. Earth Sci. Rev. 59, 49–76.
Dimitrov, L.I., 2003. Mud volcanoes: a significant source of atmospheric methane. Geo. Mar. Lett. 23, 155–161.
Ecker, C., Dvorkin, J., Nur, A., 1998. Sediments with gas hydrates: internal structure from seismic AVO. Geophysics 63, 1659–1669.
Eichhubl, P., Greene, H.G., Naehr, T., Maher, N., 2000. Structural control of fluid flow: offshore fluid seepage in the Santa Barbara Basin, California. J. Geochem. Explor. 69 (70), 545–549.
Faulkner, D., Jackson, C., Lunn, R., Schlische, R., Shipton, Z., Wibberley, C., Withjack, M., 2010. A review of recent developments concerning the structure, mechanics and fluid flow properties of fault zones. J. Struct. Geol. 32, 1557–1575. http://dx.doi.org/10.1016/j.jsg.2010.06.009.
Freeze, R.A., Cherry, J.A., 1979. Groundwater. Prentice-Hall, Inc, Englewood Cliffs, NJ.
Gay, A., Lopez, M., Berndt, C., Seranne, M., 2007. Geological controls on focused fluid flow associated with seafloor seeps in the Lower Congo Basin. Mar. Geol. 244 (1–4), 68–92. http://dx.doi.org/10.1016/j.margeo.2007.06.003.
Golmshtok, A.Y., Soloviev, V.A., 2006. Some remarks on the thermal nature of the double BSR.Marine. Geology 229 (3–4), 187–198.
Gorman, A.R., Holbrook, W.S., Hornbach, W.S., et al., 2002. Migration of methane gas through the hydrate stability zone in a low-flux hydrate province. Geology 30 (4), 327–330.
Greinert, J., Artemov, Y., Egorov, V., De Batist, M., McGinnis, D., 2006. 1300-m-high rising bubbles from mud volcanoes at 2080 m in the Black Sea: hydroacoustic characteristics and temporal variability. Earth Planet. Sci. Lett. 244, 1–15.
Guliev, I.S., 1992. A review of mud volcanism. Translation of the report by Azerbaijan Academy of Sciences Institute of Geology 65 p.
Haacke, R.R., Westbrook, G.K., Hyndman, R.D., 2007. Gas hydrate, fluid flow and free-gas: formation of the bottom-simulating reflector. Earth Planet. Sci. Lett. 261, 407–420. http://dx.doi.org/10.1016/j.epsl.2007.07.008.
Haacke, R.R., Hyndman, R.D., Park, K.P., et al., 2009. Migration and venting of deep gases into the ocean through hydrate-choked chimneys offshore Korea. Geology 37 (6), 531–534.
Haeckel, M., Suess, E., Wallmann, K., Rickert, D., 2004. Rising methane gas bubbles from massive hydrate layers at the seafloor. Geochim. Cosmochim. Acta 68, 4335–4345.
Heggland, R., 1997. Detection of gas migration from a deep source by the use of exploration 3D seismic data. Mar. Geol. 137, 41–47.
Henry, P., Thomas, M., Clennell, M.B., 1999. Formation of natural gas hydrates in marine sediments. Part 2: Thermodynamic calculations of stability conditions in porous sediments. J. Geophy. Res. B 104, 23005–23022.
Holbrook, W.S., Lizarralde, D., Pecher, I.A., et al., 2002. Escape of methane gas through sediment waves in a large methane hydrate province. Geology 30, 467–470.
Hornbach, M.J., Ruppel, C.D., Dover, V., Lee, C., 2007. Three dimensional structure of fluid conduits sustaining an active deep Marine cold seep. Geophys. Res. Lett. 34, L05601. http://dx.doi.org/10.1029/2006GL028859 2007.
Hovland, M., Judd, A.G., 1988. Seabed pockmarks and seepages – impact on geology, biology and the marine environment. Graham & Trotman Ltd, London, p. 293.
Hovland, M., Svensen, H., 2006. Submarine pingoes: indicators of shallow gas hydrates in a pockmark at Nyegga, Norwegian Sea. Mar. Geol. 228, 15–23.
Hovland, M., Gallagher, J.W., Clennell, M.B., Lekvam, K., 1997. Gas hydrate and free gas volumes in marine sediments: example from the Niger delta front. Mar. Petrol. Geol. 14, 245–255 1997.
Hubbert, M.K., Willis, D.G.W., 1957. Mechanics of hydraulic fracturing. Trans. Am. Inst. Min. Eng. 210, 153–168.
Hyndman, R.D., Spence, G.D., 1992. A seismic study of methane hydrate marine bottom simulating reflectors. J. Geophys. Res. 97, 6683–6698.

Jerosch, K., Schlueter, M., Foucher, J.P., et al., 2007. Spatial distribution of mud flows, chemoautotrophic communities, and biogeochemical habitats at Hakon Mosby Mud Volcano. Mar. Geol. 243, 1–17.

Judd, A.G., 2001. Pockmarks in the UK Sector of the North Sea Technical Report TR_002, Strategic Environmental Assessment, SEA2.

King, L.H., MacLean, B., 1970. Pockmarks on the Scotian shelf. Geol. Soc. Am. Bull. 81, 3141–3148.

Knittel, K., Boetius, A., 2009. Anaerobic oxidation of methane: progress with an unknown process. Annu. Rev. Microbiol. 63, 311–334.

Kopf, A., 2002. Significance of mud volcanism. Rev. Geophys. 40, 1005.

Kulm, L.D., Suess, E., Moore, J.C., et al., 1986. Oregon subduction zone: venting, fauna and carbonates. Science 231, 561–566.

Kurz, W., Imber, J., Wibberley, C.A.J., Holdsworth, R.E., Colletini, C., 2008. The internal structure of fault zones: fluid flow and mechanical properties. In: Wibberley, C.A.J., Kurz, W., Imber, J., Holdsworth, R.E., Collettini, C. (Eds.), The Internal Structure of Fault Zones: Implications for Mechanical and Fluid-Flow Properties. Geological Society of London, vol. 299, pp. 1–3.

Kvenvolden, K.A., Rogers, B.W., 2005. Gaia's breath: global methane exhalations. Mar. Petrol. Geol. 22 (4), 579–590.

Law, C.S., Nodder, S.D., Mountjoy, J.J., et al., 2010. Geological, hydrodynamic and biogeochemical variability of a New Zealand deep-water methane cold seep during an integrated three-year time-series study. Mar. Geol. 272, 189–208.

Levin, L.A., 2005. Ecology of cold seep sediments: interactions of fauna with flow, chemistry and microbes. Oceanogr. Mar. Biol. 43, 1–46.

Løseth, H., Wensaas, L., Arntsen, B., Hanken, N., Basire, C., Graue, K., 2001. 1000 m Long Gas Blow Out Pipes. 63rd EAGE Conference And Exhibition. Extended Abstract 524–527.

Manley, P.L., Manley, T.O., Watzin, M.C., Gutierrez, J., 2004. Lakebed pockmarks in Burlington Bay, Lake Champlain: I. Hydrodynamics and implications of origin. In: Manley, T.O., Manley, P.L., Mihuc, T.B. (Eds.), Lake Champlain: Partnerships and Research in the New Millennium. Springer, Berlin/Heidelberg/New York, pp. 299–330.

Mazzini, A., Svensen, H., Akhmanov, G.G., Aloisi, G., Planke, S., Malthe-Sørenssen, A., Istadi, B., 2007. Triggering and dynamic evolution of the LUSI mud volcano, Indonesia. Earth Planet. Sci. Lett. 261, 375–388. http://dx.doi.org/10.1016/j.epsl. 2007.07.001.

Milkov, A.V., 2000. Worldwide distribution of submarine mud volcanoes and associated gas hydrates. Mar. Geol. 167, 29–42.

Morley, C.K., 2003. Outcrop Examples of Mudstone Intrusions from the Jerudong Anticline, Brunei Darussalam and Inferences for Hydrocarbon Reservoirs. Geological Society of London Special Publication No. 216. The Geological Society, London, pp. 381–439.

Musgrave, R.J., Bangs, N.L., Larrasoaña, J.C., et al., 2006. Rise of the base of the gas hydrate zone since the last glacial recorded by rock magnetism. Geology 34 (2), 117–120. http://dx.doi.org/10.1130/G22008.1.

Nimblett, J., Ruppel, C., 2003. Permeability evolution during formation of gas hydrates in marine sediments. J. Geophys. Res. 108, 2420. http://dx.doi.org/10.1029/2001JB001650.

Ostrander, W.J., 1984. Plane wave reflection coefficients for gas sands at non normal angles of incidence. Geophysics 49, 1637–1648.

Paull, C.K., Hecker, B., Commeau, R.F., et al., 1995. Methane-rich plumes on the Carolina continental rise: Associations with gas hydrates. Geology 23, 89–92. http://dx.doi.org/10.1130/0091-7613.

Pecher, I.A., Holbrook, W.A., Sen, M.K., et al., 2003. Seismic anisotropy in gas-hydrate and gasbearing sediments on the Blake Ridge, from a walkaway vertical seismic profile. Geophys. Res. Lett. 30 (14), 1733. http://dx.doi.org/10.1029/2003 GL 017477.

Pilcher, R., Argent, J., 2007. Megapockmarks and linear pockmark trains on the West African continental margin. Mar. Geol. 244, 15–32.

Posevang, J., Mienert, J., 1999. The enigma of double BSRs: indicators for changes in the hydrate stability field. Geo. Mar. Lett. 19, 157–163.

Rajan, A., 2012. Geophysical Characterizations of Fluid Flow and Gas-hydrate Systems of the NW-svalbard and SW-barents Sea Margins PhD Dissertation. University of Tromso, UIT.

Rajput, S., Thakur, N.K., Rao, P.P., Joshi, A., 2010. AVO response for a complex double bottom simulating reflectors model. Curr. Sci. 98 (10).

Riedel, M., Collett, T.S., Malone, M.J., and The Expedition 311 Scientists (2006) Proceedings of IODP, 311: Washington, DC (Integrated Ocean Drilling Program Management International, Inc.). http://dx.doi:10.2204/iodp.proc.311.101.2006.

Ritger, S., Carson, B., Suess, E., 1987. Methane derived authigenic carbonates formed by subduction induced pore water expulsion along the Oregon Washington margin. Geol. Soc. Am. Bull. 98, 147–156.

Rueger, A., 1996. Reflection Coefficients and Azimuthal AVO Analysis in Anisotropic Media PhD thesis. Colorado School of Mines.

Ruppel, C., Kinoshita, M., 2000. Heat, fluid, and methane flux on the Costa Rican active margin off the Nicoya Peninsula. Earth Planet. Sci. Lett. 179, 153–165.

Santoso, D., Alfian, Alam, S., et al., 1995. Estimation of limestone reservoir porosity by seismic attribute and AVO analysis. Explor. Geophys. 26, 437–443.

Sassen, R., MacDonald, I.R., 1997. Hydrocarbons of experimental and natural gas hydrates, Gulf of Mexico continental slope. Org. Geochem. 26, 289–293.

Satyavani, N., Thakur, N.K., Aravind Kumar, N., Reddi, S.I., 2005. Migration of methane at the diapiric structure of the western continental margin of India: insights from seismic data. Mar. Geol. 219, 19–25.

Savini, A., Malinverno, E., Etiope, G., Tessarolo, C., Corselli, C., 2009. Shallow seep-related seafloor features along the Malta plateau (Sicily channel – Mediterranean Sea): morphologies and geo-environmental control of their distribution. Mar. Petrol. Geol. 26, 1831–1848.

Shaw, J., Courtney, R.C., Currie, J.R., 1997. Marine geology of St. George's Bay,Newfoundland, as interpreted from multibeam bathymetry and back-scatter data. Geo. Mar. Lett. 17, 188–194.

Shipley, T.H., Houston, M.H., Buffler, R.T., et al., 1979. Seismic evidence for widespread possible gas hydrate horizons on continental slopes and rises. Am. Assoc. Petrol. Geol. Bull. 63, 2204–2213.

Sloan, E.D., 1998. Clathrate Hydrates of Natural Gases. Marcel Decker, New York.

Solomon, E.A., Kastner, M., MacDonald, I.R., Leifer, I., 2009. Considerable methane fluxes to the atmosphere from hydrocarbon seeps in the Gulf of Mexico. Nat. Geosci. 2, 561–565.

Soter, S., 1999. Macroscopic seismic anomalies and submarine pockmarks in the Corinth-Patras rift, Greece. Tectonophysics 308, 275–290.

Suess, E., Torres, M.E., Bohrmann, G., et al., 1999. Gas hydrate destabilization: enhanced dewatering, benthic material turnover and large methane plumes at the Cascadia margin. Earth Planet. Sci. Lett. 170, 1–15. http://dx.doi.org/10.1016/S0012-821X(99)00092-8.

Talukder, A.R., Comas, M.C., Soto, J.I., 2003. Pliocene to Recent Mud Diapirism and Related Mud Volcanoes in the Alboran Sea (Western Mediterranean). Geological Society London, Special Publications No. 216. The Geological Society, London, pp. 443–459.

Talukder, A.R., Bialas, J., Klaeschen, D., et al., 2007. High-resolution, deep tow, multichannel seismic and sidescan sonar survey of the submarine mounds and associated BSR off Nicaragua pacific margin. Mar. Geol. 241, 33–43.

Talukder, A.R., 2012. Review of Submarine Cold Seep Plumbing Systems: Leakage to Seepage and venting. Terra Nova 24 (4), 255–272.

Taylor, M.H., Dillon, W.P., Pecher, I.A., 2000. Trapping and migration of methane associated with the gas hydrate stability zone at the Blake Ridge Diapir: new insights from seismic data. Mar. Geol. 164, 79–89.

Thakur, N.K., Rajput, S., 2010. Exploration of Gas Hydrates Geophysical Techniques. Springer Publications. Springer Berlin Heidelberg.

Thomsen, L., 1986. Weak elastic anisotropy. Geophysics 51 (10), 1954–1966. http://dx.doi.org/10.1190/1.1442051.

Tingay, M.R.P., Hillis, R., Morley, C.K., Swarbrick, R.E., Okpere, E.C., 2003. Pore pressure/stress coupling in Brunei Darussalam: implications for shale injection. In: Van Rensbergen, P., et al. (Ed.), Subsurface Sediment Mobilization. Geological Society of London Special Publication No. 216. The Geological Society, London, pp. 369–379.

Tréhu, A.M., Torres, M.E., Moore, G.F., Suess, E., Bohrmann, G., 1999. Temporal and spatial evolution of a gas hydrate-bearing accretionary ridge on the Oregon continental margin. Geology 27, 939–942.

Tsvankin, I., 2005. Seismic Signatures and Analysis of Reflection Data in Anisotropic Media. Elsevier Science. Print Book ISBN :9780080446189 USA.

Van Dover, C.L., Aharon, P., Bernhard, J.M., et al., 2003. Blake Ridge methane seeps: characterization of a soft-sediment, chemo synthetically based ecosystem. Deep-Sea Res. I 50, 281–300.

Van Rensbergen, P., Depreiter, D., Pannemans, B., et al., 2005. The El Arraiche mud volcano field at the Moroccan Atlantic slope. Gulf Cadiz. Mar. Geol. 219, 1–17.

Virs, R., 2015. A Comparative Seismic Study of Gas Chimney Structures from Active and Dormant Seepage Sites Offshore Mid-Norway and West-Svalbard, 2015, GEO-3900 Thesis. Faculty of Science and Technology UIT, Norway.

Vogt, P., Gardner, J., Crane, K., 1999. The Norwegian–Barents–Svalbard (NBS) continental margin: introducing a natural laboratory of mass wasting, hydrates, and ascent of sediment, pore water, and methane. Geo. Mar. Lett. 19, 2–21.

Webb, K.E., Hammer, Ø., Lepland, A., Gray, J.S., 2009. Pockmarks in the inner Oslofjord,Norway. Geo. Mar. Lett. 29, 11–124.

Wildish, D.J., Akagi, H.M., McKeown, D.L., Pohle, G.W., 2008. Pockmarks influence benthic communities in Passamaquoddy Bay, Bay of Fundy, Canada. Mar. Ecol. Progr. Ser. 357, 51–66.

Wood, W.T., Gettrust, J.F., Chapman, N.R., Spence, G.D., Hyndman, R.D., 2002. Decreased stability of methane hydrates in marine sediments owing to phase-boundary roughness. Nature 420, 656–660. http://dx.doi.org/10.1038/nature01263.

Xu, W., Ruppel, C.D., 1999. Predicting the occurrence, distribution, and evolution of methane gas hydrate in porous marine sediments. J. Geophys. Res. 104, 5081–5095.

Xu, W., 2004. Modeling dynamic marine gas hydrate systems. Amer. Miner 89, 1271–1279.

Zatsepina, O.Y., Buffett, B.A., 1997. Phase equilibrium of gas hydrate: implications for the formation of hydrate in the deep sea floor. Geophys. Res. Lett. 24, 1567–1570.

Zhang, Z., Han, D., Yao, Q., 2011. Quantitative interpretation for gas hydrate accumulation in the eastern Green Canyon Area, Gulf of Mexico using seismic inversion and rock physics transform. Geophysics 76, B139. http://dx.doi.org/10.1190/1.3581358.

Zoeppritz, K., 1919. Erdbebenwellen VIII B, on the reflection and penetration of seismic waves through unstable layers. Goettinger Nachr 6684.

Zuhlsdorff, L., Spiess, V., 2004. Three-dimensional seismic characterization of a venting site reveals compelling indications of natural hydraulic fracturing. Geology 32 (2), 101–104. http://dx.doi.org/10.1130/G19993.1.

CHAPTER

RESERVOIR CHARACTERISTICS

7

"There is a single light of science, and to brighten it anywhere is to brighten it everywhere."

Isaac Asimov

CHAPTER OUTLINE

7.1 INTRODUCTION

The oil and gas industry defines unconventional reservoirs as those that have high viscosity and/or low mobility and hydrocarbon reservoirs as those that have low permeability and porosity. Unconventional reservoirs are controlled by regional stratigraphy, which has poorly defined limits, and have "continuous" or "dispersed" accumulations that require stimulation for production. The common categorization of unconventional reservoirs is:

- Coal seam gas (CSG) or coal bed methane (CBM)
- Tight reservoirs
- Organic-rich shale reservoirs
- Oil shale or liquid-rich shale
- Gas hydrates
- Tar sand and heavy oil sandstones

To define a prospect in a continuous unconventional accumulation the following needs to be considered:

- Basin (regional)
- Block (tenement, what is the size of the block?)
- Geological considerations, e.g., depth of the reservoir, thickness, gas content, mineralogy, structural configuration, stress regime, facies, etc.
- Well spacing

Geological Controls for Gas Hydrates and Unconventionals. http://dx.doi.org/10.1016/B978-0-12-802020-3.00007-2

For unconventional reservoirs, conventional risking methodology is not applicable and risks factors fall into different categories for different reservoirs. In order to develop a better understanding of unconventional hydrocarbon reservoirs, the following need to be considered:

- Generation mechanism of hydrocarbon
- Storage of hydrocarbon
- Structural configuration
- Migration pathways

Unlike conventional reservoirs, which are small in volume but easy to develop, unconventional reservoirs are large in volume and difficult to develop. Unconventional resources can be categorized into unconventional oil and unconventional gas as shown in Figs. 7.1 and 7.2.

Unconventional reservoirs have low natural permeability and low productivity and this usually impacts the economic visibility. Some unconventional reservoirs cannot produce oil or gas prior to hydraulic fracturing because of very low permeability. The white pathways shown in Fig. 7.3a for a conventional reservoir would not be present in an unconventional reservoir (Fig. 7.3b)

Fractured, tight, and unconventional petroleum reservoirs, although less common and less well understood than conventional sandstone and carbonate reservoirs, have become an increasingly important resource base. Fractured reservoirs are composed of naturally fractured rock. Tight reservoirs contain no natural fractures and cannot be produced economically without hydraulic fracturing. Unconventional reservoirs include tar, bitumen, and heavy oil reservoirs as well as coalbed methane, shale, and basin-center gas reservoirs; they rely on emerging exploration strategies and new production

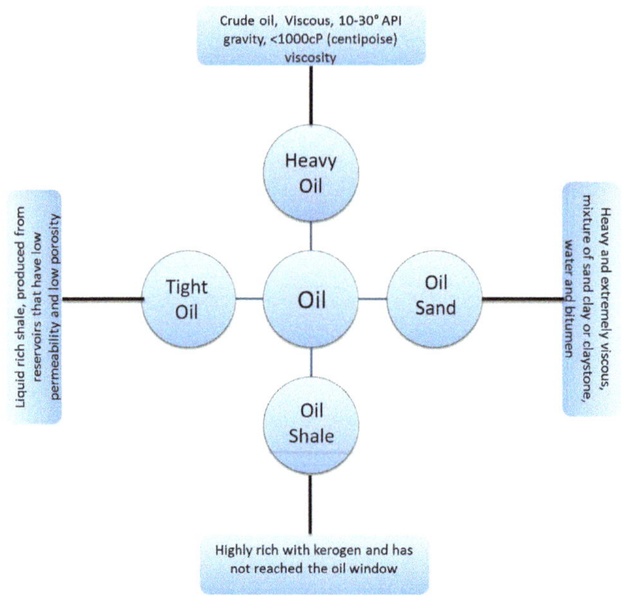

FIGURE 7.1

Categories of unconventional oil

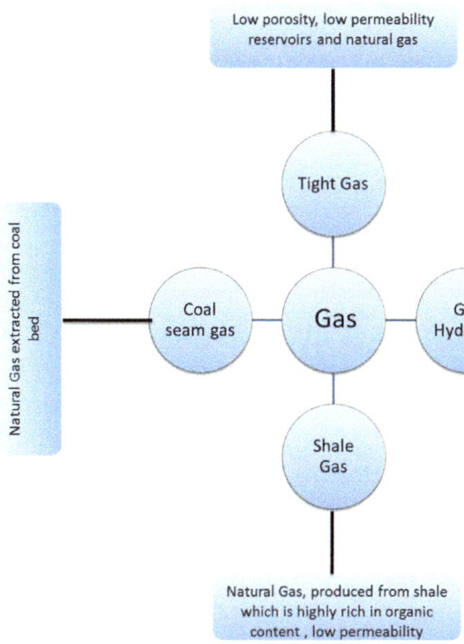

FIGURE 7.2

Categories of unconventional gas

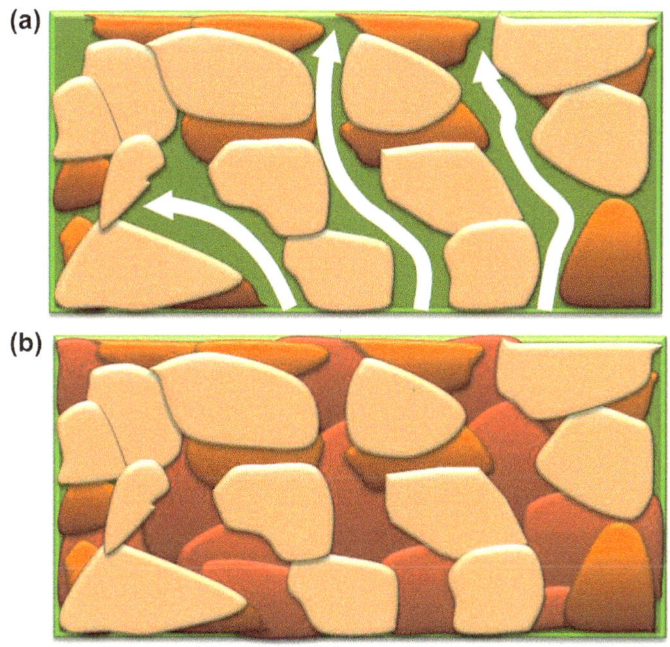

FIGURE 7.3

Conventional and unconventional reservoirs: (a) typical permeable conventional reservoir; (b) typical low permeability unconventional reservoir.

technologies to be commercially productive (Naik, http://www.pinedaleonline.com/). As a group, all of these reservoirs are increasingly important contributors to world oil and gas reserves and production. There are several ways to extract oil and gas from unconventional reservoirs:

- Surface mining – for shallow oil sands
- Horizontal wells with fracture stimulation – for tight oil or shale
- Vertical wells – for deeper reservoirs and steam
- Injection of a mixture of water, granulated solids, and chemicals

The rules for unconventional reservoirs are no different from those for other petroleum systems. The fluid movement in these reservoirs is governed by buoyancy, capillary pressure, and pressure. Petroleum is stored in source rock, which could be shale or coal, or migrates and accumulates in struc tural stratigraphic and diagenetic traps. Shales are fine-grained rocks that form from the compaction of silt and clay size particles. These rocks are formed from mud and are sometimes called mudstones. Shales are laminated, finely layered, and fissile, and are able to split along their laminations. The shift from conventional to unconventional reservoirs reflects a change in grain size from higher permeability, coarser grained rocks towards very fine-grained rocks with low permeability (Fig. 7.4). Reservoir variability, both vertically and geographically, can lead to the development of "sweet spots" of higher permeability in the finer grained reservoir rocks.

In unconventional reservoirs, shale or coal acts as a source rock, reservoir, and seal. Organic-rich shale could be high in total organic carbon (TOC) and high absorbed gas with low water saturation (S_w) and high gas saturation (S_g), and act as a mature source rock. In laminated shale, gas is stored in shale and silt with low to moderate TOC. In silty layers, high permeability can be found. In the fractured shale environment, high matrix S_w and low S_g can be found. Shale act as a source rock and gas is stored in fractures with low TOC. Most wells in unconventional reservoirs are productive but some may not meet economic thresholds. Unconventional reservoirs require a significant number of wells to achieve statistical thresholds of economic production. In unconventional reservoirs, risk is concentrated in the ability to produce economic volumes from laterally pervasive deposits of natural gas where the risk of

FIGURE 7.4

Unconventional reservoir continuum; conventional gas reservoir, tight gas reservoir, hybrid reservoir, shale gas, shale oil, and CSG reservoirs. (Courtesy of Canadian Discovery.)

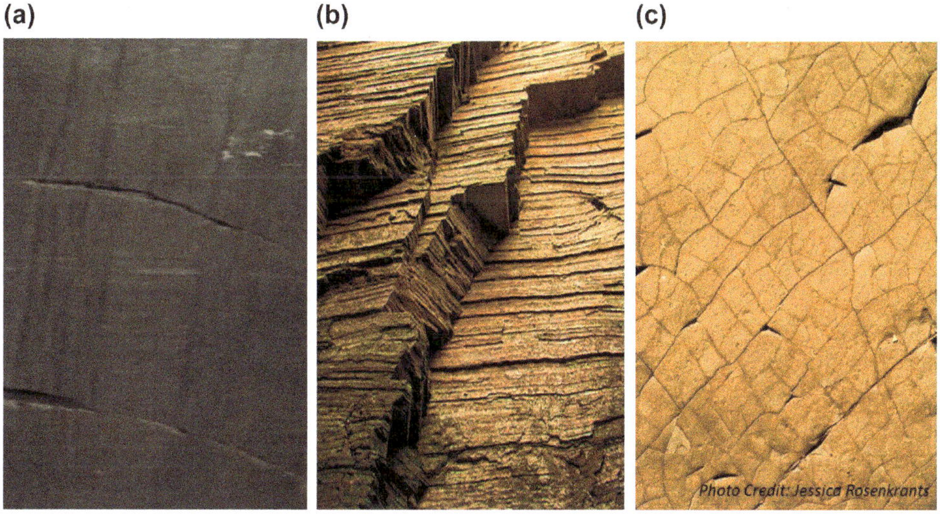

FIGURE 7.5

Describing shale: (a) Organic-rich shale. (b) Laminated shale (source: http://www.discoverbundoran.com/bundorans-coastal-geology/). (c) Fractured shale (photo credit: Jessica Rosenkrants).

finding hydrocarbons is low. Core samples of organic-rich shale, laminated shale, and fractured shale are illustrated in Fig. 7.5.

Unconventional hydrocarbon exploration is capital intensive with a large volume of investment in the early stages and return on investment throughout the lifecycle of the project. In these reservoirs, a steep decline in rates (60–80%) of production during the first 2 years is observed on a global scale. Strong commodity prices, infrastructure including pipeline and processing facilities, advanced technologies, and availability of cost-effective equipment are needed to achieve economic production.

7.2 RESERVOIR CHARACTERISTICS

The term source rock refers to an organic-rich fine-grained sedimentary rock that can produce hydrocarbons due to thermal maturation (Rezaee, 2002). Source rock is one of the main elements of a hydrocarbon system. Therefore, to identify a potential hydrocarbon system, it is necessary to first investigate the source rock and its characteristics. Thermal maturity is the primary factor that determines whether a source rock can produce oil, gas, or condensate (Lecompte et al., 2010). Source rocks are commonly shales and lime-mudstones that contain significant amounts of organic matter. Nonsource rocks also contain organic matter, but the amount is generally not significant (i.e., less than 1%). The usual method for assessing the richness and maturity of source rocks is through a variety of laboratory analyses (e.g., TOC analysis, pyrolysis, elemental analysis, vitrinite reflectance, thermal alteration index, gas chromatography, and visual kerogen description). Organic-rich rocks are composed of rock matrix, solid organic matter, and the fluids that fill the pore space at nanoscale (Herron, 1986; Hood et al., 1975;

Lawrence et al., 1984; Magara, 1978). Nonsource rocks are composed of matrix and the fluid filling the pore space. Source rocks can be categorized into two types:

- Immature source rocks – in these types of rocks, solid organic matter and rock matrix comprises the solid fraction and formation water fills the pore space.
- Mature source rocks – in these types of rocks, some percentage of the solid organic matter is transformed into liquid or gaseous hydrocarbon which moves into the pore space and displaces the formation water.

Reservoir characteristics for unconventional reservoirs can be studied by developing a thorough understanding of hydrocarbon generation and storage together with permeability pathways and stimulation limit. Understanding reservoir properties decreases the risk of completing wells that produce low volumes. The key variables that influence overall reservoir performance are shown in Fig. 7.6.

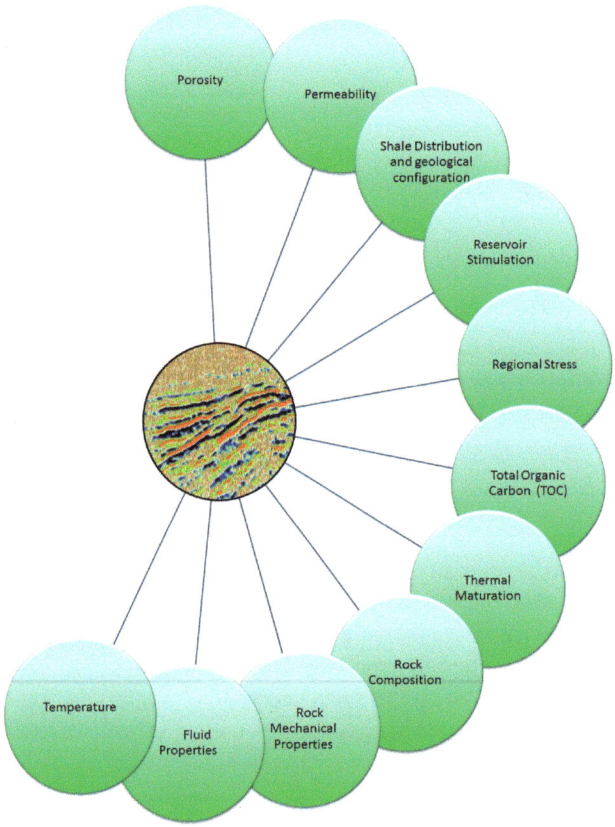

FIGURE 7.6

Key variables that influence overall reservoir performance.

Porosity and permeability are essential measurements for understanding how a reservoir will produce. Porosity is a measure of reservoir storage capacity and can be determined by measuring grain volume, pore volume, and bulk volume as defined by the following equations:

$$Ø = \frac{V_p}{V_b}$$

$$Ø = \frac{(V_b - V_g)}{V_b}$$

$$Ø = \frac{V_p}{(V_p + V_g)}$$

where Ø = porosity, V_p = pore volume, V_b = bulk volume, and V_g = grain volume.

The data source for porosity estimation includes helium gas expansion, mercury injection capillary pressure (MICP), nuclear magnetic resonance (NMR), and log analysis calibrated to core. Permeability is the measure of a rock's capability to transmit fluids. This parameter is determined by allowing a fluid of known viscosity to flow at a set rate through a core of known length and diameter and then measuring the resulting pressure drop across the core. Key characteristics of production from unconventional reservoirs include nanoscale pore sizes (10^{-9} to 10^{-6} m), tight to ultratight formations (10^{-9} to 10^{-3} mD), source rock potential, fractured environment, and hydraulically fractured horizontal wells. Porosity and permeability are created in tight shales during the organic matter maturation process, i.e., kerogen conversion to oil and gas followed by oil cracking to gas. The process involves the following:

- Kerogen conversion to gas and liquids
- Migration of gas and liquids (hydrocarbon) out of pore spaces
- Oil conversion to gas
- Creation of fractures during the process
- Pyrobitumen is formed at this stage, which leads to destruction of porosity

Permeability is considered one of the most difficult properties to quantify when characterizing shales, ranging from 0.001 to 0.0000001 mD. Shales can have a complex mixture of minerals and the relative concentrations of the constituents have the potential to make or break a potential resource play. Petrophysical measurements are used to determine mineralogy. This is a function of effective porosity, hydrocarbon saturation, and mineralogy. During the maturation process, porosity is created and destroyed but it allows a wide variation across rocks. Porosity and permeability ranges vary for conventional (porosity ≥1 μm, permeability ≥1 mD) and unconventional resources (shale gas porosity: 10^{-1} to 10^{-2} μm and permeability ≥1 to 10^{-3} μD; tight gas porosity: 10 to 10^{-3} μm and permeability 1 mD to 1 μD). A schematic of pore sizes for conventional and unconventional resources is illustrated in Fig. 7.7.

For permeability estimation the data source core analysis includes: steady-state and unsteady state, MICP, well test analysis, injection and falloff (IFOT), diagnostic fracture injection test (DFIT), post-fracture flow and buildup (FBU), production analysis: rate transient analysis (RTA), and simulation history matching.

Thermal maturity is a function of depositional history. Thermal maturity denotes diagenetic changes of organic matter that is contained in source rocks which generate hydrocarbons – from the immature phase through the main phase of oil and gas generation to the postmaturity phase. The main drivers of organic matter transformation are temperature, time, and pressure. As kerogen is exposed to

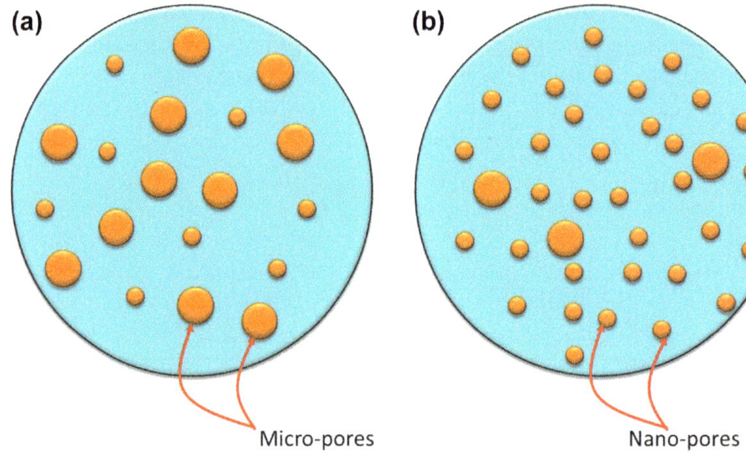

FIGURE 7.7

Pore sizes: (a) conventional oil and gas; (b) unconventional resources.

progressively higher temperatures over time, vitrinite (cell wall material and woody plant tissue preserved in the rock) undergoes irreversible alteration and develops increased reflectance. The degree of thermal maturity depends primarily on the top paleotemperature, as applied to the rocks in their geological history, and on the time of exposure to that temperature. Maturity increases with burial depth due to higher temperatures that prevail in deeper parts of the earth's crust. The increase is closely related to the geothermal gradient, which is usually expressed as °C/km. The measurement of vitrinite reflectance (R_0) was originally developed to rank coal maturity. Thermal maturity is measured on a vitrinite reflectance scale ($\%R_0$). Maturity of the rock is controlled by pressure and temperature reflecting burial conditions over time. Maturity of the organic kerogen in the rock will control the type of natural gas, natural gas liquids, or oil that is created. The value of R_0 typically ranges from 0% to 3%. Measurements in excess of 1.5% are a sign of dry gas generating source rock, which is a positive indicator for gas shale. R_0 ranging from 0.6% to 0.8% indicates oil and ranging from 0.8% to 1.1% indicates wet gas (Alexander et al., 2011). A lower value of R_0 of around 0.6% is indicative of kerogen that is immature, not having been exposed to sufficient thermal conditions over adequate time for conversion of the organic material to hydrocarbon. To quantify the relationship between porosity and matrix permeability for the key producing facies, the integrated digital rock physics (IDRP) process is recommended. IDRP begins with whole core, progresses to smaller core plugs, and finally to very high resolution 3D imaging of the pore space. This state of the art imaging technique helps in computing shale reservoir properties and offers a high resolution 3D rendering of the pore structure. IDRP involves the following steps:

- X-ray CT (computerized tomography) scan of whole rock sample:
  - Yield density and effective atomic number for the sample and can be used for facies identification
  - Density and atomic number logs provide quantitative measures to discriminate lithology, porosity, rock facies, and depositional patterns
- Core plug analysis:
  - This involves micro-CT imaging of very high resolution (~50 μm)
  - Yield porosity, kerogen, and heavy mineral volume fractions

- Special Core Analysis (SCAL):
  - From the previous stage, core plugs sampled in key reservoir facies and 3D focused ion beam scanning electron microscopy (FIB-SEM)
    - Focused ion beam (FIB) setup consists of a vacuum system and chamber, a liquid metal ion source, an ion column, a sample stage, detectors, and gas delivery system
    - The instrument is very similar to a scanning electron microscope (SEM)
    - FIB instruments may be stand-alone single beam instruments or alternatively, FIB columns can be incorporated into other analytical instruments such as an SEM, transmission electron microscope, or secondary ion mass spectrometer
    - The most common of these is a FIB/SEM dual platform instrument
  - Each layer is imaged with secondary and backscattered electrons
  - Several layers form a 3D volume, which is then segmented into separate 3D volumes of mineral, kerogen, and pore space
  - Connected and disconnected pore space and porosity associated with kerogen are then computed
  - Finally simulation of oil or gas flow using an algorithm, e.g., Lattice-Boltzmann molecular dynamics algorithms

A pore classification system proposed by Loucks et al. (2010) is shown in Fig. 7.8.

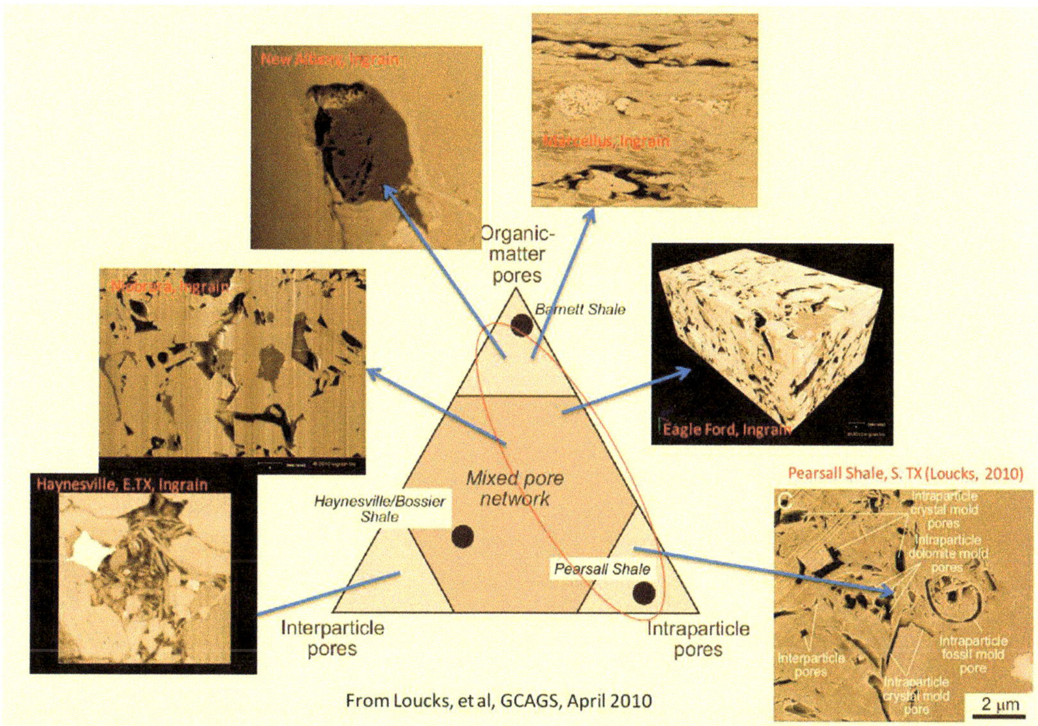

FIGURE 7.8

A pore classification system. Red oval represents range of pore types most common in Eagle Ford. (From Loucks et al., 2010.)

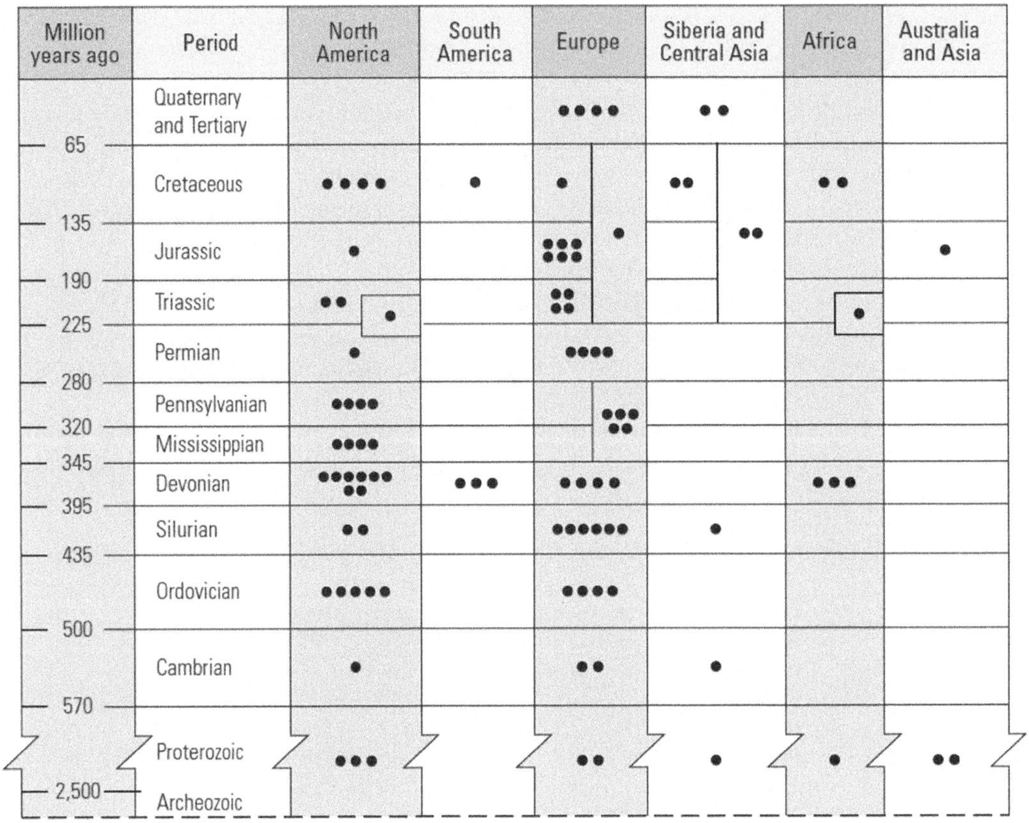

FIGURE 7.9

Global distribution of organic-rich shale units by geological period. Tectonics, geography, and climatic conditions contribute to the deposition of organic-rich sediments. Organic-rich marine shales are found across the globe. Solid black circles represent the number of occurrences for each geological age. (Compiled from Tourtelot, 1970; Fulton, 1977; Tourtelot, 1979; and Alexander et al., 2011.)

More than 60% of the earth's sedimentary crust consists of shale, and it is the primary source rock for most of the continental hydrocarbon deposits in the world (Blyth and De Freitas, 1984). Organic-rich shale (Fig. 7.5a) formations form under high levels of organic matter and low levels of oxygen. Thermal maturity of shale formation depends on the geological time range that begins in the Ordovician period and extends through the Pennsylvanian period (Alexander et al., 2011). Global distribution of organic marine shales by geological period (Fig. 7.9) has been compiled by E. B. Tourtelot (1970), H. A. Tourtelot (1979), and Fulton (1977). Recently, data from deep-sea drilling, such as the composition of pore fluids, and new methods applied to the studies of organic matter and the formation of oil and gas have clarified many aspects of the deposition and diagenesis of organic-rich sediments.

The organic material that gives shale its distinctive characteristics is derived from living things. Since organic-rich shale has accumulated throughout the entire evolution of life on earth, the composition of the organic matter in shale should reflect evolutionary development of living things at the time the shale was deposited. In deep marine environments, material accumulates and underlying ooze becomes compacted. These processes take time, for example, accumulations of a few centimeters may take several thousand years. Organic-rich shales are considered prime targets for exploration and production. However, even after passing through the maturation stage, shales do not necessarily become reservoir quality rocks.

Geochemical properties play a very important role in accessing the production potential of shales. These include TOC, gas volume and capacity, thermal maturity, permeability, and mineralogy. TOC governs the resource potential of shale; higher TOC values of shale indicate organically rich shales. If the TOC value varies from 0.5% to 1%, shale formations are considered to have poor resource potential; if the TOC ranges from 1% to 4%, shale formations are considered to have fair to good resource potential. Very good resource potential shales have a TOC ranging from 4% to 10%. Rocks with a TOC above 10% are usually too immature for development. TOC values vary across reservoir sections and need to be calibrated with petrophysical data. Another important parameter is total gas in place (GIP), which is a combination of adsorbed and free gas. Free gas volumes can be computed from log-derived effective porosity and gas saturation, after subtracting the computed pore volume occupied by the adsorbed gas.

7.2.1 RESERVOIR MODELING FOR UNCONVENTIONAL RESERVOIRS

Reservoir simulation integrates a vast amount of data from disparate sources. After the reservoir/hydraulic fracture characterization stages, reservoir modeling can be performed starting with individual well simulation. According to Dake (1978), reservoir simulations are an "estimation of hydrocarbons in place, the calculation of a recovery factor and the attachment of a time scale to the recovery."

Fig. 7.10 shows a workflow that demonstrates one method for achieving successful field development of an unconventional resource. Typically field development optimization involves a reservoir and hydraulic fracture characterization step, where key fluid, reservoir (see Fig. 7.6), and hydraulic fracture property data are estimated and integrated for the purpose of providing input to a reservoir and hydraulic fracture model. The source of information for the independent steps (see Fig. 7.6) requires the most attention for optimization of a field development plan. The information related to porosity and permeability has been explained earlier (see Section 7.2). The data sources that are required for estimating pore pressure include IFOT, "dip-in", perforation inflow diagnostic (PID), perforation inflow test analysis (PITA), and multiple open hole tests. Thermal maturity is estimated by

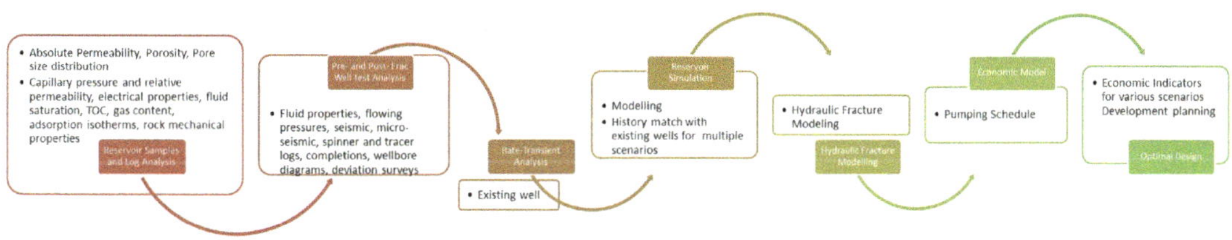

FIGURE 7.10

Illustration of a workflow used to optimize field development in unconventional gas reservoirs.

vitrinite reflectance (R_0); rock composition is estimated by X-ray diffraction (XRD), Fourier-transform infrared (FTIR) visual point count, SEM, and electron microscopy; rock mechanical properties are estimated by core measurements, log-derived (dipole sonic, Dipole shear imaging (DSI), etc.); fracture and closure stresses are computed by mini-frac tests, DFIT, and DSI calibrated to core; fluid properties are estimated by mud log, produced gas and water analysis (PVT), and for temperature, openhole logs. "Dip-in" pressure/temperature production logs are used. Rock mechanical properties are very important to optimize hydraulic fracturing scenarios. For development of prospective fields, estimation of porosity alone is insufficient, and several other parameters, including resistivity curves, need to be studied. In mature source rocks, in addition to a porosity response, the resistivity increases due to the presence of generated hydrocarbons (Passey et al., 1990). The magnitude of the curve separation in nonreservoirs should be calibrated to the TOC and maturity as it allows for depth profiling of organic richness in the absence of direct information.

For reservoir simulation models, distribution of organic content (TOC) offers very important parameters. Good quality source rocks have TOC values that vary from 1.0 to 1.5%. This parameter alone is not sufficient to access the development potential of an unconventional reservoir and additional screening is needed. Organic matter is a constituent of sedimentary rock and offers relatively low velocity, high hydrogen content and low density and source rocks have low water content. TOC is composed of extractable organic matter (EOM), convertible carbon and residual carbon. TOC can be measured by utilization of pyrolysis-gas chromatography, microscopy and pyrolysis. However, several studies (e.g. Schmoker, 1979; Fertl and Rieke, 1980; Schmoker, 1981; Dellenbach et al., 1983; Herron, 1986; Passey et al., 1990) developed new methods to estimate TOC from well logs since logs are more available than core samples and continuous recording can eliminate the statistical uncertainties of limited sampling. Dellenbach et al. (1983) developed a method using the transit-time and gamma-ray curves to provide a parameter, I-x that relates linearly to organic richness. In this method a source rock is defined by a relatively long transit time (slow velocity) and high gamma-ray intensity. Even though the derived I-x parameter has been shown in some wells to correlate roughly linearly with organic richness, calibration with sample data from specific wells is required in order to determine organic richness quantitatively. Autric and Dumesnil (1984 and 1985) developed a method in which a resistivity parameter is added to the I-x method of Dellenbach et al. (1983). For qualitatively discriminate the source rock from non-source rock, resistivity curve can be used and the organic richness is calculated by using I-x method. This method involves the selection of a variable which relates to the resistivity in non-source shales. The resistivity curve (Flower, 1983) is used only to help qualitatively discriminate potential source rocks from nonsource rocks; the organic richness then is determined using the I-x method (after calibration to direct information data). In mature source rocks, resistivity increases dramatically and this very important observation is recorded by Nixon (1973), Meissner (1978), and Schmoker and Hester (1989).

To optimize the log response to organic maturity, using sonic and resistivity log data proved very useful and a brilliant example is explained by Passey et al. (1990) in the Devonian Duvernay Formation in western Canada. Two wells (Fig. 7.11a,b) were drilled over the formation. Well A (Fig. 7.11a) was analyzed as immature with a value of 5 and well B maturity was estimated to be >10, which is considered as peak level. Maturity can be represented as "M" and was obtained from a variety of direct data analyses (e.g., vitrinite reflectance, thermal alteration index, etc.) or from estimates of burial and thermal history. Although similarity exists between the gamma-ray and transit-time curves in these two wells, the resistivity curves are very different.

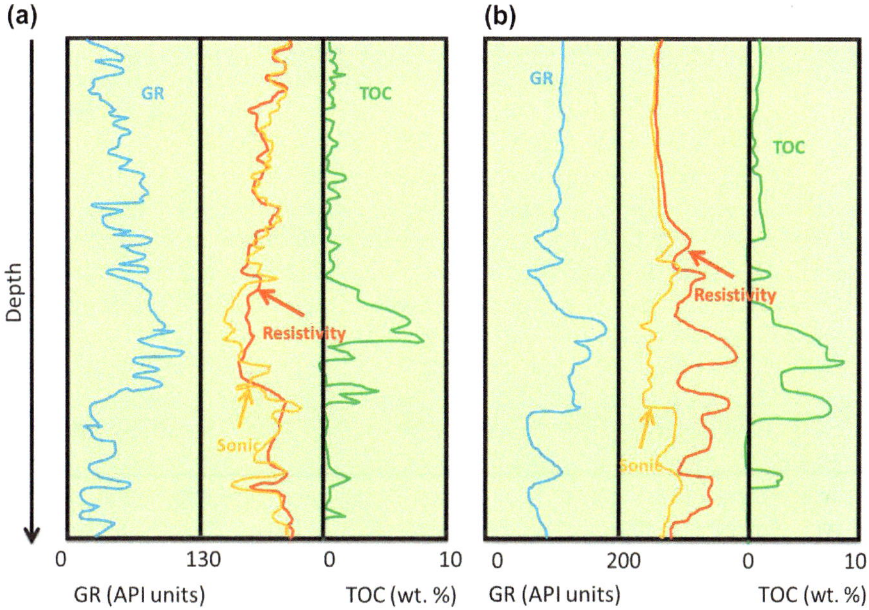

FIGURE 7.11

(a) Sonic/resistivity overlay and calculated TOC profile for an immature carbonate source rock in western Canada. (b) Sonic/resistivity overlay and calculated TOC for the same formation at peak maturity. The curves in parts (a) and (b) were hand digitized from a small-scale print. (Modified from Passey et al., 1990.)

The main steps used to perform reservoir simulations for unconventional reservoirs are as follows:

- Finer gridding near the fractures and small time steps:
  - This ensures that the pressure and saturation changes are captured in an acceptable way.
- Project-specific simulator:
  - The right simulator needs to be selected according to the reservoir behavior, e.g., dual permeability, multiphase flow, dual porosity, CSG effect, organic-rich shale, etc.
- Start with a simple model:
  - There are too many inputs required in the model that cannot be incorporated correctly due to the data quality issues, particularly in a single well scenario. For example, in a tight oil reservoir located in a Chinese basin, the primary objective was to define the horizontal target reservoir, and integrate this information with wellbore stability and fracture stimulation studies to establish the optimum well azimuth needed to maximize production. The production forecast from the integrated study yields a cumulative oil production of 25,000 m^3 after 5 years or 2.2% recovery factor (RF) for a most-likely case. The P90 case results in a 5.4% RF with 10,500 m^3 cumulative oil production. The P10 case gives 8.6% RF or 40,000 m^3 cumulative oil production. The simulation result was based on a range of reservoir properties and the frac design determined at the time of the simulation from the integrated study group. The simulation model was built in Reveal™, a numerical simulator from Petroleum Expert's Integrated Production Modeling (IPM) suite of technical software. The model was constructed without local grid refinement (LGR), which will

benefit from modeling the fracture at a smaller grid scale. The production estimate from the current model might be more optimistic than actual production.

- The simple "shoe-box" simulation model was used to understand the impact of reservoir properties in the uncertainty analysis. Owing to the size of the model, the drainage area is limited to the grid block in the model. The box model is 750 m by 1200 m wide with 30 m thickness (Fig. 7.12). The model size was chosen to accommodate the horizontal length of the well diagonally across the box model and to minimize the simulation computing time.
- Each model grid block is 25 m by 25 m wide with 1m thickness. A total of 43,200 grid blocks is used in this model. The horizontal section of the well is 1200 m long and it is positioned at N351°, which is 34° from the direction of maximum stress at N25°. This is to align each frac plane at 34° from the well as per the frac design.
- In this case there is no special core analysis (SCAL) data available and the rel-perm data are assumed values to represent tight rock. The mobile oil saturation region is only 30% with critical water saturation at 50% and residual oil saturation at 20%. The critical gas saturation used is 1% (Fig. 7.13).
- The main purpose of fracture stimulation is to create open pathways for fluid flow within the reservoir either by creation of fractures or intersection of existing fracture systems. Ideally the reservoir rock should be "brittle" so that it fractures easily. Mineral content of the shale component will determine "fracability" of reservoir – ideally a silica-rich shale is preferred.
- Eight-stage frac is uniformly distributed along the 1200 m horizontal section of the well. The frac plane is oriented at a 34° angle from the well horizontal plane. The distance between frac seed (small fractures which are considered as a flaw that acts as a fracture seed or frac seed) at the well is 150 m and along the frac axis, the perpendicular distance between frac planes is 75 m (Fig. 7.14).
- Multistage fracture stimulation effectively stimulates the entire length of the horizontal leg. Single stage stimulations tend to leave much of the intersected reservoir unstimulated due to preferential flow of fracture fluids. Isolation technologies allow the horizontal leg to be divided

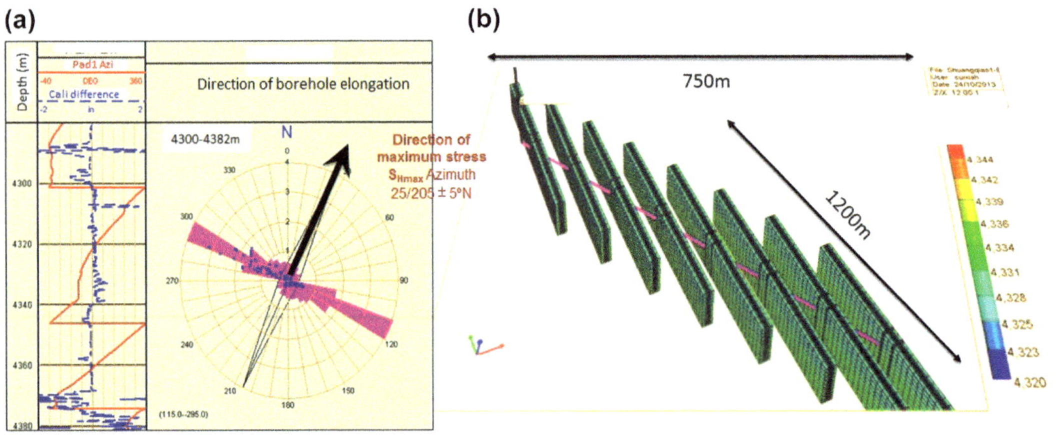

FIGURE 7.12

Reservoir model geometry. (a) Direction of maximum horizontal stress. (b) Fracture stages geometry. (Courtesy of Baker Hughes and KOC.)

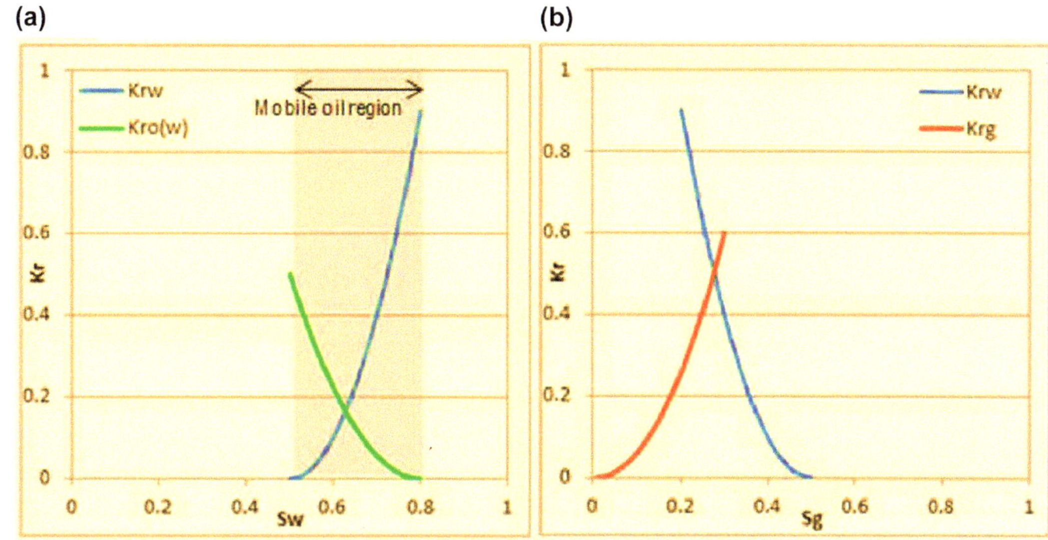

FIGURE 7.13

Oil–water (a) and gas–water (b) relative permeability. (Courtesy of Baker Hughes and KOC.)

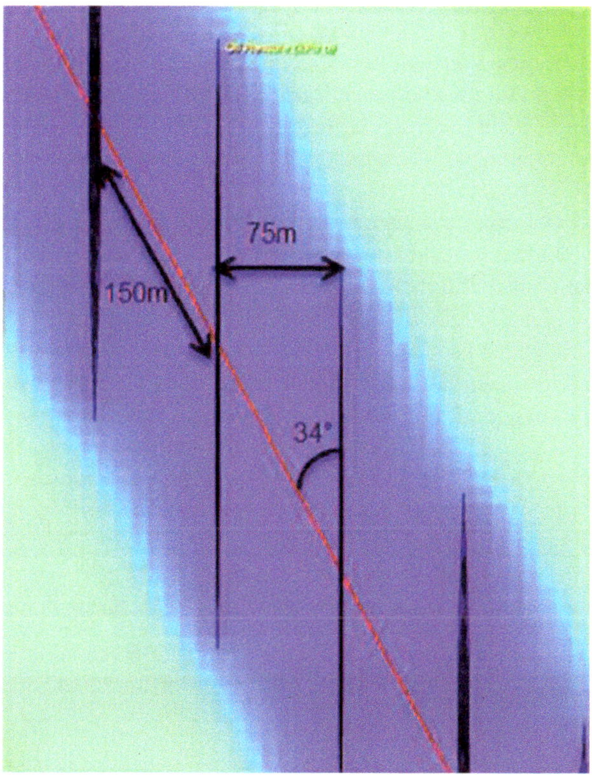

FIGURE 7.14

Fracture plane along the well axis. (Courtesy of Baker Hughes and KOC.) (Source: Bal, 2013)

into discrete stages that can be stimulated individually. Stage fracturing ensures that the maximum amount of fracture energy can be applied to a defined area thus making a more effective fracture with a greater number of fractures or a greater fracture half length.

- PVT data were limited and have come from offset wells or nearby fields. The bubble point pressure of 34,474 kPa (5000 psi) was assumed and used as a matching point to create a fluid model. Gas oil ratio (GOR) of 700 cubic meter of gas per barrel of oil is taken from a nearby well post-frac production test and used in this model. The GOR is a dimensionless ratio (volume per volume) in metric units, but in field units, it is usually measured in cubic feet of gas per barrel of oil or condensate. There are no data to indicate aquifer strength in this reservoir. Hence aquifer is not modeled and the reservoir model is producing by natural depletion.
- For uncertainty analysis nine cases of simulation are made with three reservoir permeability and three porosity values. Permeability range from low case: 0.02 mD (20% of base), base case: 0.1 mD, high case: 0.5 mD (20% of base). Porosity range from low case: 8%, base case: 10%, high case: 12%.
- Fig. 7.15 shows the corresponding cumulative oil recovery for each simulation case.
- Figs 7.16 and Fig. 7.17 show the cumulative oil recovery after 5 years. The reservoir permeability is clearly the main factor in determining oil recovery followed by porosity. At higher permeability, the porosity of the rock becomes more important in increasing the oil recovery.
- For CSG reservoirs, where dual porosity effect can be ignored, a simple tank type simulator may be adequate as explained above for a tight oil reservoir single well model (Figs. 7.12 to Fig. 7.17).
- For shale gas reservoirs there are a number of single well forecasting approaches available which combine analytical transient flow modeling (liner flow) with Arps' decline forecasts. The example of a tight oil reservoir mentioned above used a single well approach and this is equally applicable to shale gas reservoir modeling.
- Additional information including microseismic data may be required to constrain the potential drainage limits of the reservoir.

- RTA yields information useful for constraining inputs including hydraulic fracture half length and conductivity. This will add more complexity in well performance analysis.
- Surveillance information is useful including microseismic data to assist with the definition of hydraulic fracture geometry, production logs, and pressure recordings.
- Calibrate and constrain with a geocellular model to capture the uncertainty. The geocellular model should be constrained by quantitative interpretation including simultaneous inversion and angle stack and stochastic inversion using seismic data. Core-logs are also important to calibrate petrophysical models.
- Calibrate field models, which can be used for field development purposes including optimizing the producer and injector location, orientation of wells, and enhanced oil/gas recovery.
- Once fully calibrated models are developed, it is always a good idea to constrain forecasts. The model can be used for optimization of both surface and subsurface facilities. A mechanical model coupled with a flow simulation model helps to model the flow back of fracturing fluids (Clarkson et al., 2011).

Fig. 7.16 and Fig. 7.17 show the cumulative oil recovery after 5 years.

To summarize, the following key elements are critical for drilling and completion planning for an unconventional reservoir:

- Maturity (R_o) of the rock
- Total organic carbon (TOC%), which yields the hydrocarbon generation capability

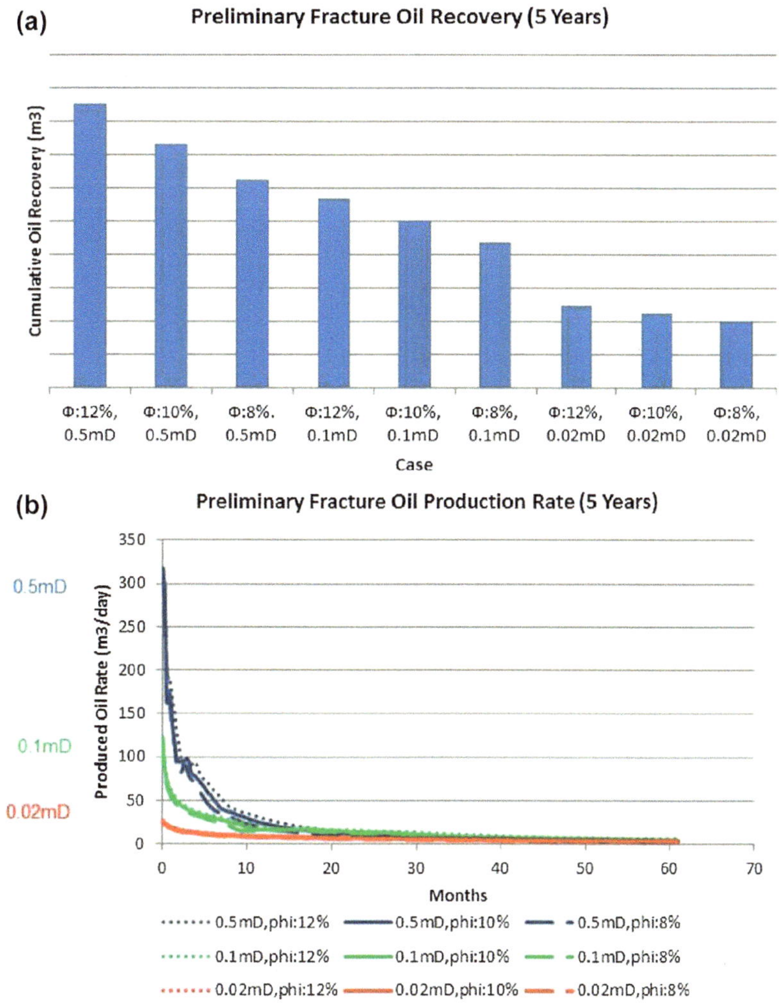

FIGURE 7.15

Initial fracture oil recovery (a) and fracture oil production rate (b). (Courtesy of Baker Hughes and KOC.)

- Understanding of lateral reservoir distribution, continuity of reservoir, and geological barriers
- Net thickness of the reservoir
- Matrix porosity
- Young's modulus and V_p/V_s ratio/Poisson's ratio
- Pressure condition of the reservoirs
- Fracability of the reservoir
  - Fracability is determined by mineral composition
  - Shale rocks that are rich in shale, particularly kaolinite and smectite, tend to be more ductile and do not fracture as effectively

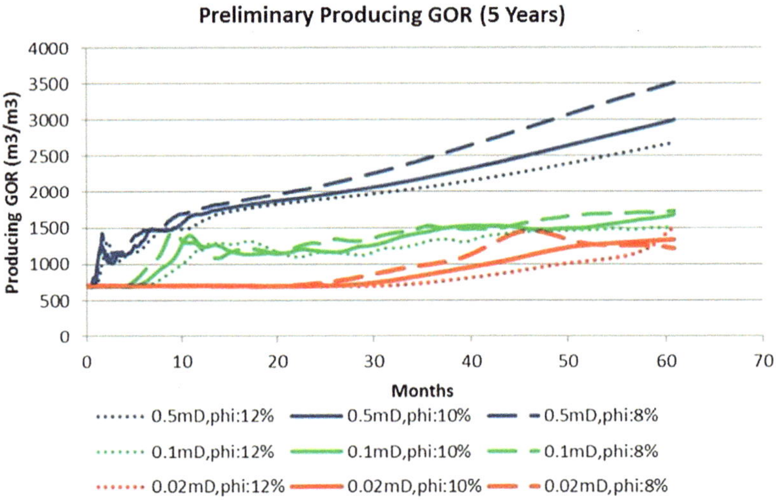

FIGURE 7.16

Production of Gas Oil Ratio (GOR) after 5 years (Courtesy of Baker Hughes and KOC.)

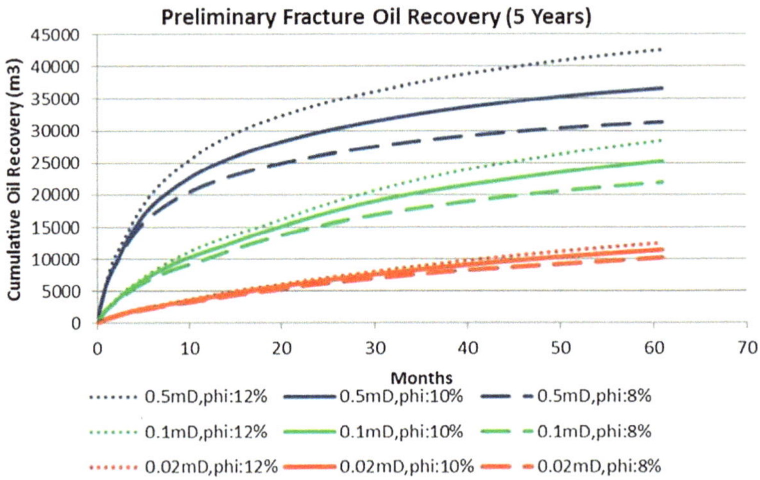

FIGURE 7.17

Cumulative oil recovery after 5 years. (Courtesy of Baker Hughes and KOC.)

  - Productive shale reservoirs have mineralogy with >45–65% quartz or carbonate
  - Reservoir rock with silica or carbonate resulting in brittle behavior
- Hydrocarbon composition
- Temperature of the reservoir
- Stress pattern of the reservoir:
  - The depth of shale reservoirs should be taken into consideration as vertical stress is always greater than horizontal stress, which results from the propagation of vertical fractures during hydraulic stimulation

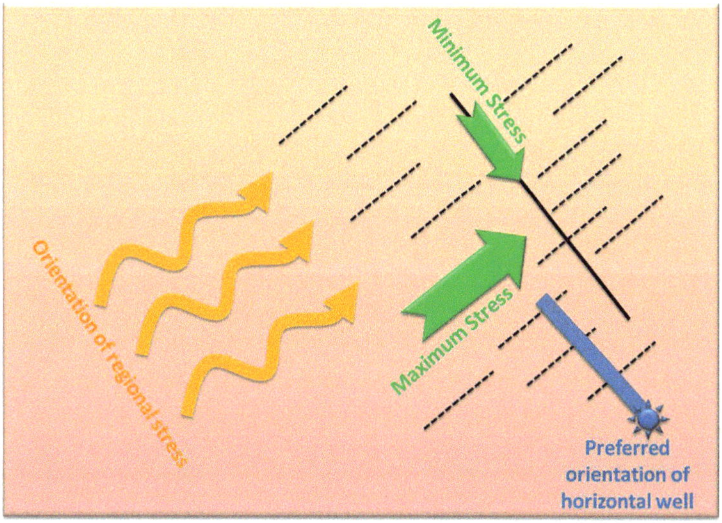

FIGURE 7.18

Schematics of the patterns of regional stress orientation, maximum and minimum stress, and decision on preferred orientation of horizontal well.

- Understanding of the orientation of the propagated fracture system
- Anisotropy distribution
- How much contribution from oil/gas in terms of reservoir producibility
- Understanding of regional stress is very important as it helps to optimize the direction of horizontal wells (Fig. 7.18)

7.2.2 GAS HYDRATE RESERVOIRS

There are large amounts of gas methane hydrate present beneath the ocean floor. This abundant energy resource is considered as a next generation unconventional resource, and many countries including southeast Asian countries, India, China, and Japan hope to become independent of energy imports by exploiting marine gas hydrate deposits. Developing economies have a strategic interest in establishing alternative energy resources, such as unconventional gas reserves. The technology for methane production from hydrate, however, is not yet fully developed. It will take another 15–20 years for industry to develop this technology. A global effort has been made to investigate the potential of gas hydrates as a potential energy resource, with significant cooperation among countries, support from different government agencies, and involvement of academic research institutions. Before these reservoirs can be exploited on a commercial scale, one needs to understand how gas hydrates are developed, their structure, and how they can be identified using a seismic scale.

Gas hydrates beneath the ocean floor in offshore environments develop in two ways:

- Biogenic methane. This is formed by microbial breakdown of biomass, which consists of dead planktonic organisms such as microalgae or krill that sink through the water column to the seafloor and build up a thick sediment structure over time. The methane is surrounded by water

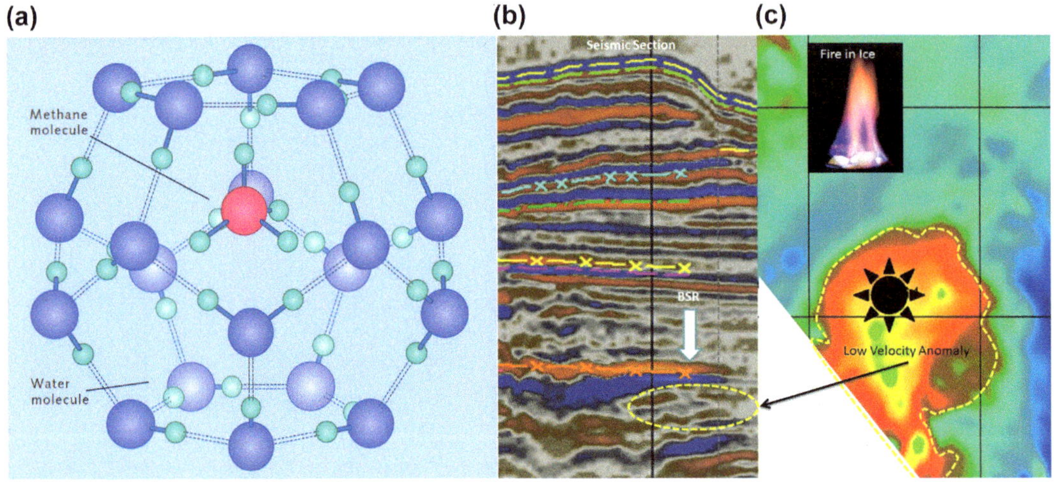

FIGURE 7.19

Chemical structure of methane molecule, seismic indicator BSR, and low velocity anomaly below BSR. (a) The water's oxygen atoms are shown in blue and the hydrogen atoms are in sky blue. Weak electrostatic forces between the atoms, called hydrogen bonds, hold the methane hydrate together. (b) Primary geophysical indicator BSR at seismic section. This example is from the southeast Asian region. (c) Map view of low velocity anomaly below the base of gas hydrates, which highlights the presence of gas below hydrates. This example is from the southeast Asian region.

molecules and trapped in a molecular cage. Chemists therefore call this kind of molecular structure a clathrate (lat. clatratus = with bars, caged). Methane hydrates develop in permafrost regions on land or beneath the seafloor. They are usually covered by a layer of sediments. Most methane hydrate occurrences worldwide lie at water depths of between 500 and 3000 m.

- Thermogenic methane. This is generated chemically in a much deeper layer of the earth's crust without the activity of microorganisms. At depths of several kilometers, under high pressures and temperatures above 100°C, the remains of biomass millions of years old in hard sedimentary rocks are transformed into methane. This is achieved by purely chemical processes driven by heat. The thermogenic methane can then rise through fissures in the rocks up to the layers where pressure and temperature conditions allow the formation of hydrates.

In methane hydrate, the methane molecule is surrounded by many water molecules (H_2O) as illustrated in Fig. 7.19a. Gas hydrates can be identified on seismic section with the help of a bottom-simulating reflector (BSR), which cross-cuts sedimentary strata and reflects opposite polarity with respect to the seafloor, caused by processes depending on sub-bottom depth and high acoustic impedance contrast between the hydrate and the free gas layer. In slope-parallel stratified sediments, i.e., where the angle of cross-cutting is small, it is difficult to determine the polarity of the BSR. An example of an identified BSR from a southeast Asian region is illustrated in Fig. 7.19b. Often, below a BSR a low seismic velocity anomaly is observed which is caused by the presence of free gas. BSR occurred at the boundary of hydrate and the free gas layer. An example of a low-velocity anomaly where a BSR has already been identified is illustrated in Fig. 7.19c. To understand the seismic character of BSR is challenging.

The zone in which gas hydrates are stable in the seafloor is called the gas hydrate stability zone (GHSZ). In this zone temperatures and pressures necessary for the formation of methane hydrates prevail. Below the GHSZ the temperature is too high due to proximity to the earth's higher temperature zone, whereas above the GHSZ the ambient pressures are too low for the methane and water to react with one another and these are the perfect conditions for the development of clathrate. In the deep upper sediments, biogeneic methane, produced by the microorganism is released. In still deeper sedimentary layers, methane is created through the chemical transformation of biomass at increasing pressure and temperatures that gives rise to thermogenic methane. An example of GHSZ from offshore eastern India is illustrated in Fig. 7.20. Thermogenic methane can rise through fractures in the various regions.

More recent estimates of worldwide amounts of gas hydrate, which attempt to consider all of these aspects, are of the order of 500 to 1500 gigatonnes of carbon. The commercialization of gas

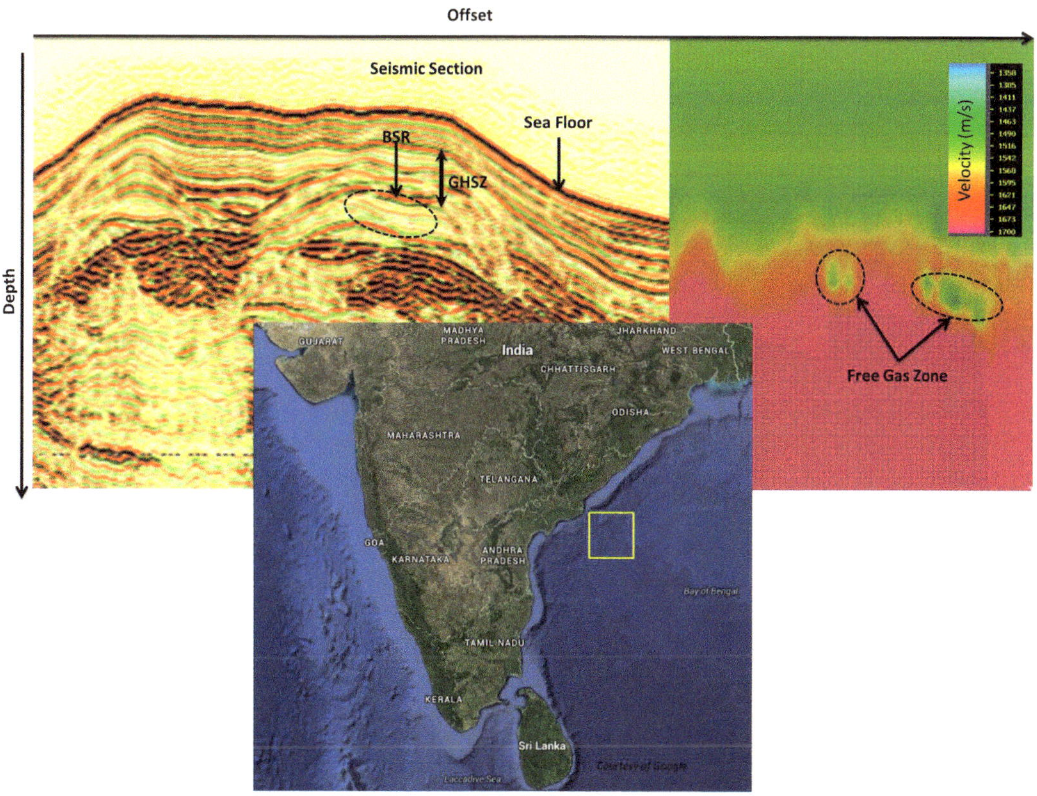

FIGURE 7.20

BSR is marked as a reverse polarity event over the seismic section (left). Low velocity anomaly representing free gas underneath BSR is marked on velocity section (right).

hydrate deposits depends most critically on the advancement of drilling technology, geopolitical stability in developing and developed countries, and a good economic climate in the energy sector (not favorable in the current scenario (2015) of declining oil and gas prices), in which there is an urgency to develop the resource potential of gas hydrates. The United States Geological Survey (USGS) summarizes post 1990 global drilling efforts that have propelled gas hydate resource studies to their present day state (Ruppel, 2011). The timeline provides few predictions about the future, other than the likely development of commercial scale production, first in permafrost-associated gas hydrates in the Arctic and then in the gas hydrates hosted in sands in the deepwater Pacific. Fig. 7.21 (courtesy of USGS) illustrates the major gas hydrate field programs and future activities.

7.2.2.1 Methane Hydrate Production Test

To prove the applicability of depressurization as a feasible method for producing methane from hydrates in deepwater sediments, the Japan Oil, Gas, and Metals National Corporation (JOGMEC) conducted the first offshore methane hydrate production test off the coast of Honshu Island (MH21 Research Consortium) in March 2013, with funding from the Ministry of Economic Trade and Industry and completion system technology from Baker Hughes. The production test site is located on the margin of the Daini-Atsumi Knoll, off the coasts of Atsumi and Shima peninsulas, in the eastern Nankai Trough, Japan. The site was selected based on seismic and well data collected from 2001 to 2008 that indicate methane hydrate-rich sedimentary layers in this area (Report from http://www.netl.doe.gov/2013). The test was devised to evaluate the ability to produce methane using an electrical submersible pumping (ESP) system to reduce the downhole pressure and dissociate the hydrate, a world first in an offshore environment. The simplest form of hydrocarbon is methane (CH4) which is made of one carbon atom and four hydrogen atoms. Methane hydrates consists of a methane molecule that is not chemically bound to the water molecules but instead are trapped within their crystalline lattice or it can be said that methane molecule surrounded by a cage of interlocking water molecules. Methane hydrates, are also known as “fire in ice” or “flammable ice” and considered as giant reservoirs of natural gas trapped in ice-like crystals. Methane hydrate contains highly compressed gas within i.e. one cubic meter of gas hydrate is almost equivalent of 164 cubic meter of natural gas. Methane hydrate is formed under very high pressure and low temperature conditions and once methane is “melted” or released, the solid crystalline lattice turns to liquid water, and the enclosed methane molecules are released as gas. This also happens when methane hydrate is exposed to unstable zone (outside the temperature and pressure conditions, which are required for methane hydrate formation). This process, called dissociation, can be demonstrated by lighting a match next to a piece of methane hydrate. The heat from the match will cause the hydrate to dissociate, and the methane molecules will ignite as they are released. This results in the curious spectacle of what appears to be burning ice. Below 100 gigapascals, methane dissociates into a mixture of hydrocarbons, and it separates into hydrogen and carbon only above 300 gigapascals (Ancilotto et al., 1997). Dissociation can be achieved by increasing the temperature (thermal stimulation), by chemical injection, or by decreasing the pressure. The first two, while feasible, are generally considered to be unrealistic for a commercially viable solution. MH21 estimates that methane hydrate formations in the Nankai Trough hold as much as 40 Tcf (1.1 Tcm) of methane in place, which is approximately 10 years of Japan’s natural gas needs. The water depth at the MH21 site varies from 857 to 1405 m. The thickness of the methane hydrate concentrated zone (MHCZ) is in tens of meters and is confirmed to contain mainly turbidite channel-type sediments within a submarine fan system in the

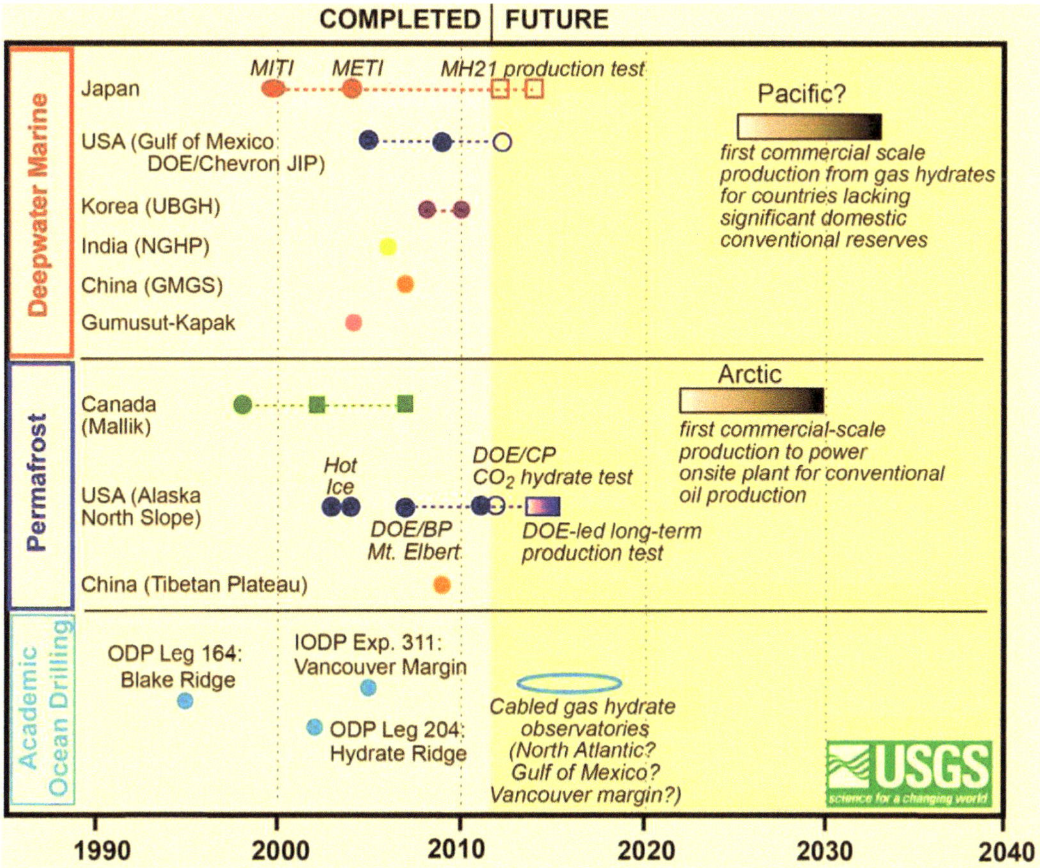

FIGURE 7.21

Timeline of major post-1990 gas hydrate field programs and future activities. Circles correspond to logging and/or coring, while squares/rectangles denote activities that included/will include production testing. Solid symbols denote completed activities, and open symbols are potential or planned activities. Rectangles filled with shaded pattern refer to longer term production activities that remain prospective. The deepwater marine programs are delineated by country and program name, with components of the Japanese activities labeled with their respective acronyms for easy cross-referencing with the literature. JIP denotes the DOE/Chevron Joint Industry Project. Permafrost activities are labeled by location, not by participants in the activities. For example, Japanese entities have been major participants and co-sponsors of some of the Mallik drilling. The DOE programs are labeled for ease of searching the associated literature, and CP and BP refer to ConocoPhillips and British Petroleum, respectively. Only the three academic drilling expeditions with an exclusive gas hydrates focus are shown in the bottom panel. Ocean Drilling Program (ODP) Legs 164 (Paull et al., 2000) and 204 (Trehu et al., 2003) investigated gas hydrates on passive and active margins, respectively. Integrated Ocean Drilling Program (IODP) Expedition 311 (Riedel et al., 2006) drilled a transect to research gas hydrate occurrences across the Vancouver margin. In the coming years, academic groups are likely to succeed in installing subsea floor gas hydrate observatories linked to seafloor cables with sufficient bandwidth to provide real-time data access to researchers. (Source: Ruppel, 2011, Courtesy of USGS.)

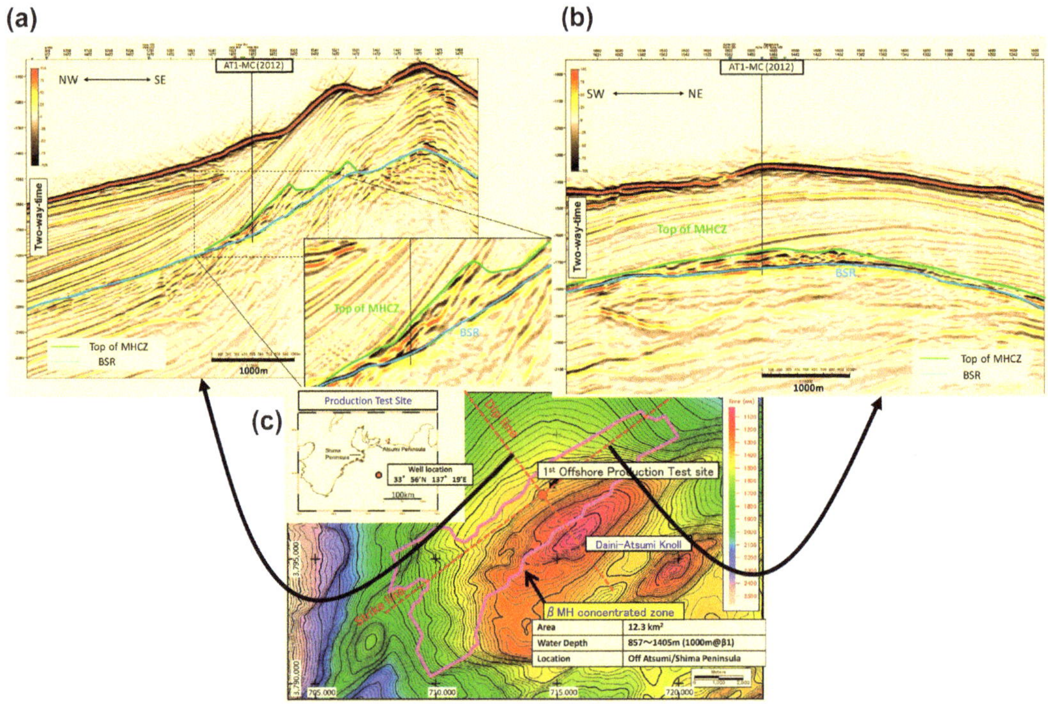

FIGURE 7.22

Seismic sections and structure maps at the production test site of MH21. (a) Dip line (Xline) and (b) strike line (Inline). On the basis of the interpretations of formation resistivity image logs, the formation dip of the MHCZ is about 20°. The MHCZ is distributed above BSRs, which are indicated by the sky-blue lines; the MHCZ is characterized by strong seismic reflectors. The top of the MHCZ is indicated by the yellow-green line. (c) Time structure map of seafloor of the Daini-Atsumi Knoll from 3D seismic survey data. Pink line outlines the β-MH-concentrated zone, as interpreted from 3D seismic data. Water depth in the b-MH-concentrated zone ranges from 857 to 1405 m. Inset (top left) locates the first offshore production test site, which is on the northwestern slope of the Daini-Atsumi Knoll. (Modified from Fujii et al., 2015.)

Ogasa Group; its age ranges from middle-to-late Pleistocene (Fujii et al., 2009, 2015; Noguchi et al., 2011). Fig. 7.22a highlights representative seismic sections from β-MHCZ (Daini-Astumi Knoll in Fig. 7.22c), a cross-section from NW to SE, which is the direction of the formation dip. Fig. 7.22b represents the cross-section from NW to SW, which is the direction of the formation strike line. The MHCZ is distributed above the BSRs.

Fujii et al. (2015) also studied the reservoir properties from well logs and core data comparisons. Total porosity was estimated at the MH21 site in the range of 40–50%. Methane hydrate pore saturation (S_h) varies at the production test site – in the upper and lower part S_h varies from 50% to 80% in sandy layers whereas in muddy layers between the sedimentary layers it ranges from 0% to 10%. The permeability was estimated in the range of 0.1 to several tens of mD in MHCZ and in seal formations.

For methane production, the following methods are considered and a lot of research has been targeted at improving these methods:

- Thermal stimulation. This refers to warming of the formation through the injection of heated fluids or direct heating of the formation (Fig. 7.23a).
- Depressurization method. This is one of the preferred and more economical methods. It does not require large energy expenditure and can be used to drive dissociation of a large volume of methane hydrate relatively rapidly. The depressurization method has been successfully tested at the MH21 production test site (Fig. 7.23b).
- Chemical inhibition method. This exploits the fact that gas hydrate stability is inhibited in the presence of certain organic (e.g., glycol) or ionic (seawater or brine) compounds. Seawater or other inhibitors might be needed during some stages of production (e.g., Bai et al., 2008) of gas from methane hydrate deposits, but would not be the primary means of dissociating gas hydrate or used for an extended period or on a large scale (Ruppel, 2011) (Fig. 7.23c).

7.2.3 COAL SEAM GAS RESERVOIRS

Coal seam gas, also known as coal bed methane, is a natural, colorless, odorless mixture of various gases (primarily methane) that is extracted from coal at a depth of ~200–1000 m. The chemical composition of CSG and conventional gas is the same. Conventional natural gas reservoirs largely consist of porous sandstone formations capped by impermeable rock. The primary difference between conventional natural gas and CSG is the type of reservoir from which they are produced. CSG is contained in natural fractures of coal seams. Coal seams possess both natural fractures and porous matrix blocks. The fractures are called "cleats", which usually occur in two sets: either perpendicular to one another or perpendicular to bedding (see Chapter 10 for more details). The cleats in one direction form first and exhibit a high level of continuity. These are called "face cleats". Cleats in the other direction are called "butt cleats"; these are discontinuous and frequently truncated by face cleats (Commonwealth of Australia, 2014). Coal seams in the ground are generally filled with water, and it is the pressure of the water that keeps the gas adsorbed as a thin film on the surface of the coal. CSG is extracted from coal deposits that are deep enough to mine economically. CSG is extracted by drilling a well vertically through rock strata until the coal seam is reached, at which point the well may also be drilled out horizontally to increase access to the methane gas. In deeper and low permeability coal seams, hydraulic fractures are used to extract methane from coal. This involves pumping a mixture of water, sand, and other additives at high pressure down the well and into the coal seam, fracturing the coal seam and providing a track for gas to flow back. CSG production involves the extraction of water and gas. The level of gas produced from a coal bed is dependent on the thickness of the coal, gas content, permeability, and the depth of the coal seam. CSG extraction can create preferential fluid flow paths. Coal seams exhibit a dual porosity system, i.e., coal contains both micropores (within the coal matrix) and macropores (fractures, coal cleats). These structures affect fluid movement by creating nonuniform flow fields with widely varying velocities. CSG exists in the following forms:

- Free gas within the pores and fractures
- Adsorbed to the coal surface
- Adsorbed within the molecular structure of coal (Rightmire et al., 1984; Rice, 1993; Shi and Durucan, 2005a,b)

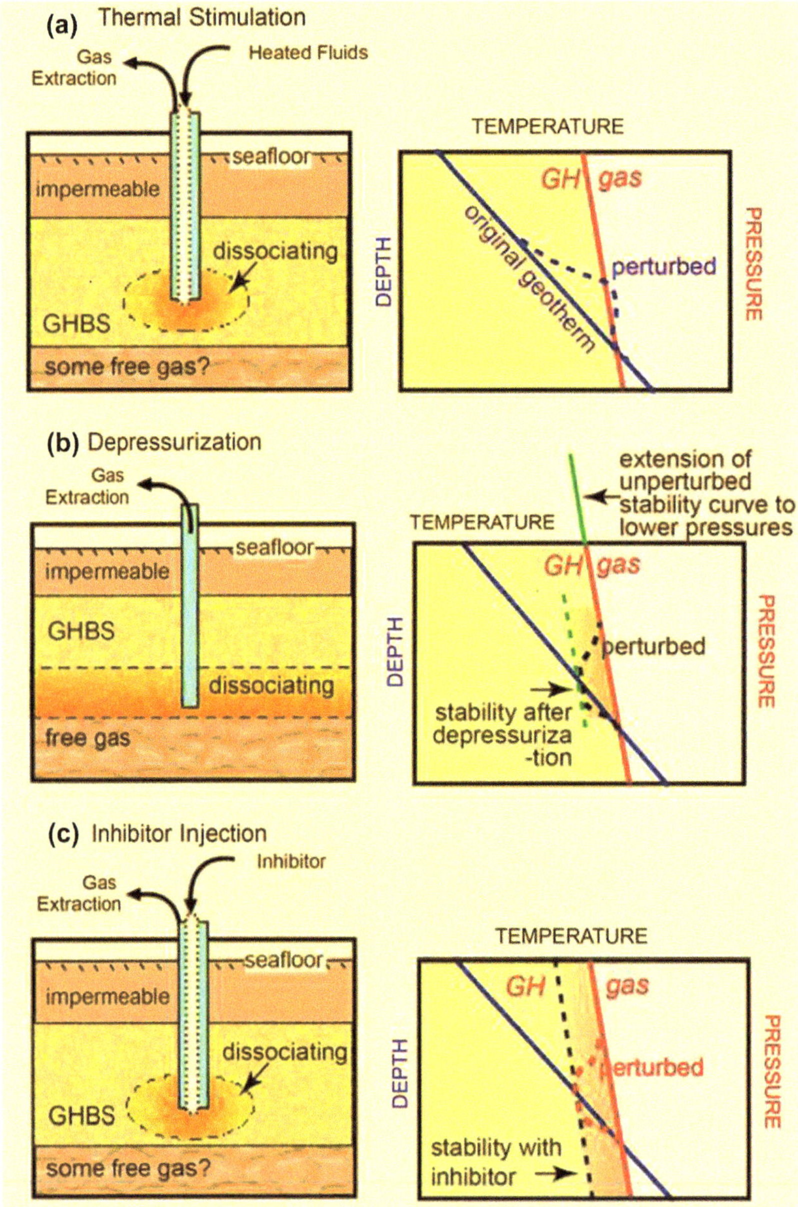

FIGURE 7.23

Possible methods for producing gas from a marine Class 1 gas hydrate deposit, which is characterized by methane hydrate-bearing sediments (yellow granular media labeled GHBS) overlying sediments with free gas (brown channeled sediments). Production techniques are shown in the left panels, modified from Collett (2002), and the impacts of the production methods on reservoir conditions are portrayed on the right. Sediments in which gas hydrate is dissociating are denoted in shaded yellow to orange. In real settings, the

The reservoir characteristics and their changes during mining can affect the production potential of the gob gas ventholes that are commonly used to control the methane emissions from the fractured zone. These holes are typically drilled from the surface to a depth that places them above the caved zone. Thus, the ability to predict changes in formation characteristics based on various factors at a given location increases the understanding of gob gas venthole production and can optimize the placement of these holes to improve their productivity (Karacan et al., 2007, 2008).

Steel-cased wells are drilled into the coal seams to release the gas, which may be 100 to 1500 m below ground. The well is equipped with a large water pump to lower the reservoir pressure within the coal seam. Water is produced to surface up the production tubing, whereas CBM gas is produced up the annulus. Produced gas is then sent to a compressor station and sold into natural gas pipelines. Produced water is either reinjected into isolated formations, released into streams, used for irrigation, or sent to evaporation ponds. Fig. 7.24 shows a schematic representation of the mining environment and the different types of supplementary controls commonly used.

The key elements in implementing reservoir engineering workflows in CSG extraction are as follows:

- Well test analysis, e.g., multi-rate drawdown gas well test analysis technique
  - Resulting venthole design for optimum production and mine safety and also for improved gas capture
- Production prediction
  - Selection of predictive method is critical for success
  - Sensitivity analysis identifies the parameter with the most influence on the performance of the gas venthole
    - Highlights mined % of the panel
    - Estimates surface elevation of the ventholes
    - Predicts distance to tailgate
    - Calculates casing diameter
    - Calculates panel length for mining
- Studies of the influence of faults on borehole modeling
  - Faults can cause intermittent production
  - Abnormal water production or gas is also caused by faults
  - Faults impact methane emission

same well would not always be used to perturb the gas hydrate stability field and to extract the gas released by hydrate dissociation. (a) Thermal stimulation introduces heat or warm fluids into the gas hydrate stability zone and dissociates gas hydrate. As shown on the right, the original geotherm (blue) will warm and dissociate gas hydrate in part of the reservoir. (b) Depressurization lowers the pressure in the GHBS. The diagram at the right shows that the part of the stability curve (green solid segment) that originally extended to pressures lower (shallower depths) than those at the seafloor applies to the gas hydrate reservoir after depressurization (dashed green). The pressure perturbation is shown schematically as the black dashed curve. (c) Inhibitors like seawater generally shift the gas hydrate stability boundary towards lower temperatures, as shown on the right with the difference between the red stability boundary and the dashed curve for stability with inhibitor. Injection of an inhibitor will dissociate gas hydrate in the vicinity of the well and result in mixed stability conditions (parts of the reservoir with and without the inhibitor), as shown by the red dashed "perturbed" stability boundary. (From Ruppel, 2011.)

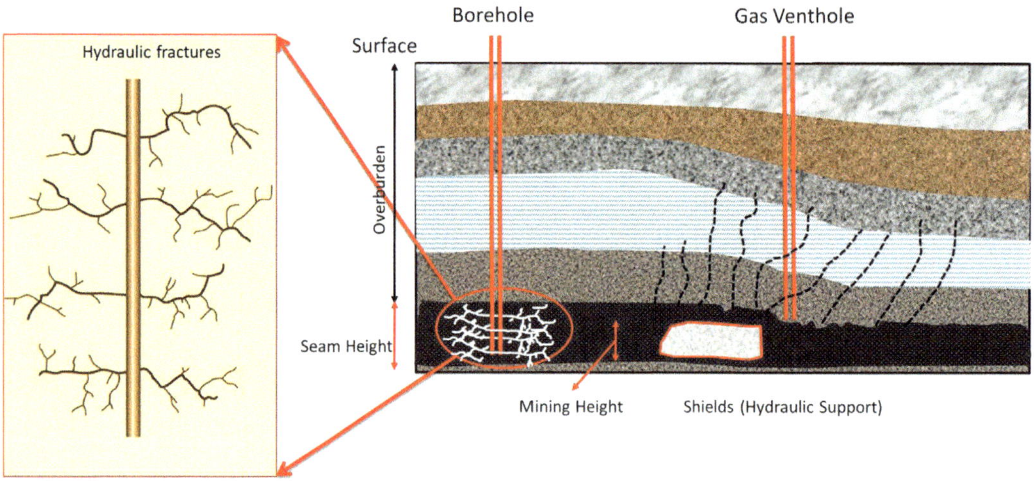

FIGURE 7.24

Schematic representation of the mining environment and the different types of degasification boreholes commonly used.

- Faults acts as barriers to methane flow
- Faults can cause gas pressure builds
- Faults can create complex compartmentalization
- Simulation of different fault configurations needs to be incorporated with geological modeling

Optimized integration of geology, geophysics, geomechanics, petrophysics, and static and dynamic modeling gives a better understanding of wellbore productivities and helps in establishing the relationships between various parameters.

REFERENCES

Alexander, T., Baihly, J., Boyer, C., et al., 2011. Shale gas revolution. Oilfield Rev. 23 (3).

Ancilotto, F., Chiarotti, G.L., Scandolo, S., Tosatti, E., 1997. Dissociation of methane into hydrocarbons at extreme (planetary) pressure and temperature. Science 275 (5304), 1288–1290.

Autric, A., Dumesnil, P., 1984. Les diagraphies de resistivite, radioactivite et temps de transit (Ts) evaluent he contenu en matieres organiques des roches a faible permeabilite: Societe pour l'Avancement de l'Interpretation des Diagraphies. Ninth International Formation Evaluation Transactions. Paper 40.

Autric, A., Dumesnil, P., 1985. Resistivity, radioactivity sonic transit time logs evaluate organic content low permeability rocks. Log Anal. 26, 36–45.

Bai, Y., Li, Q., Li, X., Du, Y., 2008. The simulation of nature gas production from ocean gas hydrate reservoir via depressurization. Sci. China Ser. E Technol. Sci. 51, 1272–1282. http://dx.doi.org/10.1007/s11431-008-0146-1.

Bal, A., September 2013. Structure and Sedimentology of Shuangqiao-1 based on EM! Borehole !mager. Baker Hughes Geoscience.

Blyth, F.G.H., de Freitas, M.H., 1984. A Geology for Engineers, seventh ed. Butterworth Heinemann, Burlington, MA. http://csl.isc.irk.ru/BD/Books/Blyth%20Freitas%20-%20Geology%20for%20Engineers.pdf.

Clarkson, C.R., Jensen, J.L., Blasingame, T.A., 2011. Reservoir Engineering for Unconventional Gas Reservoirs: What Do We Have to Consider. SPE 145080.

Collett, T.S., 2002. Energy resource potential of natural gas hydrate. Am. Assoc. Petrol. Geol. Bull. 86, 1971–1992.

Commonwealth of Australia, 2014. Coal Seam Gas Extraction: Modelling Groundwater Impacts. Prepared by Coffey Geotechnics for the Department of the Environment, Commonwealth of Australia.

Dake, L.P., 1978. Fundamentals of Reservoir Engineering. Elsevier Publications. Amsterdam, The Netherlands.

Dellenbach, J., Espitalie, J., Lebreton, F., 1983. Source Rock Logging. Transactions of the 8th European SPWLA Symposium, paper D.

Fertl, W.H., Rieke, H.H., 1980. Gama-ray spectral evaluation techniques identify fractured shale reservoirs and source-rock characteristics. Journal of Petroleum Technology 31, 2053–2062.

Flower, J.G., 1983. Use of sonic-shear-wave/resistivity overlay as a quick-look method for identifying potential pay zones in the Ohio (Devonian) Shale. J. Petrol. Technol. Mar. 638–642.

Fujii, T., Nakamizu, M., Tsuji, Y., et al., 2009. Methane-hydrate occurrence and saturation confirmed from core samples, eastern Nankai Trough, Japan. In: Collett, T., Johnson, A., Knapp, C., Boswell, R. (Eds.), Natural Gas Hydrates: Energy Resource Potential and Associated Geologic Hazards. AAPG Memoir 89, pp. 385–400.

Fujii, T., Suzuki, K., Takayama, T., et al., 2015. Geological setting and characterization of a methane hydrateres-ervoir distributed at the first offshore production test site on the Daini–Atsumi Knoll in the EasternNankai Trough, Japan, Marine and Petroleum Geology. http//:/dx.doi.org/10.1016/j.marpetgeo.2015.02.037.

Fulton, L.J.P., 1977. Stratigraphy and sedimentology of radioactive Devonian-Mississippian shales of the central Appalachian Basin: Ph.D. thesis. University of Cincinnati. 128 pp.

Herron, S.L., 1986. Derivation of a total organic carbon log for source rock evaluation. Transactions of the Twenty-Seventh SPWLA Annual Logging Symposium, Paper HH.

Hood, A., Gutjahr, C.C.M., Heacock, R.L., 1975. Organic Metamorphism and the Generation of Petroleum: AAPG Bulletin, Vol. 59, pp. 986–996.

Karacan, C.O., Diamond, W.P., Schatzel, S.J., 2007. Numerical analysis of the influence of in-seam horizontal methane drainage boreholes on longwall face emission rates. Int. J. Coal Geol. 72 (1), 15–32.

Karacan, C.O., Ulery, J.P., Goodman, G., 2008. A numerical evaluation on the effects of impermeable faults on degasification efficiency and methane emissions during underground coal mining. Int. J. Coal Geol. 75 (4), 195–203.

Lawrence, T.D., Ball, S., Harris, M., 1984. Continuous carbon/oxygen and neutron lifetime log proposed interpretation for organic and/or shaly depositional environments. Transactions of the Twenty-Fifth SPWLA Annual Logging Symposium, paper QQ.

Lecompte, B., Hursan, G., Hughes, B., 2010. Quantifying source rock maturity from logs. In: How get more than TOC Delta Log R. SPE Ann. Tech. Confer. Exhibit. Held Florence, Italy, pp. 19–22.

Loucks, R.G., Reed, R.M., Ruppel, S.C., Hammes, U., 2010. Preliminary classification of matrix pores in mudrocks. Gulf Coast Assoc. Geol. Soc. Trans. 60, 435–441.

Magara, K., 1978. Compaction and Fluid Migration. Elsevier Scientific, New York.

Meissner, F.F., 1978. Petroleum geology of the Bakken formation Williston basin, North Dakota and Montana. In: The Economic Geology of the Williston Basin. Williston Basin Symposium. Montana Geological Society, Billings, MT, pp. 207–227.

Naik, G.C., Tight Gas Reservoirs: An Unconventional Natural Energy Source for the Future. http://www.pinedale-online.com/socioeconomic/pdfs/tight_gas.pdf

Nixon, R.P., 1973. Oil source beds in Cretaceous Mowry Shale of north-western interior United States. AAPG Bull. 57, 136 161.

Noguchi, S., Shimoda, N., Takano, O., et al., 2011. 3-D internal architecture of methane hydrate-bearing turbidite channels in the eastern Nankai Trough. Jpn. Mar. Petrol. Geol. 28, 1817–1828. http://www.sciencedirect.com/science/article/pii/S0264817211000699.

Passey, Q.R., Creaney, S., Kulla, J.B., Moretti, F.J., Stroud, J.D., 1990. A practical model for organic richness from porosity and resistivity logs. AAPG Bull. 74 (12), 1777–1794.

Paull, C.K., Matsumoto, R., Wallace, P.J., Dillon, W.P., (Eds.), 2000. Proc. ODP, Sci. Results, 164: College Station, TX (Ocean Drilling Program).

Rezaee, R., 2002. Petroleum Geology. Alavi Publication. http://www-alavi.ch.cam.ac.uk/index.php.

Rice, D.D., 1993. Compositions and origins of coalbed gas. In: Rice, D.D., Law, B.E. (Eds.), Hydrocarbons from Coal: Studies in Geology 38. American Association of Petroleum Geologists, Tulsa, OK, pp. 159–184.

Riedel, M., Collett, T.S., Malone, M.J., and the Expedition 311 Scientists, 2006. Proceedings of the IODP, 311: Washington, DC (Integrated Ocean Drilling Program Management International, Inc.). http://dx.doi.org/10.2204/iodp.proc.311.101.2006 http://publications.iodp.org/proceedings/311/

Rightmire, C.T., Eddy, G.E., Kirr, J.N., 1984. Coalbed Methane Resource of the United States. In: Studies in Geology. American Association of Petroleum Geologists, Tulsa, OK.

Ruppel, C., 2011. MITEI Natural Gas Report, Supplementary Paper on Methane Hydrates. Methane Hydrates and the Future of Natural Gas. Supplement to MIT (Massachusetts Institute of Technology). In: The Future of Natural Gas An Interdisciplinary MIT Study https://mitei.mit.edu/system/files/Supplementary_Paper_SP_2_4_Hydrates.pdf.

Schmoker, J.W., 1979. Determination of organic content of Appalachian Devonian shales from formation-density logs: American Association Petroleum Geologists Bulletin 63 (9), 1504–1509.

Schmoker, J.W., 1981. Determination of organic-matter content of Appalachian Devonian shales from gamma-ray logs. AAPG Bulletin 65 (7), 1285–1298.

Schmoker, J.W., Hester, T.C., 1989. Oil generation inferred from formation resistivity – Bakken Formation, Williston basin, North Dakota. Transactions of the Thirtieth SPWLA Annual Logging Symposium, Paper H.

Shi, J.Q., Durucan, S., 2005a. CO_2 storage in deep unminable coal seams. Oil Gas Sci. Technol. Rev. IFP 60 (3), 547–558.

Shi, J.Q., Durucan, S., 2005b. A model for changes in coalbed permeability during primary and enhanced methane recovery. SPE Reserv. Eval. Eng. 291.

Tourtelot, E.B., 1970. Selected annotated bibliography of minor-element content of marine black shales and related sedimentary rocks, 1930–65. US Geol. Surv. Bull. 1293.

Tourtelot, H.A., 1979. Black Shale: its deposition and diagenesis. Clay Clay Miner. 27 (5), 313–321.

Trehu, A.M., Bohrmann, G., Rack, F.R., et al., 2003. Proceedings of the ODP. Initial Reports 204 (Online) http://www-odp.tamu.edu/publications/204_IR/.

CHAPTER

SOME FACTS, DATA ANALYSIS, AND EXAMPLES

8

"The hardest thing of all is to find a black cat in a dark room, especially if there is no cat."

Confucius

CHAPTER OUTLINE

Geological Controls for Gas Hydrates and Unconventionals. http://dx.doi.org/10.1016/B978-0-12-802020-3.00008-4

8.1 SOME FACTS ABOUT UNCONVENTIONALS

Hydrocarbon exploitation covers both conventional and unconventional hydrocarbons. Conventional hydrocarbons are oil and gas where the reservoir is sandstone or limestone; unconventional hydrocarbons refer to oil and gas that come from shale or coal seams. Professor Park A. Dickey from University of Tulsa (1958) said "We usually find oil in a new place with old ideas. Sometimes, we find oil in an old place with a new idea, but we seldom find much oil in an old place with an old idea." This statement is very much applicable in the current energy climate where the economies of many countries are dependent on fossil fuel. The demand versus supply scenarios will influence geological and geophysical theories and technologies in the future. The present day focus is on improving drilling technology and extraction processes. The discovery of unconventional hydrocarbon resources, including shale oil and gas, tight oil and gas, coal seam gas (CSG), gas hydrates, and oil shale, starts with the exploration of information to find resources. In unconventional hydrocarbon scenarios, source, seal, and reservoir rocks are same, i.e., shale or coal.

Shale is formed from muddy sediments rich in organic matter deposited in seas millions of years ago. The Wikipedia definition refers to shale as fine-grained, clastic sedimentary rock composed of mud that is a mix of flakes of clay minerals and tiny fragments (silt-sized particles) of other minerals, especially quartz and calcite. Typically having a laminate structure, shale formations are characterized by breaks along thin laminated parallel layers, known as fissility, and by low porosity and very low permeability. Another significant feature is the amount of carbonaceous material found in shale. Black shale is so called by virtue of a high degree of reduced free carbon which gives the rock its characteristic anthracite color. Shales and other mudrocks contain about 95% of all the organic matter to be found in sedimentary rocks (http://www.nortonrosefulbright.com/). The American Geological Institute defines shale as a laminated, indurated rock with >67% clay-sized minerals (Jackson, 1997). This definition clearly discriminates between shale and mudstone. The latter is a similar rock but without notable lamination. It also separates shale from siltstone, which is a mudstone in which the silt predominates over clay. Geologists believe that shale is an economically important rock as a fossil fuel, e.g., oil shale. Shale also acts as source and seal of crude oil and natural gas. Many shale samples are explained by Sandatlas (http://www.sandatlas.org/shale/) and some excellent examples from Norway, the Canary Islands, Scotland, and the coast of Estonia are illustrated in Fig. 8.1.

In recent years, the natural gas supply has improved dramatically as horizontal drilling and hydraulic fracturing technologies have made it possible to commercially develop unconventional gas reserves, particularly shale gas reserves. In North America, unconventional hydrocarbons have created the "Black Swan" event in the history of the oil and gas industry. Gas basins are located in diverse geographical areas, spanning at least 31 states in the continental United States (Fig. 8.2a–c). Europe's shale gas deposits are said almost to match those across the Atlantic (Fig. 8.2d). Europe is working very decisively to reduce its energy dependency.

There are some basis differences in unconventional hydrocarbon resource exploitation between the United States and Europe. In the continental United States, it is considered as a "revolution" whereas in Europe it is still in the "evaluation" stage. Improved production techniques, e.g., hydraulic fracturing, are unlocking huge volumes of natural gas from shale in North America. In 2000, shale gas represented only 2% of US natural gas production; by 2012 it was over 40% and this percentage is expected to increase in the near future (Hughes, 2013). Natural gas produced from shale is often referred to as

FIGURE 8.1

Shale from different parts of the world: (a) Shale is a fine-grained sedimentary rock that typically shows fine lamination (Finnmark, Norway). Width of sample 9 cm. (b) Shale as a typical sedimentary rock is clearly layered and it may be folded by a later orogenic event (shale outcrop from northern Norway). Hammer indicates scale. (c) Mud covering a dry riverbed in La Palma, the Canary Islands. The main ingredient of shale is clay which is shown here on its way from the disintegrated rock higher in the mountains to the sea, where it will be finally deposited. Rivers do the hard job of carrying all the mud to the ocean. (d) A shale outcrop in Scotland. Hammer indicates scale. (e) Pebbles of shale on the coast in Estonia. Shale is relatively easy to identify. It tends to produce flat pebbles with a dark-colored dull surface. (f) Dark brown calcareous shale associated with calcareous siltstone with a smell of oil. (Source: Photograph by R. Theriault from Globensky 1993; and http://www.sandatlas.org/shale/.)

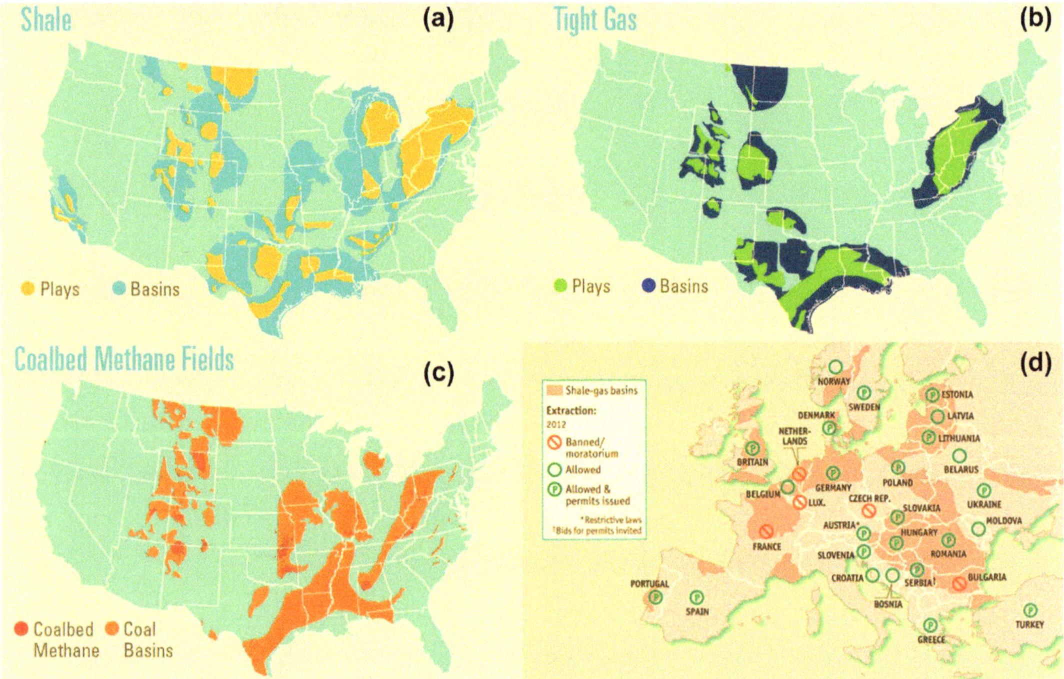

FIGURE 8.2

Unconventional hydrocarbon resources for United States (from Burwen and Flegal, 2013) and Europe (from International Energy Agency, KPMG, Press Reports). (a) Shale plays and basins in the Continental United States. (b) Tight gas plays and basins in the Continental United States. (c) Coalbed methane plays and basins in the Continental United States. (d) Europe's shale gas deposits.

"unconventional" and this refers to the type of rock in which it is found. "Conventional" oil and gas refers to hydrocarbons which have previously been sought in sandstone or limestone, instead of tight sands, shale, or coal, which are now the focus of unconventional exploration. However, the techniques used to extract hydrocarbons are essentially the same. What has changed is the advancements in technology over the last decade which have made shale gas development economically viable. Shale gas is part of a continuum (Fig. 8.3) of unconventional gas prospectivity from tight gas sands, gas shales to coalbed methane (CBM) in which horizontal drilling and fracture stimulation technology can enhance the natural fractures and recover gas from rocks with low permeability (Department of Energy and Climate Change UK).

8.2 SHALE OIL AND GAS EXTRACTION

Natural gas has formed through the thermogenic processing of carbonaceous material many millions of years old found at a significant depth in the earth's crust, where higher temperature and pressure have transformed that material into gas. Subjected to high pressures, thermogenic gas would naturally seek

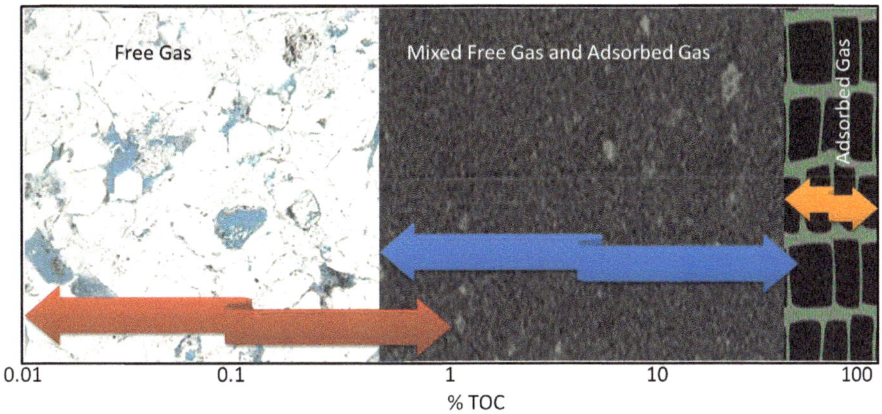

FIGURE 8.3

Ranges of total organic carbon in typical tight gas sand, shale gas, and coalbed methane prospects. (Modified from Core Lab: http://www.corelab.com/)

pathways to lower pressure environments. Shale gas is such a gas that has been trapped in shale formations that are replete with fissures arising from their laminate and fissile structures, these pathways culminating in dead-ends owing to the classic low porosity and permeability characteristics of shale formations. There are five phases in the development of a shale oil and gas field:

- Exploration: how much shale is there? This does not involve hydraulic fracturing
- Appraisal: how much oil or gas will be produced? This involves hydraulic fracturing
- Development: commercial production of oil and gas. This involves hydraulic fracturing
- Production: maintain oil and gas production rates, customize oil/gas pipeline specifications, reduce corrosion
- Rejuvenation: the primary objective is to reduce production decline and remediate subeconomic wells

The asset lifecycle approach for shale gas reservoirs, which describes the above-mentioned five steps, is shown in Fig. 8.4.

8.2.1 HYDRAULIC FRACTURING

The hydraulic fracturing technique developed in the United States to unlock the gas trapped in shale fissures has been used there for the past 65 years. It has been used to enhance coal seam gas (CSG) production from coal seams since the 1970s in the United States and since the 1990s in Australia. Increased CSG activity, mostly in Queensland and New South Wales, has caused a parallel increase in the use of hydraulic fracturing. Hydraulic fracturing is a process used in 90% of natural gas wells in the United States, where millions of gallons of water, sand, and chemicals are pumped underground to break apart the rock and release the gas. The process is illustrated in Fig. 8.5.

Hydraulic fracturing processes involve safely tapping shale and other tight rock formations by drilling below the surface before gradually turning horizontal and continuing several thousand feet or more.

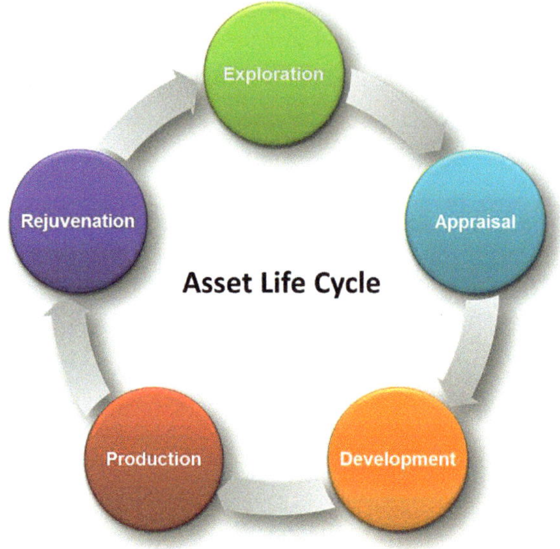

FIGURE 8.4

Shale gas reservoir asset life cycle approach that can be optimized for different types of reservoirs.

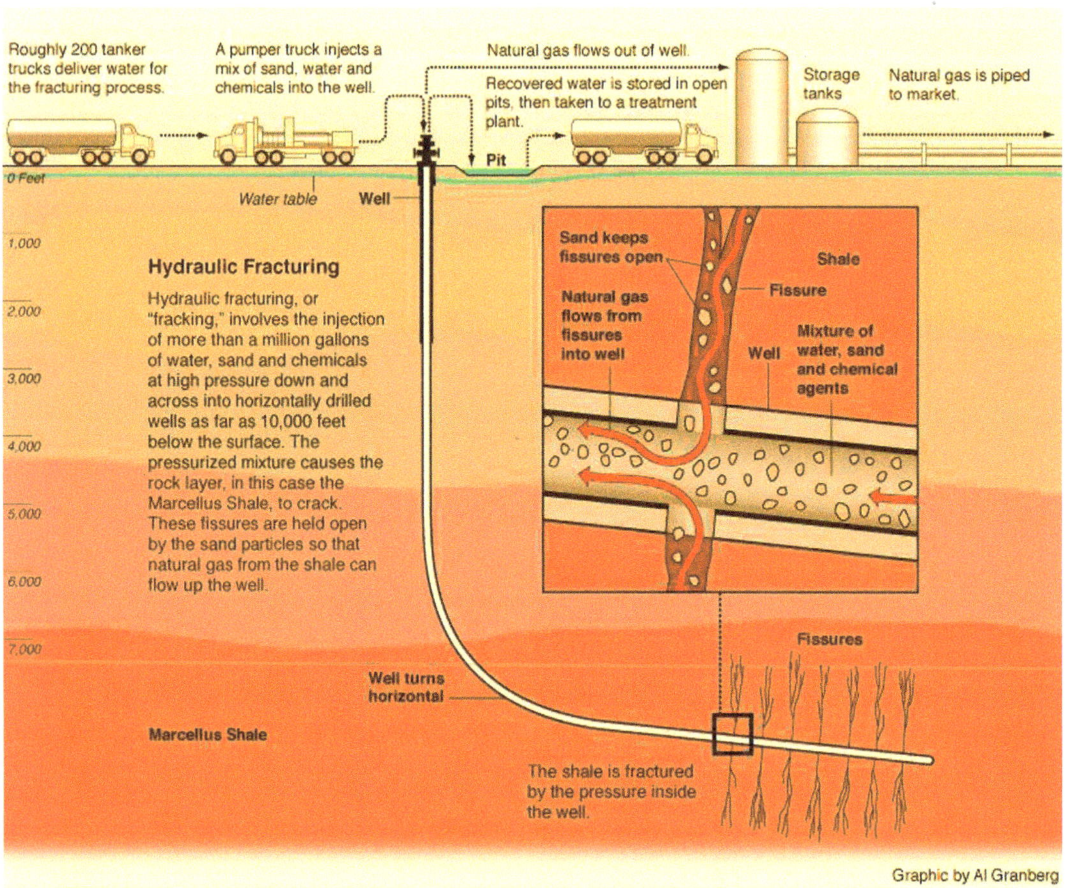

FIGURE 8.5

Hydraulic fracturing process. (Graphic by Al Granberg: https://www.propublica.org/special/hydraulic-fracturing-national)

Thus, a single surface site can accommodate a number of wells. The process can be described as follows:

- It starts with drilling of a vertical well under the nearest reservoir layer below the surface and mud (a mixture of water and additives) is pumped into the well-bore to maintain the drill-bit temperature and flush the rock cutting to the surface.
- The next step is to remove the drill-pipe and bit from the well-bore and insert the surface casing, which isolates the fresh water zone from the well-bore.
- Cement is then pumped down and out through the opening of the "shoe" at the base of the casing; this process creates a seal and prevents contamination of the aquifer by hydraulic fracturing processing and any released gas.
- Once the cement is set firm, the drill-pipe and bit is inserted again and drilling continues at the foot of the well and continues up to the "kick-off-point", which is a point 150 m above the shale formation and planned horizontal well path. This is the point at which the drill curve will begin.
- At this stage, the drill-pipe and bit are removed from the well-bore and the cementing and casing processes are repeated. An instrument which can measure the direction and angle of drill is then lowered into the well to start the horizontal drilling process.
- Approximately 400 m of angle drilling is needed from the "kick-off-point" until the drill direction is horizontal, which is known as "lateral", and drilling takes place along the shale formation.
- After drilling the target distance, the drill-pipe and bit are finally removed. At this point, the production casing is set into the full length of the well-bore including the lateral.
- Again the cement is pumped down the casing to fill in between the casing and the wall of the hole and fill in the "annulus", which is an open space between the production casing and the wall of the well-bore.
- The time has come to remove the drilling rig and place a temporary "well-head". This is the point at which hydraulic fracturing is ready to begin for the entire formation and any oil or gas is produced.
- The next step is to perforate (by using a perforating gun in the downhole) the production casing at a particular point along the lateral. This process is known as "perf". This sets up the charge that shoots small holes through the casing and the cement sleeve and a short distance into the shale formation.
- Now the perforating gun is removed from the well making it ready for hydraulic fracturing to release the oil or gas.
- Hydraulic well fracturing ("fracing") is the process of pumping fluid into a well-bore to create enough pressure to crack, or fracture, the rock layer. The fluid usually contains a "proppant", such as sand, that helps keep the fractures open to allow oil and gas to be produced to the well.
- This creates a pathway connecting the shale gas "reservoir" to the well, allowing released gas to flow into and up the well-bore.
- Water (84–96%) and proppant (3–15%), such as sand, make up around 99% of the hydraulic fracturing fluid. Proppants remain in the formation to hold the fractured fissures open.
- Some commonly used chemical additives, and their uses in hydraulic fracturing fluids, include (http://www.csiro.au/):
  - guar gum (a food thickening agent) is used to create a gel that transports sand through the fracture;
  - bactericides, such as sodium hypochlorite (pool chlorine) and sodium hydroxide (used to make soap), are used to prevent bacterial growth that contaminates gas and restricts gas flow;

  - "breakers", such as ammonium persulfate (used in hair bleach), that dissolve hydraulic fracturing gels so that they can transmit water and gas surfactants, such as ethanol and the cleaning agent orange oil, are used to increase fluid recovery from the fracture; and
  - acids and alkalis, such as acetic acid (vinegar) and sodium carbonate (washing soda), are used to control the acid balance of the hydraulic fracturing fluid.
- Most of the frac fluid is recovered from the well-bore, but there may be some losses in the formation itself. This helps the well to continue to flow for the duration of its life.
- This is the time to install a permanent well-head on the surface to regulate the production of the oil or gas flowing from the well.
- Remaining sections of the lateral are fractured sequentially; this is called multistage hydraulic fracturing. Once the entire target section of the lateral has been processed in this way, the cement plugs are bored out, allowing the freed gas to flow up the well-bore to the surface.
- This can be extended in many directions using multilateral technology. For example, the Baker Hughes S3 FracHook™ multilateral fracturing system provides operators with selective, high-pressure fracturing capabilities of all laterals (Fig. 8.6). A combination of Baker Hughes proven technologies come together to deliver a reliable multilateral fracturing solution (http://www.bakerhughes.com/). This tool can be applied in shale oil and gas, coal bed methane, hydraulic stimulation, open-hole completions, cemented liners, slotted liners, vertical wells, horizontal wells, extended-reach wells, and deviated wells. The FracHook multilateral system allows access to multiple sections of a formation for increased production, and can be commingled or selectively produced. Over the life of a well, it may become necessary for workover operations to be conducted. The diverters unique to each lateral give positive access for common remediation techniques and may even be used when restimulation is desired.

Unconventional hydrocarbons are just that, unconventional. Thus the application of integrating subsurface disciplines in their exploration and exploitation may perhaps be expected to be unconventional. Technologies such as horizontal drilling and multistage hydraulic fracturing are central to ensuring the viability of shale oil and gas resource development by maximizing contact with the most productive reservoir volumes. However, characterization efforts based on the use of well logs and cores, although very informative, may be associated with substantial uncertainty in interwell volumes. The integration of seismic data with core, petrophysical core and image logs and geomechanics plays a very important role in reducing the uncertainties during the development phase of unconventional reservoirs. A very good example of a workflow of multidisciplinary data integration and the development of a conceptual model for a horizontal well drilling program is discussed in detail in the following sections.

8.3 SUBSURFACE DATA INTEGRATION FOR HORIZONTAL WELL

Unconventional resource plays introduce challenges that are not present in the development of conventional reservoir systems. These plays are driving the technological advances necessary to develop the resource base economically, particularly with respect to drilling and completing wells. The ability to efficiently exploit ultralow-permeability plays has invigorated the oil and gas industry around the globe. The transition from vertical to horizontal wells was spurred by the development of revolutionary techniques for drilling and completion.

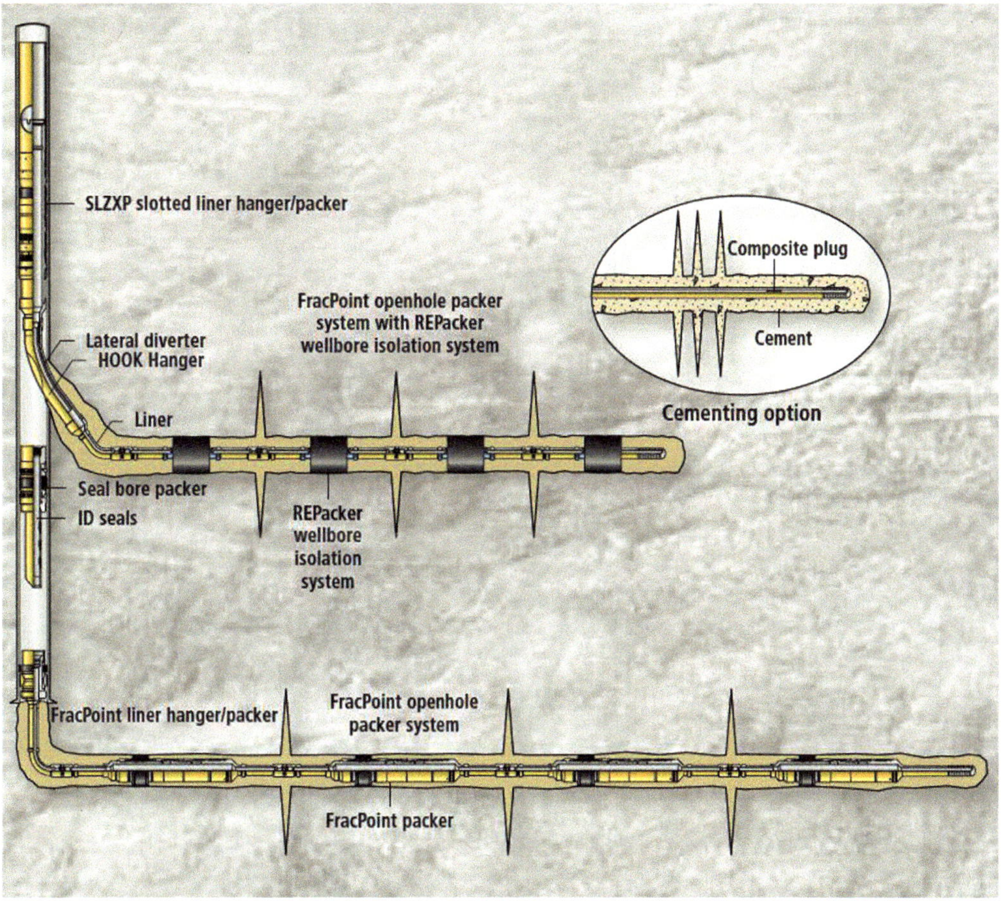

FIGURE 8.6

FrackHook Multilateral Fracturing System. (Courtesy of Baker Hughes.)

For tight oil/gas and shallow depth coalbed methane plays, a large number of wells are needed, and thus the environmental footprint is larger. To improve production from individual wells, a good understanding of the reservoir is needed so the wells can be optimally fractured and completed. Tight oil generally refers to the oil resources that have experienced short distance migration, have gathered in tight sandstone, tight limestone or tight carbonate reservoir layers, possibly interbedded with or in close proximity to the source bed, and where the overburden pressure matrix permeability is less than or equal to 0.1 mD and reservoir porosity is less than 10%. The oil contained within these reservoir rocks typically does not flow to the well-bore at economic rates without assistance from technologically advanced drilling and completion techniques.

Hydraulic fracturing of oil and gas reservoirs has been an accepted technique for enhancing production since its commercial introduction in 1949, and plays a major role in developing many oil and gas fields globally. The combination of fracturing technology and horizontal drilling has made exploiting source rock reservoirs a thriving endeavor in North America. As world economy growth has resulted in an explosion in the demand for petroleum, the horizontal well technique is being applied increasingly worldwide in the oil and gas industry. If the existing horizontal well data could be combined with seismic inversion in reservoir characterization, it would be an innovative approach. Integrated workflows, which bring together a broad range of geoscience data, have shown great potential for the comprehensive reservoir characterization required to optimize drilling and completion activities in heterogeneous unconventional resource plays.

In the following example, the main objective of the data integration and interpretation was to optimize a horizontal well placement by utilizing seismic, well, image, and geological data for a tight sandstone reservoir. This was achieved by integrating different disciplines in an innovative way of fast track (accelerated) reservoir development.

8.3.1 DATA INTEGRATION AND INTERPRETATION

A multidisciplinary approach of integrating geology, geophysics, geomechanics, and petrophysics was applied to design the optimal placement of a horizontal well in an unconventional reservoir in a Chinese basin.

8.3.1.1 Geological Settings

The area of interest is located in the central Junggar Basin, northwest China. The Badaowan Formation, comprising coals, mudstones and sandstones, the target of this horizontal well, was deposited on a lacustrine plain during the early to middle Jurassic period (Fig. 8.7).

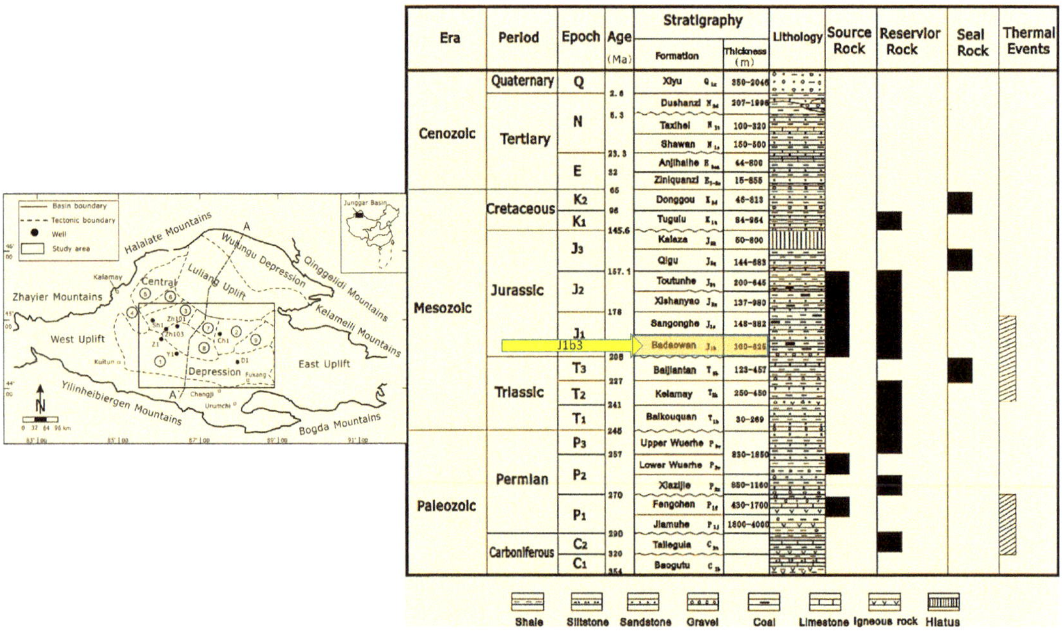

FIGURE 8.7

Generalized stratigraphic column for the Junggar Basin (right) and location map (left). Yellow arrow points out to the target interval, which represents J1b3 Badaowan Formation.

The sandstones represent a series of amalgamated meandering channels and crevasse splays, separated by layers of finer sediments that act as local seals, whereas the coals are considered to be potential source rocks (Yang et al., 2012; Yang, 2014; Xiao et al., 2010; Fan and Ma, 1991). A regional fluvial and lacustrine sandy shale seal was developed in the late Jurassic period. Source rocks are from the Permian and middle and lower Jurassic formations (Zhai et al., 2002; Wu et al., 2004). The best-developed source rocks are in the lower Permian Fengcheng Formation and the upper Permian Wuerthe Formation. These are dominated by lacustrine dark mudstones with mainly type II kerogen containing as much as 2% total organic carbon (TOC). The Jurassic source rocks consist mainly of dark shale and thin coal beds in the lower and middle Jurassic and contain mostly type III kerogen with TOC values up to 2%.

8.3.1.2 Petrophysical Understanding

Petrophysical tasks focused on selecting the best reservoir interval from a pilot vertical well and deriving suitable input parameters for basic reservoir modeling and production forecast. A zone with the highest oil saturation, best porosity, and highest permeability was chosen as the primary target interval for the horizontal section to be drilled through. Nuclear magnetic resonance (NMR) porosity from the vertical well correlated well to the core porosity. Zones with higher porosity were associated with dissolution but this relationship did not correspond with higher permeability. The NMR porosity and permeability cross-plot showed a steep porosity–permeability relationship at low porosities (Fig. 8.8). Permeability from the NMR log and air permeability from core plug data showed an acceptable agreement. Permeability was generally below 0.5 mD; this very low value was due to compaction and cementation. In summary, the reservoir has very low permeability and porosity has a patchy, random distribution controlled by dissolution diagenesis.

8.3.1.3 Understanding Sedimentology

A detailed structural and sedimentological study of the image log was performed with two main objectives:

- To quantify borehole breakout and measure the direction of maximum stress
- To measure and interpret the paleocurrents over the primary sand objective

The vertical succession of lithofacies was analyzed and grouped into four lithofacies associations, representing subenvironments of an alluvial–fluvial plain and lacustrine-delta setting, namely: lacustrine/delta plain, lower channel fill, upper channel fill, and flood-plain. The reservoir section comprises lower and upper channel fills. These sands are divided into four depositional units representing channel belts. Both the internal architecture and paleocurrent direction are different for each unit. The target sand has a predominantly west to north-west (W to NW) paleocurrent direction, representing a channel belt within a sequence of stacked channels. Once the lithofacies scheme was defined, it was calibrated to cores, open-hole logs, and seismic data (Fig. 8.9).

8.3.1.4 Geophysical and Geomechanical Analysis

The poststack statistical wavelet was extracted from well seismic ties. A check shot was used for time-depth relationships and further integrated with the P-sonic logs to ensure the highest correlation between synthetic and seismic data using the final wavelet. The wavelet and well ties were of good quality. Correlation was acceptable for the Badaowan Formation. The top of the interval was interpreted every other crossline and inline in a circle of 2 km radius from the vertical well. The horizon was

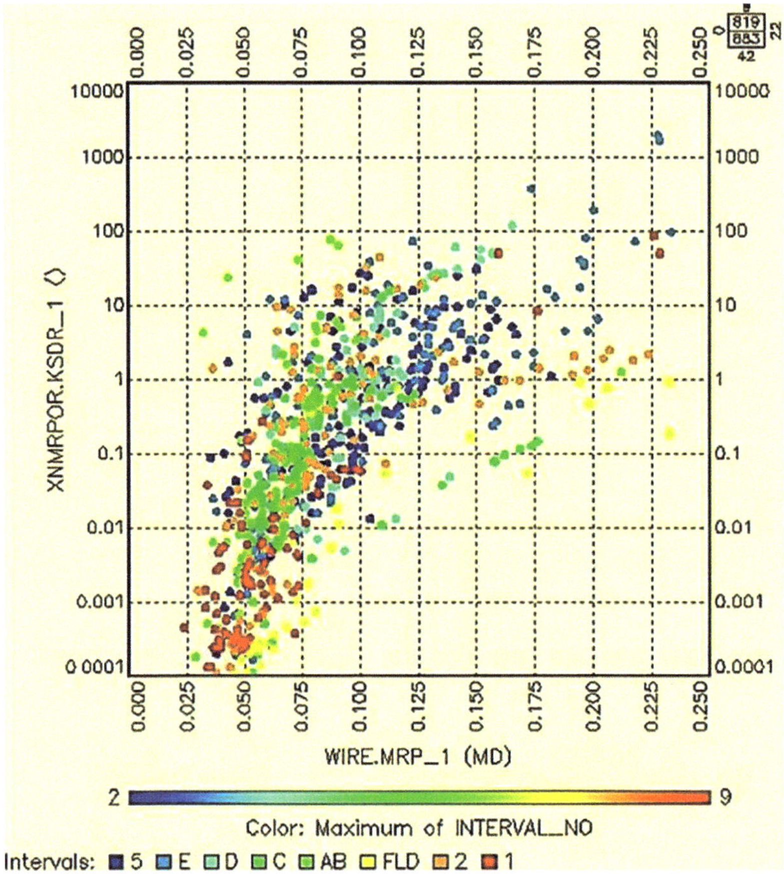

FIGURE 8.8

Porosity versus permeability from NMR data.

later flattened in order to remove any structural effect and three horizon slices were generated. Seismic amplitude time slice (Fig. 8.10a) was almost coincident with the top of the target, showing a wide zone of amplitude anomalies that were interpreted as stacked channels and crevasse splays, supported from the analysis of the resistivity image log at the well location. A model-based seismic inversion algorithm was implemented to predict acoustic impedance (AI) (Fig. 8.10b). To predict porosity (Fig. 8.10c) from AI, a linear relationship between impedance and neutron porosity was established and applied to impedance volume.

A geomechanical study was performed to optimize an operational mud weight window for drilling the planned well. A geomechanical model is comprised of a few components: rock mechanical properties (uniaxial compressive strength (UCS), internal friction coefficient, Young's modulus, and Poisson's ratio), field vertical stress S_v, formation pressure P_p, minimum horizontal stress S_{hmin}, maximum horizontal stress S_{Hmax}, and the orientation of the horizontal stresses. The historical well drilling data,

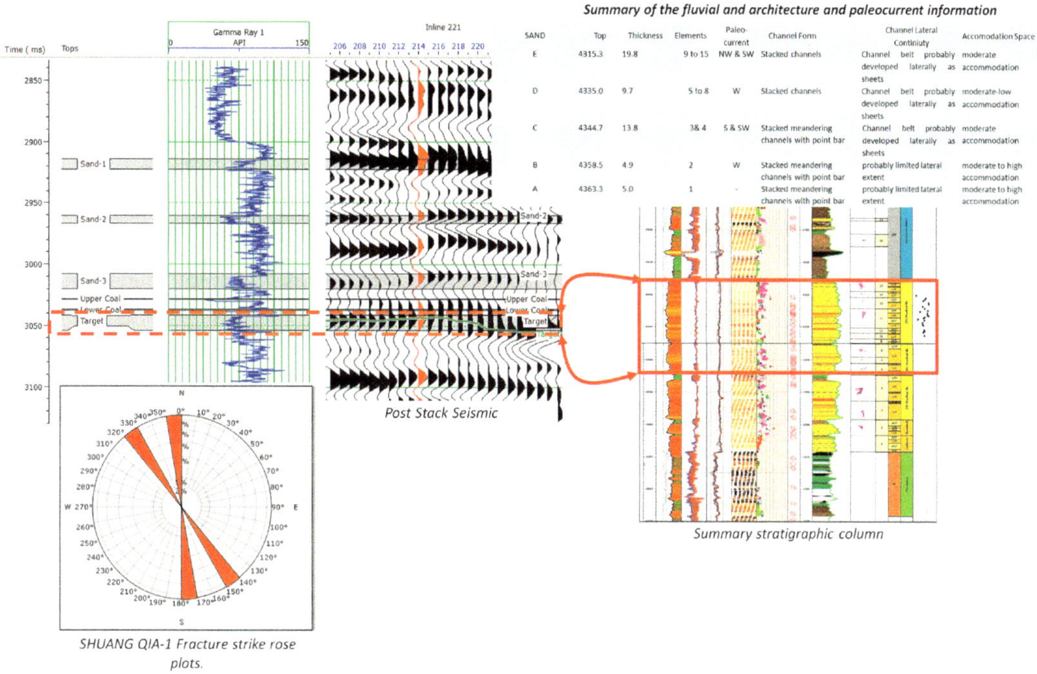

FIGURE 8.9

Image log analysis showing sand architecture and paleocurrent orientation. Target zone is highlighted by solid red lines. Dash red rectangle on log and seismic represents the target interval. Image log and seismic data are reasonably correlated. Upper right table highlights the summary of the fluvial and architecture and paleocurrents information: image log tied to seismic data. Lower left figure highlights fracture strike rose plots.

wireline logs, field tests, and newly obtained minifrac data, provided inputs for deriving a well-centric geomechanical model for the field.

The S_{hmin} profile was calibrated with a fracture closure pressure (FCP) measurement in another historical well at the minifrac test depth. An effective stress ratio principle was used to extend the profile from ground level to the target depth. This S_{hmin} profile was used for well-bore stability analysis. For hydraulic fracturing optimization, the S_{hmin} profile was further refined according to constant strain principle to model the stress contrast among various lithologies using acoustic log data available. The image data analysis introduced earlier had derived the orientation of the maximum horizontal stress S_{Hmax}, which is 025°N/205°N. The magnitudes of the S_{Hmax} were then determined by modeling the well-bore failure indicated by image data (Fig. 8.11). The derived parameters were together verified by comparing the predicted well-bore condition (breakout or not) with the observed drilling experiences of reference wells. The calibrated geomechanical model was used to predict the optimal mud weight window for the planned horizontal well-bore to avoid breakouts and mud losses, minimizing the costs.

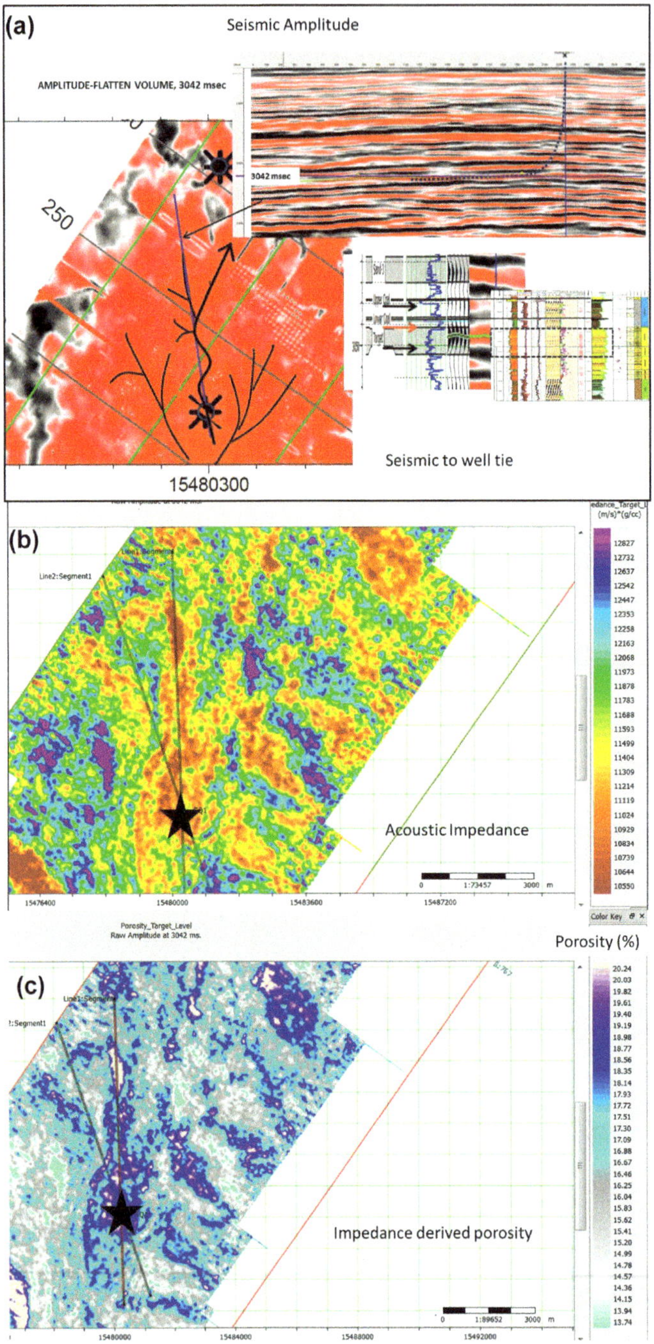

FIGURE 8.10

Seismic amplitude, inversion-derived impedance, and porosity are shown. (a) Horizon slice placed very close to the top of the target. Purple line shows the orientation of the deviated well. The seismic line at top right runs through this direction. Below the line, a composite picture showing the gamma ray (GR) Log, tops, and the interpreted image log (dashed black area is the target sand). (b) Acoustic impedance and (c) impedance-derived porosity show a similar trend. Well location is marked by a black star in b and c.

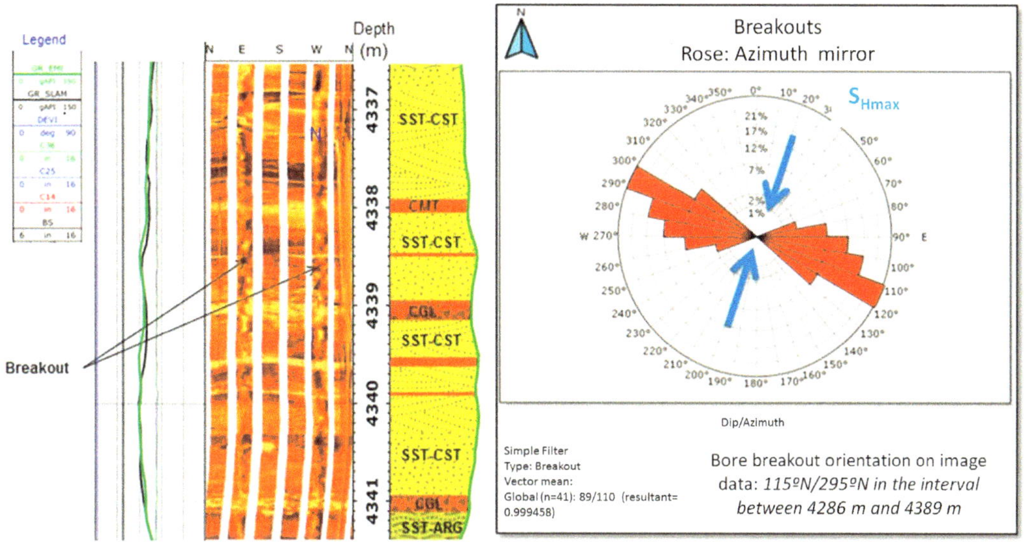

FIGURE 8.11

Interpreted image data showing breakouts (red bars) oriented 115°N/295°N in the interval between 4286 m and 4389 m of the key reference well. The S_{Hmax} orientation (blue arrows) is perpendicular to the breakout orientation.

8.3.2 SOME REMARKS

Well placement for unconventional oil reservoirs has significant economic impact for a field development prospective. Optimizing recoverable reserves from these reservoirs requires proper placement of horizontal well-bores, drilling the laterals in the right direction and keeping the lateral portion of the well-bores in the target layer. It further requires production stimulation by hydraulic fracturing of the rocks in order to induce fractures extending laterally from the horizontal well-bore. In this example, a method is presented to identify the optimum horizontal well drilling location by integrating geological, geophysical, geomechanical, and petrophysical parameters using seismic and well data from a basin in China. Available 3D seismic data and well and image logs were interpreted and integrated into a conceptual geological model. A model-based seismic inversion algorithm was implemented to predict AI. A geomechanical model was developed for the field, which provided the reservoir pressure, in situ stresses, and rock properties to predict the optimal well trajectory. The seismic amplitude time slices generated at the top of the target show a wide zone of amplitude anomalies corresponding to the meandering channels interpreted from the image log analysis. The horizontal well was successfully drilled in the reservoir sands.

To determine the optimum azimuth for the horizontal section, the information from petrophysics, geology, sedimentology, and geophysics was integrated with the results of the stress regime and well-bore stability study. A well trajectory to 351°N was expected to be optimal based on the modeled fracture propagation and sand distribution observed on the horizon slice at the top of the target. This orientation also crosses the best-developed amplitude anomalies along the trajectory. The target was located between the peak and the base of such wavelets. The real trajectory of the well was placed over the line. More than 600 m of good-quality sand were drilled following the proposed orientation with no well-bore stability problems. A few layers of shale appeared and were interpreted as pinchouts or shale-out at the edge of channels or crevasse splays.

The overall goal of this multidisciplinary study was to make the best use of available data to reduce uncertainties in drilling a horizontal well for an unconventional hydrocarbon reservoir development prospective. The primary challenges were to identify an optimal well trajectory by the integration of geological, e-log imagery, petrophysical, geophysical, and engineering information with the results of the stress regime and well-bore stability study. The horizontal section was successfully drilled in 20 days following the proposed 351°N azimuth. No geomechanical problems were encountered in the hole formed after navigating through more than 600 m of reservoir sands with hydrocarbon shows. This example demonstrated that integrating different disciplines added value in well placement. The workflow developed during this study is equally applicable in a variety of geological environments and can be applied rapidly in a fast-track mode.

8.4 COAL SEAM GAS

CSG is a form of natural gas that is found around coal deposits and can be extracted from underground coal seams. Like coal, CSG is formed by the fossilization of plant matter under pressure over millions of years. CSG is mainly composed of methane, the odorless and colorless gas used in homes and businesses. CSG collects in underground coal seams formed over tens of millions of years from compressed organic material. The naturally occurring CSG bonds to the surface of coal particles. The following aspects in particular are closely related to the production of CSG, and in many instances these features are closely aligned to the development of other unconventional gas types including shale gas:

- Well design and construction, and abandonment
- Water management (ground water and surface water)
- Hydraulic fracturing and fracture-induced seismicity
- Application and disclosure of chemicals down hole
- Social and environmental issues, including food security

CSG is sometimes called "coal seam methane" or "coal bed methane". The gas occurs in the pores and cracks in the coal seam and is held there by underground water pressure. The largest proven recoverable coal reserves, according to the latest published data, are in the United States (28.6%), followed by Russia (18.5%), China (13.5%), Australia (9.0%), and India (6.7%) (Johnston et al., 2009). Shallow coal deposits in many areas, such as in the UK and in some other European nations, have been extensively mined, yet deep coal seams beyond the reach of mining operations present opportunities for development (Fig. 8.12). Even with little minable coal remaining, the UK still ranks sixth worldwide in estimated CBM reserves.

To extract the gas, a well is drilled into the coal seam and the water is pumped out through a process called "dewatering", which allows the gas to be released from the coal and brought to the surface.

During the extraction process the quantity of water could be large. CSG has potential as an abundant clean energy supply to help replace other diminishing hydrocarbon reserves. Recent developments in technologies and methodologies are playing a large part in harnessing this unconventional resource. The utilization of hydraulic fracturing (see Section 8.2.1 Hydraulic fracturing) to exploit CSG raises

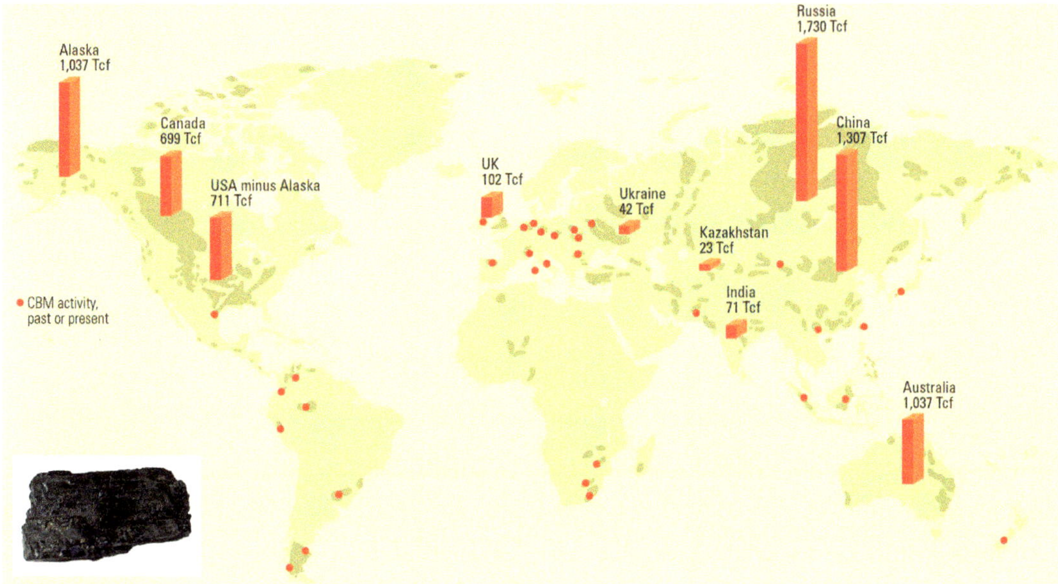

FIGURE 8.12

CBM reserves and activity. Major CBM reserves (dark blue) are found in Russia, the United States (Alaska alone has an estimated 1037 Tcf), China, Australia, Canada, the UK, India, Ukraine, and Kazakhstan. Of the 69 countries with the majority of coal reserves, 61% have recorded some form of CBM activity– investigation, testing, or production. (Source: US DOE, and BP Statistical Review and Oilfield Review Summer 2009; https://www.slb.com/~/media/Files/resources/oilfield_review/ors09/sum09/coalbed_methane.pdf.)

significant concerns about its impact on human health and the environment and can be highlighted as follows:

- Fracing chemicals can pollute drinking water aquifers by direct entry
- Pollution from the large quantity of water produced that is injected back into the ground
- Potential spill due to surface drilling and leaching of harmful waste products
- Hydraulic fracturing can cause oil and gas reservoirs to communicate with groundwater acquifer

8.4.1 OPTIMIZING FIELD DEVELOPMENT FOR COAL SEAM GAS

CSG wells are generally characterized by low production rates. To contact the maximum drainage area, a number of CSG wells are stimulated by multistage hydraulic fracturing to connect the coal cleats and natural fractures with the well-bore. Maximizing the net present value (NPV) from CSG production requires maximizing the reserves and rate of gas extraction while keeping the costs down and optimizing the well design is the key to achieving great success. Horizontal wells are often drilled into coal seams prior to mining to reduce methane levels and similarly these wells proved very useful for CSG production. To access the CSG reservoir the following inputs are needed:

- Gas content
- Gas-sorption capacity

- Permeability
- Reservoir pressure
- Reservoir geometry
- Coal chemistry

From a field development prospective, some important questions needs to be answered, e.g., how many wells are needed to supply the gas for the required purpose. Historically, CSG wells produce a considerable amount of water during the initial production and the volume generally decreases as water saturation falls and gas permeability rises. Drilling of vertical wells is common during the field development phase because it can be difficult to drill through unstable coal seams. Long-reach horizontals, where practical, can maximize contact with the reservoir, and recovery rates of between 70 and 90% within 24 to 48 months of production have been reported using multilateral wells (Coal of the Future supply, 2009). Coal has low mechanical strength and cementing in coal requires special consideration when planning the operation as coals may fail under the pressure of cement. Owing to this reason a cement of lower density should be used in the cementing operation.

8.4.2 FRACTURE STIMULATION

The formation of coal starts off as peat. After a considerable amount of time, heat, and burial pressure, it is metamorphosed from peat to lignite, which is considered as "immature coal" and remains soft and light colored. In the next stage, lignite increases in maturity and becomes darker and harder, and is known as "sub-bituminous coal". The process of burial and alteration continues and after certain physical and chemical changes it transforms into "bituminous coal", which is dark and hard. The final stage is known as "anthracite" coal, which is very hard and shiny. The order of alteration of coal is lignite (brown coal – immature), sub-bituminous, bituminous, and anthracite (mature coal) as illustrated in Fig. 8.13.

Fracture stimulation is widely used for accessing the CSG reservoir. Owing to a higher Poisson's ratio or higher compressional (V_p) and shear (V_s) ratio, it results in higher fracture gradients, often higher than those in bounding layers. The softness of coal makes the fracture propagation difficult; furthermore, the cleat system leads to a complex fracture network. In highly fractured coal with a low Young's modulus, complex fracture systems are created (Johnston et al., 2009). Coal variability from basin to basin also affects the ultimate stimulation results. From a CSG field development prospective, fracture stimulations can be divided into the following categories:

- Slickwater system
- Polymer-based gel system
- Foamed or energized system (nitrogen and carbon dioxide)

Coals are the source, the trap, and the storage media, so a different approach must be taken to evaluate them as gas reservoirs. For CSG reservoir characterization, assessment of reservoir compartmentalization relative to geological structure and in situ stress and the effect of coal properties on reservoir characteristics are very important. Seismic data play an important role in terms of assessing the reservoir, establishing the understanding of structure, optimizing the well location, and quantification of the reserves. Conventional seismic suffer from limited resolution due to its band-limited nature, whereas cross-well seismic uses a very high frequency source (>1000 Hz) and has much higher resolution as compared to the conventional high-resolution seismic data. By this method, thin coal layers

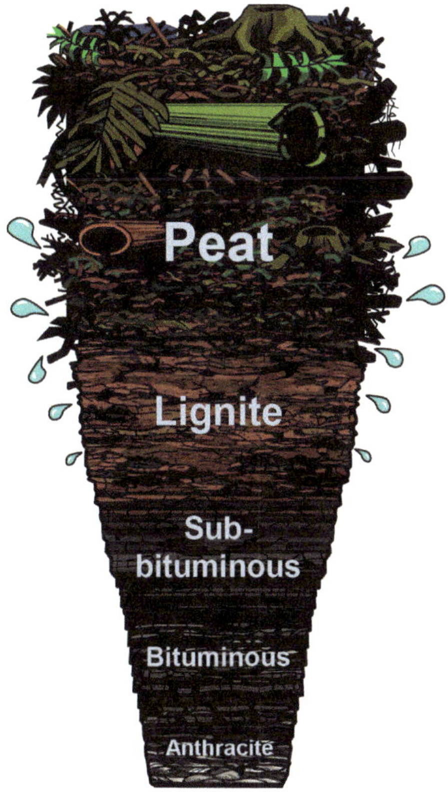

FIGURE 8.13

Classification and rank of coal. (Source: https://www.uky.edu/KGS/coal/)

can also be detected. The example illustrated in the next sections uses the application of full-waveform modeling technique to delineate thin coal layers.

8.4.3 CROSS-WELL SEISMIC MODELING

In the past decade, the upsurge of interest in the exploration, exploitation, and development of CBM in Australia represents a major opportunity to help Australia and the region move to a cleaner, less carbon-intensive future. This can be attributed to the presumably high energy resource potential of this veritable resource. The likelihood of coal seams serving as viable sites for subsurface geological storage of carbon dioxide in the foreseeable future constitutes another reason for the growing interest in CBM exploration in Australia and neighboring regions. In the past few decades, significant progress has been made in coal seismic prospecting. With the advancement of seismic methods we are able to apply them to coal seam prospecting and mining. These methods could be used in mapping the structure of the coal, discontinuities in the coal seams and measuring the thickness of coal seam, and even predicting CBM distribution (Lawrence, 1991). However, it would be worthwhile conducting a forward modeling exercise before launching any field experiments as it provides "a priori" information for subsurface mapping.

The cross-well seismic response for a realistic multilayer coal seam model from an Australian region by using 2D finite-difference simulations is evaluated. The amplitude versus offset (AVO) analysis has been performed and the characteristics of reflection coefficients for coal seams have been studied. To this end, realistic anisotropy parameters (Elapavuluri and Bancroft, 2001) were used to analyze the reflection coefficient variability of coal seams.

8.4.4 FULL-WAVEFORM MODELING

In acquiring seismic data, a seismic compressional wave (P-wave) is generated with a surface source; the wave travels at the acoustic velocity of the rock, which varies with lithology; the wave bounces off surfaces across which the impedance (product of velocity and density) varies. The strength of the reflection is quantified through a reflection coefficient, which is the difference in impedance over the sum of the impedances. The wave then returns to the surface, where geophones detect the P-waves returning vertically. The time from generation of energy to its recovery at a geophone is the travel time; it depends on the velocities of the units traversed. The amplitude of the recovered energy is governed by the contrasts in velocity and density across the interfaces.

Exploration-oriented seismic modeling requires accurate and efficient methods. There are a variety of seismic modeling algorithms incorporating analytical methods, semi-analytical methods (e.g., reflectivity method), ray-geometric methods (e.g., ray tracing method), and direct methods (finite-difference and finite-element methods). Ray tracing modeling has the advantage of being fast and easy in identifying primary reflections. It is also very efficient for obtaining travel time information but the amplitudes determined with this method are not always reliable. However, elastic wave equation modeling accounts for direct waves, primary and multiple reflection waves, converted waves, head waves, as well as diffraction waves. It therefore overcomes the shortcomings of the ray tracing approach, which fails in many cases, for example, at the edges where the calculated amplitude is infinite or in the shadow zone where the amplitude is zero.

Real rocks are not isotropic, and any such simplification in modeling the seismic response of geological structures may lead to misinterpretation or, worse, to overlooking useful information. Therefore, an accurate description of wave propagation requires a rheology, which accounts for the anisotropic behavior of the rocks, as the reservoir rocks show effective anisotropy in the seismic band. The presence of anisotropy in the earth, which manifests itself most diagnostically in terms of shear-wave splitting in multicomponent seismic data, can lead to substantial complications in the processing and interpretation of both surface seismic and vertical seismic profile (VSP) shear-wave data. However, analysis of anisotropic wave-propagation phenomena, such as shear-wave splitting or birefringence, could lead to more information, such as the strike direction and density of vertical fractures within reservoirs. Anisotropy may be caused by fine layering in sedimentary rocks, preferred crystal orientation in crystalline solids, orientation of grains in sedimentary rocks, or stress-aligned fractures or cracks. Transverse isotropy with a vertical symmetry axis is the most extensively studied type of anisotropy, which is assumed in our studies. Forward modeling plays a critical role in understanding many imaging issues in structurally complex areas, and is particularly important in designing the optimal program for a cross-hole survey. It would be worthwhile conducting forward modeling before conducting any field experiments as it provides a priori information for subsurface mapping such as identification of the top of the coal in open-cut mines.

The investigations carried out in this example are based on full-wave modeling using a finite-difference solution for the vector wave equation. In the drilling environment, the elastic wave equation modeling is attempted and seismograms are computed for a vertical source recorded on multicomponent geophones placed vertically in transversally isotropic media, and the following differential equations (Rajput et al., 2008) of 2D motion (continuous form) in velocity-stress terms are used:

$$\frac{\partial g^x}{\partial t} = a_{11}\frac{\partial f^x}{\partial x} + a_{13}\frac{\partial f^y}{\partial y} + a_{15}\left(\frac{\partial f^x}{\partial y} + \frac{\partial f^y}{\partial x}\right) \quad (1)$$

$$\frac{\partial g^y}{\partial t} = a_{13}\frac{\partial f^x}{\partial x} + a_{33}\frac{\partial f^y}{\partial y} + a_{35}\left(\frac{\partial f^x}{\partial y} + \frac{\partial f^y}{\partial x}\right) \quad (2)$$

$$\frac{\partial g^s}{\partial t} = a_{15}\frac{\partial f^x}{\partial x} + a_{35}\frac{\partial f^y}{\partial y} + a_{55}\left(\frac{\partial f^x}{\partial y} + \frac{\partial f^y}{\partial x}\right) \quad (3)$$

$$\frac{\partial f^x}{\partial t} = P\left(\frac{\partial g^x}{\partial x} + \frac{\partial g^s}{\partial y}\right) \quad (4)$$

$$\frac{\partial f^y}{\partial t} = P\left(\frac{\partial g^s}{\partial x} + \frac{\partial g^y}{\partial y}\right) \quad (5)$$

where f^x and f^y are the horizontal and vertical velocities of deformations respectively, g^x and g^y are the horizontal and vertical stresses, respectively, g^s is the shear stress, $P = \frac{1}{\rho}$ where ρ is the density, and a_{11}, a_{13}, a_{15}, a_{33}, a_{35} and a_{55}, are components of the stiffness tensor.

8.4.5 AMPLITUDE VERSUS OFFSET MODELING

Seismic AVO is a variation in seismic amplitude with change in distance between source point and the receiver. It characterizes the changes in elastic properties of fractured reservoirs such as coal beds better than the more conventional seismic techniques. The angular dependence of reflection coefficients for plane waves impinging on a plane interface between two semi-infinite, isotropic, homogeneous, elastic media is exactly described by the Zoeppritz's equation (Bortfeld, 1961; Chapman, 1976; Richards and Frasier, 1976; Aki and Richards, 1980; Shuey, 1985). These exceedingly complex equations predict that the reflection coefficient variation with the offset is dependent on the changes in medium P-wave velocity and density, thereby providing the theoretical basis for AVO analysis. For the case of normal incidence Zoeppritz equation reduces and is given as:

$$R_P = \frac{\rho_2 v_2 - \rho_1 v_1}{\rho_1 v_1 + \rho_2 v_2} \quad (6)$$

This equation states that the reflection amplitude, compared with the incident amplitude, varies directly as the change in AI. In AVO modeling, Zoeppritz equations were used to generate the exact solution. The approximation to the Zoeppritz equation was given by Aki and Richards (1980).

8.4.6 CROSS-WELL SEISMIC

Cross-well seismic is the process of acquiring seismic energy through the formation of interest, between two wells, rather than from the surface as with conventional seismic acquisition. Acquiring seismic data in this format eliminates the unconsolidated near-surface layers and results in higher frequencies at the receivers; thus providing a high-resolution profile of the reservoir between well-bores (Meyer et al., 2002; Somerville et al., 2007). The application of cross-well seismic will assist in ensuring that locations are strategically placed thus reducing the environmental footprint and streamlining the development process. This is especially important as well density increases. There are a number of unique advantages realized when using cross-well seismic modeling to delineate shallow gas reservoirs. The high bandwidth and thus high spatial resolution make it possible to image subtle features including thin zones, subseismic faulting, flow conduits, channel sands, and impermeable shale stringers (Somerville et al., 2007). In addition, cross-well seismic is referenced in depth thus eliminating the ambiguity of time to depth conversion. Here we used the 2D finite-difference method to model the cross-well seismic response for a realistic model from an Australian region. The model (Fig. 8.14) is a simplified version of the real lithological unit from an Australian region. The model parameters are listed in Table 8.1.

Transverse isotropy (TI) was considered in all the numerical simulations. TI is quantified through Thomsen parameters ε, δ, and γ. Parameter ε is a measure of P-wave anisotropy, which is the normalized difference in horizontal and vertical velocity whereas δ is more complex and has the strongest effect on

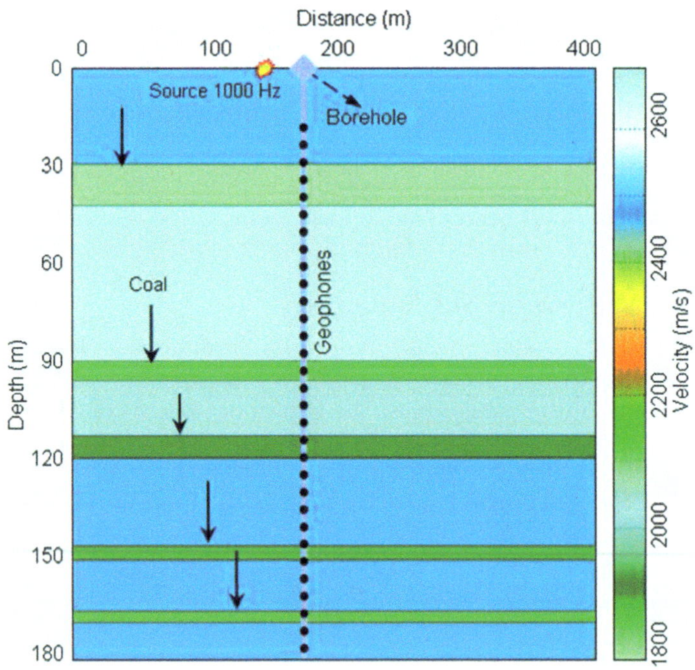

FIGURE 8.14

Flat stratigraphic layer model for vertical moving source and vertical receivers. Black arrows (solid line) show coal layers. Source receiver horizontal distance is 10 m.

Table 8.1 Model parameters for different layers including sandstone and coal. These parameters were used in full waveform modelling

S. No.	Layer Description	Compressional Velocity (m/s)	Shear Velocity (m/s)	ρ (density gm/cm^3)	Gradient
1	Sandstone	2500	1450	2.12	0.5
2	Coal	1800	1050	1.98	0.5
3	Sandstone	2650	1534	2.21	0.5
4	Coal	1850	1075	1.97	0.5
5	Sandstone	2700	1562	2.22	0.5
6	Coal	1900	1100	1.96	0.5
7	Sandstone	2500	1456	2.15	0.5
8	Coal	1875	1088	1.98	0.5
9	Sandstone	2550	1465	2.18	0.5
10	Coal	2115	1219	2.05	0.5

P- and vertically polarized shear waves at intermediate angles. In this study, we disregard γ which only affects horizontally polarized shear waves. The measured anisotropic parameters ($\varepsilon=0.12$ and $\delta=0.24$) for coal (Elapavuluri and Bancroft, 2001) are used in the model. Using the model (Fig. 8.14) synthetic seismograms have been calculated at two different source locations. Instead of considering all the source points only two scenarios were considered when the source was positioned at 1 m and 25 m depth. In both cases, all the coal layers were identified. The top of the coal was reflecting reverse polarity due to the strong AI contrast between sediments and coal. The only difference between the two different source location responses is related to the identification of the first coal layer which starts at 30 m depth.

Fig. 8.15 represents the simulated seismic responses where the source is positioned at 25 m depth below the surface. After 120 ms the signal is not clear and requires sophisticated processing workflow. The top and bottom of the first coal layer has been identified and marked on all the components. Further, wave-field propagation at two different time steps (0.022 s and 0.052 s) was analyzed to see the energy reflected from the coal layers and energy refracted/transmitted through the coal layers (Fig. 8.16). The top of the coal exhibited the reverse polarity compared to the bottom of the coal layer.

Fig. 8.16a shows the reflection from the top and the bottom of the first coal seam whereas Fig. 8.16b represents the reflected arrivals from three coal layers, i.e., 30 m, 90 m, and 113 m, which are marked on the figure. The source receiver spacing was then increased to 50 while other parameters were kept unchanged. The simulated seismic response for this scenario is illustrated in Fig. 8.17. It has been observed that vertical component is ideal for the identification of all the coal layers whereas due to the presence of converted shear waves, the horizontal component requires more specialized workflow. The first layer is clearly identified. However, detection of further layers solely on the basis of the horizontal component poses some problems and requires further signal processing tools to interpret. Fig. 8.17 shows the vertical and horizontal components from 0 to 100 ms. After careful analysis it is possible to identify the top and bottom of the coal layers.

Further, wave-field propagation at two different time steps (0.018 s and 0.050 s) was analyzed to see the level of energy traveling and reflecting off the coal layers. Fig. 8.18a and b shows a snapshot of wave-field propagation at 0.018 s and 0.050 s, respectively. At this stage, two reflections (up-going

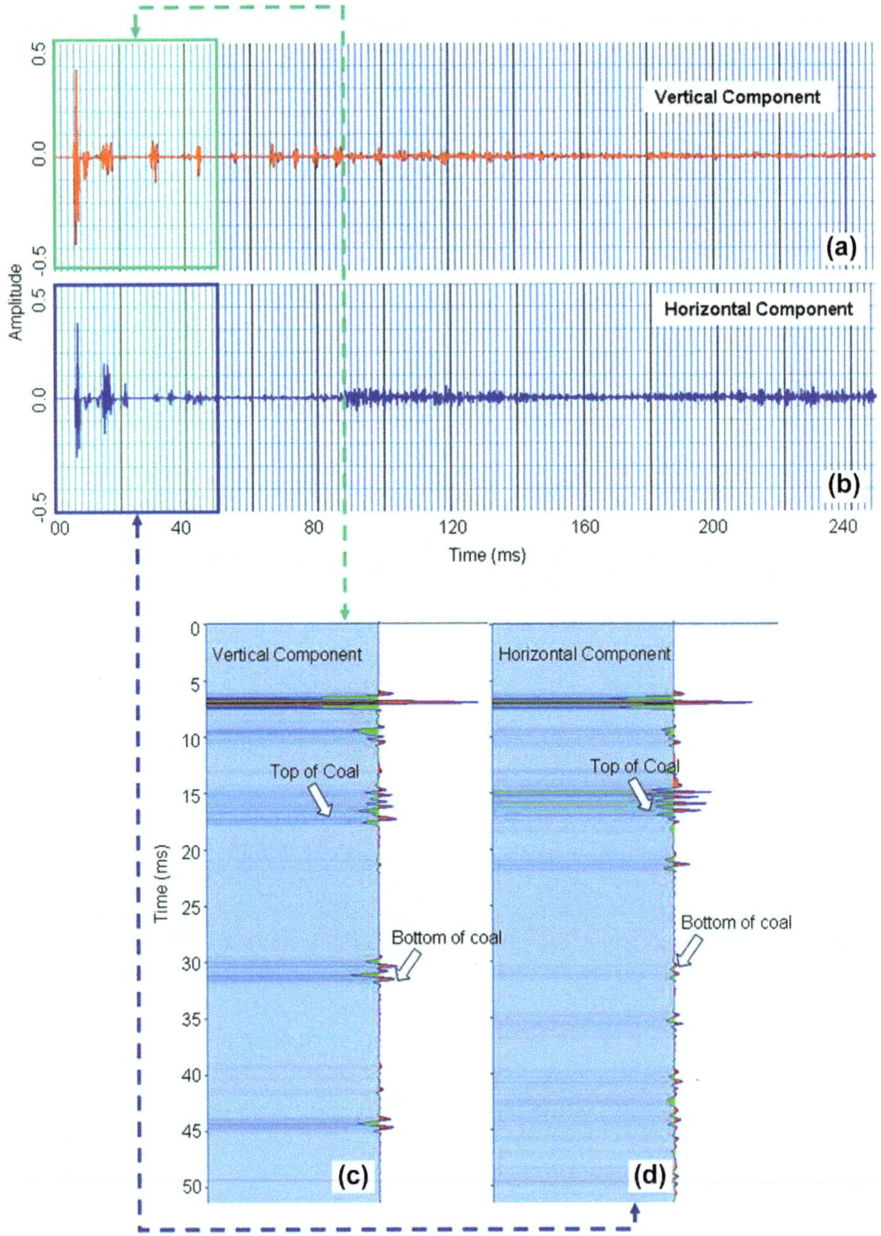

FIGURE 8.15

Single receiver response for two components (flat layer model Fig. 8.14). The source is positioned at 25 m depth. (a) Vertical component response up to 250 ms. (b) Horizontal component response up to 250 ms. (c) Represents the enlarged green box area from part (a) up to 50 ms. (d) Shows the enlarged blue box area from part (b) up to 50 ms.

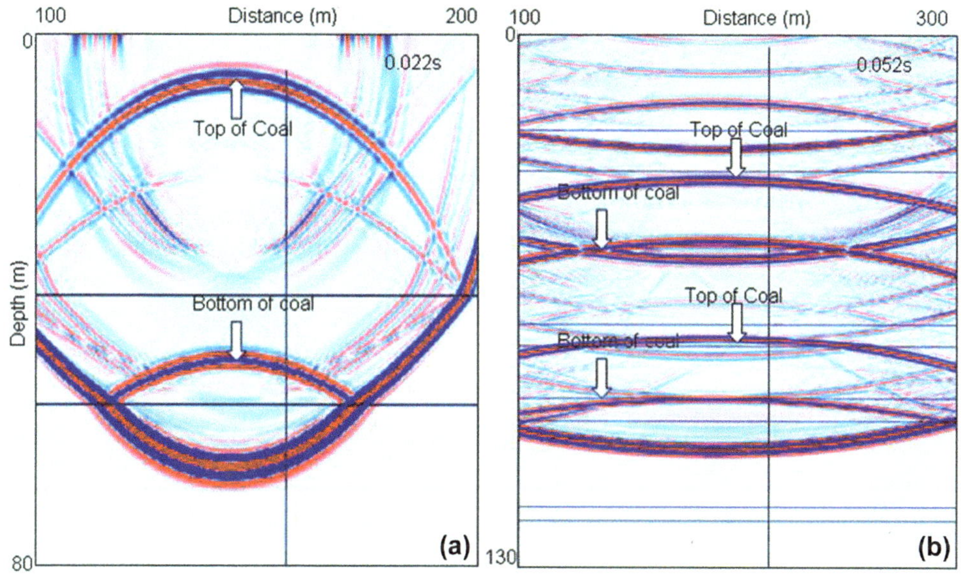

FIGURE 8.16

Wave-field propagation at two different time steps (0.022 s and 0.052 s). The elastic wave-field showing both up- and down-going waves; the top of the coal layer exhibits the reverse polarity. (a) A snapshot of wave-field at 0.022 s for flat layer model. (b) The wave-field propagation at 0.052 s.

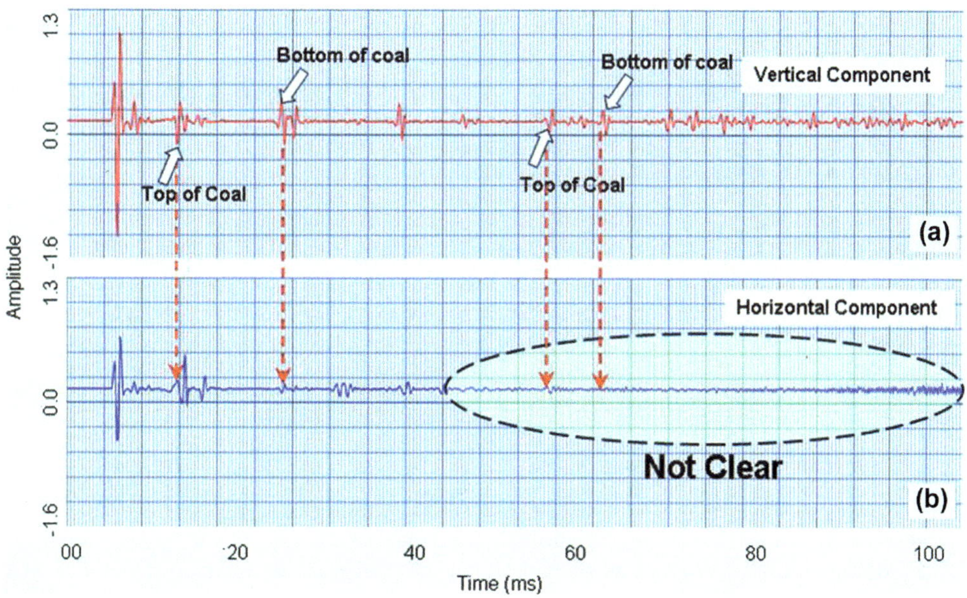

FIGURE 8.17

Simulated seismic response two components (flat layer model). The source is positioned at 1 m depth and receiver positioned at 5 m depth with 50 m source receiver spacing. (a) Vertical component response up to 100 ms. (b) Horizontal component response up to 100 ms.

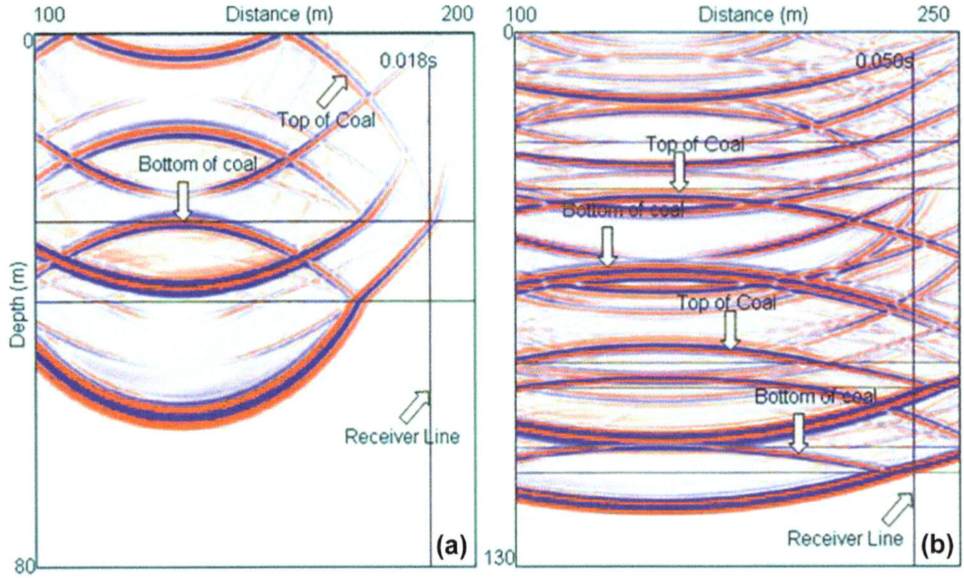

FIGURE 8.18

Wave-field propagation at two different time steps (0.018 s and 0.050 s). (a) A snapshot of wave-field propagation at 0.018 s. (b) Wave-field propagation at 0.050 s.

energy) from the top and the bottom of the first coal layer have been clearly identified. The top of the coal exhibits the reverse polarity. At 0.050 s the reflected arrivals from three coal layers, i.e., 90 m, 113 m, and 146 m, have been recorded. Further, the top and the bottom of these layers have also been identified and marked on the profile (Fig. 8.18).

8.4.7 AVO CHARACTERISTICS

To see the AVO response, the reflection coefficients were calculated for all the coal layers (top of the coal) and plotted in Fig. 8.19. The graph depicts a negative trend up to a critical angle (30°). The values of Thomson anisotropic parameters are $\varepsilon=0.12$ and $\delta=0.24$ (Elapavuluri and Bancroft, 2001). The Q-compressional value of 10 and Q-shear value of 8 were used, which is one order of magnitude less than the reported value in the literature. The magnitude of the AI anomaly suggests that a reduction in the elastic stiffness of the coal matrix frame may have been detected.

8.4.8 SOME REMARKS

Increasing resolution has always been challenging in seismic prospecting. Surface seismic provides the fundamental information for mapping and drilling decisions whereas cross-well seismic provides very high-resolution images for detection of thin and complex structures. In this example, finite-difference analyzes are attempted to simulate seismic response for a multilayered complex coal seam model, and thin coal seams are differentiated from other lithologies. Field data from the Australian region is utilized to generate a simplified model and generate synthetic seismic data for different source receiver

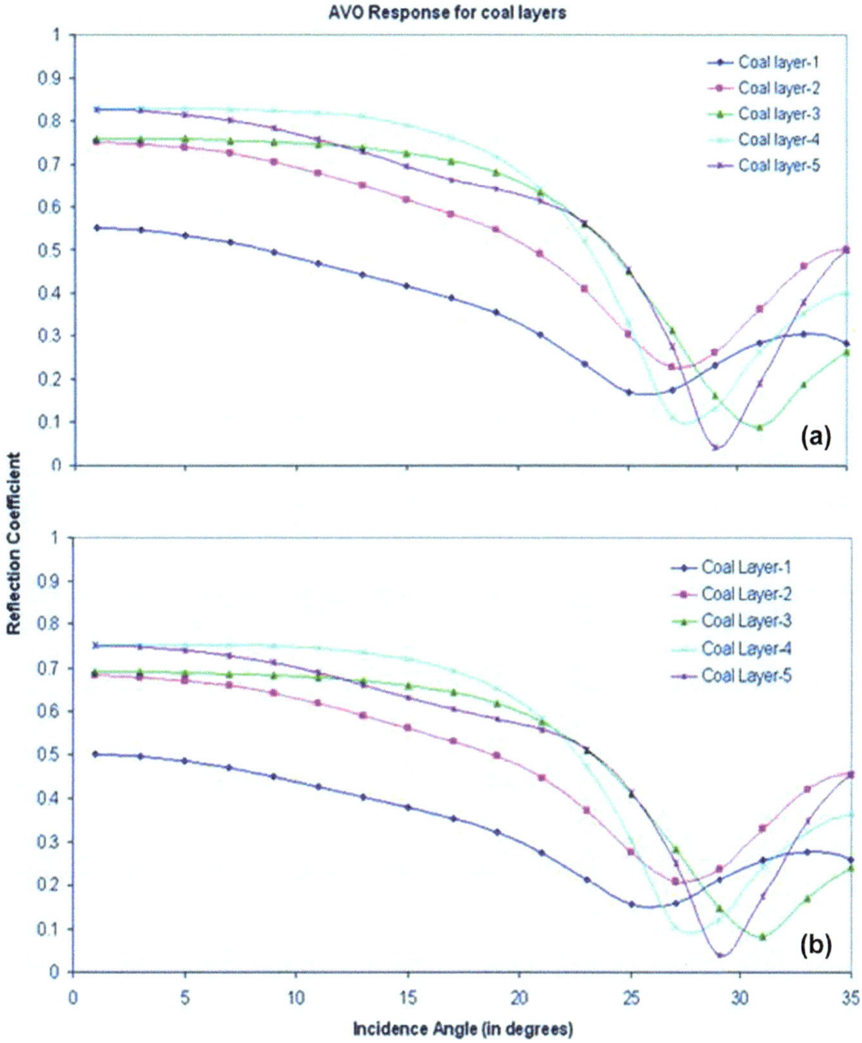

FIGURE 8.19

Comparison of AVO curves produced for a multilayered anisotropic model. The reflection coefficient is calculated for all the five coal layers. The AVO response is modeled up to 35° (a) In this case the source receiver spacing is 50 m. (b) In this case the source receiver spacing is 10 m.

configurations. The results reveal that seismic modeling with high-frequency source (1000 Hz) is feasible in delineating the top and bottom of thin coal layers for closer source and receiver configurations. The behavior of seismic amplitude could help in the detection of a sequence of thin coal layers. AVO analysis for transverse isotropic coal seam models shows a negative anomaly for coal layers with considerable variation in reflection coefficient for different configurations. Interference of thin coal layers seems to diminish the reflectivity of the interfaces. It can be concluded that 'a priori' knowledge can be gained through full-waveform seismic modeling, which can provide guidance in planning strategies for field surveys in a complex thin-layered coal seam region.

The modeling results show that the coal interval can be mapped using a seismic reflection technique. Thin coal layers can be detected on the cross-hole sections using vertical and horizontal components when the source and receiver are placed 10 m distance apart. However, only the top two coal layers were identified, when the distance between the source and receiver was 50 m. The top of the coal exhibited reverse polarity due to the strong AI contrast with respect to the host sediments. AVO response shows negative anomaly for all the coal layers, which could be acceptable up to the critical angle (~30°). For the identification of coal layers, the vertical component should be used as the processing of the horizontal component requires sophisticated tools. Forward modeling provides great insights that might help in our understanding of the field trial results.

8.5 GAS HYDRATES

The large volumes of methane stored in worldwide gas hydrate deposits represent a significant fraction of the global methane budget, and may be an almost inexhaustible resource for the future (Kvenvolden, 1993a, 1993b, 2000; Kvenvolden et al., 1993; Kvenvolden and Lorenson, 2001). Geophysical techniques such as seismic reflection/refraction, electromagnetic and geological signatures such as fluid flows, and geochemical investigations with support from logging have been extensively used to explore for the occurrence of gas hydrates on the continental margins and in permafrost regions of the world (Cherskiy et al., 1985; Hornbach et al., 2003).

The presence of higher acoustic velocity hydrate, either cementing sediment grains or pore filling in the sediments, appreciably enhances the velocity of the host medium. The impedance contrast between a high-velocity, hydrate-rich layer and a low-velocity, brine/gas-saturated layer results in the formation of a seismic reflector. These reflectors, which occur at the base of the hydrate zone, have high amplitude with reverse polarity, and are known as bottom-simulating reflectors (BSRs) (Berndt et al., 2004; Pecher et al., 1996; Singh et al., 1993). They can be considered as one of the prime seismic indicators of the presence of gas hydrates (Tucholke and Bryan, 1977; Shipley et al., 1979; Hyndman and Spence, 1992; Andreassen et al., 1997; Stoll et al., 1971; Bangs et al., 1993; Pecher et al., 1998, 2004, 2005, 2007, 2008; Mienert et al., 2001). Hydrates, by filling a portion of the pore spaces above the base of the gas hydrate stability zone (GHSZ), effectively decrease the permeability of the sediment (de Boer et al., 1985), and thereby work as a trap for the underlying free gas-bearing sedimentary layer (Holbrook et al., 1996; Ecker et al., 1998; Sain et al., 2000; Gorman et al., 2002; Holbrook et al., 1996; Hovland and Judd, 1988; Hovland and Svensen, 2006). BSRs act as the demarcation boundary of the base of the GHSZ and the top of free gas (Stoll et al., 1971; Wood et al., 2002; Tréhu et al., 2004).

To study the characteristics of seismic wave propagation in hydrate-saturated sediments, it is essential to estimate the response for different gas hydrate models (Rajput, 2009). Seismic waves propagating below the seafloor undergo transmission, reflection, and mode conversion at each interface of a layered medium. Marine seismic surveys use conventional air gun sources, which generate P-waves only; therefore, the shear waves recorded by sensors on the ocean bottom are mode-converted (PS) shear waves. The two most common mode-conversion (Fig. 8.20) types are: (1) P-wave during transmission converts to S-wave at an interface (say, at the seafloor) and then propagates as S-wave to the ocean bottom (PSS- or SS-wave type) and is recorded by the receiver; and (2) P-wave converts to

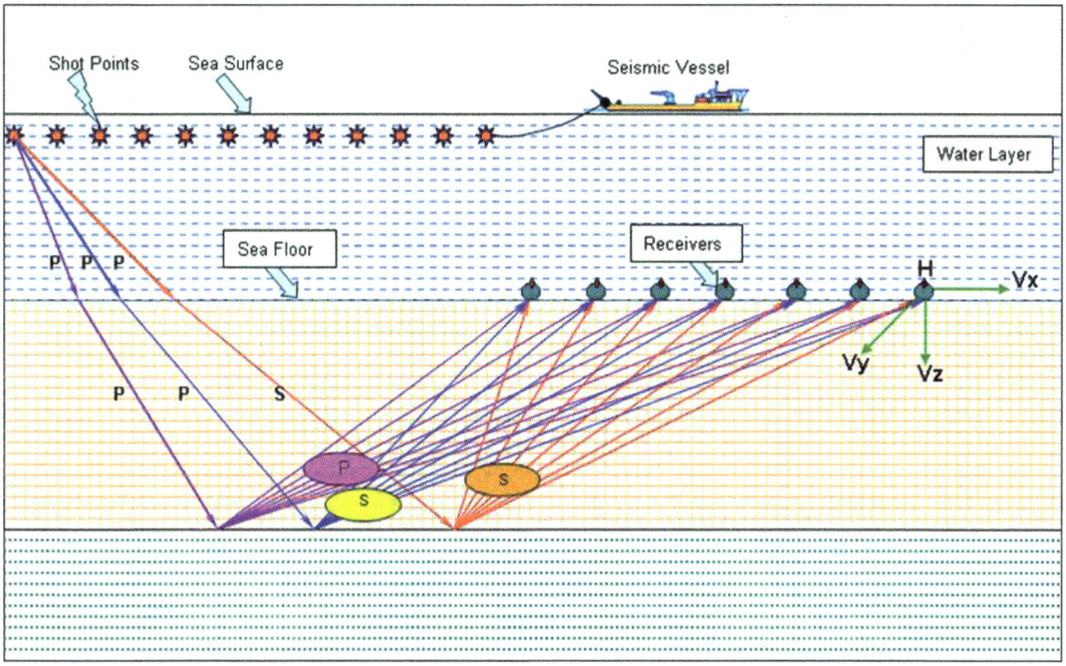

FIGURE 8.20

Converted wave survey geometry in marine environment. The green circles with a red stars on top denote the position of deployed ocean bottom receivers. The last receiver indicates all four components (H, X, Y, Z). Seismic waves (compressional and shear) are represented by P and S letters. Three possible cases are shown and indicated by different colors. First case of transmitted and reflected energy is represented by violet color arrows. Seismic energy travels as P wave in the water then transmitted as P wave in the sedimentary column and then reflected back as P wave which is recorded by receivers. This wave can be identified as PPP (transmitted P in water and sediments and reflected P wave). In the second case (blue color wave), the P-wave is converted to S-wave on the reflector (PSP); and in the third case (red color wave), the ocean bottom and the reflector are playing as a conversion boundary (PSS).

S-wave at a reflector (reflected wave) and then propagates back as S-wave to the receiver (PS-wave type). There may be many more conversions from P- to S-wave (and vice-versa) during reflection and transmission at intermediate interfaces, but they are typically too weak to be identifiable (Stewart et al., 2002; Stewart, 1991; Stewart and Ferguson, 1996; Stewart et al., 2001; Berg et al., 1994; Caldwell, 1999; Chiburies, 1987; DeVault, 1997).

For a layered earth model, SS-wave data can be analyzed in the same manner as the P-wave data (due to its symmetrical ray path). However, P- to S-wave mode conversion as a transmitted wave at the seafloor (or any other layer interface) depends on the S-wave velocity below the conversion surface (compared to the P-wave velocity above the interface). In the shallow subsurface, S-wave velocity at (or near) the seafloor is very low (50 to 250 m/s, Ritzwoller and Levshin, 2002; Gaiser, 1996; Knopoff, 1972; Xia et al., 1999) and may encounter higher values in the presence of carbonates or gas hydrates. In shallower parts of the earth, P- to S-wave mode conversion during transmission is very weak

(Mari, 1984; Cara 1973; Dziewonski et al., 1969). The PS-wave is better recorded than the SS-wave since it suffers less attenuation and is faster than the SS-wave. However, owing to lower S velocities SS reflections have better vertical resolution. Multicomponent seismic observations record both P-P and P-S converted data. The P-S waves recorded by the sensor provide additional information about the target horizons and may provide the better estimate of petrophysical properties, which cannot be determined with P-wave data alone. Detailed identification of the P-S mode in gas hydrate space has not been well documented. In this example, the seismic reflection characteristics associated with gas hydrates for different gas hydrate/free gas models have been identified and interpreted. The P-S mode characteristics and overburden effect on seismic amplitudes at the gas hydrate boundary, which provides insight into the mechanism of hydrate formation, is also studied. The focus of this attempt is to better understand mode-converted shear waves for complex gas hydrate models. In the context of the present studies, we interpret the occurrence of BSR and its behavior through discrete layers of gas hydrates and free gas in a multichannel seismic section acquired from the Green Canyon in the Gulf of Mexico, offshore United States and Kerala-Konkan (KK) basin offshore India.

8.5.1 SEISMIC DATA ANALYSIS

2D full-waveform modeling using a finite-difference solution for the electrodynamics wave equation (Rajput et al., 2005, 2010) is used to simulate the seismic response for realistic models. Synthetic seismograms have been generated for multilayer gas hydrate models. The schematic model along with model parameters is shown in Fig. 8.21.

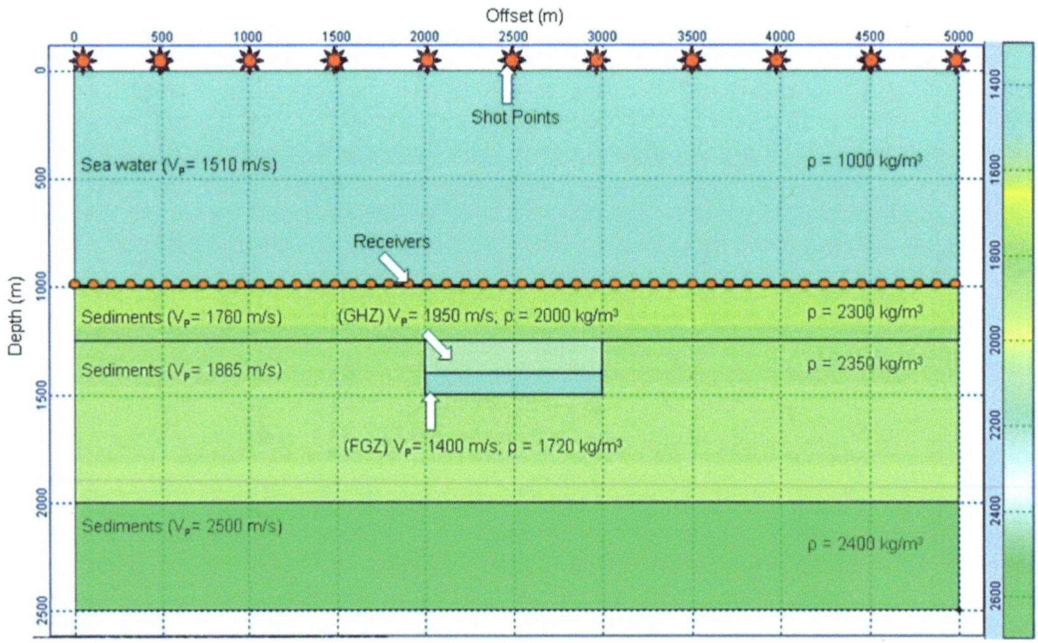

FIGURE 8.21

Key model used for simulations. Gas hydrate zone (GHZ) and free gas zone (FGZ) are located at 1250 m and 1400 m depth respectively. Receivers are deployed on the ocean bottom. 30 Hz shots are located at 15-m depth.

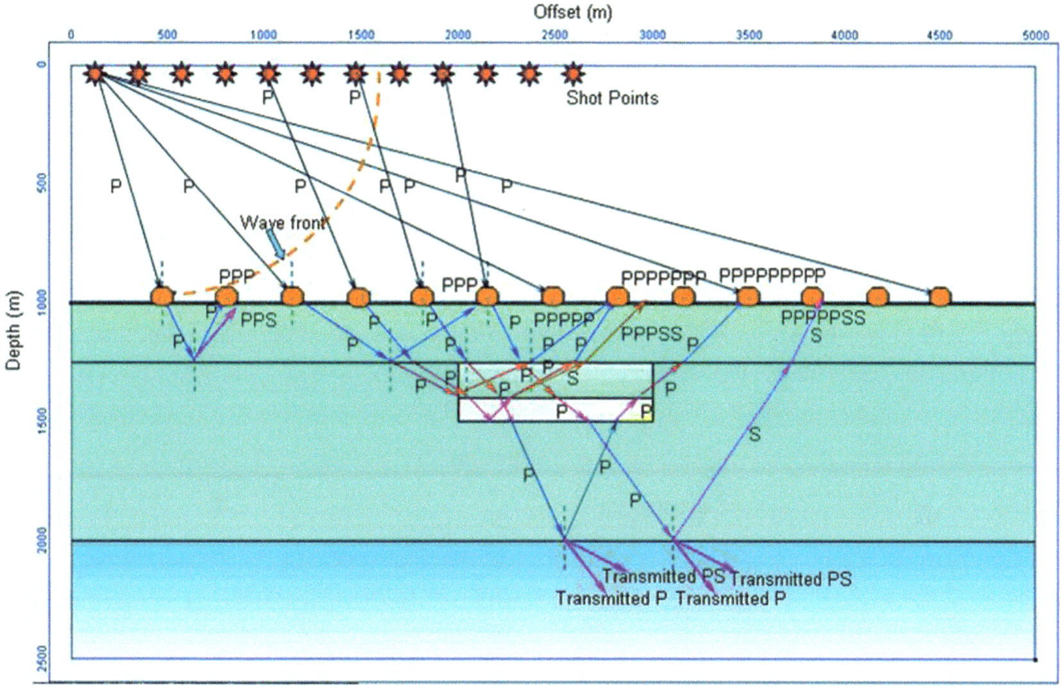

FIGURE 8.22

Converted wave ray geometry for the key model. The transmission of the seismic energy within the earth in the form of compressional and converted mode is demonstrated. Here P stands for compressional mode and PS stands for converted mode from compressional to shear. Main conversions from seafloor and gas hydrate boundary are indicated.

Water depth is considered at 1000 m with compressional wave velocity $V_p = 1510$ m/s and density $\rho = 1000$ kg/m^3. Initially five layers of different physical parameters are considered, which include gas hydrate and free gas layer. Gas hydrate layer is considered at 1250 m depth and has 150 m thickness with $V_p = 1950$ m/s and $\rho = 2000$ kg/m^3. Free gas is at a depth of 1400 m and thickness of 100 m with $V_p = 1400$ m/s and $\rho = 1720$ kg/m^3. Three different layers of sediments with $V_p = 1760$, 1865, and 2500 m/s and $\rho = 2300$, 2350, and 2400 kg/m^3 constitute the host medium. The model shown in Fig. 8.22 has been modified several times to include the effect of a low-velocity layer within the gas hydrate zone. The gas hydrate/free gas model for a number of point sources located 10 m below the sea surface with different receivers deployed at ocean is presented in Fig. 8.22. Mode conversions at different interfaces for seismic reflection trajectories have been identified and schematically represented. Synthetic seismograms have been generated and the characteristics of seismic reflections at simulating reflectors associated with gas hydrate and free gas are studied.

In order to study the mode conversions, overburden effects on the amplitude of BSRs, and various seismic indicators for the occurrence of gas hydrates and associated free gas, a detailed modeling exercise is carried out through the simulation of seismic data for different multilayered gas hydrate models. These are based on the detailed review of physical properties and sediments of the upper several meters of sediments from different parts of the world (Pecher et al., 1997; Winters et al., 2000, 2002, 2004).

In particular, the main assumptions include acoustic boundaries which separate layers with different velocities, densities, and thicknesses. To validate the considered densities and velocities in the present investigations and make them more meaningful, previous investigations of gas hydrates in sediments were reviewed (Shipley et al., 1979; Whalley, 1980; Pearson et al., 1986; Davis et al., 1990; Miller et al., 1991; Hyndman and Spence, 1992; Rajput et al., 2010; Suess, 2002; Diaz-Naveas, 1999; Morales et al., 2001; Hyndman and Spence, 1992; Hyndman and Davis, 1992; Dillon et al., 1994, 1996, 1998; Lee et al., 1993; Bouriak et al., 2000; Collett 1995, 2002, 2004; Collet et al., 2000, 2008).

8.5.1.1 Mode Conversions and Overburden Effect

Synthetic seismic reflections recorded by receivers at the ocean floor for the single point source located in the middle of the seismic profile in seawater at 10 m depth are shown in Fig. 8.23. For this computation we use the model shown in Fig. 8.20.

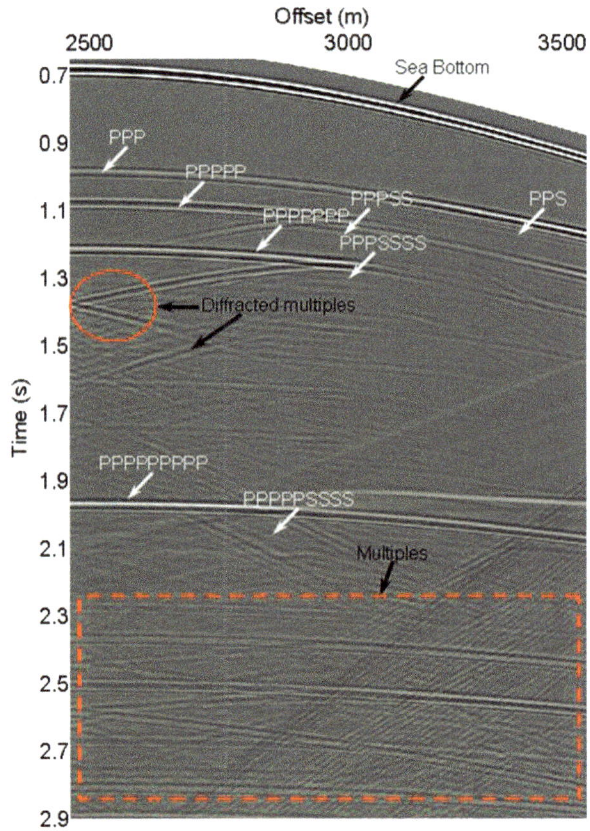

FIGURE 8.23

Synthetic seismic reflections recorded by receivers at sea bottom with obvious amplitudes. PPP, PPPPP, PPPPPPP, PPPPPPPPP, PPS, PPPSS, PPPSSSS, and PPPPSSSSS correspond to identified mode conversions in Fig. 8.22. The wave type letters P and S represent compressional wave (P-wave) and Shear-wave (S-wave) propagation, respectively. Multiples are also marked.

High-amplitude direct wave (seafloor reflection), reflected waves from the sea bottom are clearly visible on the seismic section. By the order of the arrival time seismic waves can be identified as PPP, PPPPP, PPPPPPP, PPPPPPPPP, PPS, PPPSS, PPPSSSS, and PPPPPSSSS reflection. The letter P and S represent P-wave and S-wave (converted) propagating in each layer respectively. PPP represents down-going P-wave in water, then down-going P-wave in sedimentary layer, and then reflected P-wave recorded by receivers, which are placed at the ocean bottom.

Most of the reflections (e.g., PPS, PPPSS, etc.) have partial shear-wave ray paths and experience multiple conversions at the reflectors, apart from some of the reflections (e.g., PPP, etc.). Some of the seismic reflections are due to transmitted mode conversion on the solid–gas interface. These reflections could be of higher amplitudes than those of the P-wave path reflections on the other components. Such characteristics may be given more attention during processing and interpretation of seismic data. The other characteristics of the reflection section are that the order of the arrival time of all reflections does not correspond to the depth order of the seismic interface. From dominant reflections and wave type numbers (PPP and PPS, etc.) the reflection properties and multiconverted diffractions with multiples on the seismic section can be easily identified.

Synthetic seismograms are also generated for two parallel models. In the first model, there are four discrete layers: the first gas hydrate layer, low-velocity layer, second gas hydrate layer, and a free gas layer. The model parameters remain the same as those used above. Additionally, we introduce a low-velocity layer (V_p=1350 m/s) within the gas hydrate layer. In the second model, all four layers have been considered as sedimentary layers, i.e., there are no gas hydrates or free gas, and other parameters remain unchanged. The horizontal particle velocity components (H component) for both the models are shown in Fig. 8.24.

In the converted ray geometry, we obtain all the primary reflection (e.g., PPS, PPPSS, PPPPSSS, etc.) converted at the interface. The gas hydrate model shows the main conversions within the stability zone (Fig. 8.24a), whereas the model without gas hydrate shows far fewer mode conversions at the sedimentary boundaries (Fig. 8.24b). The results indicate that the gas hydrate and free gas boundary acts as a principal conversion boundary for P-wave to S-wave.

Fig. 8.25 shows the synthetic section corresponding to a six-layer complex gas hydrate model, which includes the multiples above the half space. The multiples are not simply repeated periodically like single reflection from each interface. These types of multiples are the result of interference and overlapping between different wave types. It is difficult to identify the single reflection on the multiples reflection profile. The multiples, including the P-wave and shear-wave, introduce a complexity to the explanation and thus make data processing complicated. The direct arrivals, reflected waves from the different sedimentary layers and GHSZ, are clearly identified. Based on the modeling results it can be concluded that during data processing the first order and higher order multiples should be removed so that different quantitative interpretation procedures can be applied. Diffracted multiples should also be considered while preparing the data for reserve quantifications.

The overburden effect can be modeled in different ways like physical or synthetic modeling. However, synthetic modeling provides a more convenient and straightforward way to generate data to isolate any observed wave-propagation anomalies resulting from variations in the overburden zone. We built a simple model consisting of a 500-m-thick upper layer of water, 200-m-thick second layer of sediments, 100-m-thick layer of gas hydrates, 100-m-thick layer of gas, and 600-m-thick layer of sediments. The other parameters (V_p, V_s, and density) remained unchanged. Synthetic seismic data have been generated and the reflection coefficient of BSRs has been calculated. Computation of the reflection coefficient for BSR is done by using the Haskell–Thomson method (Haskell, 1953, 1960, 1962; Aki and Richards, 1980), and gives insight into the AVO effect in porous gas-saturated sandstones. This computation is

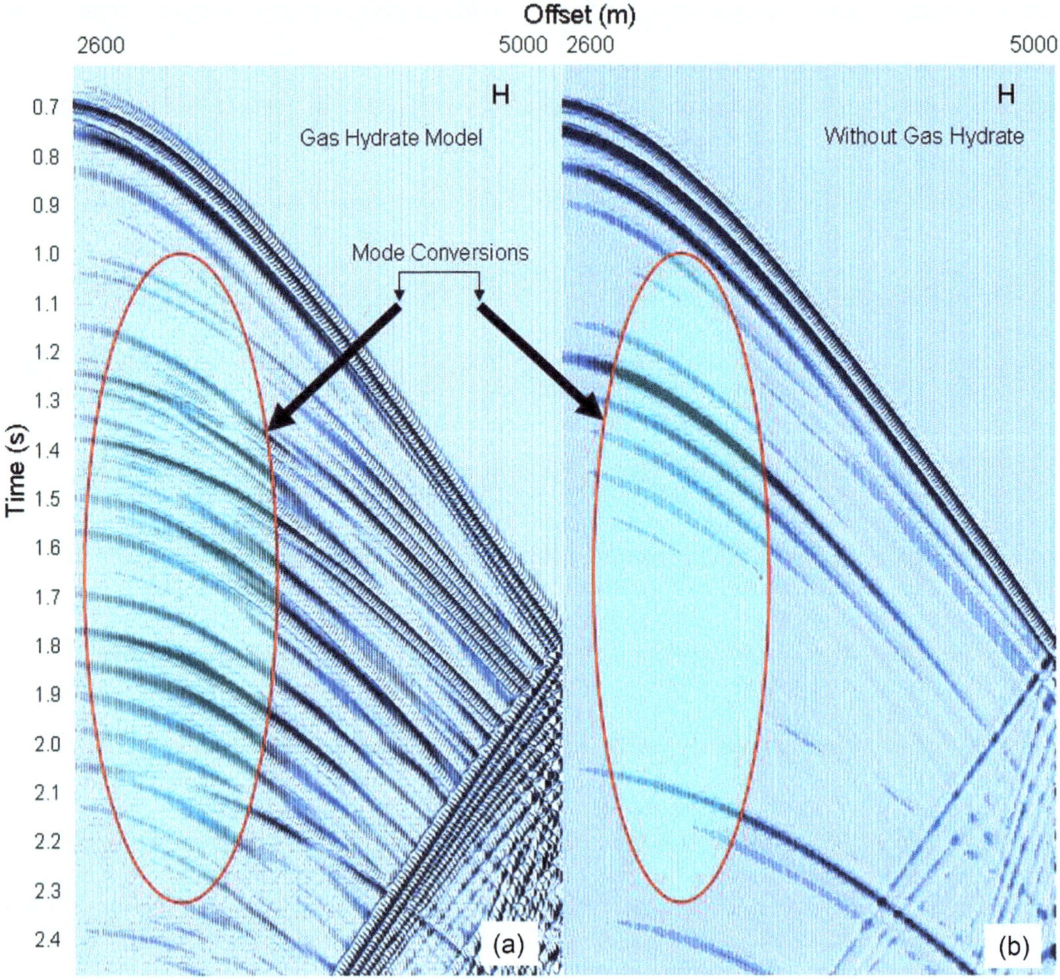

FIGURE 8.24

Synthetic seismic response (H component) for two models assuming gas hydrates and without gas hydrates. The main assumption in multicomponent surveys is that horizontal component (H) records converted waves. (a) Simulations from gas hydrate model shows principal mode conversions; (b) reflections recorded by receivers at sea bottom for the model without gas hydrate indicate very less mode conversions.

carried out in two steps. First, we determine scattering matrices for a set of plane waves with different frequencies. Then we select reflection coefficients from the scattering matrices and convolve them with a Ricker wavelet of a chosen center frequency. This approach allows the reflection and transmission coefficient vs. plane wave dip angle dependency to be stabilized. This fact is well illustrated with the linearized AVO formula for P-wave reflection coefficient (Shuey, 1985). In each step the model was modified by increasing the overburden thickness by 200 m and calculating the reflection coefficients of BSRs. In total, seven steps have been performed to see the overburden effect on seismic amplitude. The

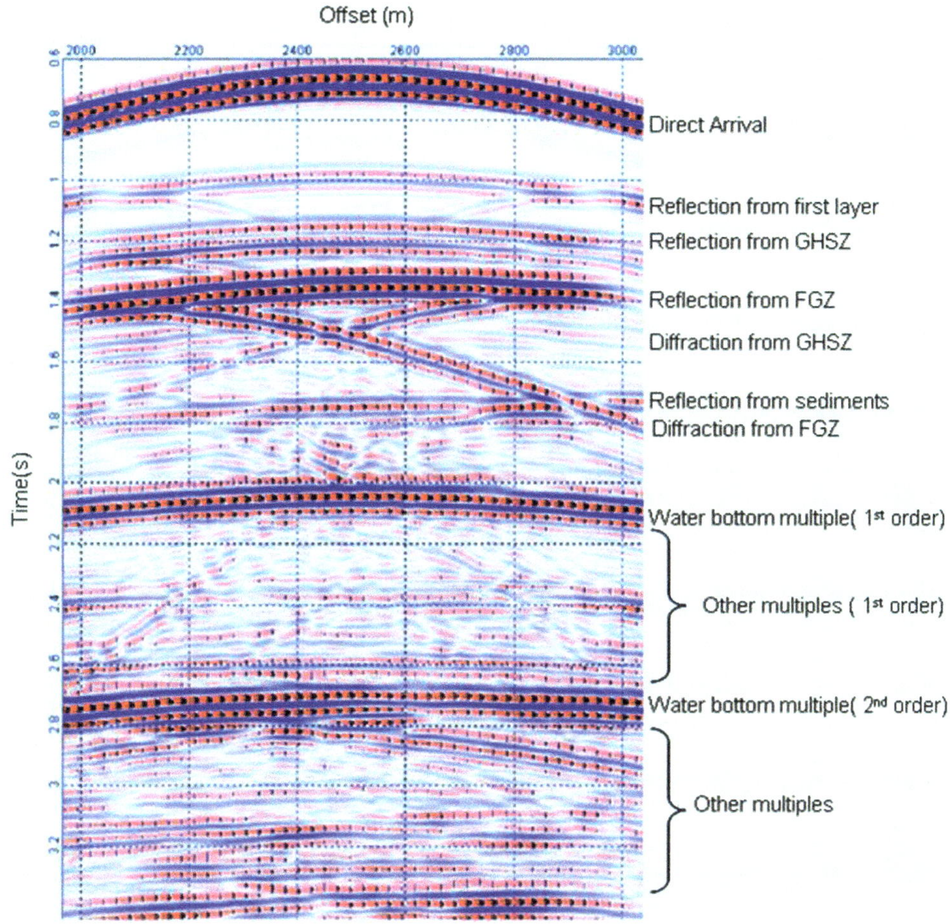

FIGURE 8.25

Vertical component synthetic section of a complex multilayer gas hydrate model which includes primary reflections from different layers and water bottom multiples (first and second order). Diffracted arrivals from gas hydrate zones are also marked.

overburden effect for synthetic data and corresponding AVO analysis is studied. A complex amplitude pattern is observed, which affects rays at all offsets as shown in Fig. 8.26.

This suggests that the overburden zone altered the signal characteristics of the simulating reflector by causing significant amplitude losses. The overburden effect is imprinted on the reflection from the basement producing a misleading interpretation. This scenario could occur in field surveys on far-offset traces that encounter variations in seismic properties along their travel paths and create anisotropy that may not come from the target horizon. This increases the uncertainty for the offset-based seismic amplitude variation. Additionally the effect of water pressure (thick water column) on seismic amplitude is studied. Three models of different water column depth (500 m, 1000 m, and 1500 m) have been taken into account and synthetic seismograms have been calculated as presented in Fig. 8.27.

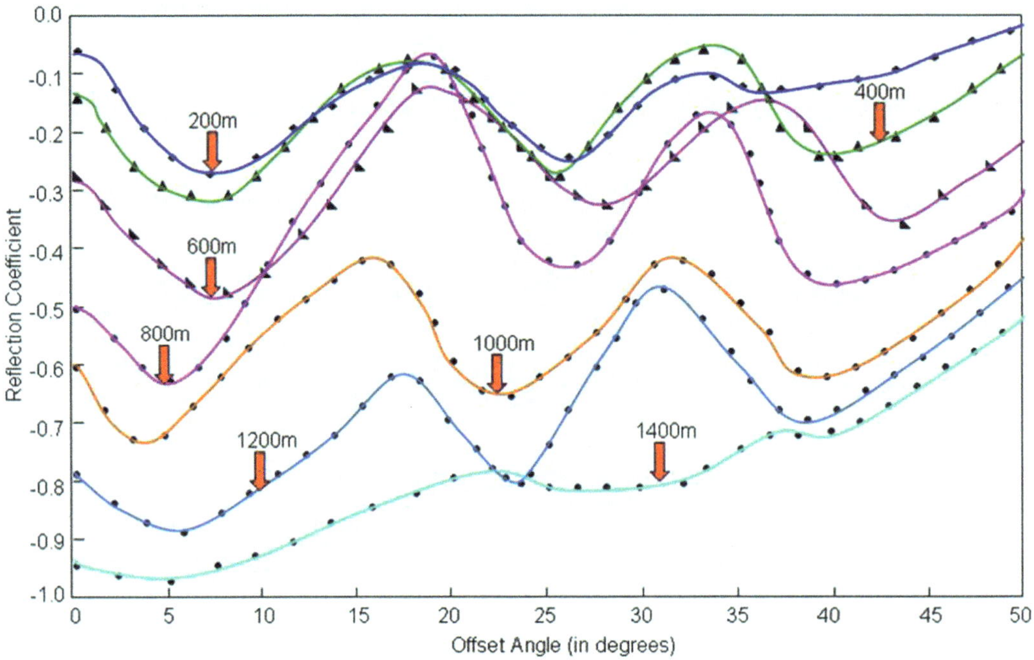

FIGURE 8.26

Amplitude verses offset analysis to analyze the overburden effect for different overburden thicknesses (200–1400 m). A complicated amplitude response suggests that the overburden zone altered the signal characteristics of simulating reflector by causing significant amplitude losses.

The model shown in Fig. 8.27a represents the simulated seismic response for a 500-m water column; the model in Fig. 8.27b represents the simulated seismic response for a 1000-m water column; and the model in Fig. 8.27c represents the simulated seismic response for a 1500-m water column. An increased water column length shows an increase in amplitude strength of different seismic reflectors.The results indicate that water pressure has a significant influence on the amplitude observed in seismic data.

8.5.1.2 Seismic Indicators

In order to study the seismic indicators related to gas hydrates and associated free gas, seismic data from Green Canyon, Gulf of Mexico have been studied. The Green Canyon area is characterized by complex geological features (migration pathways (faults and fractures) and porous sands) which promote the formation of a significant quantum of gas hydrate accumulations. To this end, Green Canyon data is used for simulating the seismic response for seismic indicators related to gas hydrates and associated free gas including BSRs, enhanced seismic reflection below the BSR, and hydrate mounds. These signatures are then correlated with real data from the same region. Different seismic indicators for gas hydrates and associated free gas are identified and discussed.

The Bottom-simulating Reflectors

Initial studies had suggested that BSRs are generated owing to impedance contrast between the high acoustic velocities of hydrate-filled sediment overlying the lower velocities of the free zone (Mienert

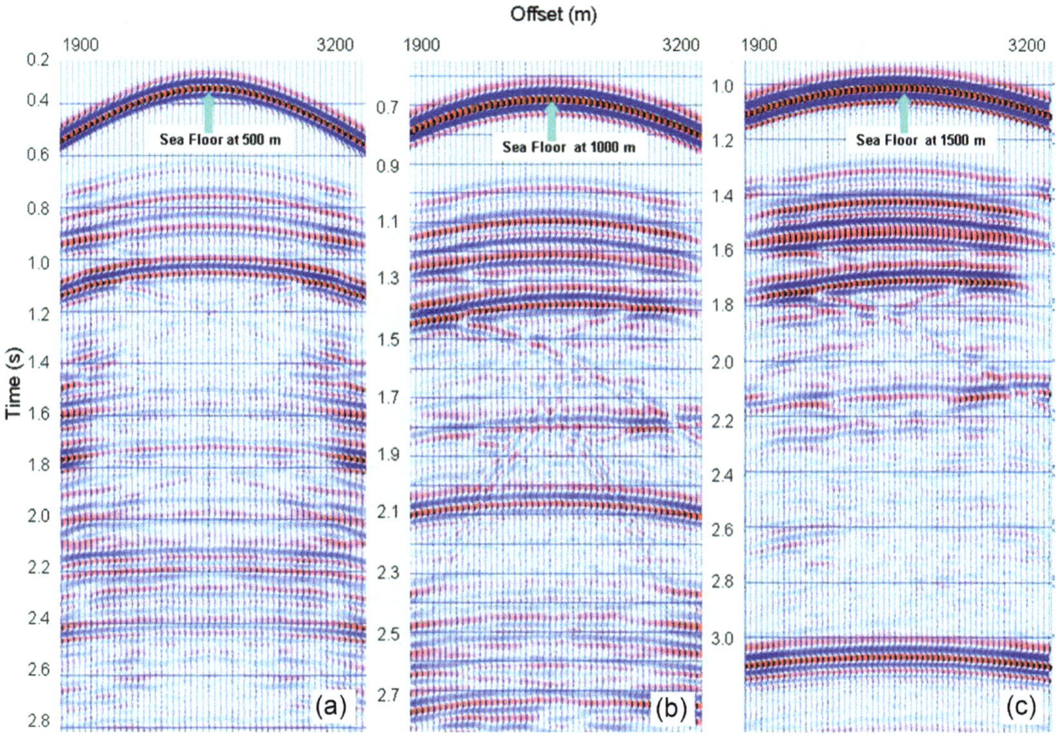

FIGURE 8.27

Synthetic seismic response for three different models of varying water pressure column. Different amplitude strengths are observed for all the cases: (a) 500-m water pressure column; (b) 1000-m water pressure column; (c) 1500-m water pressure column.

et al., 1998; Lee et al., 1993; Dvorkin and Nur, 1993; Dvorkin et al., 1991). In recent years, seismic modeling studies and high-resolution, wide-angle data have indicated that the strength of BSRs is governed by the amount of free gas beneath the hydrate layer (Chand and Minshull, 2003; Andreassen et al., 2000, 2003). Holbrook et al. (1996) argue that in the absence of free gas, a BSR may not be generated. Petrophysical data from the Blake Ridge area also support such an interpretation. The presence of a BSR may lead to an inference being drawn for the presence of gas hydrates; on the contrary, the absence of a BSR in a region does not signify the absence of gas hydrates.

The BSR is commonly observed (Fig. 8.28); it is well identifiable in the Green Canyon and KK basin, India (Fig. 8.29). In some places, the BSR exhibits low to high amplitude and with poor to moderate continuity in the region of study (Ruppel et al., 2008; Berndt et al., 2004, Foucher et al., 2002; Helgerud et al., 1999). It has been observed that BSRs in general have higher amplitude strength over the anticlines and in the vicinity of faulting, and there is appreciable reduction in amplitude strength over the basin slopes (Gei and Carcione, 2003; Andreassen et al., 2000; Hyndman and Spence, 1992; Shipley et al., 1979). One of the significant characteristics of a BSR is that it cross-cuts the sedimentary strata when the bedding planes are dipping (Wood and Ruppel, 2000; Vanneste et al., 2002a, 2002b; Markl et al., 1970). Over the Green Canyon the shallowest BSR in this region is observed at around

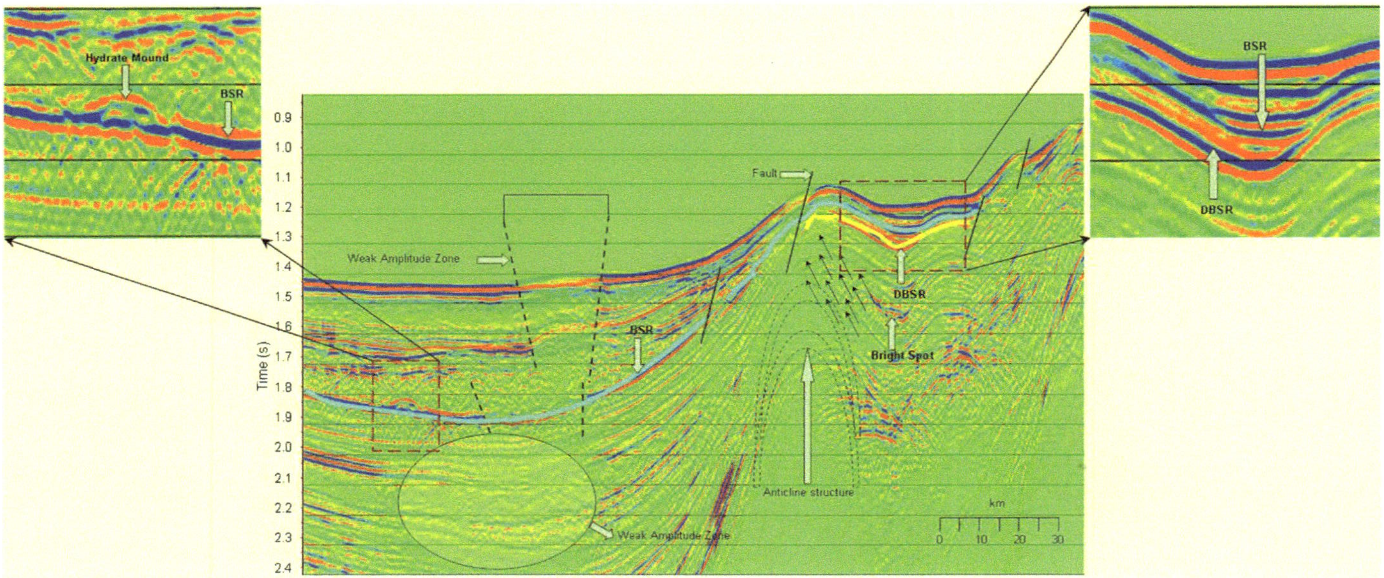

FIGURE 8.28

Real data example from Green Canyon, Gulf of Mexico. BSR and DBSR are clearly identified on Pre-Stack Time Migration (PSTM) stack section (see zoom part of the image at right-hand side). Some major faults are also marked on the profile. Week amplitude zone above the BSR can be interpreted as seismic chimney. A mound associated with the hydrate structure is shown in the zoom section (left-hand side).

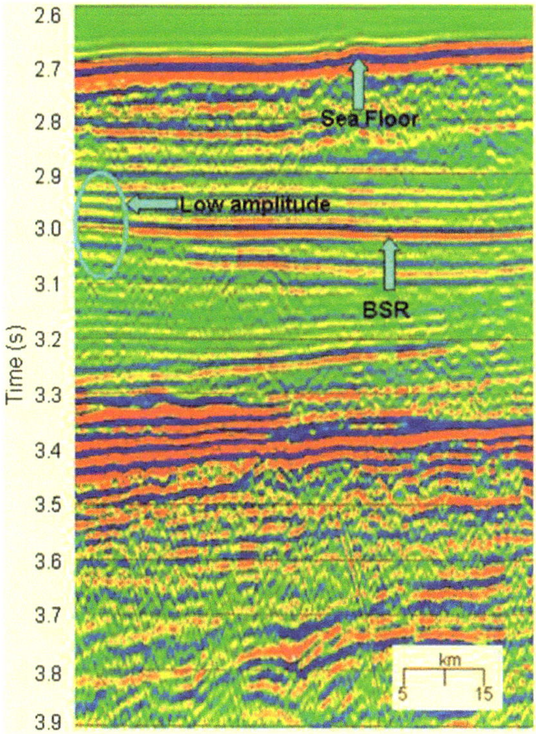

FIGURE 8.29

Real data example from Kerala-Konkan basin offshore India. BSR is marked on PSTM stack section.

1250 ms and the deepest BSR is observed at around 1900 ms. The low amplitude of the BSR at some places may be due to the low gas saturation below the GHSZ and/or due to the low hydrate saturation immediately above the base of the GHSZ. On the other hand, the relatively high amplitude and more continuous BSR in sediments may be attributed to the more favorable lithology for gas hydrate and free gas accumulations, but can also be caused by constructive interference with the strata above and below the BSR in sediments.

A double bottom-simulating reflector (DBSR)-like feature is also observed on the Green Canyon seismic section. This DBSR is possibly caused by the upward migration of gas through the discrete layers of sediments. Chand (2008) suggests that the presence of fluids in the hydrate zone appreciably modifies the amplitude pattern and attenuation characteristics of the medium. Our modeling study indicates that the presence of a low-velocity layer within the GHSZ can generate a DBSR as has been observed on the seismic section of the Green Canyon area.

Enhanced Seismic Reflections Below the BSR

Enhanced reflections below the BSR are seen in many regions and are prominent when the amplitude strength of the BSR is high (Fig. 8.30). In general, they are not laterally extensive, and fade out rapidly with depth. Termination of enhanced reflections also locally forms the BSR in uplift sediments.

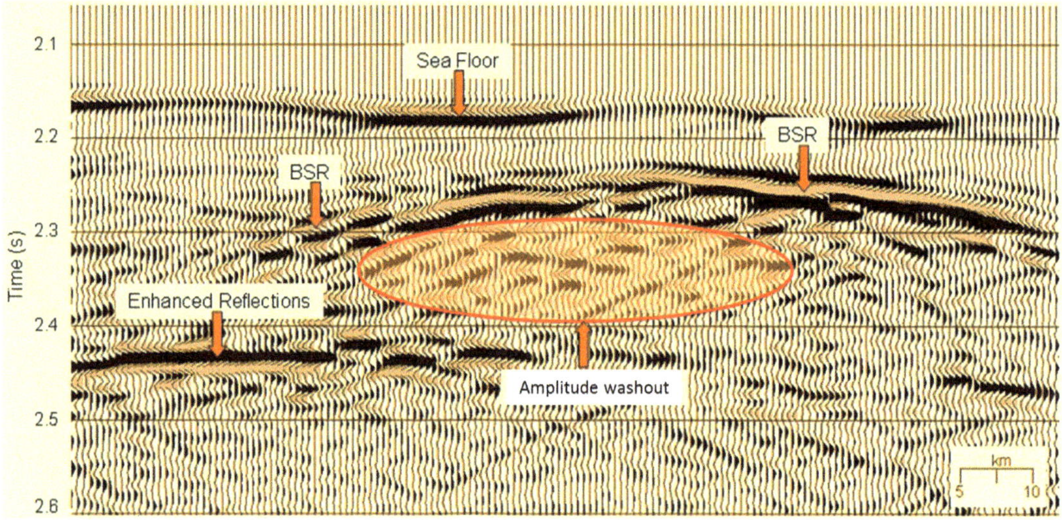

FIGURE 8.30

Real data example from Green Canyon, Gulf of Mexico. BSR is clearly identified. Enhanced seismic reflection below the simulating reflector is possibly caused by the associated free gas that acts as the source of gas hydrate occurrence in that area.

The section below the BSR shown in Fig. 8.30 includes the enhanced reflections, which are considered as a bright spot. The latter is characteristic for gas accumulations suggesting that the hydrate-bearing sediments in the GHSZ effectively seal the gas-charged layer below. Complex and attenuated seismic signals are evident at depths below the enhanced reflections. It is interpreted that most of the observed enhanced seismic reflections below the simulating reflectors are related to the free gas accumulation below the GHSZ. Some of these modeling scenarios will help us to understand the real data interpretations presented in the next section. In particular, amplitude washout hydrate mounds, seismic chimney, BSR, and double BSR.

8.5.1.3 Real Data Examples

Seismic Chimneys and Amplitude Blanking

Seismic chimneys often form a well-defined column of wipe-out of seismic signature (Walraven et al., 2004; Pekot and Gregory, 1987; O'Brein et al., 1999; White, 1975). Most seismic chimneys in the study area do not fade out upward but terminate rather abruptly before reaching the seafloor and only a few crop out on the seafloor as shown in Fig. 8.31a. They may form small mounds or pockmarks or depressions on the seafloor, which is indicated as a weak amplitude zone in Fig. 8.28. The seismic chimneys appear to be wide, owing to the scale adopted in presentation of the seismic data. In this aspect, the seismic chimneys in the study area probably have rather wide features, with their lateral extent being greater than their vertical extent. The seismic chimneys appear to originate within deep structures. Weak BSRs or BSR-like features are also seen below the seismic chimney. This weak feature is slightly up-warped and shallower than the adjacent BSR. Seismic chimneys mostly terminate within the GHSZ. We interpret these seismic chimneys as seafloor expressions that

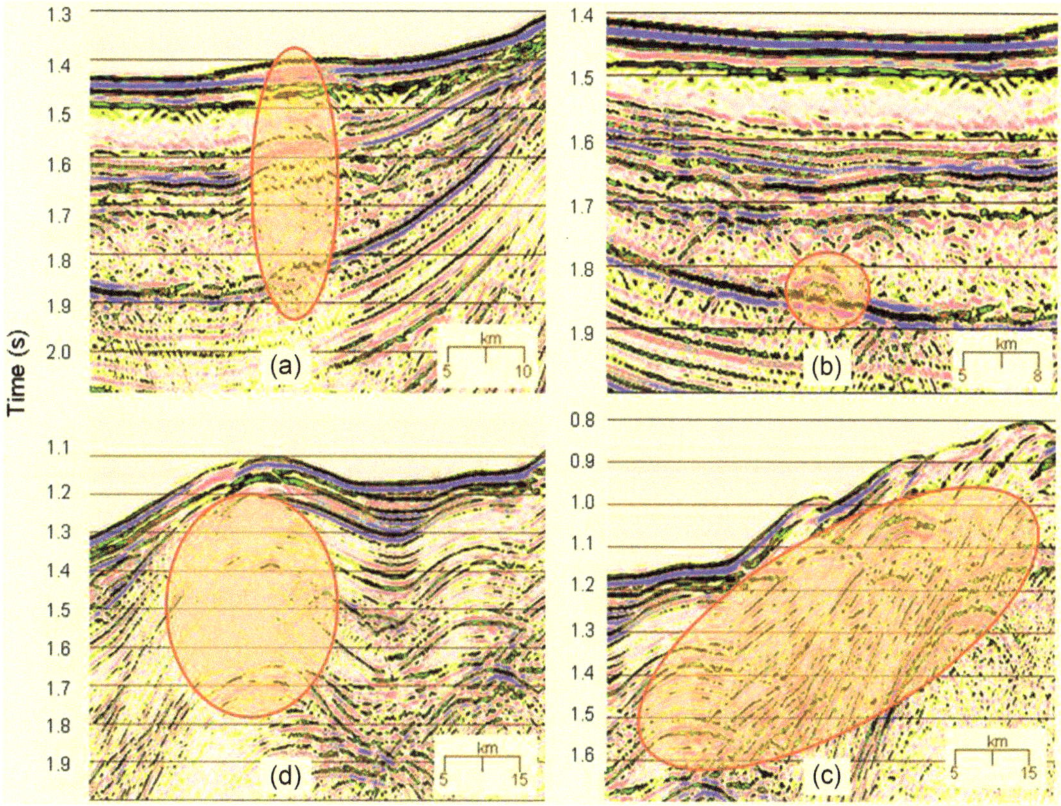

FIGURE 8.31

Seismic data example from Green Canyon, Gulf of Mexico. Key seismic indicators related to the occurrence of gas hydrate and associated free gas are observed. (a) Example of seismic chimney in gas hydrate province. (b) Example of hydrate mound found in Green Canyon area. (c) Example of amplitude blanking and weathered zone. (d) Example of weathered and seismic uplift zone. Anticline stricture can be observed.

represent vertical vents or focused conduits for gas that originates from below the GHSZ. The up-warped BSR and internal reflections may be due to the velocity effect of massive, shallow gas hydrate that causes velocity pull-ups. However, the internal reflections and BSR in seismic chimneys shown in Fig. 8.31a are also up-warped although there is no strong evidence of a high concentration of gas hydrate saturation.

The nature of amplitude blanking is an unsolved issue. On the basis of vertical seismic profiling on ODPLog 164, Holbrook et al. (1996) suggested that the amplitude blanking on Blake Ridge may be attributed to the relative inhomogeneity of the sediments rather than to hydrate cementation. Lee et al. (1994) and Dillon et al. (1996) conclude that amplitude blanking in the Blake Ridge area is caused by hydrate cementation, and the degree of blanking is proportional to the amount of hydrate in the free space. Lu and McMechan (2002, 2004) conclude that there is no one-to-one relationship between hydrate concentration and amplitude blanking. Amplitude blanking may be produced by both lithologic homogeneity and hydrate concentration. In the deep-water Gulf of Mexico study in Walker Ridge and the Green Canyon area, the seismic sections show strong amplitude blanking above the BSRs, but the Keathley Canyon 151 (JIP) and the Walker Ridge study areas (McConnell and Kendall, 2003) do not clearly indicate amplitude blanking above the BSRs. Examples of amplitude blanking from the Green Canyon area are shown in Fig. 8.30 and Fig. 8.31c. Therefore, the analysis supported the conclusion of Lu and McMechan (2002, 2004) that there is no one-to-one relationship between hydrate concentration and amplitude blanking.

Hydrate Mounds

Gas hydrate mounds form when the ideal temperature, pressure, and chemical composition conditions for the gas are achieved. Typically the mounds can be categorized by their shape, surface expression, and size. The mounds are assessed based on the environment they exist in in order to evaluate whether the mounds represent hydrates. Analysis of seismic data from Green Canyon show that the hydrate mounds are visible, as indicated in Fig. 8.31b. Generally circular to ellipsoidal in plan view, their morphologies vary from smooth, well-rounded, steep-sided mounds to rough, uneven, and gently dipping mounds. Apparently some gas is rising upward within the sediment column underneath the mound and passing through the GHSZ without being trapped as gas hydrate. However, some gas hydrate may be forming as the gas passes through this stability zone. The occurrence of these mounds on the crest of the young and apparently actively forming anticlines within the GHSZ, the presence of methane venting through one of these features, and the inherent need for expansion to form gas hydrate lenses suggest that these uplifted features may be blisters formed by expansion associated with net gas hydrate accumulation in the subsurface. The steps in a conceptual model for mound formation by subsurface gas hydrate accumulation involve focused gas migration to the crest of an actively forming anticline, formation of gas hydrate in the shallow subsurface, and upward expansion associated with continued gas hydrate growth and accumulation supported by the flux of methane through this open system, which blisters the seafloor, generating the mounds (Croker and Shannon, 1987; Gutberlet and Schenke, 1989; Hovland et al., 1994). Subsurface gas migration is presumably focused towards the crests of these anticlines. The amplitude of the BSRs varies from moderate to high near the mound, which is associated with the accumulation of the hydrates feeder structure, and decreases with distance from the mounds.

Weathered Zones and Seismic Uplift

The gas hydrate accumulation zone can be weathered out in the presence of sharp faulted structures. The amplitude of the simulating reflectors diminishes and it is very difficult to identify on the seismic section. This is illustrated on the seismic section (Fig. 8.31c & d) from the Green Canyon area, as there is no signature of BSR. Seismic uplift (Fig. 8.28) caused the anticline structure and may disrupt the continuity of the BSR. Therefore, it is very important that during the analysis of seismic data for gas hydrates, extra care should be taken with weathered zones and seismic uplift.

Simulated Seismic Response for Green Canyon Model

A synthetic seismic response from the Green Canyon model is generated to study the response from different reflectors in the gas hydrate province. Figs. 8.32 and 8.33 show an example of wave propagation through the Green Canyon model for horizontal and vertical components at different times that contain gas hydrates and associated free gas. The source used to generate waves in this example is a point source with a center frequency of 50 Hz. We can identify some of the reflection and transmission characteristics at various interfaces, especially for the snapshots corresponding to the earlier times (Fig. 8.32).

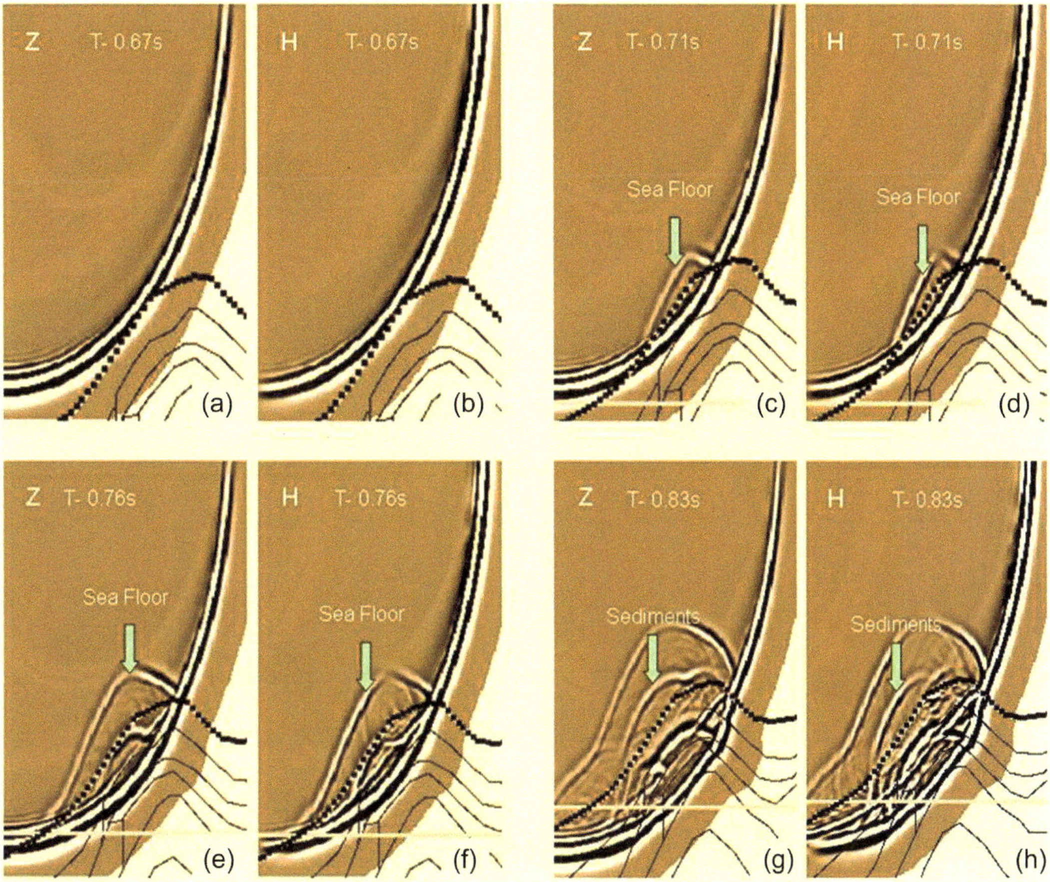

FIGURE 8.32

Synthetic seismic response for Green Canyon model. Wave-front propagation snapshots for horizontal and vertical components are recorded at different times. (a) Snapshot for vertical component recorded at time 0.67 s. (b) Snapshot for horizontal component recorded at time 0.67 s. (c) Snapshot for vertical component recorded at time 0.71 s. Seismic energy reflected from ocean floor is identified. (d) Snapshot for horizontal component recorded at time 0.71 s. Seismic energy reflected from ocean floor is identified. (e) Snapshot for vertical component recorded at time 0.76 s. Seismic energy reflected from ocean floor is identified. (f) Snapshot for horizontal component recorded at time 0.76 s. Seismic energy reflected from ocean floor is identified. (g) Snapshot for vertical component recorded at time 0.83 s. Seismic energy reflected from sediments is identified. (h) Snapshot for horizontal component recorded at time 0.83 s. Seismic energy reflected from sediments is identified.

The seafloor boundary is considered as a first conversion boundary from P to S. Fig. 8.32c represents the wave-front propagation at 0.71 s; at this time the reflected energy is seen as seafloor. While analyzing the horizontal component at the same time (i.e., 0.71 s), a converted wave (P–S) is observed as a seafloor reflection, which could be identified as a polarity change at approximately mid-point of the reflected wave-front. Fig. 8.32 e and g show the reflected compressional energy as seafloor and

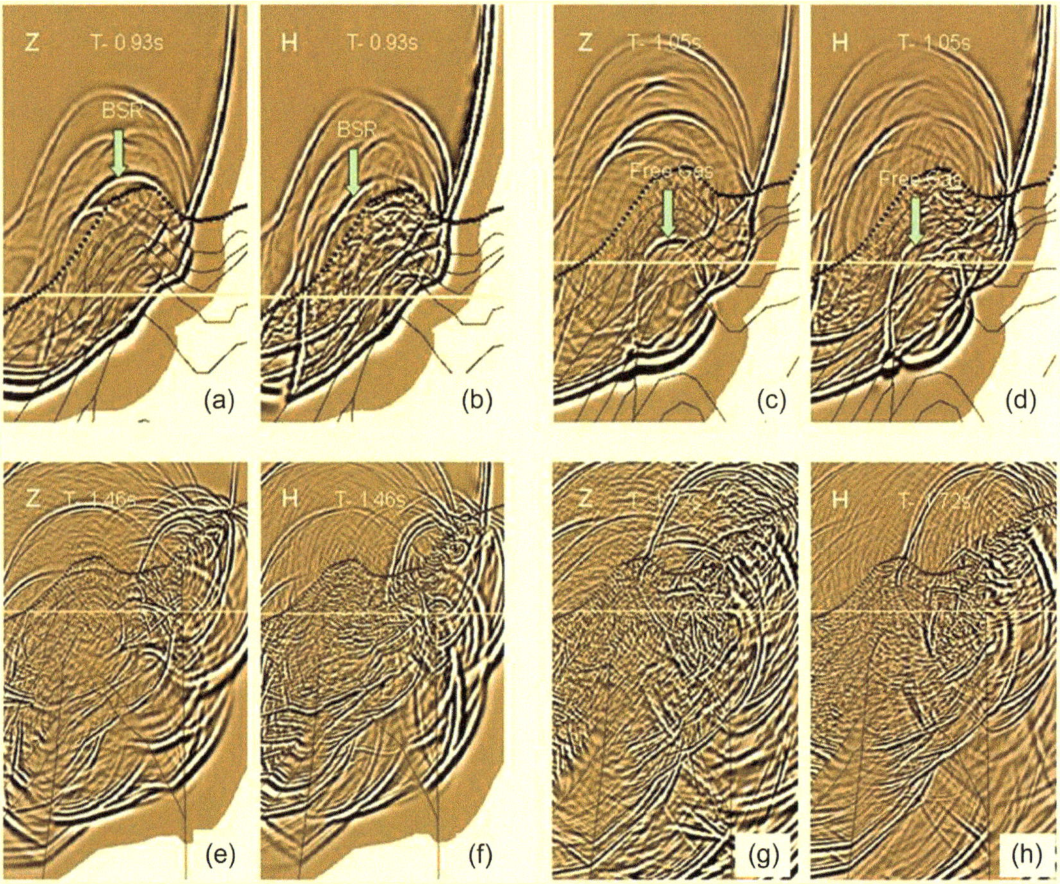

FIGURE 8.33

Synthetic seismic response for Green Canyon model. Wave-front propagation snapshots for horizontal and vertical components are recorded at different times. (a) Snapshot for vertical component recorded at time 0.93 s. Seismic energy reflected from BSR is identified. (b) Snapshot for horizontal component recorded at time 0.93 s. Seismic energy reflected from BSR is identified. (c) Snapshot for vertical component recorded at time 1.05 s. Seismic energy reflected from free gas is identified. (d) Snapshot for horizontal component recorded at time 1.05 s. Seismic energy reflected from free gas is identified. (e) Snapshot for vertical component recorded at time 1.46 s. Seismic energy reflected from different reflectors is shown. (f) Snapshot for horizontal component recorded at time 1.46 s. Seismic energy reflected from different reflectors is shown. (g) Snapshot for vertical component recorded at time 1.72 s. Seismic energy reflected from different reflectors is shown. (h) Snapshot for horizontal component recorded at time 1.72 s. Seismic energy reflected from different reflectors is shown.

sediments at time 0.76 s and 0.83 s, and Fig. 8.32f and h represent the converted waves from these two reflectors. For later times (Fig. 8.33), the picture becomes more complex and it becomes difficult to identify all the reflections and transmissions.

Despite the complexity of the model the reflections from gas hydrate and free gas layers are identified (Fig. 8.34). These figures also show the seismic data recorded by horizontal and vertical

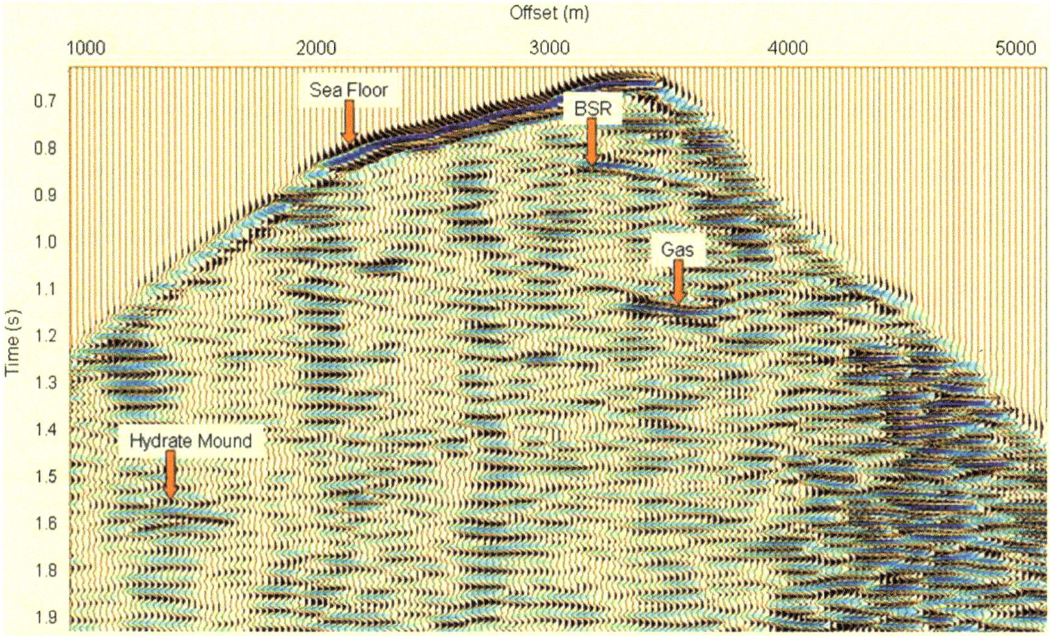

FIGURE 8.34

Synthetic seismic response example (offset gathers) for Green Canyon model. Seismic reflections from hydrate mound; BSR and associated free gas are marked.

components. To study the conversion coefficients a detailed study is recommended. The converted waves are expected to have the highest amplitude in the horizontal components since shear waves have a predominantly horizontal motion. Another advantage of measuring the horizontal direction is that multiples are not so strongly present. In this work reflected conversions are studied.

Fig. 8.34 shows the synthetic seismic response (offset gathers) for the Green Canyon model. Seismic reflections from hydrate mound, BSR, and associated free gas are marked. The seismic signatures recorded on synthetic seismic data are correlated with the real data from Green Canyon and found to be positively correlated.

8.5.2 SOME REMARKS

BSRs are the primary seismic signatures of subsurface gas hydrate accumulations. Studies of BSRs as wave conversion boundaries are relatively sparse. In view of the multicomponent seismic surveys available today, we performed two-dimensional seismic full-waveform modeling for multilayer, complex gas hydrate models. The primary focus of this modeling was to analyze seismic reflections and mode conversions for BSRs representative for the occurrence of gas hydrates. The modeling results suggest that the hydrate and free gas interface acts as the principal boundary for mode conversion from compressional (P) waves into shear waves (PS). It also appears that wave-propagation effects in the overburden zone significantly influence the P-wave amplitude of BSR. Modeling results were correlated with seismic data from the Green Canyon and the KK basin. The presence of BSRs, enhanced seismic reflection below BSRs, seismic chimneys, and hydrate mounds were found to be the key seismic indicators for the occurrence of gas hydrates. The occurrence of DBSR over Green Canyon originated from the migration of free gas through discrete layers of fractured sediments.

Simulations of seismic reflections for different gas hydrate models are carried out and some significant aspects of converted waves at the gas hydrate boundary are examined. P-wave impressions occur with a wide azimuth; in contrast, PS-wave effects occur at near-vertical propagations. The strong AI contrast (due to the presence of hydrate) boundary representing the BSR in the complex gas hydrate model is the principal conversion. Even the pressure record will include the effect of the vector wave-field, which could be considerably different from that of the scalar wave-field. The ocean bottom boundary may be treated as the natural shear-wave source in seismic prospecting. At the seafloor most energy is expected to be transmitted as P-waves. However, at the sediment/sediment and sediment/hydrate interface and for the considered model parameters, more energy seems to be reflected as S-waves in horizontal component recordings. The reflection due to shallow structures recorded on the seafloor are primarily controlled by the pattern of the shear-wave velocity below the seafloor. As the AVO phenomena translate the sharing of the energy of the incident compressional wave between the compressional and converted reflections so its relationship can be measured from a theoretical or practical standpoint. The overburden effect on seismic amplitude of BSR shows that this zone alters the signal characteristics of the simulating reflector by causing significant amplitude losses. The detection of irregularities and amplitude variations at the top of the gas hydrate reservoir can be attributed to the combined effect of transmission loss and significant decrease in seismic amplitude, and this leads to the wrong interpretation. This increases the uncertainty for the offset-based seismic amplitude variation.

A synthetic seismic response together with real data from different regions of the world reveal various seismic indicators of gas hydrates and associated free gas, such as the BSR, DBSR, enhanced seismic reflection below the BSR, seismic chimneys, and hydrate mounds. Seismic chimneys are characterized by columnar amplitude reduction that terminates at different stratigraphic levels. A few seismic chimneys crop out on the seafloor, forming very small mounds. These seismic chimneys with seafloor expressions probably represent vertical vents for gas, originating from below the GHSZ. Analysis of seismic data from the Green Canyon area reveals that gas hydrates are usually confined to permeable sand layers causing large changes in velocity across sedimentary formations. The large impedance contrasts so produced will result in high amplitude anomalies similar to those formed due to the presence of gas in sediments. The synthetic modeling outlined here could be used for further study to develop the relationship between reflected, refracted, diffracted waves and multiples, phase change phenomenon, and PS-wave studies, and to authenticate the technique of processing 4C OBS seismic data.

8.6 GAS HYDRATE MAPPING

Methane hydrates are found in areas of elevated hydrostatic pressures and low temperature regimes worldwide, and many studies have been carried out to map them using various geological proxies and geophysical anomalies. Variables such as gas composition, migration pathways for gas and water, seals and reservoir rocks, and pore water salinity are important for establishing the stability zone for hydrates. In order to identify and map the gas hydrates from regional to local scales, an integrated approach is required.

The example shown in Fig. 8.35 reveals the synthesis of a seismic and rock physics based integrated approach which should be used for the characterization of gas hydrates. The synthetic seismic response

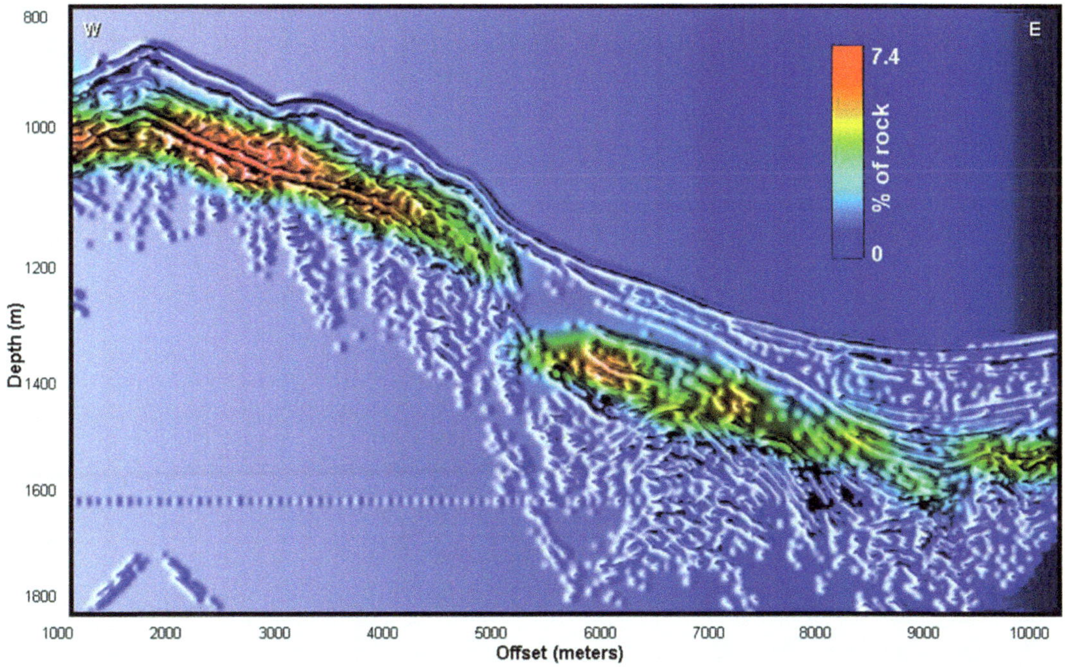

FIGURE 8.35

Depth migrated seismic section from Hydrate Ridge offshore USA. Hydrate concentrate in terms of % of rock is superimposed. Red color denotes higher concentration of gas hydrates.

together with field data from different regions of the world reveals various geological proxies, such as development of hydrate mounds, pockmark depressions, mud volcanism, fault structures and fracture zones, and geophysical indicators, such as BSR, DBSR, enhanced seismic reflection below the BSR, seismic chimneys and hydrate mounds, associated with methane hydrates and underlying free gas as discussed in previous sections. The regional identification of gas hydrates could be achieved by travel time inversion schemes. Gas hydrate saturation can be predicted by integrating seismic information with rock physics models. An example of predicting gas hydrates saturation over offshore United States (hydrate ridge) by the modified Wood equation (Kumar et al., 2007) is shown (Fig. 8.35). The integrated method of mapping gas hydrates has been used that involves integrating travel time inversion, seismic interpretation, rock physics analysis, and modeling. The seismic interpretation identifies reservoir quality sediments, migration pathways, and the availability of methane gas source for hydrate formation, and establishes the hydrate stability zone.

8.6.1 REAL DATA EXAMPLE – ALASKA

The presence of gas hydrates alters the bulk physical properties of the host sediments. The deviation of the physical property in the hydrated layer from the regional (hydrate-free background) trend can be used to estimate hydrate concentration. Elastic properties, resistivity, density, temperature, and chloride measurements of pore fluid can be used to estimate the gas hydrate concentration. Kumar et al. (2009) discussed different methods of gas hydrate concentration estimation and tested them on real field data. Their analyses are based on the above-mentioned properties.

Before going into details it is important to understand the terms "saturation" and "concentration". Hydrate saturation (S_h) is the volumetric fraction of hydrates present in the pore space, whereas hydrate concentration (S_{hyd}) is the volumetric fraction of hydrates in the sediment. Given the total porosity (ϕ) of the rock, saturation and concentration can be estimated from one another by the following relation:

$$S_{hyd} = \phi S_h$$

Hydrate saturation can be estimated by the following methods.

Electrical resistivity. Assuming that high resistivity above the BSR is caused by the presence of gas hydrate in pores (meaning pores are filled with water and/or gas hydrate), hydrate saturation can be estimated by Archie's law as proposed by Lu and McMechan (2002):

$$Sh = 1 - Sw = 1 - \left(\frac{R_o}{R_t}\right)^{\frac{1}{n}}$$

where R_o is the resistivity of the water saturated formation (background), R_t is the measured resistivity, and n is the exponent. The exponent is about 1.94 for hydrate-bearing clastic sediments and is about 1.62 in gas-bearing zones.

Chloride counts. The main challenge for this method is to accurately predict the baseline (background) chlorinity profile. Hyndman et al. (1999) showed that it is possible to estimate both the in situ pore fluid salinity and the in situ gas hydrate concentration using resistivity data. The chloride anomaly can be used to estimate gas hydrate saturation in the pore space before recovery (Yuan et al., 1996) as:

$$S_h = \frac{1}{\rho_h}\left(1 - \frac{Cl_{pw}}{Cl_{sw}}\right)$$

where Cl_{pw} and Cl_{sw} are measured chloride concentrations of pore water and normal bottom seawater (background), respectively.

Nuclear magnetic resonance – DENSITY. The NMR logging tool has the ability to measure water-filled porosity, and density porosity measures space occupied by both water and/or gas hydrate (Lee, 2008). Given the NMR-porosity and density porosity logs, gas hydrate saturation (Fig. 8.36) is estimated using the following equation (Kumar et al., 2009):

$$Sh = 1 - \frac{NMR_{porosity}}{DENSITY_{porosity}}$$

where $NMR_{porosity}$ and $DENSITY_{porosity}$ represent porosities derived from the NMR and density log respectively. Gas hydrates saturation estimation from this method is only possible at the well location.

Density: The formation of gas hydrate reduces the density of rock and this property can be used to estimate hydrate content. Assuming the original pore space was filled with water and now part of the pore space is filled with gas hydrate, the saturation of gas hydrate can be estimated by knowing the

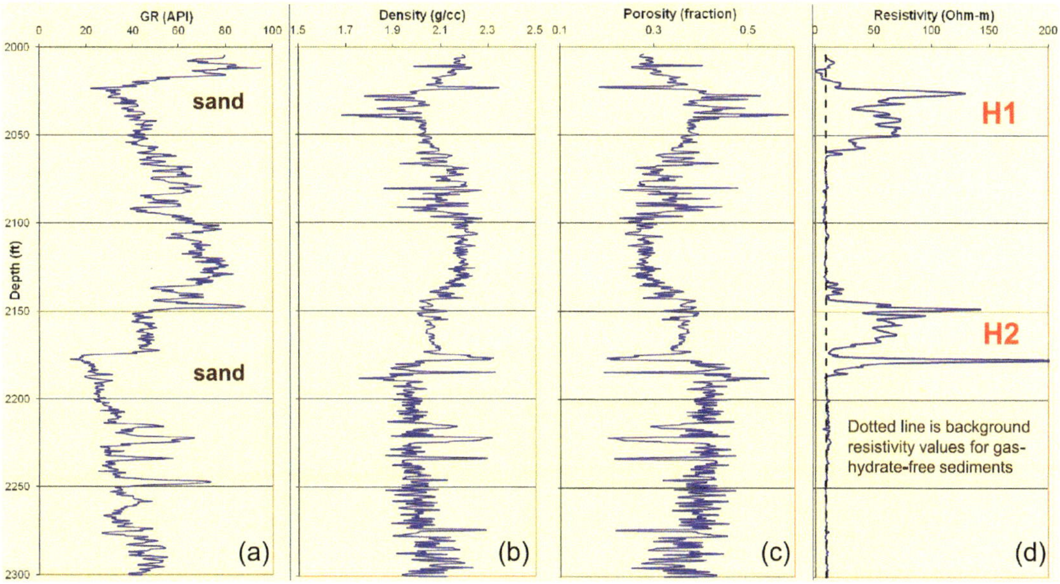

FIGURE 8.36

Well logs from Mount Elbert #1: (a) gamma ray log, (b) density log, (c) porosity log, and (d) deep resistivity log. Two possible gas hydrate zones are marked as H1 and H2. These two zones H1 and H2 are Units D and C in Inks and Agena (2008), respectively. It is evident that hydrate layers are identified in higher porosity sands because of higher seismic velocities, lower density, and higher resistivity than hydrate-free background layers. (From Kumar et al., 2009; well data source: NETL (National Energy Technology Laboratory). Available at http://www.netl.doe.gov/technologies/oil-gas/FutureSupply/MethaneHydrates/ANSLogData.html)

background rock density (ρ_0) for water-filled pores and measured density of hydrate-bearing rock (ρ) as (Kumar et al., 2009):

$$S_h = \frac{1}{\phi}\frac{\rho_o - \rho}{\rho_w - \rho_h}$$

The density measurement and therefore hydrate saturation estimation is possible from well log-derived density and three-dimensional density derived from seismic data amplitude versus angle (AVA) inversion.

In the example from Alaska shown in Fig. 8.36, the well data of hydrate-bearing sediments are shown. The BP-DOE-USA consortium conducted a well test in 2007 called the "Mount Elbert test" to study the feasibility of hydrate exploration and exploitation in Alaska (Inks and Agena, 2008; Boswell et al., 2008). An example of the well logs from the Mount Elbert site is shown in Figs. 8.36 and 8.37, where depth is measured from KB (54.80 ft above sea level). In this example, a background resistivity of 9 ohm-m is assumed. For hydrate saturation estimation using seismic velocities, constant background values of V_p (2000 m/s) and V_s (720 m/s), as shown by black dotted lines in Fig. 8.36, are assumed. Fig. 8.36c shows gas hydrate saturations estimated from V_p alone and from both V_p and V_s together. In this example, hydrate saturation varies from 40% to 80% of pore space in the GHSZ.

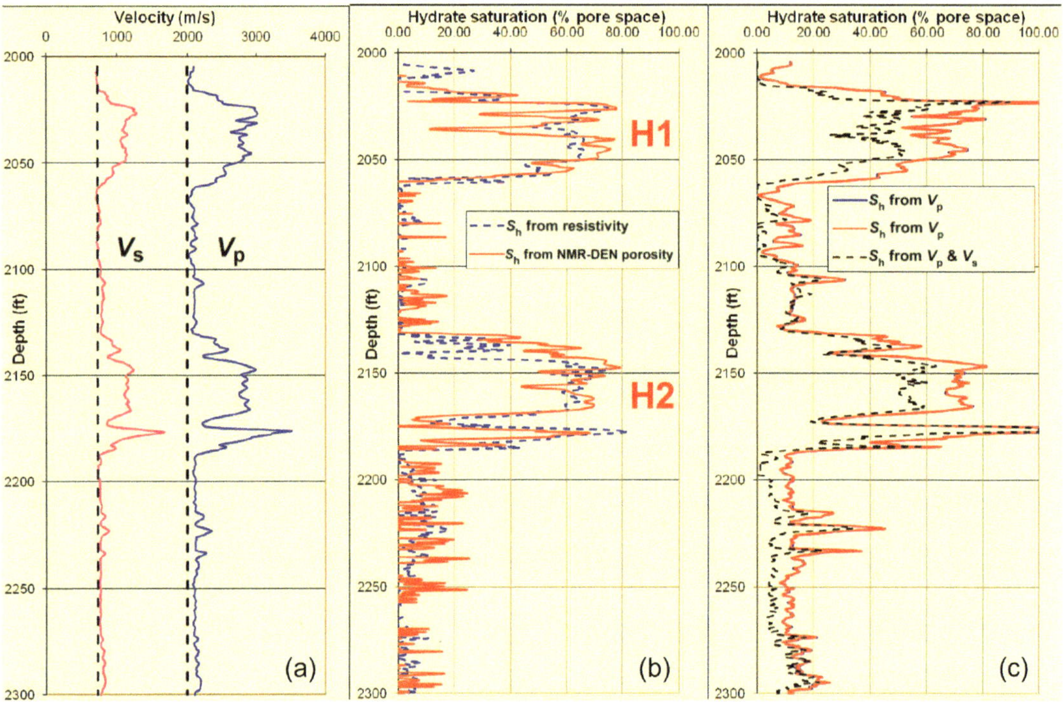

FIGURE 8.37

Gas hydrate saturation estimated from well logs at Mount Elbert #1. On the left are V_p and V_S logs with corresponding background values plotted by black dotted vertical lines (a), and in the middle panel (b) is gas hydrate saturation derived from resistivity log (blue dotted line) and from NMR-DENSITY porosity logs (red solid line). The right panel (c) shows gas hydrate saturations estimated from seismic velocities: blue and red color solid curves show saturations estimated from V_p alone, respectively (they are indistinguishable), and black dotted curve shows hydrate saturation estimated from both V_p and V_S simultaneously. (From Kumar et al., 2009; well data source: NETL (National Energy Technology Laboratory). Available at http://www.netl.doe.gov/technologies/oil-gas/FutureSupply/MethaneHydrates/ANSLogData.html)

8.6.2 REAL DATA EXAMPLE – CASCADIA MARGIN, IODP EXPEDITION 311 DATA, CANADA (KUMAR ET AL., 2009)

The northern Cascadia continental margin offshore western Canada has been under study for the last two decades and the presence of gas hydrates is confirmed from many geophysical and geological experiments. More recently, an extensive list of log data was acquired at five locations (U1325 to U1329) through IODP Expedition 311 in 2005 to study the geological models for the formation of hydrates as well as to estimate hydrate concentration (Riedel et al., 2006). The drillhole locations were chosen so as to study the formation (and dissociation) of gas hydrates within the accretionary prism at different stages of its development away from the Juan de Fuca ridge. Chen (2006) carried out an elaborative study of the different logs from IODP Expedition 311 to estimate the hydrate concentration.

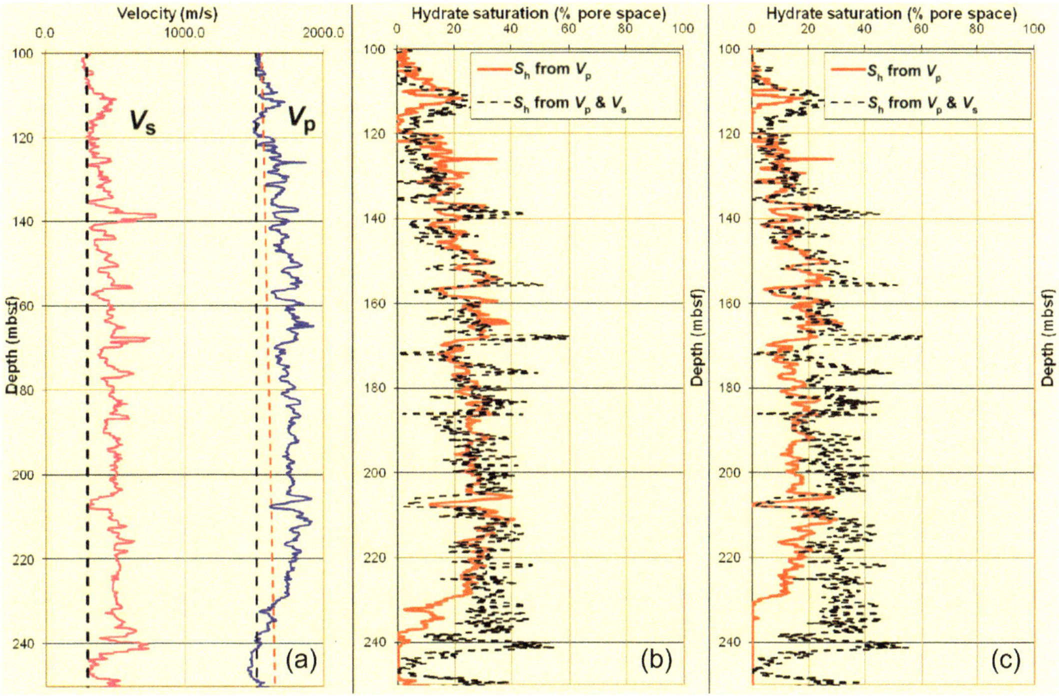

FIGURE 8.38

Gas hydrate saturation estimated from well logs at northern Cascadia margin, Expedition 311 site U1327E. On the left (a) are V_p and V_S logs with corresponding background values plotted by black dotted vertical lines and another background depth varying V_p (red dotted line) from Yuan et al. (1996). The middle panel (b) and the right panel (c) show gas hydrate saturations estimated from V_p alone (red solid curve) and V_p and V_S combined (black dotted curve) using method 1. The estimated hydrate saturation shown in the middle panel uses constant background velocities (black vertical dotted lines in part (a)), and saturation shown in the right panel was estimated using a constant background value for V_S as in the middle panel and a depth-varying VP trend (red dotted curve in part (a)). Note the change in hydrate saturation (middle vs. right panels) due to change in background V_p value only. Depth shown is in mbsf (meters below seafloor), water depth is 1300 m, and the BSR is at 225 mbsf. (From Kumar et al., 2009; well data source NETL (National Energy Technology Laboratory). Available at http://www.netl.doe.gov/technologies/oil-gas/FutureSupply/MethaneHydrates/ANSLogData.html)

A combined seismic reflection and sonic log study by Dash (2007) showed a hydrate saturation of about 15% of pore space at the IODP Site U1327, where the BSR depth is 225 mbsf (meters below seafloor). Fig. 8.38 shows hydrate saturation estimated from seismic velocities with a constant background VS trend and two different background VP trends. The estimated hydrate saturations vary by changing the background VP trend, which emphasizes the importance of a reliable background model. In this example, hydrate saturation varies from 10% to 40% of pore space in the GHSZ.

8.7 OIL SHALE OR LIQUID-RICH SHALE

Oil shale or liquid-rich shale generally refers to any sedimentary rock that contains solid bituminous material called kerogen that is released as a petroleum-like liquid when the rock is heated in the chemical process of pyrolysis. Oil shale has been used as an energy source for centuries. This type of unconventional reservoir proved to be challenging in terms of reservoir characterization, identifying pay potential from reserves that have been interbedded for long intervals in indistinguishable shale, organic mudstone, siltstone and carbonate, predicting producibility, and estimating ultimate recovery for improved economics. Oil shale or liquid-rich shale play types include Bakken and Montney. Oil shale can be mined using one of two methods: underground mining using the room-and-pillar method or surface mining. After mining, the oil shale is transported to a facility for retorting, a heating process that separates the oil fractions of oil shale from the mineral fraction. In order to understand tight oil reservoirs, the following objectives should be met:

- Geological understanding and characterization of the core interval to get information about facies, rock type, and depositional environment.
- Stratigraphic characterization for improved understanding of lithology, mineralogy, clay types and contents, fractures, etc.
- TOC measurement and characterization.
- Measurement of various petrophysical properties including permeability, grain density, and saturations (i.e., water saturation, gas saturation, oil saturation, hydrocarbon-filled porosity, and bound water).
- Mechanical properties for optimizing fracture gradient and fracture design and/or horizontal wells such as Young's modulus, Poisson's ratio, bulk modulus, horizontal stress, and proppant embedment.
- Rock physics modeling.
- Integration of rock physics modeled logs with seismic data through inversion.
- Integration of core and well data, fracture stimulation techniques, and production test results.
- Regional to local assessment for the total reserves and recoverables.

Multidisciplinary datasets should be integrated to optimize the drilling location, trajectories, well design, and production tests. An example of an oil shale reservoir from a Southeast Asian region is shown in Fig. 8.39.

Knowledge of the geophysics and Geology optimizes the performance of oil shale resource plays. Consequently, unconventional plays are driving the technological advances necessary to develop the resource base economically, particularly with respect to drilling and completing wells. Geophysics has a critical role to play in optimizing production performance by well placement, especially in tight oil reservoirs. Glenn Winters (2013) describes a very good example from Wolfberry Play where 3D seismic is used primarily as a structural guide in areas with limited well control to keep horizontal laterals extending one to two miles within their target zones. However, prestack datasets and seismic inversions were also analyzed and from these data, important variations and sweet spots within the reservoir are seen. An example of the benefits of utilizing prestack data volumes and seismic inversions can be seen in the Wolfberry Play, where fracture networks created and rejuvenated during the fracturing of a horizontal well propagated through natural fracture systems had an effect on a vertical well that was being

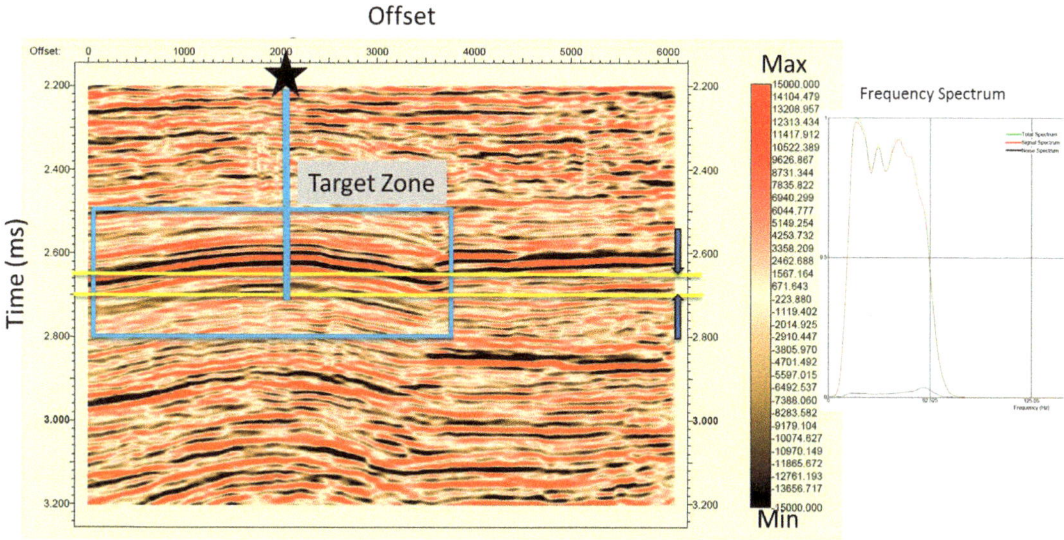

FIGURE 8.39

An example of a seismic section from a region located in Southeast Asia where a vertical well was drilled through the target zone and a tight oil discovery was made.

drilled one mile away. Fig. 8.40 shows the Strawn Formation and corresponding seismic and enhanced frequency seismic through inversion. Geophysics utilizes second order seismic data combined with first order geological well data that often results in an answer that is typically not in the order of magnitude that engineers commonly deal with.

8.7.1 PRODUCING OIL–GAS RATIO

Producing oil–gas ratio is an important characteristic for oil shale wells as it determines the richness of the produced fluids. The oil–gas ratio for liquid-rich shale or oil shale is directly proportional to the "value" of oil, i.e., a high oil–gas ratio means more valuable oil, and a low oil–gas ratio means more gas. Conventional reservoirs, however, may produce with an oil–gas ratio equal to the in-solution oil–gas ratio for a significant amount of time. The oil rate can be expressed as the sum of the oil being produced as solution oil ($r_s(P_{wf})$) and oil that flows freely in the reservoir:

$$q_{o-\text{total}} = q_{o-\text{free}} + q_g r_s \left(P_{wf}\right)$$

For many liquid-rich shale gas reservoirs the producing oil loss is as high as a factor of 2 to 50 and the producing oil–gas ratio is equal to the solution oil–gas ratio at the flowing bottomhole pressure (Whitson and Sunjerga, 2012). The producing oil–gas ratio appears to be dependent on a number of factors with the main factors being producing bottomhole pressure (P_{wf}), degree of undersaturation (initial reservoir pressure), initial oil–gas ratio of the reservoir fluid, and relative permeability as a function of oil saturation. For undersaturated gas reservoirs, with the same in situ conditions as for the

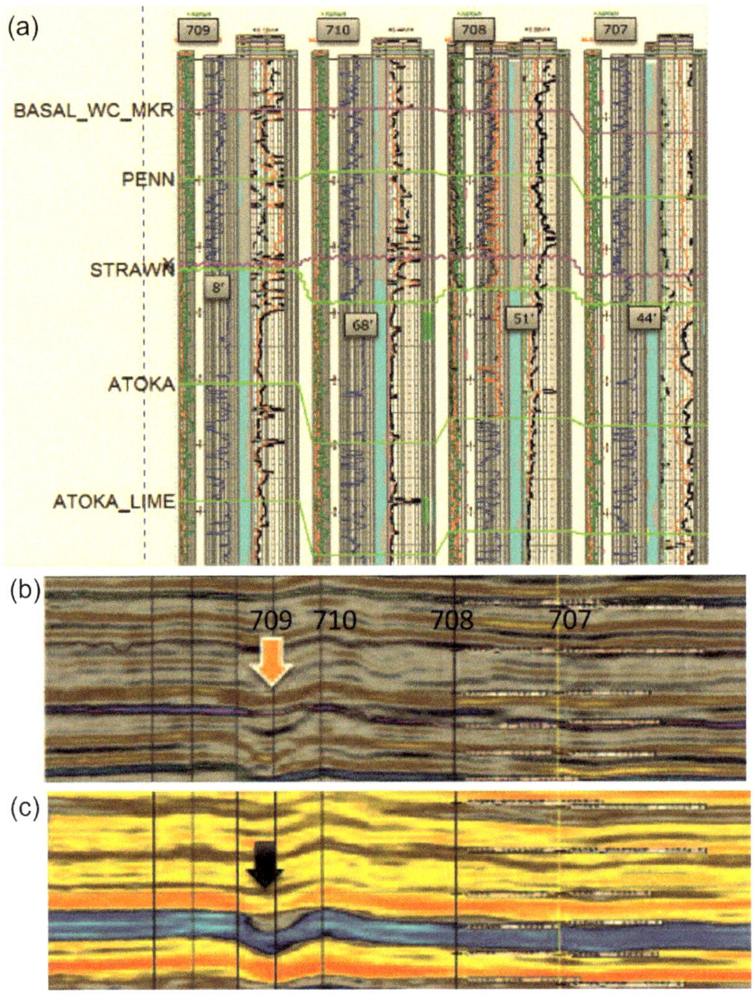

FIGURE 8.40

Wolfberry Play wells, seismic and acoustic impedance. (a) This cross-section shows the Strawn formation coming in 60 feet lower in Well 710 than in Well 709. The seismic also identifies the lower-velocity material and matches the well data. (b) One of the lessons learned from analyzing seismic data in the Wolfberry play is that the geology can change quickly, and intervals present in the stratigraphy of one well can be absent in the next well. This image compares wells 709 and 710, which are located only 550 feet apart. (c) The seismic lines are the conventional bandwidth extension in the upper section and seismic inversion in the lower section. The arrows mark the interpreted zone. Contrasted to the conventional data, the inversion shows a pod of low-velocity material. (Source: Winters, 2013).

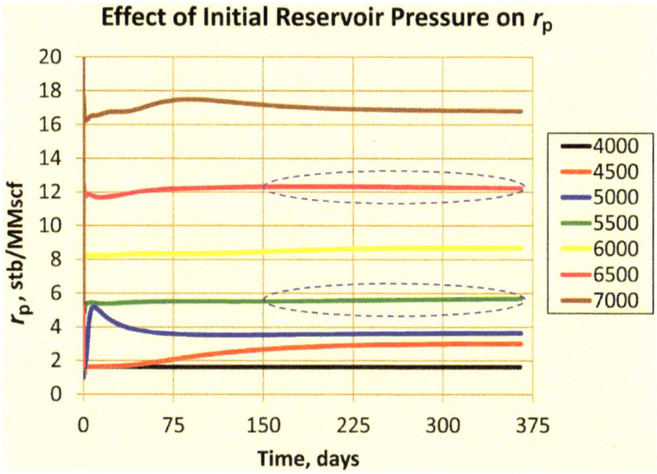

FIGURE 8.41

Different producing oil–gas ratios for different initial reservoir pressures (producing oil and gas ratio r_{si} = 100 stb/MMstb, dewpoint pressure p_d = 3800 psia, permeability $k = 1e - 4$ mD and n = 2). (From Holme, 2013.)

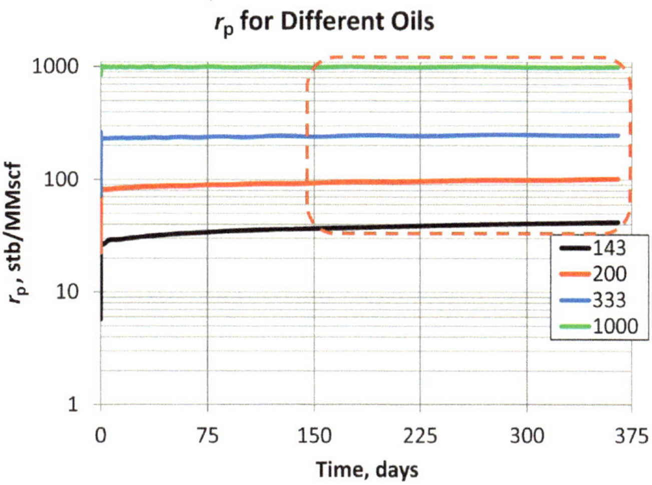

FIGURE 8.42

Different producing oil–gas ratios for different initial oil–gas ratios for reservoir saturated with oil for 2D reservoir model ($k = 1 - 4$ mD, pi = 6000 psia, and n = 2) and relative permeability exponents equal to 2. (From Holme, 2013.)

saturated reservoir except the initial reservoir pressure is higher than the dewpoint (Holme, 2013), the producing oil–gas ratio appears to be slightly higher (Fig. 8.41). The pressure gradient is steeper with a higher initial reservoir pressure, and the oil that comes out of solution does so closer to the well-bore in a more confined area, which leads to a larger build-up of oil saturation. The oil saturation in parts of the reservoir is then large enough for the oil to be mobilized and this oil is therefore produced along with the in-solution oil, increasing the total oil–gas ratio (Holme, 2013).

For liquid-rich reservoirs saturated with oil ($S_g = 0$), the producing oil–gas ratios behave differently. The oil will always be mobile in the reservoir and problems with insufficient oil saturation are not encountered (Holme, 2013). However, as the pressure drops below the bubble point pressure (P_b), gas starts coming out of solution. The gas phase mobility is far greater than the oil phase mobility. For gas-heavy oil systems (near critical oil) a large amount of gas will come out of solution. The gas will flow more easily towards the fracture because of a higher mobility and "block" some of the oil, so reducing the producing oil–gas ratio (Fig. 8.42). For richer oil systems (less gas) the amount of gas that comes out of solution is not that significant and will not affect the producing oil–gas ratio in the same way, and the producing oil–gas ratio is approximately equal to the initial oil–gas ratio (Fig. 8.42).

REFERENCES

Aki, K., Richards, P.G., 1980. Quantitative Seismology: Theory and Methods. W.H. Freeman and Sons. San Francisco.

Andreassen, K., Berteussen, K.A., Sognnes, H., et al., 2003. Multicomponent ocean bottom cable data in gas hydrate investigation offshore of Norway. J. Geophys. Res. 108, 2399.

Andreassen, K., Hart, P.E., MacKay, M., 1997. Amplitude versus offset modeling of the bottom-simulating reflection associated with submarine gas hydrates. Mar. Geol. 137, 25–40.

Andreassen, K., Miniert, J., Bryan, P., Singh, S.C., 2000. A double gas hydrate related bottom simulating reflector at the Norwegian continental margin. In: Holder, G.D., Bishnoi, P.R. (Eds.), Gas Hydrates: Challenges for the Future Ann. N. Y. Acad. Sci., vol. 912, pp. 126–135.

Bangs, N.L.B., Sawyer, D.S., Golovchenko, X., 1993. Free gas at the base of the gas hydrate zone in the vicinity of the Chile triple junction. Geology 21, 905–908.

Berg, E., Svenning, B., Martin, J., 1994. SUMIC: Multi-component sea-bottom seismic surveying in the North Sea; data interpretation and application. In: 64th Annual International Meeting of the Society of Exploration Geophysicists. Expanded Abstracts, pp. 477–480.

Berndt, C., Bünz, S., Clayton, T., Mienert, J., Saunders, M., 2004. Seismic character of bottom simulating reflectors: examples from the mid-Norwegian margin. Mar. Petrol. Geol. 21 (6), 723–733.

Bortfeld, R., 1961. Approximations to the reflection and transmission coefficients of plane longitudinal and transverse waves. Geophys. Prosp. 9, 485–502.

Boswell, R., Hunter, R., Collett, T., Digert, S., Hancock, S., Weeks, M., Mount Elbert Science Team, 2008. Investigation of gas hydrate-bearing sandstone reservoirs at the "Mount Elbert" Stratigraphic test well, Milne Point, Alaska. In: Proceedings of the 6th International Conference on Gas Hydrates Vancouver, BC, Canada, July 6–10, 2008.

Bouriak, S., Vanneste, M., Saoutkine, A., 2000. Inferred gas hydrates and clay diapers near the Storegga Slide on the southern edge of the Voring Plateau, offshore Norway. Mar. Geol. 163, 125–148.

Burwen, J., Flegal, J., 2013. Unconventional Gas exploration & production: case studies on the Government's role in energy technology innovation. http://americanenergyinnovation.org.

Caldwell, J., 1999. Marine multi-component seismology. Lead. Edge 399, 1274–1282.

Cara, M., 1973. Filtering of dispersed wave trains. Geophys. J. R. Astr. Soc. 33, 65–80.

Chand, S., 2008. Gas hydrate anomalies in seismic velocities, amplitudes and attenuation: what do they imply? In: Proceedings of the 6th International Conference on Gas Hydrates Vancouver, BC, Canada.

Chand, S., Minshull, T.A., 2003. Seismic constraints on the effects of gas hydrate on sediment physical properties and fluid flow: a review. Geofluids 3, 275–289.

Chapman, C.H., 1976. Exact and approximate generalized ray theory in vertically inhomogeneous media. Geophys. J. R. Astron. Soc. J6, 201–233.

Chen, M., 2006. Northern Cascadia Marine Gas Hydrate: Constraints from Resistivity, Velocity, and AVO MSc thesis. University of Victoria.

Cherskiy, N.V., Tsarev, V.P., Nikitin, S.P., 1985. Investigation and prediction of conditions of accumulation of gas resources in gas-hydrate pools. Petrol. Geol. 21, 65–89.

Chiburies, E.F., 1987. Studies of amplitude versus offset in Saudi Arabia. In: 57th Annual International Meeting Society of Exploration Geophysicists. Expanded Abstracts, pp. 614–616. New Orleans, USA.

Coal of the Future (Supply Prospects for Thermal Coal by 2030–2050). 2009. European Commission Joint Research Centre, Report EUR 22644 EN, February 2007. http://ie.jrc.ec.europa.eu/publications/scientific_publications/2007/EUR22644EN.pdf (accessed April 6, 2009).

Collett, T.S., 2002. Energy resource potential of natural gas hydrates. AAPG Bull. 86, 1971–1992.

Collett, T.S., Reidel, M., Cochran, J.R., et al., 2008. Indian National Gas Hydrate Program: Expedition 01-Initial Report, vol. 1. Directorate General of Hydrocarbon, New Delhi, India.

Collett, T., 2004. Gas hydrates as a Future Energy Resource. Geotimes. November, 2004. http://www.agiweb.org/geotimes/nov04/feature_futurehydrates.html.

Collett, T., Riedel, M., Cochran, J., Boswell, R., Presley, J., Kumar, P., Sathe, A., Sethi, A., Lall, M., Siball, V., the NGHP Expedition 01 Scientific Party, 2008. Indian national Gas hydrate program Expedition 01 initial Reports. In: Prepared by the United States geological survey. The Directorate General of Hydrocarbons, Ministry of Petroleum and Natural Gas, India.

Collett, T.S., 1995. Gas hydrate resources of the United States. In: Gautier, D.L., et al. (Ed.), National Assessment of United States Oil and Gas Resources on CD-ROM. US Geological Survey Digital Data Series, 30. USGS, USA.

Collett, T.S., Lewis, R., Uchida, T., 2000. Growing interest in gas hydrates. Oilfield Rev. 86, 1971–1992.

Croker, P.F., Shannon, P.M., 1987. The evolution and hydrocarbon prospectivity of the Porcupine Basin, offshore Ireland. In: Brooks, J., Glennie, K.W. (Eds.), Petroleum Geology of NorthWest Europe. Graham and Trotsman, London, pp. 633–642.

Dash, R., 2007. Crustal Structure and Marine Gas Hydrate Studies Near Vancouver Island Using Seismic Tomography PhD thesis. University of Victoria.

Davis, E.E., Hyndman, R.D., Villinger, H., 1990. Rates of fluid expulsion across the northern Cascadia accretionary prism: constraints from new heat flow and multichannel seismic reflection data. J. Geophys. Res. 95, 8869–8889.

de Boer, R.B., Houbolt, J.J.H.C., Lagrand, J., 1985. Formation of gas hydrates in a permeable medium. Geol. Mijnb. 64, 245–249.

DeVault, B.C., 1997. Prestack Multi-component Amplitude Analysis, Vacuum Field, Lea County, New Mexico PhD thesis. Colorado School of Mines.

Diaz-Naveas, J., 1999. Sediment Subduction and Accretion at the Chilean Convergent Margin between 35 and 40S. Dissertation zur Etlangung des Doktorgrades. Christian-Albrechts-Universitat zu Kiel, p. 130.

Dillon, W.P., Danforth, W.W., Hutchinson, D.R., et al., 1998. Evidence for faulting related to dissociation of gas hydrate and release of methane off the southeastern United States. In: Henriet, J.P., Mienert, J. (Eds.), Gas Hydrates:Relevance to World Margin Stability and Climate Change. Geological Society Publishing House, London, pp. 293–302. Geological Society Special Publication 137.

Dillon, W.P., Hutchinson, D.R., Drury, R.M., 1996. Seismic Reflection Profiles on the Blake Ridge Near Sites 994, 995, and 997. In: Paull, C.K., Matsumoto, R., Wallace, P.J., et al. (Eds.), Proceedings ODP, Initial Reports. Ocean Drilling Program, 164 College Station, TX, pp. 47–56.

Dillon, W.P., Lee, M.W., Coleman, D.F., 1994. Identification of marine hydrates in situ and their distribution off the Atlantic coast of the United States. In: Sloan Jr., E.D., Happel, J., Hnatow, M.A. (Eds.), Ann. N.Y. Acad. Sci., Int. Conf. Natural Gas Hydrates, vol. 715, pp. 364–380.

Dvorkin, J., Mavko, G., Nur, A., 1991. The effect of cementation on the elastic properties of granular material. Mech. Mater. 12, 207–217.

Dvorkin, J., Nur, A., 1993. Rock physics for seismic characterization of gas hydrates. In: The Future of Energy Gases. Professional Paper 1570. US Geological Survey, Virginia, USA, pp. 293–298.

Dziewonski, A.M., Bloch, S., Landisman, M., 1969. A technique for the analysis of transient seismic signals. Bull. Seism. Soc. Am. 59, 427–444.

Ecker, C., Dvorkin, J., Nur, A., 1998. Sediments with gas hydrates: internal structure from seismic AVO. Geophysics 63, 1659–1669.

Elapavuluri, P., Bancroft, J., 2001. Estimation of Thomsen's anisotropy parameter, delta, using CSP gathers. CREWES Research Report 13, 479–490.

Fan, J., Ma, B., 1991. Oil-and-gas geology of Tarim. In: Biostratigraphy and geologic evolution of Tarim, Vol 4. Science Press, Beijing. 466 p.

Foucher, J.P., Nouzé, H., Henry, P., 2002. Observation and tentative interpretation of a double BSR on the Nankai slope. Mar. Geol. 187 (1–2), 161–175. http://dx.doi.org/10.1016/S0025-3227(02)00264-5.

Gaiser, J.E., 1996. Multicomponent Vp-Vs correlation analysis. Geophysics 61, 1137–1149.

Gei, D., Carcione, J.M., 2003. Acoustic properties of sediments saturated with gas hydrate, free gas and water. Geophys. Prospect. 51 (2), 141–158. http://dx.doi.org/10.1046/j.1365-2478.2003.00359.x.

Globensky, Y., 1993. Lexique stratigraphique canadien Volume V-B. Région des Appalaches, des Basses-Terres du Saint-Laurent et des îles de la Madeleine. Direction générale de l'Exploration géologique et minérale, Direction de la Recherche géologique, Service géologique de Québec. DV 91-23. 327 p.

Gorman, A.R., et al., 2002. Migration of methane gas through the hydrate stability zone in a low-flux hydrate province. Geology 30, 327–330.

Gutberlet, M., Schenke, H.W., 1989. Hydrosweep: new era in high precision bathymetric surveying in deep and shallow water. Mar. Geod. 13, 1–23.

Haskell, N.A., 1953. The dispersion of surface waves on multilayered media. Bull. Seismol. Soc. Am. 43, 17–34.

Haskell, N.A., 1960. Crustal reflection of plane SH waves. J. Geophys. Res. 65, 4147–4150.

Haskell, N.A., 1962. Crustal reflection of plane P and SV waves. J. Geophys. Res. 67, 4751–4769.

Helgerud, M., Dvorkin, J., Nur, A., Sakai, A., Collett, T., 1999. Elastic-wave velocity in marine sediments with gas hydrates: effective medium modeling. Geophys. Res. Lett. 26 (13), 2021–2024. http://dx.doi.org/10.1029/1999GL900421.

Holbrook, W.S., Hoskins, H., Wood, W.T., Stephen, R.A., Lizzaralde, D., the Leg 164 Science Party, 1996. Methane hydrate and free gas on the Blake Ridge from vertical seismic profiling. Science 273, 1840–1843.

Holme, A., 2013. Optimization of Liquid-rich Shale Wells Thesis. Petroleum Geoscience and Engineering, Norwegian University of Science and Technology, Department of Petroleum Engineering and Applied Geophysics.

Hornbach, M.J., Holbrook, W.S., Gorman, A.R., et al., 2003. Direct Seismic Detection of Methane Hydrate on the Blake Ridge. Geophysics 68, 92–100.

Hovland, M., Crocker, P.F., Martin, M., 1994. Fault associated seabed mounds (carbonate knolls?) off western Ireland and northwest Australia. Mar. Pet. Geol. 11, 232–246.

Hovland, M., Judd, A.G., 1988. Seabed Pockmarks and Seepages: Impact on Geology, Biology, and Marine Environment. Graham and Trotman, London.

Hovland, M., Svensen, H., 2006. Submarine pingoes: indicators of shallow gas hydrates in a pockmark at Nyegga, Norwegian Sea. Mar. Geol. 228, 15–23.

Hughes, D.J., 2013. A reality check on the shale revolution. Nature 494, 307–308.

Hyndman, R., Davis, E., 1992. A mechanism for the formation of methane hydrate and sea floor bottom-simulating reflectors by vertical fluid expulsion. J. Geophys. Res. 97, 7025–7041.

Hyndman, R., Spence, G., 1992. A seismic study of methane hydrate marine bottom simulating reflectors. J. Geophys. Res. 97, 6683–6698.

Hyndman, R.D., Yuan, T., Moran, K., 1999. The concentration of deep sea gas hydrates from downhole electrical resistivity logs and laboratory data. Earth Planet. Sci. Lett. 172, 167–177. http://dx.doi.org/10.1016/S0012-821X(99)00192-2.

Inks, T.L., Agena, W.F., 2008. Successful Gas Hydrate Prospecting Using 3D Seismic: A Case Study for the Mt. Elbert Prospect, Milne Point, North Slope Alaska. SEG expanded abstract. pp. 473–477.

Jackson, J.A., 1997. Glossary of Geology, fourth ed. American Geological Institute. Published by Elsevier Science, ISBN 10: 0913312894. Alexandria, VA.

Johnston, A.A.S., Lambert, C.B.S., Bustos, O.A., Pashin, J.C., Wary, A., 2009. Coalbed methane: clean energy for the world. Oilfield Rev. 21 (2).

Knopoff, L., 1972. Observation and Inversion of Surface Wave Dispersion. Tectonophysics 13, 497–519.

Kumar, D., Dash, R., Dewangan, P., 2009. Methods of Gas Hydrate Concentration Estimation with Field Examples. GEOHORIZONS December 2009/76.

Kumar, D., Sen, M.K., Bangs, N.L., 2007. Gas hydrate concentration and characteristics within Hydrate Ridge inferred from multicomponent seismic reflection data. J. Geophys. Res. 112. http://dx.doi.org/10.1029/2007JB004993 B12306.

Kvenvolden, K.A., 1993a. A primer on gas hydrates. In: Howell, D.G. (Ed.), The Future of Energy Gases. US Geological Survey Professional Paper 1570, pp. 279–291.

Kvenvolden, K.A., 1993b. Gas hydrates: geological perspective and global change. Rev. Geophys. 31 (2), 173–187.

Kvenvolden, K.A., 2000. Gas hydrates and humans. In: Holder, G.D., Bishnoi, P.R. (Eds.), Gas Hydrates: Challenges for the Future. Ann NY Acad Sci., vol. 912, pp. 17–22.

Kvenvolden, K.A., Ginsburg, G.D., Soloviev, V.A., 1993. World-wide distribution of subaquatic gas hydrates. Geo. Mar. Lett. 13, 32–40.

Kvenvolden, K.A., Lorenson, T., 2001. The global occurrence of natural gas hydrate. In: Paull, C.K., Dillon, W.P. (Eds.), Natural Gas Hydrates: Occurrence, Distribution, and Detection. American Geophysical Union, Washington, DC, pp. 3–18.

Lawrence, M.G., 1991. Advances in seismic reflection profiling for US coal exploration. Lead. Edge 10, 24–29.

Lee, M.W., 2008. Models for Gas Hydrate-bearing Sediments Inferred from Hydraulic Permeability and Elastic Velocities. USGS Scientific Investigations Report 2008-5219. Online at http://pubs.usgs.gov/sir/2008/5219/pdf/SIR08-5219_508.pdf.

Lee, M.W., Hutchinson, D.R., Collett, T.S., Dillon, W.P., 1993. Seismic velocities for hydrate-bearing sediments using weighted equation. J. Geophys. Res. 101, 20347–20358.

Lee, M.W., Hutchinson, W.F., Agena, W.F., et al., 1994. Seismic character of gas hydrates on the southeastern US continental margin. Mar. Geophys. Res. 16, 163–184.

Lu, S., McMechan, G.A., 2002. Estimation of Gas Hydrate and Free Gas Saturation, Concentration, and Distribution from Seismic Data. Geophysics 67, 582–593.

Lu, S., McMechan, G.A., 2004. Elastic impedance inversion of multi-channel seismic data for estimation of gas hydrate and free-gas distribution and concentration. Geophysics 69, 164–179.

Mari, J.L., 1984. Estimation of static correction for shear-wave profiling using the dispersion properties of Love waves. Geophysics 49, 1169–1179.

Markl, R.G., Bryan, G.M., Ewing, J.I., 1970. Structure of the Blake-Bahama outer Ridge. J. Geophys. Res. 75 (24), 4539–4555. http://dx.doi.org/10.1029/JC075i024p04539.

McConnell, D.R., Kendall, B.A., 2003. Images of the base of gas hydrate stability in the deepwater Gulf of Mexico: Examples of gas hydrate traps in northwest Walker Ridge and implications for successful well planning. The Leading Edge 22, 361–367.

Meyer, J., Evans, R., Martin, R.L., Welch, C.L., 2002. Crosswell seismic: review of field operations, survey planning and case history. In: Hunt, T.J., Lufholm, P.H. (Eds.), The Permian Basin: Preserving Our Past – Securing Our Future. West Texas Geological Society, Inc. Texas, USA.

Mienert, J., Posewang, J., Baumann, M., 1998. Gas hydrates along the northeastern Atlantic Margin: possible hydrate-bound margin instabilities and possible release of methane. Geol. Soc. Spec. Publ. 137, 275–291.

Mienert, J., Posewang, J., Lukas, D., 2001. Changes in hydrate stability zone on the Norwegian margin and their consequences for methane and carbon releases into the oceanosphere. In: Schaefer, P., Ritzau, W., Schlueter, M., Thiede, J.J. (Eds.), The Northern North Atlantic. Springer Verlag, Berlin, Germany, pp. 259–280.

Miller, J.J., Lee, M.W., von Huene, R., 1991. An analysis of a seismic reflection from the base of gas hydrate zone, offshore Peru. Am. Assoc. Petrol. Geol. Bull. 75, 910–924.

Morales, E., Vera, E., Diaz, J., et al., 2001. Hidratos de gas submarinos: una nueva fuente de energía para el siglo XXI. Proyecto FONDEF D00I 1104 19.

O'Brien, M., Whitmore, D.N., Brandsberg-dahl, S., Etgen, J.T., Murphy, G.E., 1999. Multicomponent modelling of the Valhall field. In: Presented at the 61st Annual Conference and Exhibition, EAGE.

Pearson, C.F., Murphy, W.F., Hermes, R., 1986. Acoustic and resistivity measurements on rock samples containing tetrahydrofuran hydrates: laboratory analogous to natural gas hydrate deposits. J. Geophys. Res. 91, 132–138.

Pecher, I., Coffin, R., Henrys, S., CHARMNZ working Group, 2007. Tangaroa TAN0607 Cruise Report: Gas Hydrate Exploration on the East Coast, North Island, New Zealand. GNS Science Report, Lower Hutt, New Zealand.

Pecher, I.A., et al., 2008. Erosion of seafloor ridges at the top of the gas hydrate stability zone, Hikurangi Margin, New Zealand: new insights from research cruises between 2005 and 2007. In: 6th International Conference on Gas Hydrates, Vancouver, BC, Canada.

Pecher, I.A., Henrys, S.A., Ellis, S., Chiswell, S.M., Kukowski, N., 2005. Erosion of the seafloor at the top of the gas hydrate stability zone on the Hikurangi Margin, New Zealand. Geophys. Res. Lett. 32 (L24603).

Pecher, I.A., Henrys, S.A., Zhu, H., 2004. Seismic images of gas conduits beneath vents and gas hydrates on Ritchie Ridge, Hikurangi Margin, New Zealand. N. Z. J. Geol. Geophys. 47, 275–279.

Pecher, I.A., Holbrook, W.S., Stephen, R.A., et al., 1997. Offset-vertical seismic profiling for marine gas hydrate exploration; is it a suitable technique? First results from ODP Leg 164. In: Proceedings Offshore Technology Conference, vol. 29, pp. 193–200.

Pecher, I.A., Minshull, T.A., Singh, S.C., von Huene, R., 1996. Velocity structure of a bottom simulating reflector offshore Peru. Earth Planet. Sci. Lett. 139, 459–469.

Pecher, I.A., Ranero, C.R., von Huene, R., Minshull, T.A., Singh, S.C., 1998. The nature and distribution of bottom simulating reflectors at the Costa Rican convergent margin. Geophys. J. Int. 133, 219–229.

Pekot, L.J., Gregory, A.G., 1987. Ekofisk. In: Spencer, A.M. (Ed.), Geology of the Norwegian Oil and Gas Fields. Graham&Trotman Ltd., pp. 73–87.

Rajput, S., Thakur, N.K., Rao, P.P., Joshi, A., 2010. AVO response for a complex double bottom simulating reflectors model. Curr. Sci. 98, 10.

Rajput, S., 2009. Analysis of Ocean Bottom Seismometer (OBS) Data for Gas Hydrate Studies and Subsurface Models PhD thesis. Department of Geophysics, Kurukshetra University, Kurukshetra, India.

Rajput, S., Rao, P.P., Thakur, N.K., 2005. Amplitude Strength of Simulating Reflectors by Full Waveform Modeling. SEG Expanded Abstracts, pp. 1787–1790.

Rajput, S., Thakur, N.K., Joshi, A., 2008. Full waveform seismic modelling for Gas hydrate studies. In: 70th EAGE Conference & Exhibition, Rome, Italy.

Richards, P.G., Frasier, C.W., 1976. Scattering of elastic waves from depth-dependent inhomogeneities. Geophysics 458.

Riedel, M., Willoughby, E.C., Chen, M.A., et al., 2006. Proceedings of the Integrated Ocean Drilling Program, vol. 311.

Ritzwoller, M.H., Levshin Anatoli, L., 2002. Estimating shallow shear velocities with marine multicomponent seismic data. Geophysics 67, 1991. http://dx.doi.org/10.1190/1.1527099.

Ruppel, C., Boswell, R., Jones, E., 2008. Scientific results from Gulf of Mexico gas hydrates joint industry project Leg 1 drilling: introduction and overview. Mar. Pet. Geol. 25 (9), 819–829. http://dx.doi.org/10.1016/j.marpetgeo.2008.02.007.

Sain, K., Minshull, T.A., Singh, S.C., Hobbs, R.W., 2000. Evidence for a thick free gas layer beneath the bottom simulating reflector in the Makran accretionary prism. Mar. Geol. 164, 3–12.

Shipley, T.H., Houston, M.H., Buffler, R.T., et al., 1979. Seismic evidence for widespread possible gas hydrate horizons on continental slopes and rises. Am. Assoc. Pet. Geol. Bull. 63, 2204–2213.

Shuey, R.T., 1985. A simplification of Zoeppritz equations. Geophysics 50, 609–614.

Singh, S.C., Minshull, T.A., Spence, G.D., 1993. Velocity structure of a gas hydrate reflector. Science 260, 204–207.

Somerville, K., Cassinat, J., McCallum, M., 2007. Crosswell seismic applied to shallow coalbed methane and tight gas sands, SEG Extended Abstract, pp. 1257–1260.

Stewart, R.R., 1991. Rapid map and inversion of P-SV waves. Geophysics 56, 859–862.

Stewart, R.R., Ferguson, R., 1996. Shear-wave interval velocity from P-S stacking velocity. Can. J. Expl. Geophys. 32, 139–142.

Stewart, R.R., Gaiser, J.E., Brown, J.R., Lawton, D.C., 2002. Converted-wave seismic exploration: Methods. Geophysics 67, 1348–1363.

Stewart, R.R., Lu, H., Bland, H., Mewhort, L.E., 2001. A lake bottom cable seismic survey: Acquisition and processing. Can. Soc. Expl. Geophys. Rec. 8, 37–40.

Stoll, R.D., Ewing, J., Bryan, G.M., 1971. Anomalous velocities in sediments containing gas hydrates. J. Geophys. Res. 76, 2090–2094.

Suess, E., 2002. The evolution of an idea: from avoiding gas hydrates to actively drilling for them. Spec. Issue Joides J. 28, 45–50.

Tréhu, A.M., et al., 2004. Feeding methane vents and gas hydrate deposits at south Hydrate Ridge. Geophys. Res. Lett. 31, 1–4.

Tucholke, B.E., Bryan, G.M., 1977. Gas-hydrate horizons detected in seismic-profiler data from the western North Atlantic. Am. Assoc. Pet. Geol. Bull. 61, 698–707.

Vanneste, M., Mienert, J., Guidard, S., HYDRATECH-INGGAS Partners, 2002a. "Arctic" gas hydrates offshore Western Svalbard, Norway. In: Mori, Y.H. (Ed.), Proceedings of the 4th International Conference on Gas Hydrates. International Conference on Gas Hydrates, Yokohama, Japan, pp. 222–227.

Vanneste, M., Poort, J., De Batist, M., Klerk, J., 2002b. Atypical heatflow near gas hydrate irregularities and cold seeps in the Baikal Rift Zone. Mar. Pet. Geol. 19 (10), 1257–1274. http://dx.doi.org/10.1016/S0264-8172(03)00019-9.

Walraven, D., Aminzadeh, F., Connoly, D.L., 2004. Predicting seal risk and charge capacity using chimney processing: three Gulf of Mexico case histories. In: 74th Annual International Meeting, SEG, Expanded Abstracts.

Whalley, E., 1980. Speed of longitudinal sound in clathrate hydrates. J. Geophys. Res. 85, 2539–2542.

White, J.E., 1975. Computed seismic speed and attenuation in rocks with partial gas saturation. Geophysics 40, 224–232.

Whitson, C., Sunjerga, S., 2012. PVT in Liquid-rich Shale Reservoirs. SPE 155499.

Winters, G., 2013. Geophysics Optimize Performance of Tight Oil Resource Plays. The American Oil & Gas Reporter. http://seisware.com/wp-content/uploads/2014/08/AOGR_GlennWinters.pdf.

Winters, W.J., Dillon, W.P., Pecher, I.A., Mason, D.H., 2000. GHASTLI: determining physical properties of sediment containing natural and laboratory-formed gas hydrate. In: Max, M.D. (Ed.), Coastal Systems and Continental Margins: Natural Gas Hydrate in Oceanic and Permafrost Environments, vol. 5. Kluwer Academic Publishers, Dordrecht, Netherlands, pp. 311–322.

Winters, W.J., Waite, W.F., Mason, D.H., Gilbert, L.Y., 2004. Physical properties and rock physics models of sediment containing natural and laboratory-formed methane gas hydrate. Am. Min. 89, 1221–1227.

Winters, W.J., Waite, W.F., Mason, D.H., Dillon, W.P., Pecher, I.A., 2002. Sediment properties associated with gas hydrate formation. In: 4th International Conference on Gas Hydrates, Yokohama, Japan, pp. 722–727.

Wood, W.T., Ruppel, C., 2000. Seismic and thermal investigations of the Blake Ridge gas hydrate area: a synthesis. Proc. Ocean Drill. Program. Sci. Results 164, 253–264. http://dx.doi.org/10.2973/odp.proc.sr.164.203.2000.

Wood, W.T., Gettrust, J.F., Chapman, N.R., Spence, G.D., Hyndman, R.D., 2002. Decreased stability of methane hydrates in marine sediments owing to phase-boundary roughness. Nature 420, 656–660.

Wu, X.H., Yi, H.S., Wang, C.S., 2004. Petroleum geochemical characterization in Shinan oil field of Junggar Basin (in Chinese with English abstract). Nat. Gas Ind. 24 (12), 24–27.

Xia, J., Miller, R., Park, C., 1999. Estimation of shear-velocity by inverting Rayleigh wave phase velocity. Geophysics 64, 691–700.

Xiao, F., Xu, C.Y., Zhang, F.J., Lin, C.X., 2010. Major breakthrough in the Hatu gold deposit, western Junggar, XinJiang. Xinjiang Geol. 409–412.

Yang, Y.-T., Guo, Z.-X., Song, C.-C., Li, X.-B., Sheng, H., 2014. A short-lived but significant Mongol–Okhotsk collisional orogeny in latest Jurassic–earliest Cretaceous. Gondwana Res. 28, 1096–1116. Journal ISSN: 1342-937X. http://dx.doi.org/10.1016/j.gr.2014.09.010.

Yang, G.X., Li, Y.J., Gu, P.Y., et al., 2012. Geochronological and geochemical study of the Darbut Ophiolitic Complex in the West Junggar (NW China): implications for petrogenesis and tectonic evolution. Gondwana Res. 21, 1037–1049. http://dx.doi.org/10.1016/j.gr.2011.07.029.

Yuan, T., Hyndman, R.D., Spence, G.D., Desmons, B., 1996. Seismic velocity increase and deep-sea hydrate concentration above a bottom-simulating reflector on the northern Cascadia continental slope. J. Geophys. Res. 101, 13655–13671.

Zhai, G.M., Song, J.G., Jin, J.Q., Gao, W.L., 2002. Plate Tectonic Evolution and its Relationship to Petroliferous Basins (In Chinese). Petroleum Industry Press, Beijing.

CHAPTER

GEOLOGICAL HAZARDS

9

"If you thought that science was certain well, that is just an error on your part."

Richard Feynman

CHAPTER OUTLINE

9.1 INTRODUCTION

Gas hydrates occur in different geological and tectonic environments and their stability is delicately balanced under suitable thermobaric conditions. Gas hydrates as a potential unconventional resource may provide a viable alternative for future energy requirements. It is estimated that the shallow subsurface gas hydrate reserve on the continental margins and in permafrost regions combined holds twice the amount of carbon of all other hydrocarbon resources put together (Sloan, 2006). Enormous amounts of methane are believed to be trapped by hydrates, both in the hydrate crystal structure itself as well as free gas in sediments beneath the hydrate deposits. Primarily, gas hydrates occupy the pore spaces in the sediments and cement them together. They are also formed in secondary voids in fractures and joints. During growth, hydrates create their own space by disturbing the sediment matrix (Suess et al., 1999a). Pore-filling morphologies of gas hydrates replace pore fluid between grains of sediment; this gas hydrate may or may not cement grains together. Grain-displacing gas hydrate does not occupy the pore volume between grains; instead it forces the grains apart, forming veins, layers, and lenses of pure gas hydrate. Grain-displacing hydrates may cover a vast range of sizes, from thin veins possibly only a few microns thick to nodules of tens of centimeters or even meters in diameter (Fig. 9.1).

Grain-displacing and pore-filling are not equivalent to the terms "massive" and "disseminated" but these terms apply to cores which have already undergone gas hydrate dissociation, where massive gas

Geological Controls for Gas Hydrates and Unconventionals. http://dx.doi.org/10.1016/B978-0-12-802020-3.00009-6

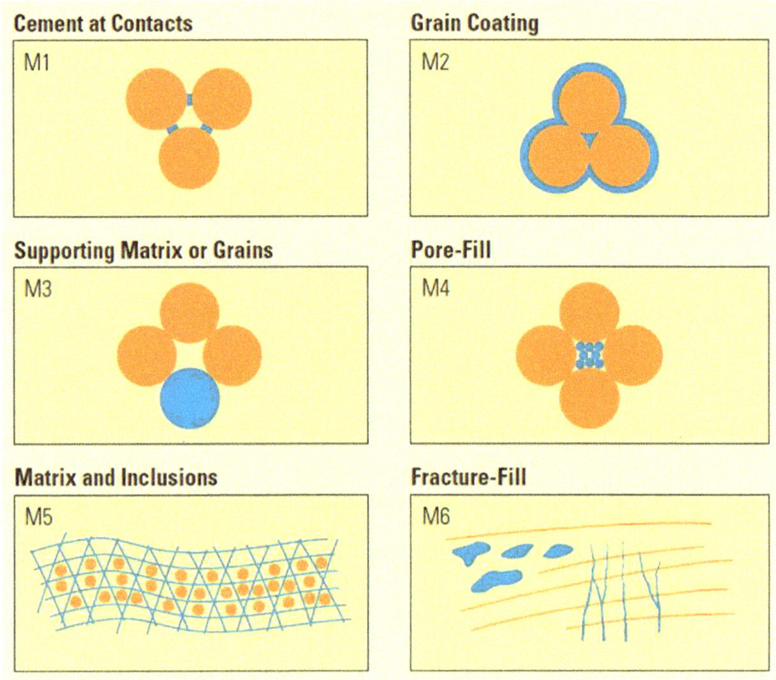

FIGURE 9.1

Microstructural models of hydrate-bearing sediments. In the first five of the six models, gas hydrates (blue) are evenly distributed throughout the sedimentary grains (tan) to a first approximation. Hydrate may occur as cement at grain contacts (top left), as coating on grains (top right), as a component of the grain matrix (middle left), or as pore-filling material (middle right). The fifth model considers sedimentary grains as inclusions in a hydrate matrix (bottom left). The sixth model (bottom right) depicts hydrates as nodules or fracture-fill on fine-grained, low-permeability sediments. These models are used to simulate the response of hydrate-bearing sediments to logging and seismic measurements. (Source Dai et al., 2008.)

hydrate is still visible, and disseminated gas hydrate is invisible to the naked eye, and may have already been completely dissociated (Holland et al., 2008). Disseminated hydrates have a tendency to occupy coarse-grained clastic sediments with higher porosity and permeability and contain fine-grained sediments in remotely connected geological structures. In subsurface earth they may exist as individual grains or particles disseminatedthrough the sediments but also may cement the sediment grains as nodules and laminate, layers of different shapes and sizes with dimensions of the order of a few centimeters or less (Collett, 2003). Massive gas hydrate samples have been recovered during drilling in the Malik gas hydrate research well in the Mackenzie Delta formed in high porosity coarse-grained and gravel rocks (Dallimore et al., 1999). Drilling over the Hydrate Ridge in the Cascadian margin has shown that gas hydrates are found to occur in fine-grained sediments confined to the fractures and joints mostly running parallel to the bedding planes and in some places in the fractures oblique to bedding planes (Suess et al., 1999b). Gas hydrate deposits of different shapes and sizes, such as veins and layers of comparable size and those with spherical to oblate features, sometimes with edges (considered as nodules) and massive lens-shaped hydrate accumulations, were obtained during drilling over the Blake Ridge (Site 570: Shipboard Scientific Party, 1985).

Low molecular weight gases, such as hydrogen and helium, which are smaller than the small cavities (3.8 Å), and high molecular weight gases, such as pentane, hexane, and paraffin hydrocarbons, which are bigger than the large cavities (>9 Å), are unsuitable for the formation of hydrates under natural conditions. However, in recent times attempts have been made to synthesize hydrates with low molecular weight gases in the laboratory. Most low molecular weight hydrocarbon and nonhydrocarbon gases such as oxygen, hydrogen, nitrogen, carbon dioxide, methane, hydrogen sulfide, argon, and krypton as well as some higher hydrocarbons and freons are trapped in the cavities and form hydrates at suitable temperatures and pressures. Without the support of the trapped molecules, the lattice structure of hydrate clathrates would collapse into a conventional ice crystal structure or liquid water. When the cavities are fully occupied by guest molecules of single chemical components then the composition can be expressed in terms of a definite formula. However, in nature, depending upon the physical conditions (pressure, temperature) and the availability of suitable gases, only parts of the voids are filled with guest molecules (degree of saturation).

Earlier studies had suggested that methane, the main gaseous component of gas hydrate, is generated locally due to the decomposition of organic content and reduction of carbon dioxide in sediments. Methane is also produced due to the thermochemical, i.e., thermal degradation and cracking, process at elevated temperatures (50°C to 150°C) and extended periods of heating (Rice, 1993). Fracturing and other weak zones in the subsurface earth provide conduits to bring thermogenic methane to shallower levels, where the conditions are suitable for gas hydrate formation. Methane gas hydrate was assumed to be uniformly distributed wherever pressure and temperature conditions were appropriate and generated methane was present in large quantities (Trofimuk et al., 1973). The higher quantum of methane required for formation of gas hydrates has led to the conclusion that methane has to be remotely brought into the hydrate stability zone to enhance methane concentration, which is a prerequisite for the formation of gas hydrates (Hyndman and Davis, 1992). Subsequent studies have revealed a heterogeneous global distribution of gas hydrates and this may be due to the advective transport in supplying sufficient methane-rich fluids for hydrate formation (Buffett and Archer, 2004; Wood and Jung, 2008). Geological weak zones such as faulting or fracturing provide conduits for the migration or seepage of methane-rich fluids in the gas hydrate stability region. Thus it is expected that hydrates not only exist under suitable temperature conditions but are also likely to get confined to weak geological (faulting/fracture/joints) environments.

The estimates of gas hydrate as a resource potential have been quite speculative and uncertain ever since first attempts were made to quantify them in the 1970s (Kvenvolden, 1999). The resource estimates were primarily based on the identification of the bottom-simulating reflector (BSR) in the multichannel seismic reflection data and the observation of geological proxies such as pockmarks and venting gas through the sedimentary layers and the water column. Parameters such as the thickness of the hydrated layer, porosities, saturation in sediment pores, areal extent, volume, and hydrate and gas yield were considered for estimation of the global gas content in hydrates (Collett, 2003; Milkov, 2004). These estimates are being modified in view of new information about their formation processes, development of new techniques, high-resolution data acquisition, and ground truth by well log data acquisition or interpretation and obtaining hydrate samples under pressurized conditions. The global estimates of methane gas in hydrate shows large variations. During the 1970s and 1980s, the gas content (mainly methane gas) was estimated to be of the order of 1017–1018 m^3; this fell in the 1980s and 1990s to the order of 1016 m^3. Recent estimates (Milkov, 2004) suggest that the gas hydrates reserves may be the order of 1014–1015 m^3 (Fig. 9.2, Milkov, 2004). The global estimates of methane gas contained in gas hydrates have substantially reduced from the 1970s to the present, a significant reduction by a factor of 2–4 The estimated quantum of gas hydrates exceeding 1000 Gt (high probability) and in

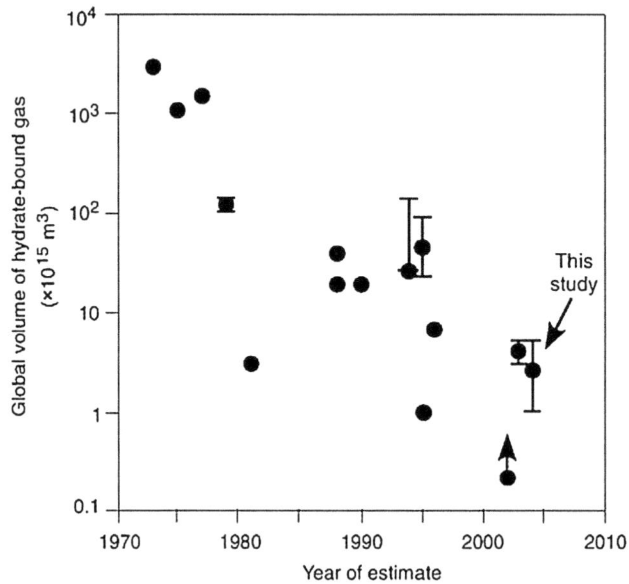

FIGURE 9.2

Global estimates of the volume of hydrate-bound gas in marine sediments versus the year in which the estimate was made. (Data from Milkov, 2004.)

the range of 1000–10,000 Gt, equivalent to ~ 2000–20,000 trillion cubic meters (medium probability), was calculated at a recently held workshop (Bohannon, 2008).

Suitable physical conditions, enormous primary productivity, and influx of terrestrial organic matter make continental margins a most attractive proposition for the commercial exploitation of gas hydrates to balance supply and demand of oil and gas. However, owing to their accumulation in deepwater, lower saturation (17–20%), which is about 3% concentration in sediments, increases the likelihood of geological hazards that come about as a result of their dissociation (Maslin et al., 2010). Furthermore, limitations of the present-day technology mean that commercial exploitation of gas hydrates over the continental margins does not look feasible in the immediate future. Gas hydrate morphology describes the relationship between gas hydrates and the surrounding marine sediments. The morphology of gas hydrates determines the basic physical properties of the sediment–hydrate matrix.

9.2 PHYSICAL PROPERTIES

The presence of gas hydrates in sediments dramatically alters some of the normal physical properties of the sediments. An understanding of the physical properties of the hydrate-bearing sediments is necessary for interpretation of geophysical data collected in the field setting, including borehole and slope stability analyses, reservoir simulation, and production models. The geomechanical properties of hydrate-bearing sediments are not yet fully understood. Laboratory experiments, small scale physical modeling, and theoretical analyses have been undertaken to estimate their behavior with regard to degree of saturation, grain size, porosity, and other parameters; such attempts in turn may provide "a priori" knowledge to assess the impact of geological processes that govern submarine slope failure and mitigate the risks during oil and gas exploitation.

Gas hydrates cement between sediment grains or act as cement around contacting grains, and provide grain-to-grain cohesion and frictional resistance to the shallow soft sediments, where gas hydrates are likely to occur. The strength of methane hydrate-bearing sediments is a fundamental parameter required for the assessment of methane production from a hydrate accumulation and assessment of submarine slope stability and its productivity. Studies undertaken so far demonstrate that the presence of hydrates will increase strength and stiffness and decrease permeability of sediments. In fact, laboratory experiments indicate that the strength of pure hydrates can be 20 times that of pure ice, a contrast that increases with lower temperatures (Durham et al., 2003). However, when contained within the sediments, laboratory results show increases in the strength of hydrate-bearing sediments over hydrate-free sediments, with hydrate- and ice-bearing sediments having similar strengths (Masui et al., 2005, 2008; Edinuma et al., 2005). The strength of hydrate-bearing sediments is the function of hydrate saturation, strain rate, temperature, consolidation stress, grain size, density, and cage occupancy (Winters et al., 2004). Even at low concentrations, intergranular hydrate may thus influence the formation strength depending on its articulation within the sediment host. At higher concentrations, the strengthening effects from pore-filling gas hydrate can be obtained by analogy to strength enhancement in frozen soils or from general theory and experiments with rheologies of aggregate mixtures. Hydrated sediment specimens show increase in strength with hydrate saturation (Durham et al., 2005).

The formation of hydrates between the sand particles contributes to the increases in cohesion but has little impact on the friction angle (Masui et al., 2005; Suzuki et al., 2008). Testing to date indicates that stiffness, cohesion, and dilation increase with increase in saturation, while the friction angle remains unaffected. Modeling results on synthetic hydrates (Yun et al., 2007) show that the stress–strain behavior of hydrate-bearing sediments is a complex function of particle size, confining pressure, and hydrate concentration. The mechanical properties of hydrate-bearing sediments at low hydrate concentrations (probably <40% of pore space) appear to be determined by stress-dependent soil stiffness and strength. At high hydrate concentrations (>50% of pore space), the behavior becomes independent of stress because the hydrates control both the stiffness, strength, and possibly the dilative tendency of sediments by effectively increasing interparticle coordination, cementing particles together, and filling the pore space. The contribution of cementation to the shear strength of hydrate-bearing sediments decreases with increasing specific surface of the soil minerals. Hydrates contained in the pore spaces accentuate the pore pressure response, shear decrease in the coarse-grained sediments, and shear increase in the fine-grained sediments. The presence of a gas phase dampened the pore pressure response (Winters et al., 2007).

The thermal properties of hydrate-bearing rocks are a controlling factor for all processes involving the formation and decomposition of gas hydrates in nature which are inevitably coupled with the transport of heat within the formation. The stability of gas hydrates is controlled more by temperature than pressure. In this context, detailed knowledge about the thermal properties of rocks containing gas hydrates is required in order to quantify processes involving the formation and decomposition of gas hydrates in nature. The measured heat flow in a region is the product of thermal conductivity and the geothermal gradient. Thermal conductivity describes the material's ability to transfer heat. Measurement of in situ thermal conductivity provides information about the rock matrix where hydrates are likely to be present. The mineralogical composition of the rock matrix determines the thermal conductivity of the host rock. Geological features such as salt-bearing structures (e.g., diapirs) have good thermal conductivity and shale diapirs have lower thermal conductivities. For a constant heat flow, the isotherm has to undulate by increasing or decreasing the geothermal gradients due to changes in thermal conductivities of the host rock (detailed in Chapter 4). Such variations in geothermal gradients affect the thickness of the gas hydrate stability zone. The thermal conductivity of pure methane hydrate was determined as 0.45 W/mK at 216.2 K (Cook and Leaist, 1983). The thermal conductivity of pure hydrate is therefore about 20% lower than the thermal conductivity of water, and up to 80% lower than that of ice. Owing to the low thermal conductivity of pure hydrate, the presence of gas hydrate in sediments may significantly alter the bulk rock thermal conductivity and the geothermal gradient within the hydrate-bearing formations (Ruppel, 2000).

At low saturations (S_h ~30–35%), thermal conductivity of investigated gas saturated sediments varied slightly and did not exceed 1–2% when compared to the host rock. In contrast there were significant changes in thermal conductivity of gas-saturated sediments above 35–40%. Thus, in the sand sample with the increase of (S_h) from 0 to 38% the thermal conductivity increased about 8% (from 1.85 W/mK to almost 2.0 W/mK). Unlike samples without gas hydrates, an abnormal fall in thermal conductivity was observed in hydrate-saturated sediments. Thus, the difference in the values of thermal conductivity of hydrate-containing sediments and samples without hydrates can reach 10% or even higher; this distinction was 20% in the sand sample, and more than 100% in samples of sandy loam sand. Such behavior of thermal conductivity of hydrate-saturated sediments is caused not only by a difference in the values of thermal conductivity of porous ice and porous hydrates but also by structurally textural changes in hydrate-saturated sediments during freezing. Decrease of thermal conductivity was observed during hydrate accumulation in sediments. Thus, in the sand sample (sand% ~22 in the sediment mix) at $S_h = 0\%$ thermal conductivity was 2.00 W/mK, and at Sh = 38% thermal conductivity was 1.73 W/mK. In this case, decrease of the thermal-physic parameter was 14% (Chuvilin and Buhanov, 2011).

The hydrate target occurs in the shallow section (hundreds of meters in depth), and is manifested by subtle resistivity contrasts (a few ohm-m). Electrical sounding provides an alternate method of locating hydrate in the absence of a BSR (Yuan and Edwards, 2000). Hydrates form from pore fluid but usually exclude salt from the pore fluid, so they have high electric resistivity just as ice and sediments containing hydrates have a higher resistivity compared to sediments without gas hydrates (Judge, 1982). Laboratory experiments have shown the electrical conductivity (σ) of methane hydrate at 0°C to be 5×10^{-5} S/m; for ice it is of the order of 10^{-5} to 10^{-6} S/m. The resistivity of hydrate-saturated sediments approximately varies from 103 to 106 ohm-m (Tzirita, 1992), whereas water has a resistivity of 10 ohm-m and for sand it is of the order of 102 to 103 ohm-m. The significant contrast in electric parameters due to the presence of hydrates in sediments make electromagnetic (EM) surveys a good tool to map hydrate occurrences (Du Frane et al., 2011). The unconsolidated sediments in the upper several hundred meters of the marine sediment section (50% porosity) normally have a very low resistivity of about 10 ohm-m. For 15–20% hydrate saturation in the pore space (7–10% of sediment), the resistivity increases by a factor of about 2. Calculation of hydrate content using downhole electrical measurements based on Archie's law requires knowledge of the saturation exponent. The saturation exponent is an empirical parameter that includes influences from the internal rock structure such as pore shape, size, connectivity of the pore network, and the distribution of the conducting phase. Spangenberg (2004) reported that, for the parameters used in his model calculations, the saturation exponent varied between 0.5 and 4. The saturation exponent is an empirical parameter that includes influences from the internal rock structure such as pore shape, size, connectivity of the pore network, and the distribution of the conducting phase. Seismic measures can delineate gas hydrates in the sediments provided their saturation in the pore spaces is of the order of ~40%, as such percentage rises can significantly alter the seismic velocities to cause anomalous behavior. With lower saturations of gas hydrates in pore spaces, seismic analysis may not have a greater advantage. This means that some seismic techniques may miss a significant amount of methane hydrate in areas where the saturation is less than ~40%. As the lack of appreciable impedance contrast has not been clearly brought out in many seismic studies, apart from the top of the hydrated zone, the resource estimate of gas hydrate is speculative. Laboratory studies show that electrical methods are more sensitive to lower saturations of gas hydrates and may also delineate the top of the hydrated layer. This has fueled interest in the application of electromagnetic (EM) methods for regional characterization of gas hydrate deposits or the joint application of EM and seismic techniques. The sensitivity of electrical properties to a wide range of hydrate saturations is also manifest by the widespread reliance of borehole resistivity logging to identify hydrate-bearing sediments in both marine and permafrost associated settings.

Well-planned strategies and technological advancement are the prerequisites for exploitation of gas hydrates for the production of methane and its recovery in commercial quantities, taking into consideration the likely geological hazards one may encounter during such operations.

9.3 DECOMPOSITION OR DISSOCIATION

Gas hydrates are not chemical compounds as the sequestered molecules are never bound to the lattice. The formation and decomposition of hydrates are first order phase transitions, not chemical reactions (Maslin et al., 2010). Hydrate decomposition is a sequence of lattice destruction and gas desorption processes. The formation and decomposition of gas hydrates is complex and not fully understood (Bishnoi and Natarajan, 1996; Gao et al., 2005). The stability conditions are likely to be altered with the presence of varying ionic content of water and the admixture of different constituent gases (Fig. 9.3). The presence of gases other than methane in the hydrate frame significantly alters the stability conditions. Hydrates stabilize at appreciably higher temperatures and lower pressures when hydrogen sulfide

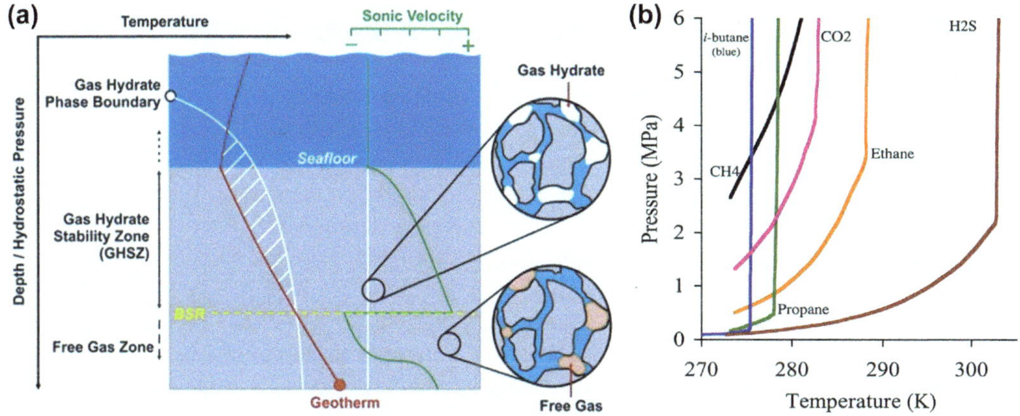

FIGURE 9.3

Gas hydrate stability conditions. (a) The hydrate stability zone in subsea sediments (source: http://www.pet.hw.ac.uk/research/hydrate/hydrates_where.cfm). (b) Temperature versus pressure plot for different gases shows the variation of stability conditions (source: knol.google.com/k/gas-hydrates).

(H_2S), carbon dioxide (CO_2), ethane (C_2H_6), and propane (C_3H_8) are part of the hydrate-forming gas mixture (Carroll and Mather, 1991; Sloan, 2006; Kvenvolden, 1998). The presence of hydrogen sulfide (H_2S) and carbon dioxide gases with methane make the hydrate phase equilibrium stable at higher temperatures and lower pressures. On the contrary, the presence of nitrogen with methane may cause hydrate formation at higher pressures and lower temperatures.

Hydrates stiffen the sediment in which they exist, and this semiconsolidated mass overlies the highly fluidized and frequently gassy silts and muds at the base of the gas hydrate stability zone (GHSZ). Thus, as a general principle, either creation or dissociation of hydrate can potentially cause problems, though dissociation is the more immediate concern. Dissociation of the hydrate induces fluid and gas production, loss of strength and stiffness, and elevated pore pressures (Grozic, 2010). Factors affecting gas hydrate decomposition in sediments include sediment type, mineral composition, pore size distribution, particle size, pore water composition, hydrate saturation distribution, initial formation pressure, and temperature and cement characteristics. Furthermore, the decomposition rate of hydrate sediments decreases with decreasing particle size and increasing pore size (Yu et al., 2013). Where hydrate exists near the bonds of stability, even small increases in temperature may destabilize it; absorbing heat during decomposition of hydrates can give rise to formation temperature changes. However, the formation temperature change will limit the decomposition rate of hydrates (Cheng and Li, 2012). Hydrate dissociation starts when the pressure–temperature (PT) state reaches the equilibrium boundary. In reality, the hydrate stability curve should be expressed as a broad band that reflects different boundaries for different pore sizes rather than a single phase boundary. A single boundary causes instantaneous dissociation as temperature increases; in contrast a broad margin across the PT boundary is expected to cause gradual dissociation. This is due to hydrates in smaller pores dissociating at lower temperatures compared to those in larger pores (Anderson et al., 2003; Kwon et al., 2008).

Any variations of these factors alter the stability conditions and gas hydrate may decompose and liquefy to produce liquid and large quantities of methane gas. Gases are highly concentrated in the solid hydrate phase, thus the decomposition of hydrates can release large volumes of gas. One cubic meter of methane hydrate can contain the equivalent of 164 m^3 of methane at standard temperature and pressure (Sloan, 2006). However, the immediate volume increase that occurs on hydrate decomposition is dependent on temperature and pressure conditions in the sediment column. The volume expansion factor increases as water depth (hydrostatic pressure) decreases. For example, at a water depth of 1000 m (and typical seafloor temperatures), the volume of released methane and water is a little over twice the original gas hydrate volume, whereas at 500 m water depth, the volume increases by a factor of 3.5, and in very shallow hydrate deposits (<300 m water depth), expansion factors may exceed by a factor of 6.0 (Fig. 9.4).

When solid hydrate turns into gas and water, the volume of pore-filling material can increase significantly and the sediment becomes fluidized, compromising the strength and stiffness of the sediment column (Birchwood et al., 2010). Strengthening effects of the hydrate are not only lost owing to loss of material and cohesion but this weakening effect is in addition to the pore pressure effect due to the release of a high quantum of gas. These many properties and effects, in addition to those introduced by the environmental pore pressure conditions, place many limits on the conditions under which gas hydrates may be stable in nature (Durham et al., 2005). Modeling of a gas hydrate system during tectonic uplift revealed that significant pore pressure may build up beneath shallow gas hydrates easily reaching lithostatic pressure and thus may lead to fracturing of the seafloor.

The presence of free gas in the GHSZ leads to two different problems with respect to hydrate thermodynamic stability. First, it reduces the lithostatic pressure by decreasing the density of the sediment

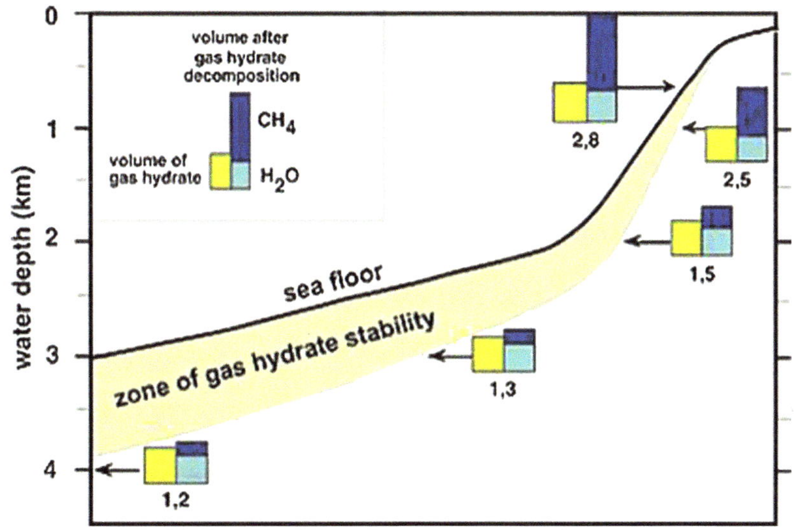

FIGURE 9.4

Thickness of the zone of gas hydrate stability in the seafloor (assuming a geothermal gradient of 30°C/km) depending on water depth (bold line). The bar charts show the equally depth-dependent increase in volume (number = factor by which the volume increases) if a given volume of gas hydrate dissociates. (From Paull et al., 2000.)

column. Secondly, and probably more importantly, it steepens the geothermal gradient due to its small thermal conductivity. As the seafloor constitutes a nearby heat reservoir of essentially infinite capacity, steepening the gradient causes the temperature at the base of the GHSZ to increase. Both pressure and temperature effects tend to destabilize the hydrate deposit and this effect in the sediment can be exacerbated by a pore-filling growth habit (Kleinberg et al., 2003).

Accumulation of gas under the base of the GHSZ due to the destabilization of hydrate or migration from below is also clearly problematic. Increasing gas saturation reduces the shear strength of the formation. With a huge quantum of methane gas preserved in hydrate or free gas zone in shallow subsurface earth under delicate PT conditions, its dissociation may cause a significant geological hazard in terms of climatic impact and margin slope stability during conventional oil and gas drilling and production operations. In this chapter, we mainly discuss the latter aspect of geological hazards due to the dissociation of gas hydrates in the continental margins.

9.4 HYDRATE IMPACT ON CLIMATE

Gas hydrates in the sediments are in a state of metastable equilibrium with the environment. Methane as a guest molecule is captive in a structural frame of hydrates and provides stability to hydrates over a higher range of temperatures. PT regimes make them stable in shallow subsurface depths of the earth. Changes brought about in temperature and pressure by any mechanism leads to destabilizing or stabilizing of hydrates over the timescale to the order of hundreds to thousands of years. The stabilizing or destabilizing timescales depend on the thermal properties of sediments (MacDonald, 1990). The amount of methane trapped in the hydrates is almost 3000 times the quantum of methane in the atmosphere (Kvenvolden, 1998). The methane concentration in the atmosphere is increasing at the rate of about 1% every year (Watson et al., 1990). Release of large amounts of methane from hydrates either as unaltered methane or oxidized to carbon dioxide could have a large impact on the composition of atmospheric gases. Methane releases from hydrates that could be most significant to climate change are more likely to be chronic in nature. Methane, being radioactively active, is a greenhouse gas and a major contributor to global climatic change (McDonald, 1990). It is suggested that methane released as a greenhouse gas from hydrates has a 20 times larger effect on global warming as compared to carbon dioxide integrated into the atmosphere over the last 100 years (Shine et al., 1990).

The shallow hydrate stability zone makes it vulnerable to near-surface changes. Any changes brought about in PT conditions on the surface or near surface by means of sea-level fluctuations during periods of glaciations, atmospheric warming due to climatic change, and tectonic activities in terms of faulting and earthquakes may lead to disturbances in the stability of hydrates. These phenomena alter the temperature and/or pressure regime thereby altering the stability conditions. During the global warming period, the glaciers and ice caps melt; increase water flow in the oceans and polar shelves. The thermal expansion of oceans also causes rise in sea levels. The rise in sea levels results in an increase in the hydrostatic pressure and stabilizes hydrates in continental margins and in polar continental shelves (Kvenvolden, 1998). During global cooling cycles, the whole system reverses. Glaciers and ice caps grow resulting in a fall in sea levels over the continental shelves reducing the hydrostatic pressure over the continental shelves. The reduction in hydrostatic pressure due to the sea-level drop of 120 m during the last Glacial Maximum (~18,000 years ago) may have resulted in decreasing the hydrate stability zone by about 20 m, assuming the temperature to be constant (Dillon and Paull, 1983). The reduction in the thickness of the stability zone depends on the rate of sea-level fall and thermal properties of near-seafloor sediment. A global map of the thickness of the GHSZ under steady-state conditions for the present-day climate (Fig. 9.5a) shows values of 600–900 m in polar regions due to low bottom water temperatures (<0°C) and along continental margins where bottom water temperatures are low and a thick sediment cover exists (Kretschmer et al., 2015). Thinner GHSZs (<150 m) are predicted for most of the deeper ocean, where the vertical extent is restricted by thin sediment thicknesses. The existence of abundant Early Paleogene marine gas hydrates despite warm deep-ocean temperatures was studied by Gu et al. (2011). Numerical simulations of gas hydrate accumulations were used at Paleogene seafloor temperatures to show that near-present-day values of gas hydrates could have been hosted in the Paleogene. A modified methane production profile with respect to depth below the seafloor at different seafloor temperatures is illustrated in Fig. 9.5b. Changes in the thickness of GHSZ are quite pronounced for higher geothermal gradients (Fig. 9.5c).

Fluctuation in sea level and changes in thickness of sediment column manifests in hydrate stability, thereby releasing about 200–400 Gt of carbon in the atmosphere over the time scale 1000s years. The fall in sea levels also changes the near-surface temperature and warms up the oceanic waters. A few degrees of warming in the deep ocean can have a significant impact on the stability of the hydrate, and it is known that the temperature of the deep ocean responds to changes in surface climate, albeit with a lag of centuries to millennia (Schiermeier, 2008). Different mechanisms have been proposed to explain the cyclic link between the dissociation of gas hydrates (release of methane) and climate change or global warming. Nisbet (1990) suggested that the methane released from the continental gas hydrates has led to a rapid rise in the global temperatures, which in turn may have almost stopped glacial activity

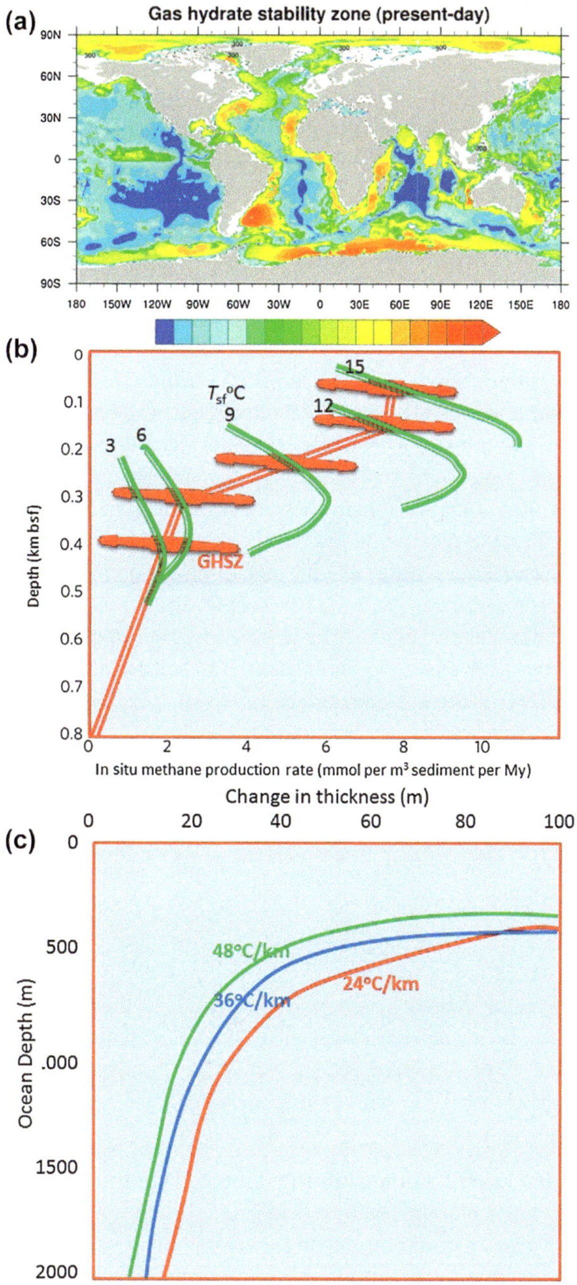

FIGURE 9.5

Effect of change in gas hydrate thickness, global gas hydrate stability zone, and methane production profile for a model. (a) Global map of the gas hydrate stability zone thickness (GHSZ in m) under present-day climate conditions (mean 1988–2007). The white shadings indicate areas where GHSZ = 0 m. The dashed contour line marks the 300 m isobath. (From Kretschmer et al., 2015). (b) Methane production profiles with respect to depth below the seafloor at different seafloor temperatures. D_{sf} = 2.0 km. The effects of higher organic carbon input and elevated methanogenesis rates at higher sediment temperatures have been included. The black bars show the base of the GHSZ for each T_{sf} °C. (c) Change in thickness and ocean depth effect on temperature for gas hydrate stability zone.

about 13.5 ka ago. Termination of a glacial period makes higher latitude regions get exposed to higher temperatures and results in melting ice sheets or caps and thereby reduces pressure. The change in thermobaric conditions leads to destabilization of gas hydrates and the release of uncontrolled methane into the atmosphere.

Conversely, Paull et al. (1991) postulated that during periods of global cooling, water inputs over the outer continental margins are curtailed causing sea levels to fall. These falling sea levels reduce the pressure over the margins leading to dissociation of gas hydrates and the release of methane into the atmosphere, resulting in global warming triggering deglaciation. Both the proposed models suggest the interconnectivity between different regions (high latitude – tropical) of gas hydrate occurrences through the atmosphere. The two proposed models either due to release of methane by continental hydrates, altering global temperatures or global cooling during periods of glaciations over the polar regions and determining the fate hydrates in tropical gas hydrates, envisages interconnectivity between different regions (high latitude-tropical) through the atmosphere in relevance to gas hydrate to gas hydrates occurrences. Physical changes brought about in one region lead to perturbing the physical conditions in another region via atmospheric connectivity. According to the latest discussion by the US Geological Society (USGS),which refers to the numbered type locals or sectors shown in Fig. 9.6a of gas hydrate deposits, currently, gas hydrates are most likely dissociating in shallow arctic shelves (sector 2 in Fig. 9.6a) and the upper edge of stability (sector 3 in Fig. 9.6a), contributing to release of methane into the atmosphere. As per the USGS (www.USGS.gov), these sectors are as follows:

- Thick onshore permafrost. Gas hydrates that occur within or beneath thick terrestrial permafrost will remain largely stable even if climate warming lasts hundreds of years. Over thousands of years, warming could cause gas hydrates at the top of the stability zone, about 625 ft (190 m) below the earth's surface, to begin to dissociate.
- Shallow arctic shelf. The shallow water continental shelves that circle parts of the Arctic Ocean were formed when sea-level rise during the past 10,000 years inundated permafrost that was at the coastline. Subsea permafrost is thawing beneath these continental shelves, and associated methane hydrates are likely dissociating now.
- Upper edge of stability. Gas hydrates on upper continental slopes, beneath 1000 to 1600 ft (300 to 500 m) of water, lie at the shallowest water depth for which methane hydrates are stable.
- Deepwater. Most of the earth's gas hydrates, about 95%, occur in water depths greater than 3000 ft (1000 m). They are likely to remain stable even with a sustained increase in bottom temperatures over thousands of years.

An example of methane bubble plumes off Washington, Oregon suggests warmer ocean may be releasing frozen methane (www.phy.org) as shown in Fig. 9.6b. New University of Washington research suggests that subsurface warming could be causing more methane gas to bubble up off the Washington and Oregon coast. However, due to the long ventilation times of the deep ocean (~100–1000 years), where the bulk of methane hydrates reside, and the slow propagation of the temperature signal into the sediment column (~180 m/1000 years), a new equilibrium is only reached on a timescale of 1000–10,000 years. Moreover, the fraction of methane from the bottom of the ocean that reaches the atmosphere is uncertain and dependent on the transport mechanism, e.g., bubbles, dissolved (Lamarque 2008). However, the susceptibility of hydrate dissociation to near-surface natural or human-induced changes in pressure and temperature might have played an important role in past climates and may also bring about significant global climatic change in the future due to activities for extraction of gas

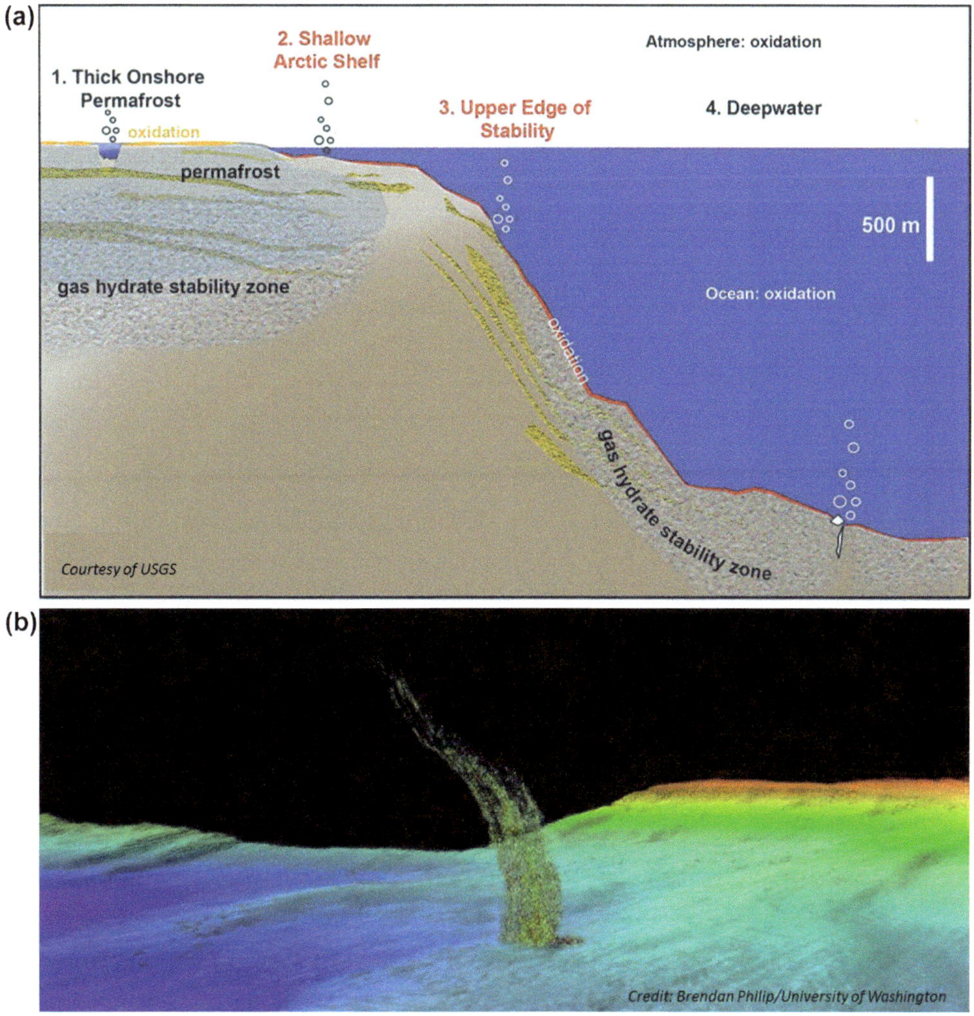

FIGURE 9.6

Gas hydrate deposit over continental margins and gas bubble rising. (a) Gas hydrate deposits by sector. Currently, gas hydrates are most likely dissociating in sectors 2 and 3. Only sector 2 is likely to release methane that could reach the atmosphere (source: http://soundwaves.usgs.gov/2012/06/). (b) Sonar image of bubbles rising from the seafloor off the Washington coast. The base of the column is 1/3 of a mile (515 m) deep and the top of the plume is at 1/10 of a mile (180 m) depth. (Credit: Brendan Philip/University of Washington.)

hydrates as an energy resource. Therefore, there is an increasing interest in assessing the potential of methane release in a warmer world and its consequences for future climate change (Brook et al., 2008; Schiermeier, 2008; Westbrook et al., 2009).

Interpretation of identified BSR in seismic reflection data suggests that the causative mechanism of BSR requires free gas, trapped-gas as bubbles in interconnected voids in sediments. The thickness of the free gas zone and the quantum of methane preserved in this zone are still speculative. In particular, the size and role of the free gas zone below the gas hydrate provinces remain relatively unconstrained, largely because the base of the free gas zone is not a phase boundary and has thus defied systematic description. It is suggested that the maximum thickness of an interconnected free gas zone is mechanically regulated by valving caused by fault slip in overlying sediments as seen in Fig. 9.7. Modeling results (Hornbach et al., 2004) show that such critical gas columns exist below most hydrate provinces in basin settings, likely poised for mechanical failure and are therefore highly sensitive to changes in ambient physical conditions. These studies indicate that the estimated global free gas reservoir may contain from one-sixth to two-thirds of the total methane trapped in hydrate. If gas accumulations are critically thick along passive continental slopes, a 5°C temperature increase at the seafloor could result in a release of ~2000 Gt of methane from the free gas zone, offering a mechanism for rapid methane release during global warming events (Hornbach et al., 2004).

9.5 GEOLOGICAL HAZARDS

Tectonic forces and sea-level changes have brought in many changes in seafloor morphology by deforming the soft unconsolidated sediments on the continental margins. The shape and scale length of these features are governed by smoothness and gradients in the seafloor relief (Fig. 9.8). Over the continental shelf or slope regions, under normal circumstances these slumps are gravitationally driven when the mechanical shear strength of sediments does not overcome the gravitational force. Gravity-driven sediment mass transport can be classified as slide slump and debris flow. Movement of a large coherent mass of sediments with no internal deformation over the planar glide plane in contact bedding is termed as slide. The slump occurs as a coherent rotational mass movement on the concave up glide plane with internal deformation of the slumped mass. Debris flows are incoherent viscous masses in which intragranular movement produces a chaotic end product. Increase in water content makes the debris flow into turbidity currents (Shanmugam et al., 1995). The gravity-driven mass transport of sediments on the continental slope has quite distinct characteristics: slides occur on large planar surfaces, no internal deformation is seen, and the end member is debris flow.

When the seafloor surface is inclined, the gravitational force tends to move the sediment mass downward. This movement is resisted by the shear strength of sediments along the rupture surface. When the gravitational force exceeds the resistance offered by the sediment strength, slope failure is likely to occur (Siriwardane and Smith, 2006). Sub-aerial and submarine landslides occur when the stresses exerted on the sediments exceed the sediment strength. Increase in shear stress or decrease in shear strength can lead to shear failure in sediments, which in turn leads to slope failure. These stresses are from external sources or from changes brought about within the sediment matrix itself. In the marine environment, slope failure may result from over-steepening of slopes, faster sediment deposition, tectonics, and earthquakes. Submarine gas hydrate deposits are found on continental slopes, where average slope grades are only a few degrees. Basic fracture mechanics theory suggests that low angle failures cannot result from slow changes of earth stresses or mechanical properties. Low angle failures, including those found in geological records, can only result from (1) large transient stresses, such as those associated with earthquakes, (2) sudden changes in material properties, such as those associated with rapid decomposition of gas hydrate, or (3) sudden changes of pore pressure, which might be due to hydrate decomposition or migration of gas from other sources (Kleinberg, 2004). The mechanisms

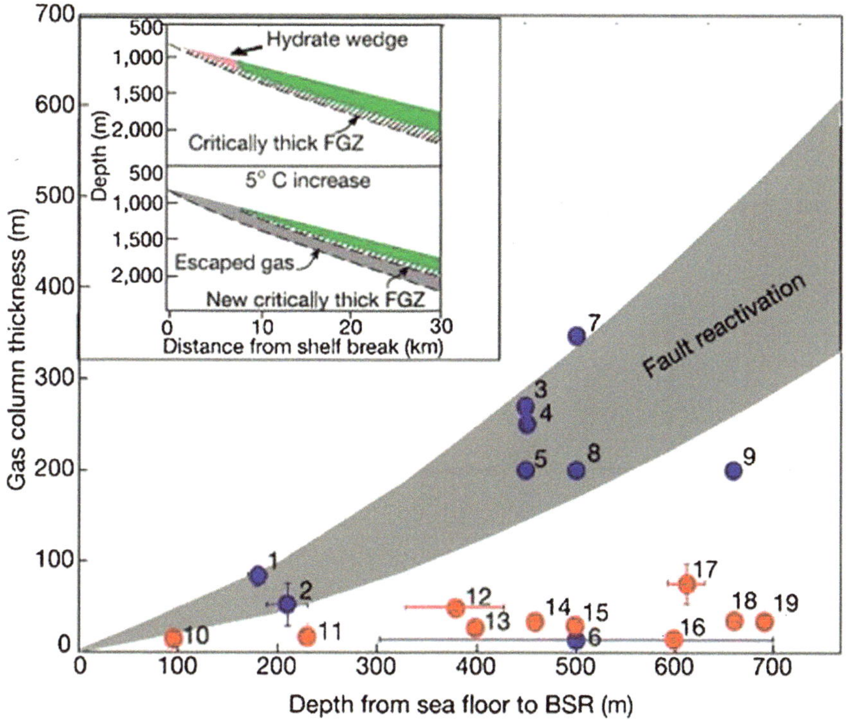

FIGURE 9.7

Blue symbols (numbered 1–9) show gas column thickness below BSRs at several hydrate provinces in basin settings where 1 is near vertical. Red symbols (10–19) show gas column thickness below BSRs at hydrate provinces in compressional settings where 1 is probably near horizontal. Note that gas column thickness generally increases with depth for the blue points, but remains nearly constant for the red points. The gray area indicates the interconnected FGZ thickness required for fault reactivation. The upper bound of the gray area corresponds to values of $\alpha = 0.77$ (zero hydrate at the BSR) and $\upsilon = 0.45$; the lower bound corresponds to low end values of $\alpha = 0.67$ (25% bulk hydrate) and $\upsilon = 0.41$. Point locations: 1, Barents Sea (Andreassen et al., 1990); 2, the Norwegian margin (Mienert et al., 2000); 3, Bering Sea basin (Scholl and Hart, 1993); 4, 5, Blake Ridge (Holbrook et al., 1996); 6, Beaufort Sea (Andreassen et al., 1995); 7–9, basin in Makran accretionary prism that has experienced rapid sedimentation (Sain et al., 2000; Grevemeyer et al., 2000); 10, offshore Cascadia (Hyndman et al., 2001); 11, offshore Chile (Bangs et al., 1993; Miller, et al., 1996), Niger delta (Hovland et al., 1997); 13, 15, offshore Colombia (Minshull et al., 1994); 14, Gulf of Oman anticline (Minshull and White, 1989); 16, offshore Peru (Pecher et al., 1996); 17, South Shetland islands (Tinivella et al., 1998); 18, 19, Oman accretionary toe (White, 1979). Top inset, model of the hydrate stability zone and critically thick FGZ below for a bottom water temperature of 0°C. The BSR is located at the interface between the sediment overburden where hydrate can exist (green) and the FGZ (black and white stripes). The pink area marks the hydrate wedge. Bottom inset, new BSR depth and critical FGZ thickness for a 5°C temperature rise at the seafloor. The gray area denotes the area over which gas (and dissociated hydrate) must escape. Note the thinner critical FGZ due to bottom water warming, which results in less sediment overburden above the BSR. (From Matthew et al., 2004.)

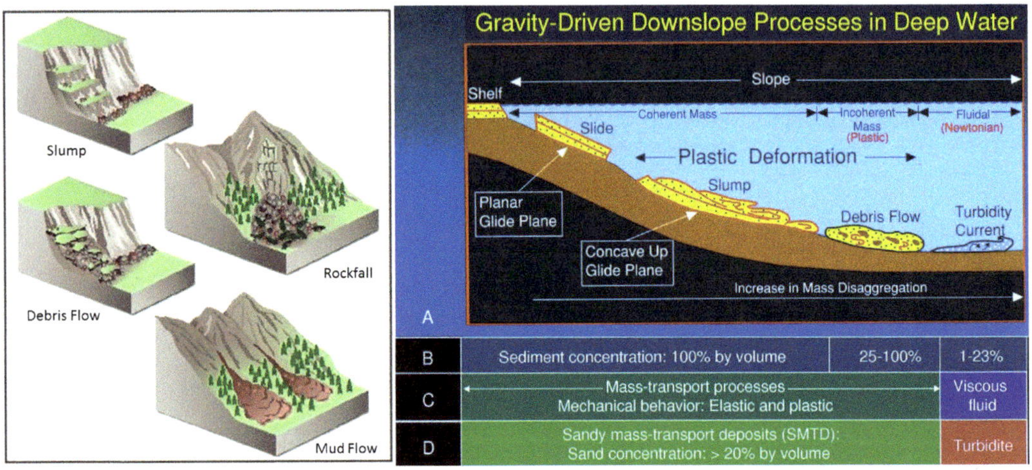

FIGURE 9.8

(a) Schematic diagram showing four common types of gravity-driven downslope processes (slides, slumps, debris flows, and turbidity currents) that transport sediment into deep-marine environments (from Shanmugam et al., 1994).(b) Sediment concentration (% by volume) in gravity-driven processes. Note that turbidity currents are low in sediment concentration (i.e., low-density flows) (after Shanmugam, 2000). (c) Based on mechanical behavior of gravity-driven downslope processes, mass-transport processes are considered to include slide, slump, and debris flow, but not turbidity currents (Dott, 1963). (d) The prefix "sandy" is used for mass-transport deposits (SMTDs) that have grain (>0.06 mm: sand and gravel) concentration value equal to or above 20% by volume. The 20% value is adopted from the original field classification of sedimentary rocks by Krynine (1948).

that potentially cause shear failure also include (a) removal of lateral supports, (b) increase in the weight of overlying sediment, (c) upward fluid flow (Davies and Clark, 2006), and (d) frictional heat generated in the subduction zone (Jahren et al., 2005).

Geological hazards produced by gas hydrates are related to the instability of the sediment pile due to the decomposition of hydrates or possible violent releases of methane in the water column, and further in the atmosphere. The decomposition of hydrates has to be rapid enough in order to trigger hazard-producing phenomena. At geological scale, "rapid enough" means dynamics of the decomposition processes able to suddenly influence the consolidation process of marine sediments. For causing deformation on the scale length of slope failure, gas hydrate accumulation has to be widespread, the slide has to occur in the hydrate stability zone, and there has to be reduction in permeability at the base hydrate stability zone to accentuate pore pressure, which may lead to slope instability due to any change in external physical conditions (Dillon and Max, 2003).

Parameters affecting gas hydrate formation include temperature, pore pressure, gas chemistry, and pore water salinity. During gas hydrate formation, methane and water become immobilized as a solid, restricting pore space and retarding the migration of fluids. Hydrates occupy pore spaces and inhibit consolidation due to sedimentation and mineral cementation. However, hydrates themselves can act as metastable cementation for sediment grains. The continuous process of hydrate formation impedes the permeability of the sediment as gases and liquids combine to form hydrates. Eventually, gas hydrate may

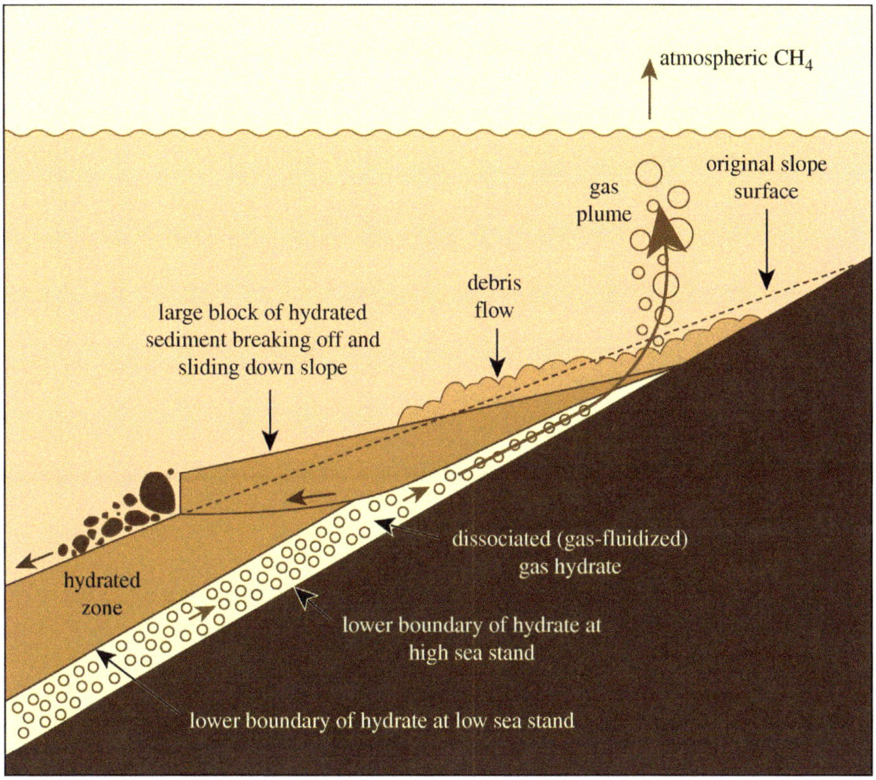

FIGURE 9.9

Champagne cork effect: illustration of what happens when a slope failure occurs above a gas hydrate layer and how large quantities of gas can be released. (From http://rsta.royalsocietypublishing.org/content/368/1919/2369Kvenvolden, 1998.)

occupy much of the pore space within the GHSZ. Continued sedimentation leads to deeper burial of the gas hydrate. Finally, the gas hydrate is buried so deep that temperatures at the base of the stability zone are those at which the gas hydrate is no longer stable. Thus, the basal zone of the gas hydrate becomes underconsolidated and possibly overpressured because of the large quantum of released gas, leading to a zone of weakness. This volume expansion of gases causes the pore fluids to become highly pressurized and push the sediment apart resulting in the loss of sediment compaction, formation of large pore spaces leading to fracturing in sediments, and potentially triggering submarine landslides. Failure follows on moderate slopes unless the increased fluid pressures can be adequately vented (Kvenvolden, 1999).

Submarine landslide phenomena and gas hydrate dissociation is a linked, auto-feedback process (Fig. 9.9). A landslide may be triggered either by gravitational or sea-level changes, tectonic or earthquake activities, or due to processes over a small segment of continental slope, which may be under the gas hydrate-bearing layer. Transportation of sediment mass in the upper sediment layer lowers the lithostatic pressure over the hydrate layer, thereby disturbing the hydrate stability equilibrium. During decomposition, the mechanical strength of sediments is considerably reduced and the sediments can no longer support the load of the overlying strata. Destabilizing the gas hydrate will release a large quantum of gas and liquid and cause an increase in pore pressure, decrease in shear stress, and loss of shear strain of the sediment layer by increasing the pore size. Reduction of sediment strength and thinning of the hydrated layer makes the upper layer slide further down along the slope due to gravity and release more gases and fluids (McIver, 1982).

Near-surface changes in PT conditions lead to the destabilization of gas hydrates. Over the continental margins, pressure being the dominant component regulates the thickness of the GHSZ, thickening with increase in the level of the water column and in its reduction during sea-level fall. When gas hydrates dissociate due to any of the aforementioned causes, the released methane either gets partially dissolved in the overlying seawater or can also remain in gaseous phase with the seawater. The dissolution of methane in seawater reduces the density of seawater (Xu and Ruppel, 1999), leading to mass transfer, which is a slow process (Zhang and Xu, 2003). The estimated values of methane release due to the dissociation of unit volume of gas hydrate in an oceanic regime exhibit a distinct trend (Max and Dillon, 1998; Paull et al., 2003). The pore pressure on sediments that are close to the up-dip end of the GHSZ are significantly higher when compared to deeper parts of the ocean resulting in larger quantities of methane release near the up-dip side in the shallow regions, leading to a reduction in sediment strength. The exposure of hydrated sediments to warmer ocean conditions, higher gradients on the upper side of the slope, and the tendency of free gas to move to the flanks weaken the sediment strength and make them susceptible to sliding (Dillon and Max, 2003).

Gas hydrates occupy pore spaces and also cement sediments to increase the shear strength and enhance the slope stability. On the one hand, hydrates may strengthen the sediments by cementing grains; on the other hand, if hydrates impede fluid flow due to reduced permeability, they may weaken the underlying sediment by trapping fluids and free gas. Coarse-grained sediments are favorable for gas hydrate growth but impede the bulk permeability of the host sediments. The internal mechanisms which weaken the sediment strength are related to increased porosity, pore pressure, and the presence of gas bubbles in the pores. Any causative mechanism that alters the stability equilibrium may result in dissociation (gas hydrate turns into free gas or water mixture) and/or dissolution (gas hydrate becomes a mixture of water and dissolved gas) of the gas hydrate. During the formation of gas hydrates, the water and the gas content in the sediment pores is converted to solid form and increases sediment strength. On the contrary, the dissociation of gas hydrate will freshen the sediment with water and gases, which may reduce the sediment strength. The generation of excess pore pressure due to hydrate dissociation is known to be the most relevant process responsible for destabilizing hydrate-bearing sediments (Xu and Germanovich, 2006; Moridis and Kowalsky, 2007). The excess pore pressure reduces the effective stress of sediments and the shear strength, which results in shear failure. The excess pore pressure generated by decomposition of gas hydrates depends on the hydrate saturation in the pore, the gas solubility, and medium compressibility (Sultan et al., 2004) and the pore pressure increases with temperature. The magnitude of excess pore pressure depends on the initial methane concentration and hydrostatic pressure. For a hydrostatic pressure of 4 MPa, the dissociation of hydrate will generate a pore pressure of 0.1 MPa for the initial hydrate saturation of 2.6% and pore pressure increases to 0.3 MPa for the hydrate concentration of 52% in the pore spaces as seen in Fig. 9.10 (Sultan et al., 2004).

Gas hydrate decomposition at the upper shelf edge can result in an enormous pore pressure that leads to a massive loss of compactness while the large pore space makes the sediment highly deformable. As would be expected, the steeper the sediment slope, the higher the probability that a slide will occur, where slides are more likely to be promoted by gas hydrate breakdown on slopes of more than 4°. Laboratory experiments, small-scale physical modeling, and theoretical slope stability analyses indicate that dissociation of even a small amount of hydrate can cause a significant loss of sediment strength. Hydrate

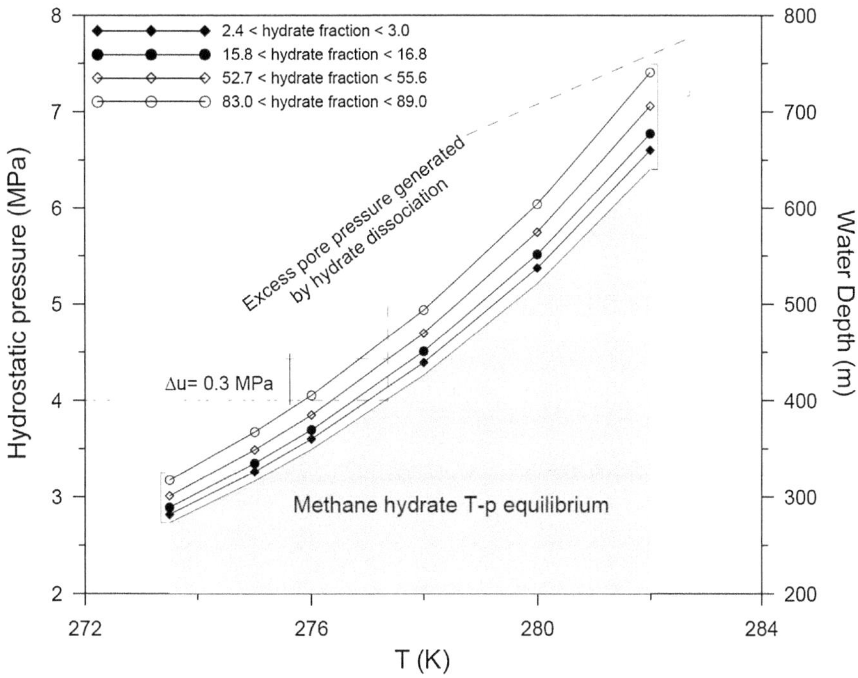

FIGURE 9.10

Excess pore pressure generated by the melting of methane hydrate at different gas hydrate pore saturations. For a hydrostatic pressure of 4 MPa, the melting of the methane hydrate will generate around 0.3 MPa for an initial hydrate fraction of 52%. (From Sultan et al., 2004.)

dissociation could be critical in inciting slope failures for low-permeability sediments in shallower water depths (Nixon and Grozic, 2006). Based on the characteristics of the medium, the pore pressure response differs significantly. The dissipation pore pressure in clayey rocks is not appreciable, which makes this medium more susceptible to deformation and possible slope failure. In sandy rocks, the pore pressure dissipates faster, although pore pressure buildup is rather slow (Kayen and Lee, 2001). In coarse-grained rocks, pore pressure in conjunction with other external factors such as earthquakes and faster rate of sedimentation may lead to seafloor collapse or failure (Kayen and Lee, 1991). The most important factor for controlling pore pressure generation is hydraulic permeability. Generation of excess pore pressure is a function of the amount of gas hydrate present initially, the rate of gas hydrate dissociation, sediment permeability, background fluid flux, capillary pressure, and the depth of the base of the GHSZ (Crutchley et al., 2010; Nimblett and Ruppel, 2003). Fine-grained sediments are more susceptible to hydraulic fracturing due to the high gas pressure needed to overcome capillary entry pressure. This process reopens the fluid pathway and locally increases the bulk permeability of the sediments (Scholz et al., 2011). A number of oceanic gas hydrate deposits such as the Ulleung Basin sediments (Kwon et al., 2011), hydrate deposits in the Gulf of Mexico (Francisca et al., 2005), and the Krishna–Godavari basin in India (Yun et al., 2010) are characterized by fine-grained low-permeability sediments.

The top of the GHSZ is generally a diffused (hydrate or seawater) boundary due to a lack of noticeable contrast in physical properties and is therefore difficult to identify in geophysical field datasets.

Gas hydrate dissociation at the base of the GHSZ is often considered a major cause of sediment deformation and submarine slope failures. Based on theoretical considerations it is suggested that excess pore pressure and shear discontinuities generated by hydrate dissociation are unlikely to be a hazardous factor. In the natural environment, the excess pore pressure generated by hydrate dissociation is bound by the gas hydrate stability law inducing a natural temperature increase, a limited amount of excess pore pressure, and limited shear discontinuities at the base of the GHSZ. However, under natural temperature changes hydrate dissolution at the top of the GHSZ, (which can occur at a regional scale) is a hazardous process that can lead to catastrophic landslides (Sultan et al., 2004). During the process of hydrate dissociation, gas hydrates liquefy and sediments unconsolidate. The free-gases trapped beneath the remaining gas hydrates would constitute a weak layer of overpressurized sediments. The combined effect of these processes may lead to slope failure (Bünz et al., 2003).

Periodic Pleistocene eustatic sea-level transgressions and regressions provide a mechanism to account for the waxing and waning of submarine gas hydrates. Sea-level changes, especially the lowering of sea levels during the falls, have given rise to large-scale slumping along the margins. Fluctuations in global climate, reflected in Pleistocene sea-level falls, likely caused these submarine slides and perhaps caused other slides on other continental margins where gas hydrates are present (Kayen and Lee, 1991).

During the last glacial period between 45 ka and 16 ka about nine continental slope failures confined to the Heinrich event are reported in the middle and low latitudes, and these failures were supposedly triggered due to lowering of the sea levels. About 70% of the slope failure occurred during the past 45 ky was displaced in two periods, between 15 and 13 ka and between 11 and 8 ka. Both these intervals correlate with rising sea levels and peaks in the methane record during the Bølling-Aollerød and Preboreal periods (Fig. 9.11).

The glaciation and deglaciation period also changes the thermal pattern of continental margins. During the period of lower sea levels, hydrates are exposed to warmer surface temperatures. Also the global scale current patterns in the oceans periodically alter the sub-bottom temperature of the oceans. This is an important consideration as it is the intermediate ocean waters between about 200 and 1500 m which bathe the sediment most likely to contain gas hydrates. This predicted warming of the intermediate ocean depth tends to destabilize marine gas hydrates. The small predicted rise of future sea levels of up to 1 m may increase the hydrate stability slightly but is completely insufficient to counter the

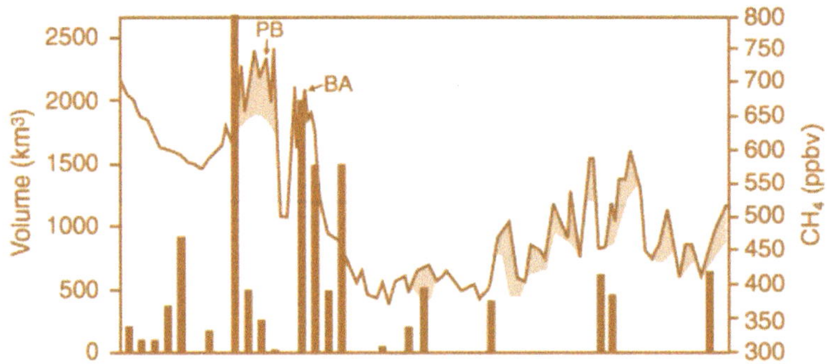

FIGURE 9.11

Methane record during the Bølling-Aollerød (BA) and Preboreal (PB) periods.

warming. However, to destabilize gas hydrates through increased temperatures, the increased warmth must penetrate through the gas hydrate layer to the phase transition at the hydrate–gas boundary. Thermal diffusivity is a relatively slow process and related to the temperature gradient, the hydrate depth, and sediment composition (Clennell et al., 1999; Henry et al., 1999).

Paleocene–Eocene thermal maximum (PETM) is the most intensely studied high temperature event that occurred ~54.95 Ma. The large, negative carbon isotopic excursion (CIE) recorded in both marine and terrestrial sediments during the PETM has been interpreted as reflecting widespread release of isotopically-light (microbial) carbon from dissociating marine methane hydrates (Dickens et al., 1995; Zachos et al., 2005). It has also seen a number of climate and carbon cycle modeling studies attempting to integrate the available proxy data into a coherent view of the earth-system behavior across the event (e.g., Bice and Marotzke, 2002; Panchuk et al., 2008; Zeebe et al., 2009). The current synthesis (although with significant continuing uncertainties) appears to consist of an approximately 4% negative CIE in the oceanic dissolved inorganic carbon pool, a global temperature anomaly of 5–6°C, a rapid (less than 10 ky) onset, a total PETM duration of approximately 170 ky, and a carbon input within the range of 4000–7000 Gt with some contribution but perhaps not the dominant one, from methane hydrates as shown in Fig. 9.12 (Shipboard Scientific Party, 2002).

It has been reported that during the past 55 years (1955–2010) the 0–2000 m layer of oceans worldwide has been warmed by 0.09°C because of global warming. The global warming has latitude dependence in the increase of temperature, more pronounced over higher latitudes when compared to lower latitude regions. This raises the scientific concern that if warming of the bottom water of deep oceans continues, it would dissociate natural gas hydrates and could eventually trigger massive slope failures (Kwon and Cho, 2012). For example, a 1°C increase in the temperature of hydrate-bearing sediments results in the pore fluid pressure increasing by the order of several megapascals under the no mass flux condition (Jang and Santamarina, 2011; Holtzman and Juanes, 2011). Therefore, any thermal change can stimulate hydrate dissociation and thus mechanically destabilize sediments.

The presence of BSR, in the vicinity of deformed structural configuration, has led to the inference that many submarine slope failures are due to reduction in the shear strength. Seafloor slope failures occur in the hydrate-bearing sedimentary deposits and might be directly triggered, or at least primed by gas hydrate decomposition. Submarine slope failures have been reported from many regions of the world and the different causative mechanisms have been proposed for such features (Kayen and Lee, 1991; Maslin et al., 1998; Rothwell et al., 1998; Dillon et al., 2001; Holbrook, 2001;Vogt and Jung, 2002; Canals et al., 2004; Kwon and Cho, 2012). For almost three decades, the link between gas hydrates and slope failure has been examined and debated extensively. Most of the studies have attempted to link gas hydrate decomposition and slope failures. The inference about connectivity between the gas hydrate and slope failure is drawn on the basis of identification of slope failure in the vicinity of occurrence of gas hydrate-bearing sediments either by means of the presence of BSRs or other geophysical, geochemical, and geological signatures (Summerhayes et al., 1979; McIver, 1982; Nisbet and Piper, 1998; and others).

9.5.1 STORREGA SLIDE-NORWEGIAN SEA

The Storrega Slide in the Norwegian Sea, Norway, is one the most impressive scars preserved on the continental slope; it is dated at around 7000 years old. The scar is about 290 km long and is comprised of three segments. More than 5500 km^3 of its sediments have been eroded during their transportation

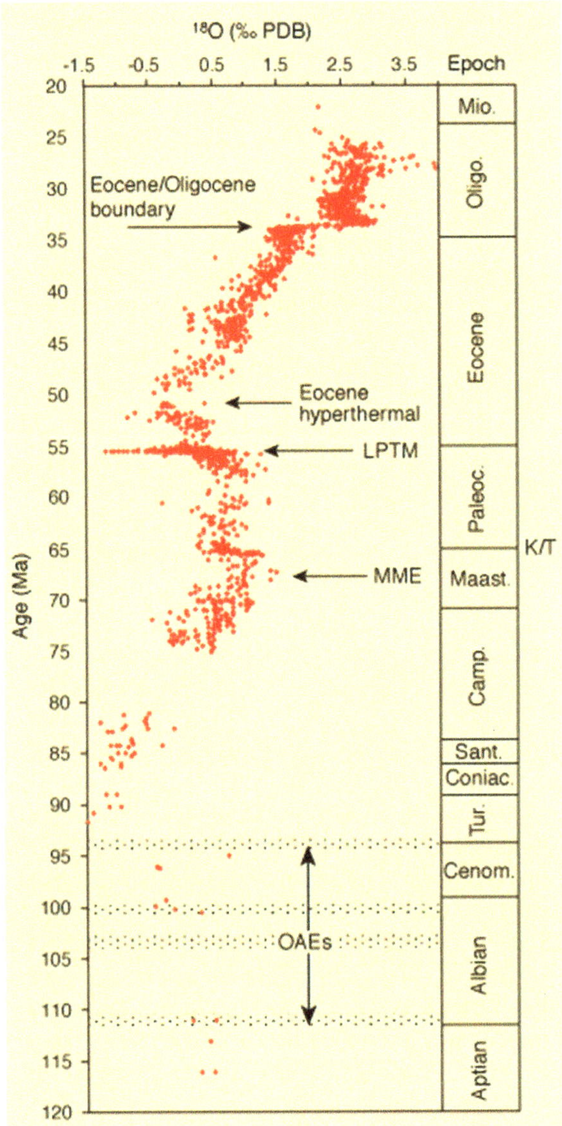

FIGURE 9.12

A stacked record of temperatures and ice volume in the deep ocean through the Mesozoic and Cenozoic periods. LPTM, Paleocene–Eocene thermal maximum; OAEs, oceanic anoxic events; MME, Mid-Maastrichtian event

over 800 km towards the Atlantic Ocean (Bugge et al., 1987; Paull et al., 2000; Berndt et al., 2002). The Storegga Slide is bound west-north-west by the Voring Plateau, further northwest by the Norwegian basin, and in the south by North Sea Fan (Fig. 9.13).

This massive slide is a compound of three events. Recent studies indicate that all these events took place at ~8.15 ka (Hafidason et al., 2001). The headwall of the slide is approximately 300 km

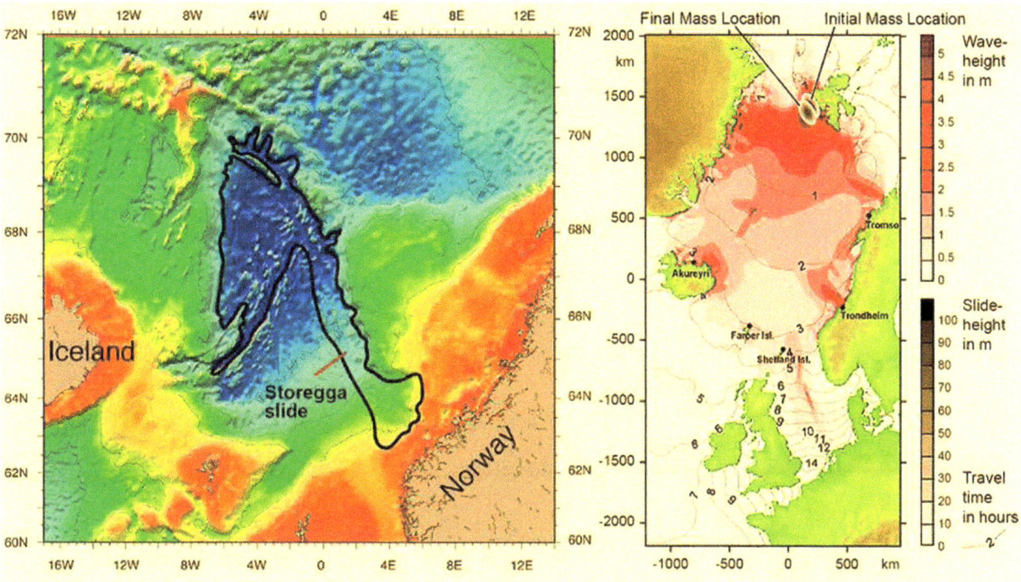

FIGURE 9.13

Bathymetry of the Storegga Slide, Norway illustrating the enormous size of mass-transport processes. Right: modeling of the tsunami that would be caused if a similar but one-third smaller landslide than the Storegga Slide happened in the Fram Strait. (From Berndt et al., 2009.)

wide and the slide narrows to a 60-km gateway in the opening between the Møre and Vøring volcanic highs (Bryn et al., 2005). The slide runs from the shallow depths of the shelf of Norway to the abyssal plain with a slope angle of 0.6°; major areas of the slide have an inclination <0.4° (Kvalstad et al., 2002). The genesis of the Storegga Slide is debated extensively and consensus about the mechanism for its formation has not emerged so far (Bugge et al., 1987; Berndt et al., 2002; Maslin et al., 2004; Sultan et al., 2004; Mienert et al., 2005; Kvalstad et al., 2002, 2005; Bryn et al., 2005). In general, excess pore pressure is considered the main causative mechanism for the genesis of the slide; however, the interpretation for the generation of excess pore pressure differs significantly. This region has experienced seismic activity in the past and is disposed to earthquakes, the most recent being in 1998.

Two seismic profiles were shot across the Storegga Slide and the data obtained indicated that the headwall slide is characterized by rollover structures under the undisturbed sediment section, suggesting that the mass has been transported from the base of rollover structures. The local extent of disturbance in the sediment pattern is confined to the fault region which rules out the possibility of earthquake-generated liquefaction to account for the observed sediment pattern. The identification of a BSR (Fig. 9.14) in the vicinity lent credence to the dissociation of gas hydrate as the most likely mechanism for the Storegga Slide (Berndt et al., 2002). The Storegga composite slump scar runs to about 290 km with run off distance of about 800 km (Fig. 9.14). Seafloor collapse within the Storegga Slide and above the dissolved gas hydrate suggests that the hydrate reservoir is highly dynamic, and methane released due to landslide may contribute significantly to a greenhouse effect (Berndt et al., 2002). A large amount of methane is released into the atmosphere as a result of submarine sediment failures. A

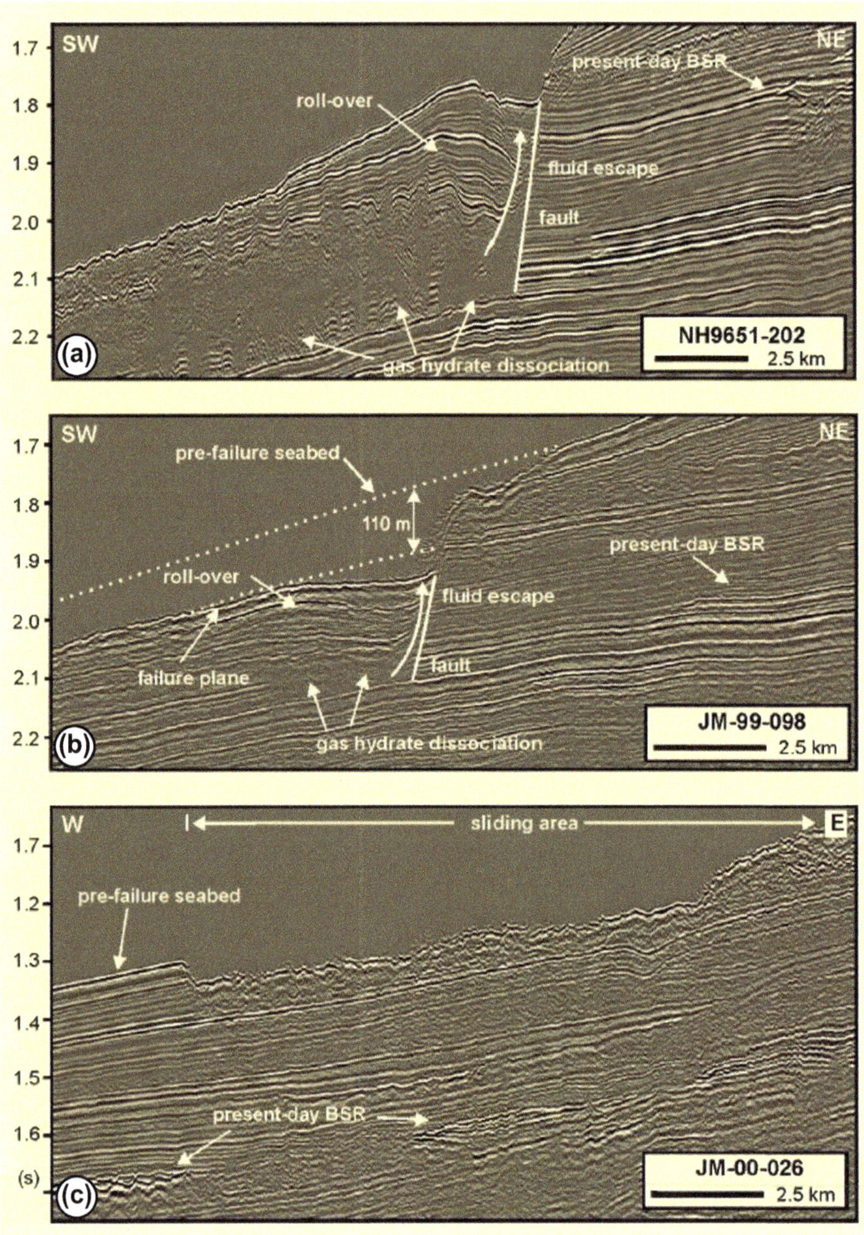

FIGURE 9.14

Migrated high-resolution seismic data. (a) Multichannel data. (b,c) Single-channel data. Note that the bottom-simulating reflector (BSR) within the slide area in (c) mimics the seafloor indicating that it has adapted to the new pressure/temperature conditions. (From Berndt et al., 2002)

link between releases of methane due to dissociation of hydrate and its climatic impact can be ascertained through the clathrate gun hypothesis. More than 70% of slope failures in the North Atlantic occurred during two time spans, between 15 to 13 ka and between 11 to 8 ka, and correlate with rising sea levels and peaks in the methane record during the Bølling-Ållerød and Preboreal periods. These data support the clathrate gun hypothesis for glacial–interglacial transitions (Maslin et al., 2004).

Sea-level changes associated with glaciation and interglaciations have been attributed to many slope failures across the globe due to variations in PT on the earth's surface. The last episode of this activity was confined to the Quaternary era and most of the slope failure was found to have occurred during the Pleistocene–Holocene epoch (Vogt and Jung, 2002). Maslin et al. (2004) have shown a good correlation between slope failure and sea-level changes. These slope failures correlated well with sea-level rises and falls and enhancement in methane in the northern hemisphere. Most of the slope failures over the North Atlantic occurred during the Preboreal period. The Storegga Slide and Andoya failures are the main events that seem to have formed during this period. The isostatic rebound results in the reduction of lithostatic pressure and dissociation of gas hydrates leading to slope failure. Isostatic rebound also produces a significant increase in seismics and earthquakes, which may also give rise to slope failure (Mörner, 1991). Such prevailing conditions may have formed the Storegga Slide during the Preboreal period (Maslin et al., 2004).

Seismic reflection data delineated the BSR in the Storegga Slide area and its vicinity. Relatively faster increases in temperatures in ocean waters was shown during the Younger Dryas (~12.5–10 ka) and warmer ocean temperatures have prevailed since then. Owing to warm ocean water conditions, there was a substantial reduction in the gas hydrate stability zone along the upper slope of the mid-Norwegian margin, even with the rise in sea-level during this period. Additionally, the BSRs within the slide complex seem to have more or less adjusted to the new equilibrium conditions, highlighting the dynamics of hydrate stability in continental margin sediments under environmental changes (Mienert et al., 2005).

Sultan et al. (2004) theoretically studied the thermodynamic, chemical equilibrium of gas hydrates in soil by considering the effects of pressure, temperature, pore pressure, and pore water chemistry. The modeling results indicated that increase in temperature and pressure results in dissociation at the top portion of the GHSZ. Modeling of the Storegga Slide under the assumption of unique shear strength profile and unique gas profile suggests that pressure reduction and temperature increase during deglaciation leads to dissociation of gas hydrates and the origin of retrogressive failure of the Storegga Slide. Moreover, due to the gas solubility, the failure interface is initiated at the top of the hydrate layer not the bottom of the hydrate stability zone as proposed earlier by Kvalstad et al. (2002).

Late Jurassic early Cretaceous fault activation and earthquakes generated from the sediment load and deglaciation add dynamic load and excess pore pressure which may lead to the initiation of slope failure in marine clayey sediments. The Storegga Slide was most likely triggered by a strong earthquake in the area 150 km from the slide 8200 years ago. Modeling results show that excess pore pressure generated in the North Sea Fan area is transferred to the Storegga area with reduction in the slope stability in the downslope from the Orman Lange gas field and developed as a retrogressive slide (Bryn et al., 2005; Kvalstad et al., 2005).

9.5.2 CAPE FEAR SLIDE – ATLANTIC MARGIN

About 200 slide scars have been mapped on one of the most extensive gas hydrate locals on the Atlantic continental margin of the United States (Booth et al., 1994). The occurrence of slides in the vicinity of identified BSRs and other proxies of gas hydrates provided opportunities to validate the relationship between them (Carpenter, 1981; Cashman and Popenoe, 1985; Rothwell et al., 1998; Paull et al., 2003). The Cape Fear Slide and the Cape Lookout Slide are two prominent slope failures situated east of Cape Fear on the Carolina rise off the eastern continental margin of the United States. The amphitheater-shaped headwall scarp running 50 km in length is elevated by 120 m at 2600 m water depth on the continental slope (for location map see Fig. 9.14). Two salt diapirs appear within the slide to disturb the slope topography as the diapirs protrude above the seafloor. The mass wasting associated with the slope extends 250 km across the rise and into the Hatteras Abyssal Plain (Embley 1980). The Cape Fear Slide, the largest known submarine landslide on the U.S. Atlantic continental margin, is a recent (~20 ka) mass wasting event that generated potentially large tsunamis.

Multiparameter geophysical surveys along with drilling over the headwall of the slide were undertaken to ascertain the triggering mechanism of the slide (Paull et al., 1996; Popenoe et al., 2001; Hornbach et al., 2007a,b). Initially it was proposed that the triggering mechanism may be linked to the presence of diapirs (Cashman and Popenoe, 1985) and dissociation of gas hydrates either due to the lowering of hydrostatic pressure during glacial low stands or due to global warming. The higher thermal conductivity of salt diapirs leads to an increase in the geothermal gradients; thinning the GHSZ over upwarp and sharp geothermal gradients at the flanks of diapir may promote slope failure.

Identification of BSRs in the Cape Fear Slide area infers the presence of gas hydrates in this region. Shallow sediment samples collected during the drilling on the headwall of the slide have provided an estimate of ages. Carbon-14 dating estimates that the ages range from 9 to 14.5 ka suggesting that the Cape Fear Slide may have formed during the last glacial sea-level low stand, which may have occurred ~18 ka (Paull et al., 1996). Sediment lithology and accumulation rates in the cores believed to be continuous over the Carolina Rise have not changed much during sea-level changes over 35 ka (Haskell, 1991). With the absence of much change in sediment rate pattern and lower seismicity, the Carolina Rise and Blake Ridge are some of the most prospective gas hydrate areas of the world.

Owing to the processes of formation and mass movement of the Cape Fear Slide, the thick salt layer has been displaced and compressed into a line of diapirs. BSRs can easily be identified to the south of the slide area and the crest of Blake Ridge. A BSR is conspicuously not seen under the slide, suggesting that saturation of gas hydrates and free gas in the sediments is quite low. Upslope of the lower headwall scar, a series of slip planes that flatten with depth and stretch into a stratigraphic layer that is characterized by irregular bedding and thickening and thinning suggest zone weak strata. The zone of weakness coincides with the discontinuous base of the gas hydrate layer (Popenoe et al., 2001).

A significant number of faults have been identified in the region on the seismic reflection data; these faults appeared to be near the BSR and the headwall of the slide. The base of the gas hydrate zone, i.e., the BSR, coincides with the plane of slide over which the mass movement may have taken place. A BSR is conspicuously absent under the slide and the BSR upwarps at the flanks of the slide and is found to be present beyond the extent of the slide. Absence of the BSR under the slide suggests dissociation of hydrate and escape of gas resulting in weakening of the sediment strata (Dillon et al., 1993; Paull et al., 2003). It is hypothesized that gas hydrate dissociation reduces the sediment strength and increases the chances of continental slope failure. Circumstantial evidence strongly suggests that gas hydrate breakdown is instrumental in triggering sediment mass movements on the seafloor (Paull et al., 2003).

Multibeam and chirp data were collected and processed to produce the highest resolution bathymetric and subsurface images of the Cape Fear Slide. High-resolution images constrain shallow subsurface structures of the region and a single massive normal fault that intersects the main Cape Fear Slide headwall was delineated (Hornbach et al., 2007a). The fault strikes parallel to a linear chain of salt diapirs and extends ~100 km south of the Cape Fear Slide.

An M7.5 earthquake located up to 102 km from the upper slope could cause failure on slopes of 6°, but such high slope angles are confined to only small portions of the continental margin (Fig. 9.15). Failures on the more typical slopes of 2° require the epicenter to be within 62 km of the shelf edge. An M6.5 earthquake could cause a landslide only if located within 28 km of the continental slope of 2° and within 42 km of a 6° slope, and an M5.5 only if located within 7 and 14 km, respectively (Brink et al., 2009). Analysis of the interactions with the gas hydrate phase boundary and the generations of various

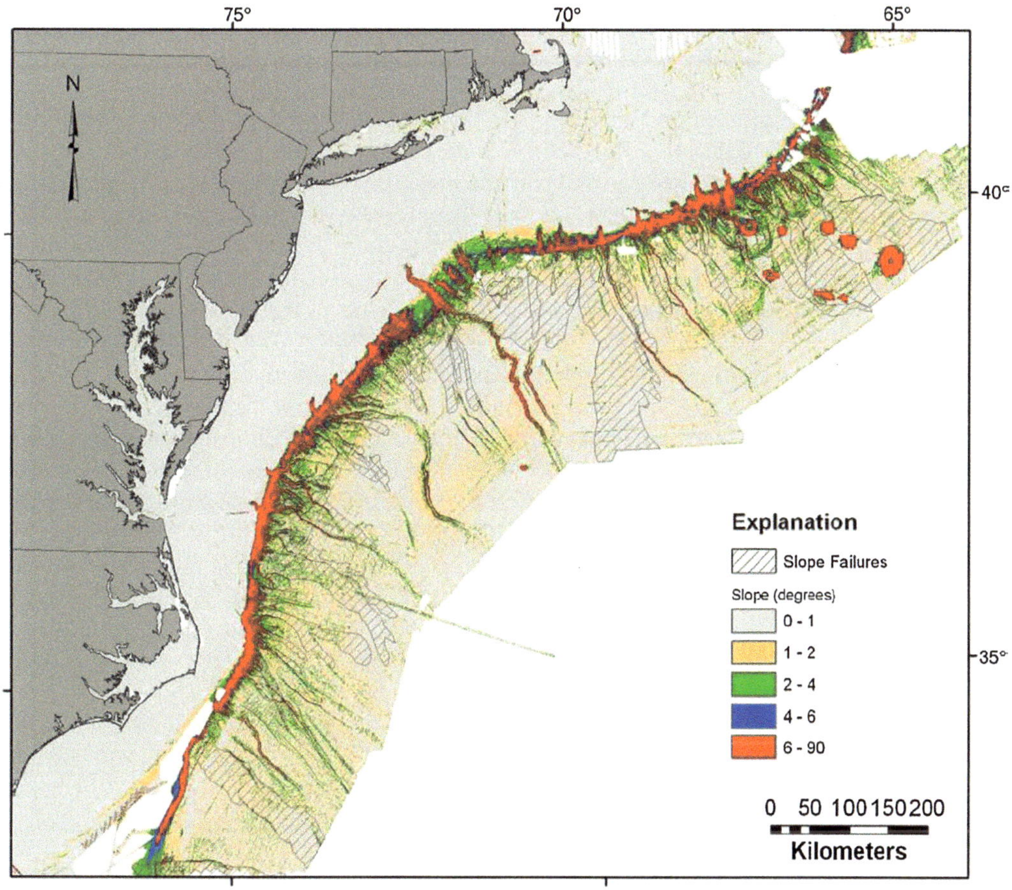

FIGURE 9.15

Slope angle of the US Atlantic continental margin and overlays of interpreted slope failures. (From Twichell et al., 2009.)

slides indicates that only the most landward slide likely intersected the phase boundary and inferred high gas pressures below it. Interpretation of the acquired data suggests slide failure initiated along this fault. Studies indicate that this fault is active and continuous salt diapirism over this area suggests that slide failure may likely occur again along the fault. Ongoing salt intrusion activates this fault and augments slope instability in the region. Hydrodynamic modeling suggests that mass wasting events due to the slide and faulting generates ocean waves which may create large tsunamis that could significantly impact the US coast (Hornbach et al., 2007b).

Most submarine landslides on the continental margin occur on the continental slope and upper rise; a lack of detailed maps for parts of this region has hampered efforts to produce a quantitative assessment of tsunami hazard. During May 11–25, 2009, a team of scientists conducted a 15-day survey aboard the National Oceanic and Atmospheric Administration (NOAA) ship *Ronald H. Brown* to provide a complete seafloor map of the continental slope and upper rise from Cape Hatteras in the south to the eastern end of Georges Bank in the north, a distance of 1200 km (750 miles).

9.5.3 BEAUFORT SEA

The continental shelf of the Beaufort Sea is composed of complex marine and nonmarine sequences of clay, silt, and sand (Fig. 9.16, Brink, 2009). Seismic surveys show that the area of the continental slope across which the seismically inferred gas hydrates occur exceeds 7500 km and may encompass a significant part of the Arctic Ocean basin where the water depth is deeper than 400 m to 600 m at different locations. In many areas of the shelf these sediments contain occurrences of ice-bonded permafrost and associated pressure and temperature conditions that are conducive to the occurrence of methane gas hydrates. The most extensive offshore occurrence of gas hydrates in the North American Arctic lies north of Alaska, beneath the Beaufort Sea (Grantz and Dinter, 1980). The Beaufort Sea continental slope is disrupted by a belt of massive bedding plane slides and rotational slumps (Kayen and Lee, 1991). USGS seismic surveys have shown clear evidence of landslides on the Beaufort Sea continental slope at depths between 200–400 and 2000 m. The shelf to slope transition is thought to be an area of extensive regional instability with acoustic records indicating there is upwards of 500 km of slumps and glides extending over the entire Beaufort margin. Some of these slide regions are coincident with the up-dip limit of the permafrost gas hydrate stability (Grozic and Dallimore, 2012). BSRs from the sediment of the Beaufort Sea are most strongly developed beneath the bathymetric highs. The base of the gas hydrate zone coincides with a region of sediment-containing gas hydrates.

This complex environment is undergoing dramatic warming, where changes in sea level, ocean bottom temperatures, and geothermal regimes are inducing permafrost thawing and gas hydrate decomposition. Eustatic sea-level falls during Pleistocene times caused reduced pressures acting on the seafloor sediment. In oceanic areas underlain by sediment with gas hydrate, the reduction of sea level initiates dissociation along the base of the gas hydrate, which, in turn, causes the release of large volumes of gas into the sediment and creates excess pore fluid pressures and reduced slope stability. In such settings, the overlying permafrost cap may act as a permeability barrier, which could result in significant excess pore pressures and reduction in sediment stability. The deformed sediments lay above a smooth BSR, suggesting that overpressures due to gas hydrate dissociation may have contributed to or caused the slides. A quantitative approach was taken to predict excess pressures at the base of the gas hydrate zone, and the stability of the overlying sediment for a variety of sediment types. The studies suggest that eustatic sea level falls and fluid diffusion properties dominate the

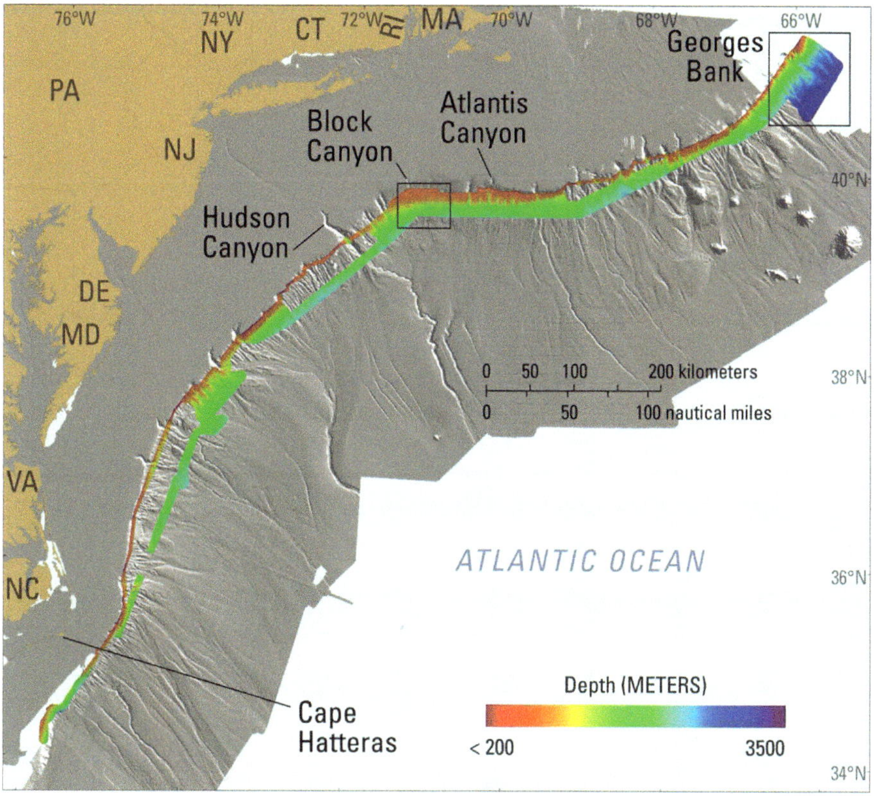

FIGURE 9.16

Shaded-relief map of US Atlantic margin (gray), gridded from single-beam bathymetric soundings, and new bathymetric data (color-coded for depth) collected with a multibeam echosounder during the May 2009 cruise. (From Brink, 2009.)

dissociation process in fine-grained marine sediment. Considering the physical properties and the pattern of sedimentation, two-dimensional modeling was carried out to ascertain the pore pressure required for initiation of triggering of slope failure. These models give credence to the theory that fluctuations in global climate, manifest in periodic Pleistocene eustatic sea-level regressions, likely triggered seafloor landslides on the continental slope of the Beaufort Sea and other margins where gas hydrate is present in the seafloor sediment (Kayen and Lee, 1991).

By combining mechanical and thermal loading of the sediment, a more accurate indication of slope stability was obtained. The stability analysis results indicate a relatively low factor of safety for the Beaufort sediments without the presence of permafrost and gas hydrate, owing to the relative slope steepness compared to other submarine failures. Including the effects of the permafrost and gas hydrate in the sediments can result in an increase of the factor of safety under static conditions. However, modeling of the temporal effects of transgression of the Beaufort Shelf (considering change in pressure and temperature), indicates that, for a reasonable assumption of between 5 and 35% hydrate content, the factor of safety reduces to below unity and failure occurs (Grozic and Dallimore, 2012).

9.5.4 CASCADIAN MARGIN

The Cascadian margin is one of the most extensively studied gas hydrate provinces confined to convergent continental margins of the world. The entire length of the Juan de Fuca oceanic plate is steeply subducting beneath the North American plate at a rate of about 46 mm/a. This subducting plate led to the accumulation of a 3- to 4-km-thick accretionary prism onto the American plate marked by mega thrusts associated with intense earthquake activity with a return period of about 300–700 years. On the northern Cascadia accretionary margin off Vancouver Island, Canada, there are numerous sedimentary slide features near the base of the slope; notable among them are the Orca Slide and Slipstream Slide. The presence of marine gas hydrate beneath the ridge is based on a widespread hydrate BSR, high velocities determined by ocean bottom seismograph data, and sediment core samples and downhole logs collected by the Integrated Ocean Drilling Program at Site U1326. The depth of the BSR at 255 (±15) m coincides closely with the estimated depth of the glide plane beneath the slide. Isolation of the causative mechanism for slides in terms of steep dip angles, earthquakes, and localized excess pore pressure either due to local geological configuration or due to the dissociation of gas hydrate in this complex region is quite challenging. This suggests that the base of the slope failure is related to the contrast between strong hydrate-cemented sediments above the BSR and underlying weak sediments containing free gas. Strong earthquake shaking on this convergent margin likely provides the trigger for the slide.

Gas hydrate concentration estimates are constrained by seismic, logging, and coring data and available high-resolution bathymetry helps to define slope geometries. As an important input for slide age estimation the local sea-level history is used to calculate the boundary of the GHSZ over time. With reduced pressure due to a sea-level low stand near the end of the Last Glacial Maximum, BSR subseafloor depths were shallower by ~10 m compared to the current depths. Effects on the top of the gas hydrate occurrence zone appear to be negligible. Rapid sea-level changes occurred around 9000–10,000 BP and 13,000–14,500 BP, which compares well with the estimated slide ages from 14C radiocarbon dating, sulfate reduction rate modeling, and from the comparison of turbidity sequences used to date the occurrence of past megathrust earthquakes. As a first step a simplified geometry and sedimentation history has been used to calculate overpressure generation. Sediment column thermal models, based on observed water column warming trends offshore Washington (USA), show that substantial volume of gas hydrate along the entire Cascadia upper continental slope is vulnerable to modern climate change. Dissociation along the Washington sector of the Cascadia margin alone has the potential to release 45–80 Tg of methane by 2100 (Hautala et al., 2014).

9.5.5 HYDRATE-ASSOCIATED RISKS FOR OIL AND GAS EXPLOITATION

The potential rewards of extracting methane from gas hydrate fields must be balanced with the risks. Most of the oil and gas over the continental margins are located in shallow waters mostly confined to continental shelves. Conventional oil and gas are located about 1–4 km below the seafloor. Gas hydrates on the continental margins are found to occur >500 m below the seafloor, most likely on the continental slope or rise. At these seafloor depths hydrates are formed in shallow subsurface layers <1 km below the seafloor. Exploration and extraction of oil and gas from deeper water may encounter risk due to destabilization of gas hydrates as extraction of oil and gas from deeper levels has to pass through the hydrated layer. Any small perturbance in the temperature or pressure regime due to deep-sea activity may dissociate gas hydrates causing unknown changes in the host rock matrix and thereby to the wellheads: blow out preventers, pipe lines, installations and anchoring support, and to the platform itself (Hovland and Gudmestad, 2001). Decomposition of gas hydrate reduces the sediment strength, releases excessive gas and water, and increases the pore pressure at shallow depth leading to mass wasting, such as slope failures and deformation of sediment strata. Thus, as a general principle, either creation or dissociation of hydrate can cause problems, though dissociation is the more immediate concern. To meet such eventualities associated with gas hydrate dissociation, some precautionary measures have to be undertaken before the deep-sea explorations are launched. The formation and dissociation of gas leads to substantial changes in the physical properties of subsurface sediments. Observance of BSRs in seismic data has been considered to be a good inference of gas hydrate and free gas below the hydrated layer. Seafloor collapse (pockmarks), gas venting, transparency of sediment, gas hydrate molds, and mud volcanoes are considered major indicators of gas hydrate formation and dissociation in the region. Any exploratory activity for oil and gas in deepwater may consider these as "a priori" for precautionary measures to be undertaken.

The fluidization of sediment due to decomposition of gas hydrates and release of excessive gas and water specifically in shallower depths leads to compaction of sediments in the producing zone and over burden, destabilization of faults, sand production, and mass wasting as debris flow due to slope failure or other causes that may potentially damage the infrastructure (Birchwood et al., 2010).

Most of the methane hydrates are found in shallow subsurface sediments. That means over the continental margins drilling rigs must be able to reach down through more than 500 m of the water column and then through the hydrates layer, which is generally located hundreds of meters below the seafloor before the extraction of oil or gas can begin. Drilling through the layer has to be done with caution as any temperature rise or pressure alteration due to drilling mud may make the gas hydrate dissociate, leading to reduction in sediment strength and thereby causing damage to drilling installations. To overcome this aspect, the entire sediment column of the hydrate layer has to be covered with casing, which works out to be an expensive proposition.

Submarine gas hydrate deposits are found on continental slopes, where gradients in slope may vary significantly from region to region. Convergent margin slope topography may be quite irregular when compared to smooth slope topography of the divergent margins. Drilling in these oceanic deposits could destabilize the seabed, causing vast swaths of sediment to slide for miles down the continental slope. Evidence suggests that such underwater landslides have occurred in the past, with devastating consequences. The movement of so much sediment would certainly trigger massive tsunamis similar to those seen in the Indian Ocean tsunami of December 2004. The change in gradients across the slope and abyssal plane make laying of the pipeline a challenging task. Mass movement due to slope failure and debris flow could rupture the pipeline laid on the sea bottom and the problem is more serious for buried pipelines. If gas hydrates dissociate suddenly and release expanded gas during offshore drilling, it could disturb the marine sediments and compromise the pipelines and production equipment on the seafloor. Even if you can situate a rig safely, methane hydrate is unstable once it is removed from the high pressures and low temperatures of the deep-sea. Methane begins to escape even as it is being transported to the surface. Unless there is a way to prevent this leakage of natural gas, extraction will not be efficient. It will be a bit like hauling up well water using a pail riddled with holes. The decomposition of the hydrate may cause a pressure increase in the well-bore, gasification, and a possible blowout. The presence of gas hydrates has hindered attempts to plug the Deepwater Horizon oil blowout in the Gulf of Mexico, and may have had some role in contributing to the anomalous gas pressure in the well-bore that caused the blowout itself.

Gas hydrate production may lead to uncontrolled gas release during the drilling and damage the well casing during and after the installation of the well. Offshore drilling that disturbs the gas hydrate-bearing marine sediments could fracture or disturb the bottom sediments and compromise the well-bore, pipelines, rig support, and other equipment used in oil and gas production from the seafloor.

Bily and Dick (1974) showed that hydrate decomposition during, and subsequent to, penetration of the hydrate zone can be controlled by lowering the temperature of the circulating mud. It remains important, however, to detect the presence of a hydrate zone since, in deeper drilling, higher mud temperatures cannot be avoided and such zones require adequate casing. Davidson et al. (1978) reported several examples of gas kicks associated with hydrate decomposition. Chilling of the drilling mud, close density control, and continuous monitoring of mud gases are now routine procedures in drilling hydrate-prone areas of northern Canada.

Briaud and Chaouch (1997) propose a model of gas hydrate dissociation beneath the oil platform due to the heat released around pipes where hot oil travels from the well to the platform. They report that the melting process generates a large amount of gas that can endanger the stability of the foundation. Delisle et al. (1998) propose a model of gas hydrate formation due to the thermal re-equilibration occurring after slumps. The main result suggests that the structure does not regain complete thermal equilibrium after slumping in the course of several tens of years. Both Briaud and Chaouch (1997) and Delisle et al. (1998) emphasize that the latent heat greatly impedes gas hydrate formation and dissociation: additional heat source and heat sink are produced as the gas hydrate forms and dissociates respectively.

Destabilization of natural hydrates has occasionally affected the integrity of the seafloor or boreholes, led to well control problems, or contributed to shallow water flows (e.g., Dutta et al., 2010). Gas hydrates in near-mud-line subsea sediments present significant challenges in the production of underlying hydrocarbons, impacting well-bore integrity and placement of subsea equipment. As the fluids of an underlying reservoir flow to the mud-line, heat carried by the fluids warms near-well sediments and dissociates hydrates which release gas that can displace and fracture near-well soil (Peters et al., 2008).

The long-time industry practice of simply avoiding areas with known gas hydrates during production activities that target deeper, conventional hydrocarbons has become increasingly impractical with the push for more deepwater operations.

REFERENCES

Anderson, R., Llamedo, M., Tohidi, B., Burgass, R.W., 2003. Experimental measurement of methane and carbon dioxide clathrate hydrate equilibria in mesoporous silica. J. Phys. Chem. B 107, 3507–3514.

Andreassen, K., Hart, P.E., Grantz, A., 1995. Seismic studies of a bottom simulating reflection related to gas hydrate beneath the continental margin of the Beaufort Sea. J. Geophys. Res. 100, 12659–12673.

Andreassen, K., Hogstad, K., Berteussen, K.A., 1990. Gas hydrate in the southern Barents Sea, indicated by a shallow seismic anomaly. First Break 8, 237–245.

Bangs, N.L.B., Sawyer, D.S., Golovchenko, X., 1993. Free gas at the base of the gas hydrate zone in the vicinity of the Chile triple junction. Geology 21, 905–908.

Berndt, C., Brune, S., Nisbet, E., Zschau, J., Sobolev, S.V., 2009. Tsunami modeling of a submarine landslide in the Fram Strait. Geochem. Geophys. Geosyst. 10 (4).

Berndt, C., Mienert, J., Vanneste, M., Bünz, S., Bryn, P., 2002. Submarine slope-failure offshore Norway triggers rapid gas hydrate decomposition. In: Proceedings of the 4th International Conference on Natural Gas Hydrates, Yokohama, Japan, pp. 71–74.

Bice, K.L., Marotzke, J., 2002. Could changing ocean circulation have destabilized methane hydrate at the Paleocene/Eocene boundary? Paleoceanography 17, 1018. http://dx.doi.org/10.1029/2001PA000678.

Bily, C., Dick, J.W.L., 1974. Naturally occurring gas hydrates in Mackenzie Delta. NWT Bull. Can. Pet. Geol. 32, 340–352.

Birchwood, R., Dai, J., Shelander, D., 2010. Developments in Gas Hydrates. Schlumberger. Oilfield Review Spring 1, 22.

Bishnoi, P.R., Natarajan, V., 1996. Formation and decomposition of gas hydrates. Fluid Ph. Equilib. 117, 168–177.

Bohannon, J., 2008. ENERGY: weighing the climate risks of an untapped fossil fuel. Science 319, 1753.

Booth, J.S., Winters, W.J., Dillon, W.P., 1994. Circumstantial evidence of gas hydrate and slope failure associations on the United States Atlantic Continental Margin. Ann. N. Y. Acad. Sci. 715, 487–489.

Briaud, J.L., Chaouch, A., 1997. Hydrate melting in soil around hot conductor. J. Geotech. Geoenviron. Eng. 123, 645–653.

Brink, U., 2009. Assessment of tsunami hazard to the U.S. East Coast using relationships between submarine landslides and earthquakes. Mar. Geol. 264, 65–73.

Brink, U.S.T., Lee, H.J., Geist, E.L., Twichell, D., 2009. Assessment of tsunami hazard to the U.S. East Coast using relationships between submarine landslides and earthquakes. Mar. Geol. 264, 65–73.

Brook, E., Archer, D., Dlugokencky, E., Frolking, S., Lawrence, D., 2008. US Climate Change Science Program. Synthesis and Assessment Report 3.4: Abrupt Climate Change US Geological Survey.

Bryn, P., Berg, K., Forsberg, C.F., Solheim, A., Kvalstad, T.J., 2005. Explaining the Storegga Slide. Mar. Petrol. Geol. 22, 11–19.

Buffett, B., Archer, D., 2004. Global inventory of methane clathrate: sensitivity to changes in deep ocean. Earth Planet. Sci. Lett. 227, 185–199. http://dx.doi.org/10.1016/J.epsl.2004.09.005.

Bugge, T., Befring, S., Belderson, R.H., et al., 1987. A giant three-stage submarine slide off Norway. Geo. Mar. Lett. 7, 191–198.

Bünz, S.J., Mienert, J., Berndt, C., 2003. Geological controls on the Storegga gas-hydrate system of the mid-Norwegian continental margin. Earth Planet. Sci. Lett. 209, 291–307.

Canals, M., Lastras, G., Urgeles, R., et al., 2004. Slope failure dynamics and impacts from seafloor and shallow sub-seafloor geophysical data: Case studies from the COSTA project. Mar. Geol. 213, 9–72.

Carpenter, G.B., 1981. Coincident sediment slump/clathrate complex on U.S. Atlantic slope. Geo. Mar. Lett. 1, 29–32.

Carroll, J.J., Mather, A.E., 1991. Phase equilibrium in the system water hydrogen sulphide:Hydrate-forming condition. Candian Journal Chemical Engineering 69, 1206–1212.

Cashman, K.V., Popenoe, P., 1985. Slumping and shallow faulting related to the presence of salt on the continental slope and rise off North Carolina. Mar. Pet. Geol. 2, 260–272.

Cheng, Y., Li, L., 2012. Fluid-Solid Coupling Numerical Simulation on Natural Gas Production from Hydrate Reservoirs by Depressurization. In: Al-Megren Dr., H. (Ed.), Advances in Natural Gas Technology 6, 147–192.

Chuvilin, E.M., Buhanov, B.A., 2011. Change of thermal conductivity of gas-saturated sediments during hydrate formation and freezing. In: Proceedings of the 7th International Conference on Gas Hydrates (ICGH 2011), July 2011, Edinburgh, Scotland, United Kingdom, pp. 17–21.

Clennell, M.B., Hovland, M., Booth, J.S., Henry, P., Winters, W.J., 1999. Formation of natural gas hydrates in marine sediments – Part 1: Conceptual model of gas hydrate growth conditioned by host sediment properties. J. Geophys. Res. 104, 22985–23003.

Collett, T.S., 2003. Natural gas hydrate as a potential energy resource. In: Max, M.D. (Ed.), Natural Gas Hydrate in Oceanic and Permafrost Environment. Kluwer Academic Publishers, London, pp. 123–136.

Cook, J.G., Leaist, D.G., 1983. An exploratory study of the thermal conductivity of methane hydrate. Geophys. Res. Lett. 10, 397–399.

Crutchley, G.J., Pecher, I.A., Gorman, A.R., 2010. Seismic imaging of gas beneath seafloor vent sites in a shallow marine gas hydrate province. Hikurangi Margin, New Zealand, Mar. Geol. 272, 114–126 http://dx.doi.org/110.1016/j.margeo.2009.1003.1007.

Dai, J., Banik, N., Gillespie, D., Dutta, N., 2008. Exploration of gas hydrates in the deep water northern Gulf of Mexico: Part II, Model validation by drilling. Marine Petroleum Geology 25, 830–844.

Dallimore, S.R., Uchida, T., Collett, T.S., 1999. Scientific results from JAPEX/JNOC/GSC Mallik 2L-38 gas hydrate research well, Mackenzie Delta, Northwest Territories, Canada. Geol. Surv. Can. Bull. 544.

Davidson, D.W., El-Defrawy, M.D., Fulgem, M.O., Judge, A.S., 1978. Proceedings of the 3rd International Conference on Permafrost. National Research Council of Canada 1, pp. 938–943.

Davies, R.J., Clark, I.R., 2006. Submarine slope failure primed and triggered by silica and its diagenesis. Basin Res. 18, 339–350.

Delisle, G., Beiersdorf, H., Neben, S., Steinmann, D., 1998. The geothermal field of the North Sulawesi accretionary wedge and a model on BSR migration in unstable depositional environments. In: Henriet, J.P., Mienert, J. (Eds.), Gas Hydrates: Relevance to World Margin Stability and Climate Change. Geological Society Special Publication, No. 137. The Geological Society, London, UK, pp. 267–274.

Dickens, G.R., O'Neil, J.R., Rea, D.K., Owen, R.M., 1995. Dissociation of oceanic methane hydrate as a cause of the carbon isotope excursion at the end of the Paleocene. Paleoceanography 10, 965–971.

Dillon, W.P., Max, M.D., 2003. Oceanic gas hydrates. In: Max, M.D. (Ed.), Natural Gas Hydrates in Oceanic and Permafrost Environment. Kluwer, London, pp. 61–76.

Dillon, D.P., Nealon, J.W., Taylor, M.H., et al., 2001. Seafloor collapse and methane venting associate with gas hydrate on the Blake Ridge – causes and implications to seafloor stability and methane release. In: Paull, C.K., Dillon, W.P. (Eds.), Natural Gas Hydrates: Occurrence, Distribution, and Detection. American Geophysical Union, Washington, DC, pp. 211–233.

Dillon, W.P., Lee, M.W., Felhaber, K., Coleman, D.F., 1993. Gas hydrates on the Atlantic continental margin of the United States: control and concentration. In: Howell, D.G. (Ed.), The Future of Energy Gases. U.G. Geological Survey, Prof. Paper, 1570, 313–330.

Dillon, W.P., Paull, C.K., 1983. Marine gas hydrates-II: geophysical evidences. In: Cox, J.I. (Ed.), Natural Gas Hydrates: Properties, Occurrence and Recovery. Butterworth, 1983, Boston, pp. 73–90.

Dott Jr., R.H., 1963. Dynamics of subaqueous gravity depositional processes. AAPG Bull. 47, 104–128.

Durham, W., Kirby, S., Stern, L., Zhang, W., 2003. The strength and rheology of methane clathrate hydrate. J. Geophys. Res. 108 (B4), 2182–2193.

Durham, W., Stern, L., Kirby, S., Circone, S., 2005. Rheological comparisons and structural imaging of sI and sII end members gas hydrate and hydrate/sediment aggregates. In: Proceedings of the 5th Internation Conference on Gas Hydrates, Trondheim, Norway, June, 2005.

Dutta, N.C., Utech, R.W., Shelander, D., 2010. Role of 3D seismic for quantitative shallow hazard assessment in deepwater sediments. Lead. Edge 29, 930–942.

Ebinuma, T., Kamata, Y., Minagawa, H., et al., 2005. Mechanical properties of sandy sediment containing methane hydrate. In: Proceedings of the 5th International Conference on Gas Hydrates, Trondheim, Norway, June 12–16, 2005.

Edinuma, T., Kamata, Y., Minagawa, H., 2005. Mechanical properties of sandy sediments containing methane hydrates. In: 5th Int. Con. on Gas Hydrates, June 12–16. Tromdheim, Norway.

Embley, R.W., 1980. The role of mass transport in the distribution and character of deep-ocean sediment with reference to North Atlantic. Mar. Geol. 38, 28–50.

Du Frane, W.L., Stern, L.A., Weitemeyer, K.A., et al., 2011. Electrical conductivity of laboratory-synthesized methane hydrate. In: Proceedings of the 7th International Conference on Gas Hydrates (ICGH 2011), pp. 1–11.

Francisca, F., Yun, T.S., Ruppel, C., Santamarina, J.C., 2005. Geophysical and geotechnical properties of near-seafloor sediments in the northern Gulf of Mexico gas hydrate province. Earth Planet. Sci. Lett. 237, 924–939.

Gao, S., House, W., Chapman, W.G., 2005. NMR/MRI Study of Clathrate Hydrate Mechanisms. J. Phys. Chem. B 109, 19090–19093.

Grantz, A., Dinter, D.A., 1980. Constraints of geologic processes on Beaufort Sea oil development. Oil Gas J. 78, 304–319.

Grevemeyer, I., Rosenberger, A., Villinger, H., 2000. Natural gas hydrates on the continental slope off Pakistan: constraints from seismic techniques. Geophys. J. Int. 140, 295–310.

Grozic, J.L.H., Dallimore, S.R., 2012. Submarine Slope Failures in the Beaufort Sea: Influence of Gas Hydrate Decomposition, AGU Fall Meeting, San Francisco, USA.

Grozic, J.L.H., 2010. Interplay between gas hydrates and submarine slope failure. In: Mosher, D.C. (Ed.), Submarine Mass Movements and Their Consequences. Vol. 28. Advances in Natural and Technological Hazards Research, pp. 11–30.

Gu, G., Dickens, G.R., Bhatnagar, G., et al., 2011. Abundant Early Palaeogene marine gas hydrates despite warm deep-ocean temperatures. Nat. Geosci. 4, 848–851 http://www.nature.com/ngeo/journal/v4/n12/abs/ngeo1301.htm.

Haflidason, H., Sejrup, H.P., Bryn, P., 2001. The Storegga Slide; chronology and flow mechanism. In: paper presented at the XI European Union of Geoscientists Meeting, 8–12 April, Strasbourg, France.

Haskell, B.J., 1991. The Influence of Deep Western North Atlantic Circulation on Late Quaternary Sedimentation on the Blake Outer Ridge. PhD Thesis. Duke University, USA.

Hautala, S.L., Solomon, E.A., Johnson, H.P., 2014. Dissociation of Cascadia margin gashydrates in response to contemporary ocean warming. Geophys. Res. Lett. 41. http://dx.doi.org/10.1002/2014GL061606.

Henry, P., Thomas, M., Clennel, M.B., 1999. Formation of natural gas hydrate in marine sediments, Part 2: Thermogenic calculation. J. Geophys. Res. 104, 23005–23020.

Holbrook, W.S., 2001. Seismic studies of the Blake Ridge: Implications for hydrate distribution, methane expulsion, and free gas dynamics. Geophys. Monogr. Ser. 124, 234–256.

Holbrook, W.S., Hoskins, H., Wood, W.T., 1996. Metahne gas hydrates and free gas on the Blake Ridge from Vertical Seismic Profiling. Science 273, 1840–1843.

Holland, M., Schultheiss, P., Roberts, J., Druce, M., 2008. Observed gas hydrate morphologies in marine sediments. In: Proceedings of the 6th International Conference on Gas Hydrates (ICGH '08), Vancouver, Canada, July, 2008.

Holtzman, R., Juanes, R., 2011. Thermodynamic and hydrodynamic constraints on pore pressure caused by hydrate dissociation: a pore scale mode. J. Geophys. Res. 38, L14308.

Hornbach, M.J., Saffe, D.M., Holbrook, W.S., 2004. Critically pressured free-gas reservoirs below gas-hydrate provinces. Nature 427, 142–144.

Hornbach, M.J., Lavier, L.L., Ruppel, C.D., 2007a. Active Faulting Coincident with the Cape Fear Slide Headwall: Implications for Slope Stability and Tsunamis along the U.S. East Coast. Northeastern Section – 42nd Annual Meeting, 12–14 March, 2007.

Hornbach, M.J., Lavier, L.L., Ruppel, C.D., 2007b. Triggering mechanism and tsunamogenic potential of the Cape Fear Slide complex, U.S. Atlantic margin. Geochem. Geophys. Geosyst. http://dx.doi.org/10.1029/2007GC001722.

Hovland, M., Gudmestad, O.T., 2001. Potential influence of gas hydrates on seabed installations. In: Paull, C.K., Dillon, W.P. (Eds.), Natural Gas Hydrate Occurrence, Distribution, and Detection. Geophys. Mongr. Ser. AGU, Washington D.C, pp. 307–315.

Hovland, M., Gallagher, J.W., Clennell, M.B., Lekvam, K., 1997. Gas hydrate and free gas volumes in marine sediments: example from the Niger Delta front. Mar. Petrol. Geol. 14, 245–255.

http://www-odp.tamu.edu/publications/164_SR/.

Hyndman, R.D., Spence, G.D., Chapman, R., 2001. Geophysical studies of marine gas hydraes in Northern Cascadia. In: Paul, C.K., Dillon, W.P. (Eds.), Natural Gas Hydrates: Occurrence, Distribution, and Detection. American Geophysical Union, Washington DC, pp. 273–295.

Hyndman, R.D., Davis, E.E., 1992. A mechanism for the formation of methane hydrate and seafloor bottom-simulating reflectors by vertical fluid expulsion. J. Geophys. Res. 97 (B), 6683–6698.

Jahren, A.H., Conrad, C.P., Arens, N.C., Mora, G., Lithgow-Bertelloni, C., 2005. A plate tectonic mechanism for methane hydrate release along subduction zones. Earth Planet. Sci. Lett. 236, 691–704.

Judge, A., 1982. Natural gas hydrate in Canada. In: Proceedings of the 4th Canadian Permafrost Conference, pp. 320–328.

Jang, J., Santamarina, J.C., 2011. Recoverable gas from hydrate-bearing sediments: pore network model simulation and macroscale analyses. J. Geophys. Res. 116, 2156–2202.

Kayen, R.E., Lee, H.G., 1991. Pleistocene slope instability of gas hydrate laden sediment on the Beaufort Sea margin. Mar. Geotechnol. 10, 125–141.

Kayen, R.E., Lee, H.J., 2001. Pleistocene slope instability of gas hydrate laden sediment on the Beaufort Sea margin, Mar. Geotechnol. 10, 125–141.

Kleinberg, R.L., 2004. Quantitative Assessment of Submarine Slope Stability. Abstracts: AAPG Hedberg Research Conference on Gas hydrates: Energy Resource Potential and Associated Geologic Hazards, September 12–16, 2004, Vancouver, BC, Canada.

Kleinberg, R.L., Flaum, C., Griffin, D.D., et al., 2003. Deep sea NMR: Methane hydrate growth habit in porous media and its relationship to hydraulic permeability, deposit accumulation, and submarine slope stability. J. Geophys. Res. 108, 2508.

Kretschmer, K., Biastoch, A., Rupke, L., Burwicz, E., 2015. Modeling the fate of methane hydrates under global warming. Glob. Biogeochem. Cycles 29 (5), 610–625.

Krynine, P.D., 1948. The megascopic study and field classification of sedimentary rocks. J. Geol. 56, 130–165.

Kvalstad, T.J., Andresen, L., Forsberg, C.F., et al., 2005. The Storegga slide: evaluation of triggering sources and slide mechanics. Mar. Petrol. Geol. 22, 245–256.

Kvalstad, T.J., Gauer, P., Kaynia, A.M., Nadim, F., 2002. Slope stability at Ormen Lange. In: Proceedings of Offshore Site Investigation and Geotechnics. Diversity and Sustainability, London, UK, pp. 233–250.

Kvenvolden, K., 1998. A primer on geological occurrence of gas hydrates. In: Menriet, J.P., Mienert, J. (Eds.), Gas Hydrates in Relevance to World Margin Stability and Climate Change. Geological Society Special Publications, London, pp. 9–30. No. 137.

Kvenvolden, K.A., 1999. Potential effects of gas hydrate on human welfare. Proc. Natl. Acad. Sci. U. S. A. 96, 3420–3426.

Kwon, T.H., Cho, G.C., 2012. Submarine slope failure primed and triggered by bottom water warming in oceanic hydrate-bearing deposits. Energies 5, 2849–2873. http://dx.doi.org/10.3390/en5082849.

Kwon, T.-H., Cho, G.-C., Santamarina, J.C., 2008. Gas hydrate dissociation in sediments: Pressure-temperature evolution. Geochem. Geophys. Geosyst 9, 1–14. http://dx.doi.org/10.1029/2007GC001920.

Kwon, T.H., Lee, K., Cho, G.C., Lee, J.Y., 2011. Geotechnical properties of deep oceanic sediments recovered from the hydrate occurrence regions in the Ulleung Basin, East Sea, Offshore Korea. Mar. Pet. Geol. 28 (10), 1870–1883.

Lamarque, J.F., 2008. Estimating the potential for methane clathrate instability in the 1%-CO2 IPCC AR-4 simulations. Geophys. Res. Lett. 35, L19806.

Macdonald, G.J., 1990. Role of methane clathrates in past and future climates. Clim. Change 16, 247–281.

Maslin, M., Mikkelsen, N., Vilela, C., Haq, B., 1998. Sea-level and gas-hydrate-controlled catastrophic sediment failures of the Amazon Fan. Geology 26, 1107–1110.

Maslin, M., Owen, M., Betts, R., et al., 2010. Gas hydrates: past and future geohazard? Phil. Trans. R. Soc. A 368. http://dx.doi.org/10.1098/rsta:2010.0065.

Maslin, M., Owen, M., Day, S., Long, D., 2004. Linking continental-slope failures and climate change: testing the clathrate gun hypothesis. Geology 32, 53–56.

Masui, A., Haneda, H., Ogata, Y., Aoki, K., 2005. The effect of saturation degree of methane hydrate on the shear strength of synthetic methane hydrate sediments. In: Proceedings of the 5th International Conference on Gas Hydrates, 12–16 June, Trondheim, Norway. Paper 2037, pp. 657–663.

Masui, A., Miyazaki, K., Haneda, H., Ogata, Y., Aoki, K., 2008. Mechanical characteristics of natural and artificial gas hydrate bearing sediments. In: Proceedings of the 6th International Conference on Gas Hydrates, Vancouver, Canada.

Max, M.D., Dillon, W.P., 1998. Oceanic methane hydrate: The character of the Blake Ridge hydrate stability zone, and the potential for methane extraction. J. Pet. Geol. 21, 343–358.

McDonald, G.J., 1990. The future of methane as energy resource. Annual Review of Energy 15, 53–83.

McIver, R.D., 1982. Role of naturally occurring gas hydrates in sediment transport. AAPG Bull. 66, 789–792.

Mienert, J., Andreassen, K., Posewang, J., 2000. Gas hydrates: Challenges for the future. New York Academy of Sciences, New York, pp. 200–210.

Mienert, J., Vanneste, M., Bünz, S., et al., 2005. Ocean warming and gas hydrate stability on the mid-Norwegian margin at the Storegga slide. Mar. Pet. Geol. 22 (1–2), 233–244.

Milkov, A.V., 2004. Global estimates of hydrate-bound gas in marine sediments: how much is really out there? Earth Sci. Rev. 66, 183–197. http://dx.doi.org/10.1016/j.earscirev.2003.11.002.

Miller, K.G., Liu, C., Feigenson, M.D., 1996. Oligocene to middle Miocene Sr-isotopic stratigraphy of the New Jersey continental slope. In: Mountain, G.S., Miller, K.G., Blum, P., Poag, C.W., Twichell, D.C. (Eds.), Proc. ODP, Sci. Results, 150: College Station. TX (Ocean Drilling Program), pp. 97–114. http://dx.doi.org/10.2973/odp.proc.sr.150.011.1996.

Minshull, T.A., Singh, S.C., Westbrook, G.K., 1994. Seismic velocity structure at a gas hydrate reflector, offshore Western Columbia, from full waveform inversion. J. Geophys. Res. 99, 4715–4734.

Minshull, T., White, R., 1989. Sediment compaction and fluid migration in the Makran accretionary prism. J. Geophys. Res. 94, 7387–7402.

Moridis, G.J., Kowalsky, M.B., 2007. Response of oceanic hydrate-bearing sediments to thermal stresses. SPE J. 2007 (12), 253–268.

Mörner, N.A., 1991. Intense earthquakes and seismotectonics as function of glacial isostasy. Tectonophysics 117, 139–153.

Nimblett, J., Ruppel, C., 2003. Permeability evolution during the formation of gas hydrates in marine sediments. J. Geophys. Res. 108 (B9), 2420.

Nisbet, E.G., 1990. The end of the ice age. Can. J. Earth Sci. 27, 148–157.

Nisbet, E.G., Piper, D.J.W., 1998. Giant submarine landslides. Nature 392, 329–330.

Nixon, M.F., Grozic, L.H., 2006. A simple model for submarine slope stability analysis with gas hydrates. Norw. J. Geol. 86, 309–316.

Panchuk, K., Ridgwell, A., Kump, L.R., 2008. Sedimentary response to Paleocene–Eocenethermal maximum carbon release: a model-data comparison. Geology 36, 315–318. http://dx.doi.org/10.1130/G24474A.1.

Paull, C.K., Brewer, P.G., Ussler III, W., et al., 2003. An experiment demonstrating that marine slumping is a mechanism to transfer methane from seafloor gas-hydrate deposits into the upper ocean and atmosphere. Geo. Mar. Lett. 22, 198–203. http://dx.doi.org/10.1007/s00367-002-0113-y.

Paull, C.K., Matsumoto, R., Wallace, P.J., 1996. In: Proceedings of the Ocean Drilling Program Initial Reports, vol. 164. Ocean Drilling Program, College Station, TX, pp. 142–144.

Paull, C.K., Ussler III, W., Dillon, W.P., 2000. Potential role of gas hydrate decomposition in generating submarine slope failures. In: Max, M.D. (Ed.), Natural Gas Hydrate in Oceanic and Permafrost Environments. Kluwer Academic Publishers, Dordrecht, The Netherlands, pp. 149–156.

Paull, C.K., Buelow, W.J., Ussler III, W., Borowski, W.S., 1996. Increased continental-margin slumping frequency during sea-level lowstands above gas hydrate bearing sediments. Geology 24, 143–146. http://dx.doi.org/10.1130/0091-7613(1996).

Paull, C.K., Ussler III, W., Dillon, W.P., 1991. Is the extent of glaciation limited to marine gas hydrates? Geophys. Res. Lett. 18, 432–434.

Pecher, I.A., Minshull, T.A., Singh, S.C., von Huene, R., 1996. Velocity structure of a bottom simulating reflector offshore Peru: Results from full waveform inversion. Earth Planet. Sci. Lett. 139, 459–469.

Peters, D., Hatton, G., Mehta, A., Hadley, C., 2008. Gas hydrate geohazards in shallow sediments and their impact on the design of subsea systems. In: Proceedings of the 6th International Conference on Gas Hydrates (ICGH 2008), Vancouver, BC, Canada, June 6–10, 2008.

Popenoe, P., Schmuck, E.A., Dillon, W.P., 2001. The Cape Fear landslide: slope failure associated with salt diapirism and gas hydrate dissociation. In: Submarine Landslides Selected Studies in US Exclusive Economic Zone, pp. 40–53.

Rice, D.D., 1993. Biogenic gas: controls, habitats and resource potential. In: Howell, D.G. (Ed.), The Future of Energy Gases. USGS Professional Paper 1570, pp. 583–606.

Rothwell, R.G., Thomson, J., Kahler, G., 1998. Low-sea-level emplacement of a very large Late Pleistocene 'megaturbidite' in the western Mediterranean Sea. Nature 392, 377–380.

Ruppel, C., 2000. Thermal state of the gas hydrate reservoir. In: Max, M.D. (Ed.), Natural Gas Hydrate in Oceanic and Permafrost Environments, Kluwer Academic Publishers, Dordrecht, The Netherlands, pp. 29–42.

Sain, K., Minshull, T.A., Singh, S.C., Hobbs, R.W., 2000. Evidence for a thick free gas layer beneath the bottom simulating reflector in the Makran accretionary prism. Mar. Geol. 164, 3–12.

Schiermeier, Q., 2008. Fears surface over methane leaks. Nature 455, 572–573.

Scholl, D.W., Hart, P.E., 1993. The Future of Energy Gases. Prof. Paper, US. Geological Survey 331–351.

Scholz, N.A., Riedel, M., Spence, G.D., 2011. Do dissociating gas hydrates play role in triggering submarine slope failures, a study from Northern. In: Proc. 7th Int. Conf. On gas Hydrates, Edinburg, U.K. July 17–21.

Shanmugam, G., Bloch, R., Mitchell, S.M., et al., 1995. Basin-floor fans in the North Sea: sequence stratigraphic models vs sedimentary facies. AAPG Bull. 79, 477–512.

Shanmugam, G., 2000. 50 years of the turbidite paradigm (1950s–1990s): deep-water processes and facies models – a critical perspective. Mar. Pet. Geol. 17, 285–342.

Shanmugam, G., Lehtonen, L.R., Straume, T., et al., 1994. Slump and debris flow dominated upper slope facies in the Cretaceous of the Norwegian and Northern North Seas (61º–67º N): implications for sand distribution. AAPG Bull. 78, 910–937.

Shine, K., Derwent, R., Wubbles, D., 1990. Radiative forcing of climate. In: Houghton, J., et al., (Eds.), Climate Change: The IPCC Scientific Assessment. Cambridge University Press, Cambridge, pp. 41–68.

Shipboard Scientific Party, 1985. Site 570 (Leg 84). In: Huene, R., et al. (Ed.), Proceedings Deep Sea Drilling Project, Initial Reports, vol. 96. US Government Printing Office, Washington DC, pp. 3–424.

Shipboard Scientific Party, 2002. Leg 198 Preliminary Report. Ocean Drilling Program, College Station, TX. http://www-ODP.tamu.edu/publication/prel/198-prel/198PRELPDF.

Siriwardane, H.J., Smith, D.H., 2006. Gas Hydrate Induced Seafloor Stability Problems in the Blake Ridge. In: Proceedings of the 16th International Offshore and Polar Engineering Conference, San Francisco, California, USA, May 28–June 2, 2006.

Sloan Jr., E.D., 2006. Clathrate Hydrates of Natural Gases, third ed. Marcel Dekker, New York, NY.

Spangenberg, E., 2004. Modeling of the influence of gas hydrate content on the electrical properties of porous sediments. J. Appl. Geophys. 56, 73–78.

Suess, E., Bohrmann, G., Lausch, E., 1999a. Flammable ice. Sci. Am. 281, 76–83.

Suess, E., Torres, M.E., Bohrmann, G., et al., 1999b. Gas hydrate destabilization: enhanced dewatering, benthic material turnover and large methane plumes at the Cascadia convergent margin. Earth Planet. Sci. Lett. 170, 1–15.

Sultan, N., Cochonat, P., Foucher, J.P., Mienert, J., 2004. Effect of gas hydrates melting on seafloor slope instability. Mar. Geol. 213, 379–401.

Summerhayes, C.P., Bornhold, B.D., Embley, R.W., 1979. Surficial slides and slumps on the continental slope and rise of South West Africa: a reconnaissance study. Mar. Geol. 31, 265–277.

Suzuki, K., Ebinuma, T., Narita, H., 2008. Shear strength of natural gas hydrate bearing sediments of Nankai Trough. In: Proceedings of the 6th International Conference on Gas Hydrates, 6–10 July, Vancouver, Canada.

Tinivella, U., Lodolo, E., Camerlenghi, A., 1998. Seismic tomography study of a bottom simulating reflector off the South Shetland Island. In: Henriet, J.P., Mienert, J. (Eds.), Gas Hydrates: Relevance to World Margin Stability and Climate Chance. The Geological Society, London, pp. 141–151.

Trofimuk, A.A., Cherskiy, N.V., Tsarev, V.P., 1973. Accumulation of natural gases in zones of hydrate formation in the hydrosphere. Dokl. Akad. Nauk. SSSR 212, 931–934 (in Russian).

Twichell, D.C., Chaytor, J.B., ten Brink, U.S., Buczkowski, B., 2009. Morphology of late Quaternary Submarine Landslides along the U.S. Atlantic Continental Margin. Mar. Geol. 264, 4–15.

Tzirita, A., 1992. In Situ Detection of Natural Gas Hydrates Using Electrical and Thermal Properties. Offshore Tecnology Research Center, p. 220.

Vogt, P.R., Jung, W.Y., 2002. Holocene mass wasting on upper non-Polar continental slopes: due to post-glacial ocean warming and hydrate dissociation? Geophys. Res. Lett. 29, 1341.

Watson, R.T., Rodhe, H., Oeschger, H., Siegenthaler, U., 1990. Greenhouse gases and aerosols. In: Houghton, J.T., Jenkins, G.J., Ephraums, J.J. (Eds.), Climate Change: The IPCC Scientific Assessment, Intergovernmental Panel on Climate Change (IPCC). Cambridge, University Press, Cambridge, pp. 1–40.

Westbrook, G.K., Thatcher, K.E., Rohling, E.J., et al., 2009. Escape of methane gas from the seabed along the West Spitsbergen continental margin. Geophys. Res. Lett. 36, L15608.

White, R.S., 1979. Gas hydrate layers trapping free gas in the Gulf of Oman. Earth Planet. Sci. Lett. 42, 114–120.

Winters, W.J., Pecher, I.A., Waite, W.F., Mason, D.H., 2004. Physical properties and rock physics models of sediment containing natural and laboratory-formed methane gas hydrate. Am. Mineral 89, 1221–1227.

Winters, W.J., Waite, W.F., Mason, D.H., Gilbert, L.Y., Pecher, I.A., 2007. Methane gas hydrate effect on sediment acoustic and strength properties. J. Pet. Sci. Eng. 56, 127–135.

Wood, W., Jung, W., 2008. Modeling the Eextent of Earth's Mmarine Mmethane Hhydrate Cryosphere. In: Proceedings of the 6th International Conference on Gas Hydrates (ICGH 2008), July 6–10, 2008, Vancouver, BC, Canada, p. 8.

Xu, W.Y., Germanovich, L.N., 2006. Excess pore pressure resulting from methane hydrate dissociation in marine sediments: A theoretical approach. J. Geophys. Res. Solid Earth.111, B01104.Bull.

Xu, W.Y., Ruppel, C., 1999. Predicting the occurrence, distribution, and evolution of methane gashydrate in porous marine sediments. J. Geophys. Res. 104, 5081–5095.

Yu, X.C., Li, G., Li, Q.P., et al., 2013. Experimental simulation of gas hydrate decomposition in porous sediment. Sci. China Earth Sci. 56 (4), 588–593.

Yuan, J., Edwards, R.N., 2000. The assessment of marine gas hydrates through electrical remote sounding: hydrate without a BSR? Geophys. Res. Lett. 27, 2397–2400.

Yun, T.S., Fratta, D., Santamarina, J.C., 2010. Hydrate-bearing sediments from the Krishna-Godavari basin: physical characterization, pressure core testing and scaled production monitoring. Energy Fuels 24, 5972–5983.

Yun, T.S., Santamarina, J.C., Ruppel, C., 2007. Mechanical properties of sand, silt, and clay containing tetrahydrofuran hydrate. J. Geophys. Res. 112, B04106.

Zachos, J.C., Rohl, U., Schellenberg, S.A., et al., 2005. Rapid acidification of the ocean during the Paleocene–Eocene thermal maximum. Science 308, 1611–1615.

Zeebe, R.E., Zachos, J.C., Dickens, G.R., 2009. Carbon dioxide forcing alone insufficient to explain Palaeocene–Eocene thermal maximum warming. Nat. Geosci. 2, 576–580.

Zhang, Y., Xu, Z., 2003. Kinetics of convective crystal dissolution and melting, with application to methane hydrate dissolution and dissociation in seawater. Earth Planet. Sci. Lett. 213, 133–148.

CHAPTER

THE ROAD AHEAD AND OTHER THOUGHTS

10

To raise new questions, new possibilities, to regard old problems from a new angle, require creative imagination and marks real advance in science.

Albert Einstein

CHAPTER OUTLINE

10.1 INTRODUCTION

The history of human civilization can be viewed as the progressive development of energy resources and their associated conversion technologies. The development of new methods and the digital world has increased the comfort, longevity, and affluence of humans. The US Energy Information Administration (EIA) presented an analysis of global energy consumption by fuel type, which shows that hydrocarbons account for the vast majority of our energy consumption, i.e., ~84% (Fig. 10.1).

Hydroelectricity is the most significant renewable energy source at 6% of the global total. Other renewables include solar, wind, ethanol, biomass, biodiesel, and geothermal, which

Geological Controls for Gas Hydrates and Unconventionals. http://dx.doi.org/10.1016/B978-0-12-802020-3.00010-2

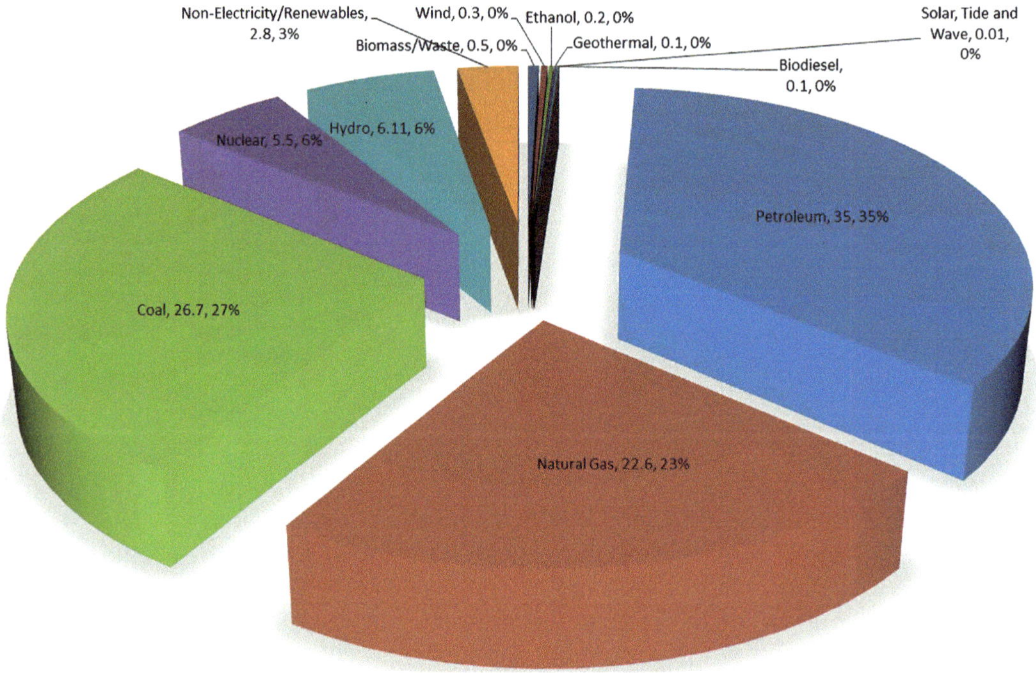

FIGURE 10.1

Global energy consumption by source (on an energy-equivalent BTU basis). (Data Source: EIA, AltaCorp Capital Inc. Non-electricity renewables are those used for heat generation.)

collectively make up just 4% of the total. What do we know about the future of hydrocarbon resources? Predictions of impending oil shortages are as old as the industry itself, and the literature is full of arguments between "optimists" and "pessimists" about how much oil there is and what other resources might be available. To understand the availability of hydrocarbon resources the industry professionals need to focus on quality, quantity, and available patterns of exploitation from the resources with the alignment of shareholders. Industry has exploited the easiest available large reservoirs first because these resources are easier to explore, appraise, and develop and required less energy to extract and refine. These fields are in the mature stage now and are called mature assets. Technological progress is in a race with the depletion of these resources and a lot of optimized workflows are being developed by the research and service industry. Hydrocarbon resources are of mainly two types and the following section explains the difference between these resource types.

10.2 CONVENTIONAL VERSUS UNCONVENTIONAL RESOURCES

Human progress has required the exploitation of energy resource, and it seems that interest has shifted recently from renewables such as solar energy back to fossil hydrocarbons. To define the age of oil we need to look back about 150 years to when the first deep oil wells were sunk and just over 200 years when the viable steam engine was developed. The connection between these two events can be understood as steam engines made deep drilling of oil wells possible. Classification of hydrocarbon resources is based on techniques/methodology for production, not hydrocarbon product. According to Petroleum Resources Management System (PRMS) classifications, conventional hydrocarbons include conventional reservoirs, tight reservoirs, heavy oil, extra-heavy oil, shale oil, and shale gas, and unconventional hydrocarbons include bitumen, coal seam gas (CSG), gas hydrates, and oil shale.

10.2.1 CONVENTIONAL RESOURCES

The term "conventional" implies these resources are easy to develop and do not need specialized technologies to unlock their potential. As they are easier and less expensive to produce, conventional oil and gas were the first targets of industry activity. The conventional oil drilling process includes the drilling of a well, a reservoir having pressure and oil flowing out of the ground. Conventional oil drilling is cheaper than unconventional oil drilling.

10.2.2 UNCONVENTIONAL RESOURCES

The term "unconventional" implies these resources are more challenging to develop. Any hydrocarbon resource that is different in terms of origin, storage, exploitation technique, and properties falls under the category of unconventional resources. In contrast to conventional resources, unconventional resources are trapped in reservoirs with low permeability, meaning little to no ability for the oil or natural gas to flow through the rock and into a well-bore. In order to produce from unconventional reservoirs, industry uses a stimulation technique like hydraulic fracturing to create cracks in the underground rock that allow the oil or natural gas to flow. In the current climate, the term "unconventional" refers mostly to shale or tight gas and liquids, leaving out other "super" unconventional resources, such as gas hydrates, extra-heavy oil, and coal bed methane (CBM), among others. Unconventional formations are fine-grained, organic-rich, sedimentary rocks, usually shales and similar rocks. The shales and rocks are both the source of and the reservoir for oil and natural gas, unlike conventional petroleum reservoirs. Unconventional gas is trapped in formations that are atypical in terms of their geological location and characteristics. Recovering these resources requires the use of techniques designed to "untrap" the gas. The term "unconventional gas" actually covers three main types of natural gas resources: shale gas, tight gas, and coal bed methane (also known as coal seam gas). Conventional and unconventional gases do not differ by their chemical composition (they are all natural gas), as shown in Fig. 10.2 (http://www.total.com). Unconventional drilling is a lengthy process. Usually the well is tight and the movement of oil takes longer geological time. To alleviate the slow process, Chestnut Exploration and Production drills the oil horizontally and fracs it. The fracking process opens up the rock and lets the oil flow more quickly. Unconventional oil drilling is a much more expensive process than conventional oil drilling.

A hydrocarbon field has a long life cycle. From the discovery of a field to the first oil or gas, exploration and production activities are spread over several years. In a conventional reservoir, the gas is in interconnected pore spaces, much like a kitchen sponge, that allow easier flow to a well; but in an unconventional reservoir, like shale, the reservoir must be mechanically "stimulated" to create additional permeability and free the gas for collection. The road map for future production can be designed on the basis of the five main steps in the life cycle of an oil and gas field described in Fig. 10.3.

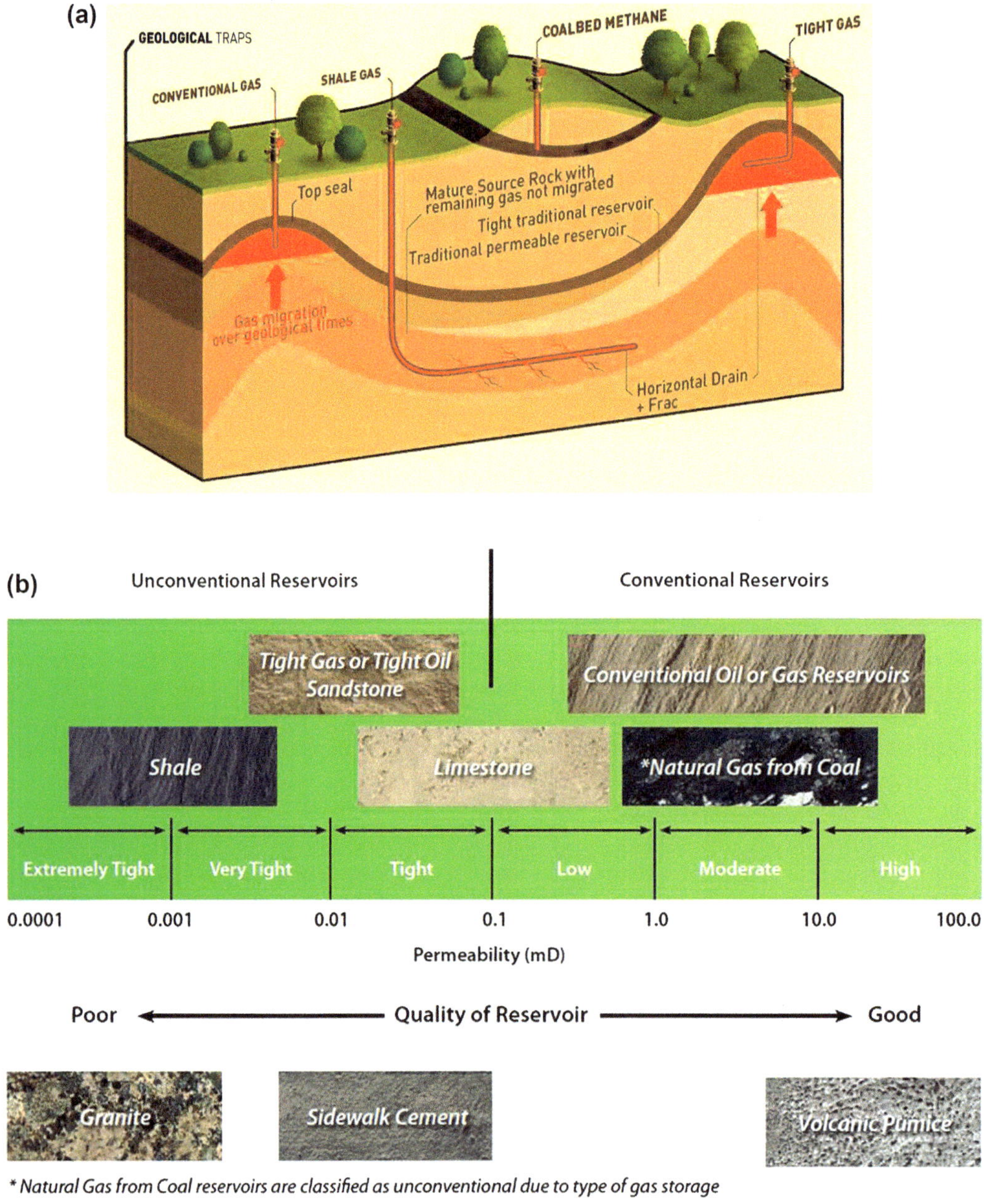

Modified from US Department of Energy

FIGURE 10.2

(a) Main sources of unconventional gas. (Source: http://www.total.com/en/energies-expertise/oil-gas/exploration-production/strategic-sectors/unconventional-gas/presentation/three-main-sources-unconventional-gas.) Oil and gas industry has explored and developed conventional oil reservoirs for several decades and most of the oil production is coming from these reservoirs. Conventional oil and gas reservoirs have diminished and industry has expanded the search for new sources of oil that includes shale oil or other tight oil reservoirs. (b) US Department of Energy has classified conventional and unconventional reserves. (Source: http://www.csur.com)

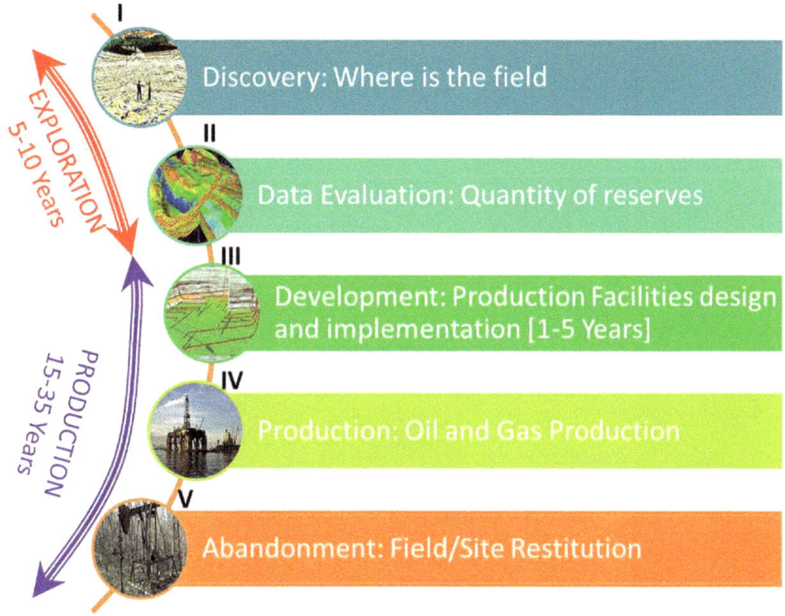

FIGURE 10.3

Five main steps of an oil or gas field development project.

Conventional oil and gas are not the main focus of this book, which mainly discusses the geological controls of hydrates and unconventional resources. The future prospects for unconventional resources are discussed in the following section.

10.3 UNCONVENTIONAL RESOURCES: FUTURE PROSPECTS

Unconventional hydrocarbon resources have been classified under the following categories.

10.3.1 TIGHT GAS RESERVOIRS

Tight gas refers to natural gas reservoirs produced from reservoir rocks with such low permeability that considerable hydraulic fracturing is required to harvest the well at economic rates. These resources are sealed in extremely impermeable, hard rock, making the underground formation extremely “tight”. Tight gas reservoirs are generally defined as having less than 0.1 millidarcy (mD) matrix permeability and less than 10% matrix porosity. Tight gas can also be trapped in sandstone or limestone formations that are atypically impermeable or nonporous, also known as tight sand. Many of the low-permeability reservoirs that have been developed in the past are sandstone, but significant quantities of gas are also produced from low-permeability carbonates, shales, and coal seams. For successful production from tight gas reservoirs, a vertical well drilled and completed must be successfully stimulated to produce at commercial gas flow rates and produce commercial gas volumes. Compared with a conventional gas

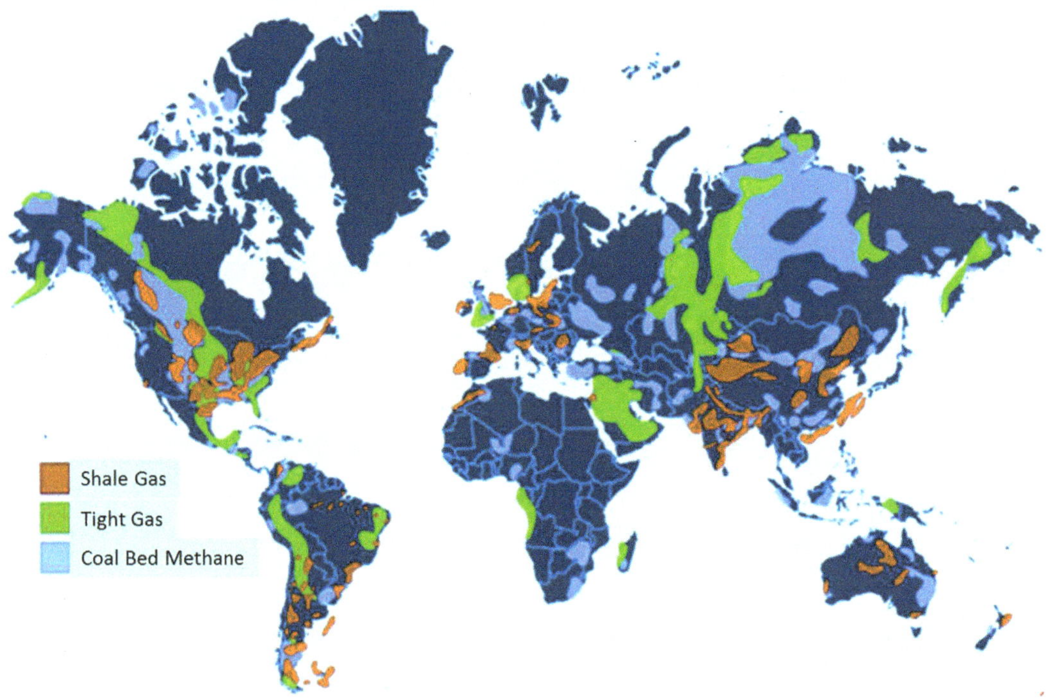

FIGURE 10.4

Global Map of tight gas potential. (Source: http://pacwestcp.com/education/shaleunconventional-resources/)

production scenario, more wells needs to be drilled in tight gas reservoirs to recover the gas-in-place. These reservoirs do not have depth constraints, they can be developed deep or shallow, high or low pressure and temperature, stacked multilayered or single layer and homogeneous, or naturally fractured. According to the department of Energy and Climate Change UK (www.gov.uk), wells produce from low-porosity sandstones and carbonate reservoirs. The gas is sourced outside the reservoir and migrates into the reservoir over geological time. Some tight gas reservoirs have also been found to be sourced by underlying coal and shale source rocks, in the so-called basin centerd gas (BCG) accumulations. A global map of of tight gas resources is shown in Fig. 10.4.

10.3.2 SHALE GAS RESERVOIRS

A shale gas reservoir starts with "shale", which is a sedimentary rock that is predominately comprised of fine-grained clay particles deposited in a thinly laminated texture. In these reservoirs, the organic-rich shale acts both as the source and the reservoir rock. Shale is one of the most abundant forms of sedimentary rock on earth and serves as a source for the hydrocarbons migrating into permeable reservoirs by acting as a seal for trapping oil and gas in underlying sediments. Shales were originally deposited as mud in low-energy depositional environments such as tidal flats and swamps together with organic matter, which is measured as total organic content (TOC). Different shale gas reservoirs have different properties. For shale gas reservoirs to become a successful shale play, the following important characteristics should be taken into account: depth of shale gas formation, organic matter content (TOC), maturation, thickness of the reservoir zone, gas-in-place, permeability, mineralogy, brittleness, and pore pressure. Shale gas reservoirs rely on natural fractures for porosity and permeability as the matrix porosity or permeability is low. The TOC is a measurement of kerogen, oil, and gas in a rock. These three different phases can be difficult to distinguish, and the TOC is therefore used as a measurement of the hydrocarbon potential of the rock. When buried deep in the earth's crust, the rock is heated and kerogen releases oil or natural gas depending on the conditions and the kerogen type. In the absence of natural fractures, these reservoirs need stimulation by way of hydraulic fracturing. According to the Department of Energy and Climate Change UK (http://www.gov.uk), matrix permeabilities of typical shales (the ability of fluids to pass through them) are very low (often termed ultra-low) compared to conventional low-permeability oil and gas reservoirs (nanodarcy 10^{-9} Darcy in shales versus millidarcy 10^{-2} in conventional sandstones) which means that hydrocarbons are effectively trapped and unable to flow under normal circumstances in shale, and are usually only able to migrate out over geological time. The slow migration of hydrocarbons from shales into shallower sandstone and carbonate reservoirs has been the source of most conventional oil and gas fields; hence shales have historically been thought of as source and seal rocks, rather than potential reservoirs, but much of the hydrocarbon still remains bound in the shale. A map of global shale gas resources is shown in Fig. 10.5. Producing

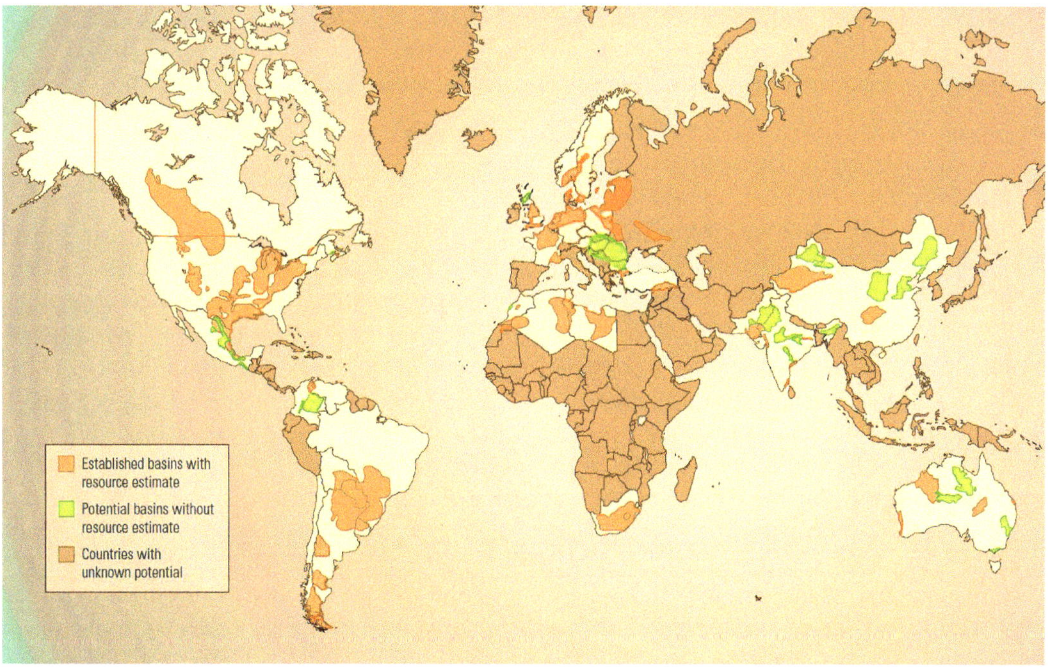

FIGURE 10.5

Global shale gas resources. The US EIA studied 14 regions for shale gas potential. Vast land masses in Russia, the Middle East, and Africa were not included in the report (gray shade). Reasons cited for not including these regions in the report were scarcity of exploration data or the presence of abundant reserves in conventional reservoirs, which make shale gas unattractive for the present. (Adapted from Kuuskraa et al., 2011 and Boyer et al., 2011.)

hydrocarbons from shale deposits is nothing new; the practice predates the modern oil industry. In 1821, decades before the first oil well was drilled, William Hart, a local gunsmith, drilled the first commercial natural gas well in the United States in 1821 in Fredonia, Chautauqua County. The Hart well was first dug to a depth of 27 feet in the shale that outcropped in the area. A 1.5 inch diameter borehole was then drilled to a depth of 70 feet. Hart built a simple gas meter and piped the natural gas to an innkeeper on the stagecoach route from Buffalo to Cleveland. By the 1920s, the world most prolific natural gas production came from similar shale deposits in the nearby Appalachain Basin.

Oil shale or liquid-rich shale refers to any source rock that contains kerogen which upon heating releases a similar product to crude oil. Kerogen requires a maturation process for oil extraction that involves heating the shale, for there is no other practical method for extracting a significant portion of contained oil or gas. In these unconventional reservoirs, oil is mobile and can be extracted through horizontal wells that are fracture stimulated ("fraced"). A liquid-rich unconventional play targets a reservoir which requires significant stimulation to provide economic liquid production rates. In these reservoirs, it is considered that source is the same as reservoir or there is an external charge (and migration) and reservoir is "shale" (fine-grained) or low-permeability (tight) (Fig 10.6) sedimentary rock with fluid type ranging from black oil, volatile oil, condensate, to wet gas (Abrams et al., 2014). Some of the tight oil reservoirs fall into the mature oil shale category, which is distinguished from immature oil shale by the fact that liquid hydrocarbons are available within the formation. The best examples of oil shale or liquid-rich shale plays can be found in the United States, where there are presently 13 oil and gas shales of consequence. Nine of them are liquid-rich and the other four essentially produce natural gas only. Producing from shale depends on the quantity of the resource, whether it is oil or gas, and how much can be extracted and at what cost. For resource assessment, the following areas need to be addressed:

- Establishment of a geological understanding of shale basins and formations
- Preliminary reservoir characterization
- Areal mapping of the major oil shale formations
- Definition of prospective area for each liquid-rich shale or oil shale formation
- Estimation of risk for oil shale in place
- Calculation of the recoverable oil shale resource

Each area has its own merit and needs to be studied in detail by applying state of the art workflows. The recovery of hydrocarbons from shale plays depends on volume and maturity of the TOC and the ability to create an effective fracture network that will conduct the hydrocarbons to each borehole. To understand this in an effective way a detailed study of mineralogy, lithology, relative rock brittleness, natural fracturing and the directionality of in situ rock stresses is needed.

During the processing of liquid-rich shale, its kerogen is converted into oil shale, which is similar to crude oil. This process involves several steps including pyrolysis, hydrogenation, or thermal dissolution. The most effective method of extraction involves pyrolysis, which is a thermochemical decomposition of organic material at elevated temperatures in the absence of oxygen (or any halogen). This is also known as retorting or destructive distillation. During the processing stage, liquid-rich shale produces spent shale or spent oil shale (also known as retorted shale), which is a solid residue from the oil shale extraction process of producing synthetic shale oil from oil shale. It consists of inorganic compounds (minerals) and remaining organic matter known as char, which is a carbonaceous residue formed from kerogen. In conventional processing of oil shale, shale is crushed into smaller pieces to increase surface area for improved extraction. The steps in the extraction of liquid-rich shale are shown in Fig. 10.7.

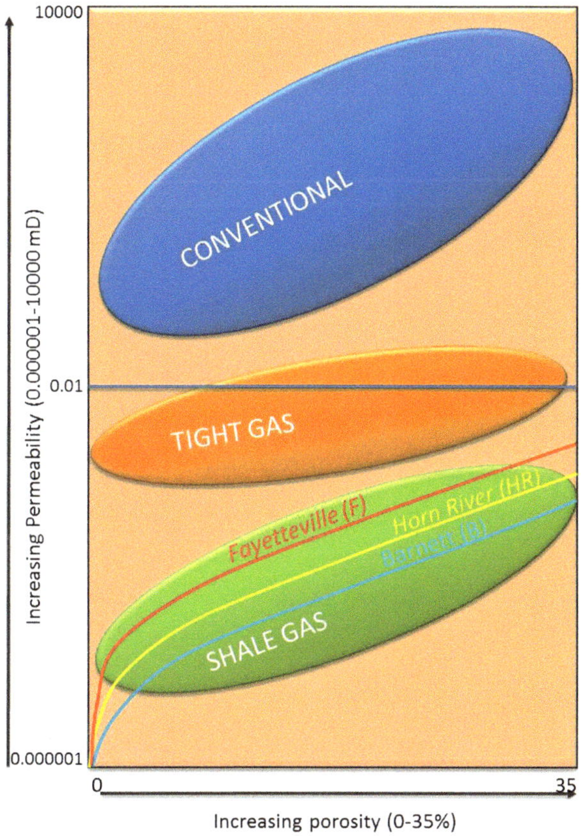

FIGURE 10.6

Permeability versus porosity plot for shale to conventional reservoirs where samples from Fayetteville (F), Horn River (HR), and Barnett (B) are plotted together with correlations between porosity, permeability, and pore throat size. (Modified from Aguilera, 2014 and Holmes et al., 2013.)

10.3.3 TIGHT OIL OR SHALE OIL RESERVOIRS

Tight oil generally refers to the oil resources that have experienced short distance migration, gathered in either tight sandstone, tight limestone or tight carbonate reservoir layers, may be interbedded with or in close proximity to the source bed, have an overburden pressure matrix permeability less than or equal to 0.1 mD, and have reservoir porosity less than 10%. Tight oil, also known as shale oil or light tight oil, is petroleum that consists of light crude oil contained in petroleum-bearing formations of low permeability, often shale or tight sandstone. The oil contained within these reservoir rocks typically will not flow to the well-bore at economic rates without assistance from technologically advanced drilling and completion technologies. Today, horizontal drilling coupled with multistage fracturing is commonly used to produce the reservoir. Many tight oil plays are siltstones without a lot of clay in the reservoir. Tight oil is considered to be an "unconventional" reservoir, requiring horizontal wells and massive hydraulic fracture jobs to perform economically. The important point to remember is that tight

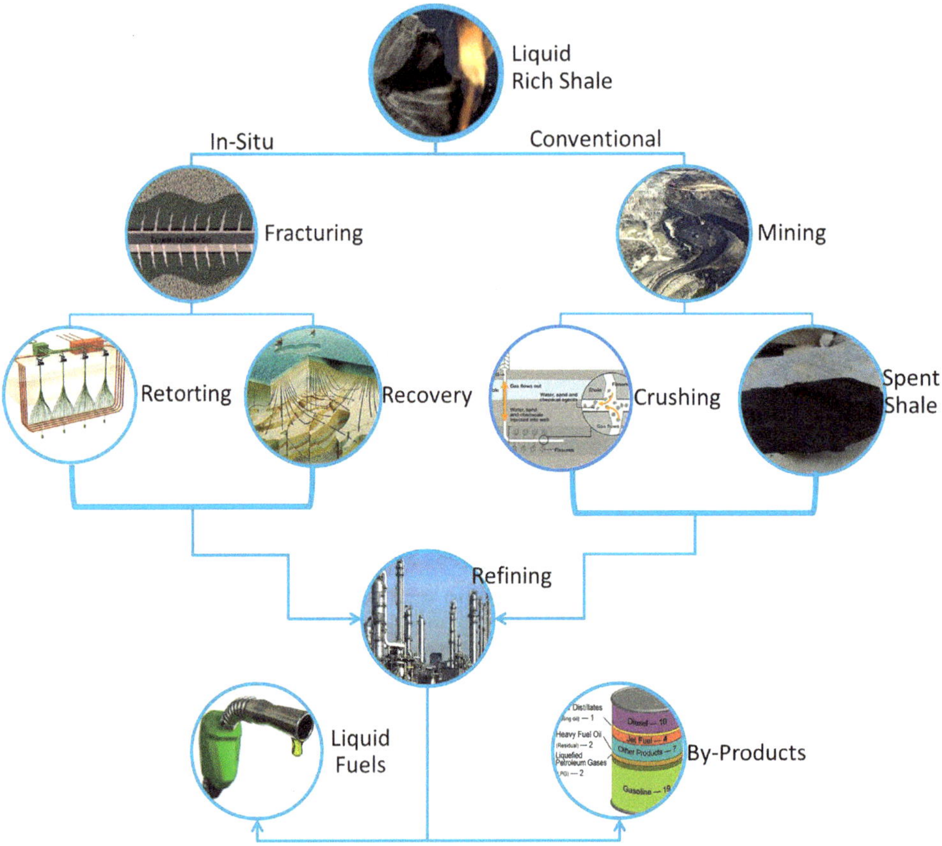

FIGURE 10.7

Liquid-rich shale extraction steps starting from fracturing and mining to liquid fuels and by-products.

oil is not limited to shale. The flow of oil from rock to well-bore is limited by the largely impermeable fine-grained nature of the oil-hosting rock, the cause for the term "tight". While some tight light oil plays produce oil directly from shales, most tight oil is produced from low-permeability siltstones, sandstones, limestones, and dolostones that are associated with the shales from which oil has been generated (www.nrcan.gc.ca). These reservoirs are unique and the simulation methods should be optimized for individual physical properties. The production cycle for a single well is long enough for the exploitation of measurable recoverables. Most of the tight oil produced is of the medium to light density, with a lower viscosity. In some of the existing oil fields, the fringe regions, or halos, surrounding the areas of historical production, are known to have oil. Applying new technologies including horizontal drilling in combination with multistage fracturing, allows oil to be recovered from the halo or fringe regions. Examples of this type of play are the Cardium and Viking formation in Western Canada (http://www.csur.com) (Fig. 10.8).

The world average depth of tight oil formation varies from 1000 m to 3000 m with the deepest well taking up to 6–7 weeks to drill. Tight oil or shale oil is different from oil shale or liquid-rich shale, which

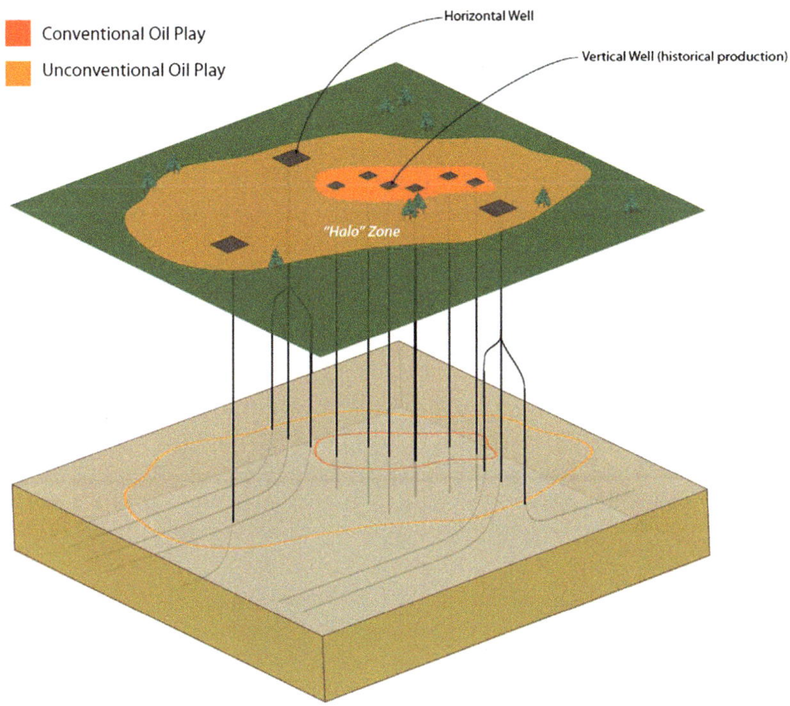

FIGURE 10.8

Schematics for halo play for tight oil reservoirs. (Source: http://www.csur.com/)

contains kerogen, a heavy mixture of organic chemical compounds, the soluble part of which is known as bitumen. Oil shale is a primitive form of oil that has not yet been converted into crude, because the sediment has experienced great heat or high pressure. It is typically found in shallow deposits and is mined, instead of drilled, with a consequent impact on the environment. The main challenges associated with tight oil reservoir exploration and developments are associated with environment-related issues. Environmental concerns over fracking may cast a shadow over the development of tight oil extraction. Fracking can cause surface scarring as production on tight formations requires a large number of wells to maximize contact with the reservoir rocks. Water pollution is another concern. Chemical-laced fracking fluid can seep through rock fissures into water wells, contaminating local supplies. The world's super majors, including BP, are seeing the exploration of tight oil play in and around the world's giant fields and source rock and convergence with enhanced oil recovery as a future exploration trend.

10.3.3.1 Some Facts about Tight Oil

- Tight oil reservoirs have very low porosities, often less than 10%, which makes discovery and evaluation of these reservoirs very challenging. This can lead to large uncertainties in calculations of reserves.
- "Tight" refers to the extremely low permeability of the reservoir rock, which impedes the ability of the oil in the formation to flow freely.

- Tight oil is conventional oil which is found in rock formations associated with organically rich shale at unminable depths.
- Some oil reservoirs produce directly from shales, but much tight oil production is from low-permeability siltstones, sandstones, and carbonates that are closely associated with oil-rich shale.
- A thorough understanding of reservoir geology, mineralogy, and lithological variation is important when considering enhanced oil recovery methods.
- Recent advancements in the technology have spurred tight oil production, including:
  - Horizontal drilling, smart completion and intelligent well systems
    - To optimize horizontal drilling, well needs to be drilled vertically until well-bore reaches a point above the target reservoir.
    - From target reservoir "kick-off" and begin to drill at an increasing angle until the well-bore runs horizontally through the targeted tight oil or shale oil reservoir.
    - Drilling should be continued horizontally to desired length.
  - Multistage hydraulic fracturing with a mixture of fluids and proppants. This makes reservoir rock more permeable and oil can flow more freely into the well-bore
    - Tight oil or shale oil reservoirs require a form of stimulation to flow the oil from already drilled well. Multistage hydraulic fracturing creates additional pathways to enable the flow of hydrocarbons.
    - Each fracturing stage is isolated using packers or plugs to contain the fracturing fluids and ensure that the fracture grows in the direction and distance that was planned.
- To produce tight oil though multistage fracking requires millions of gallons of water and frac fluid containing chemicals. Flaring is also a part of tight oil production. This opens the door of environmental criticism.
- Fracking involves the construction of thousands of well pads and drilling rigs, and the heavy truck traffic required causes road damage, creates traffic jams in villages, and increases dust levels, thereby creating public perception issues. Drilling of multiple wells and the use of advanced technologies can efficiently reduce the overall footprint of drilling operations.

10.3.3.2 Tight Oil or Shale Oil Reservoir Development Stages

Tight oil reservoir development can be optimized into five different stages. For the mature fields where existing horizontal conventional wells have already been drilled, the new activities can be classified as "infill" or "halo", where the development begins from Stage 3 (Fig. 10.9). For an infill scenario, unconventional technologies are applied to a known tight oil reservoir. For the green-field tight oil reservoir development, where the resource potential of oil-bearing formations has not been established and requires a structured approach for exploration and appraisal planning, Stages 1 and 2 are undertaken before proceeding with the pilot project, which comes under Stages 3 and 4 (Fig. 10.9).

Stage 4 mainly involves drilling multiple wells with optimization of completion techniques and multistage hydraulic fracturing. One of the very important steps in this stage is implementation of microseismic monitoring as a diagnostic tool for hydraulic fracture operations. Microseismic events are very small earthquakes of generally negative moment magnitude that are often associated with hydraulic fracturing or fluid flow in reservoirs. Microseismicity refers to discrete rock-deformation events, analogous to tiny earthquakes, which are generally of moment magnitude <0. For reference, magnitude 0.2 is the equivalent of the energy released by a large hand grenade (30 g TNT equivalent), whereas a typical small mining blast has a magnitude of around 1–1.5, corresponding to 2–2.5 kg of TNT (Bann et al., 2013). Since

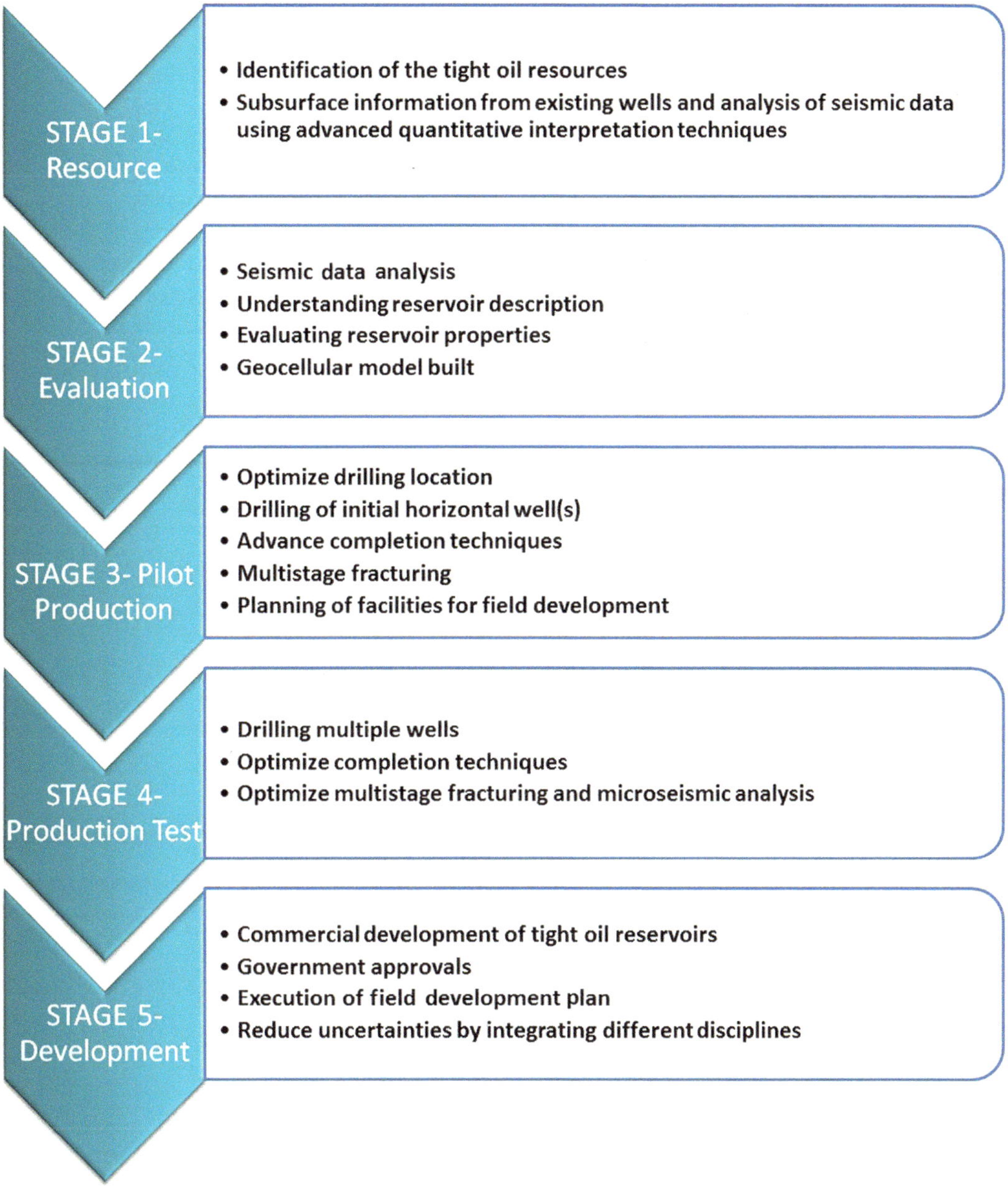

FIGURE 10.9

Tight oil or shale oil reservoir development stages.

magnitude scales are logarithmic, negative magnitude events thus correspond to the energy yield equivalent of milligrams or even micrograms of TNT. Microseismic monitoring techniques are being used increasingly by the oil and gas industry to monitor hydraulic stimulation of "tight" (very low permeability) hydrocarbon reservoirs and steam injection into heavy oil fields. Hydraulic fracturing (also known as fracking) leads to brittle failure inside a reservoir, which is typically accompanied by microseismicity. An example of a multistage hydraulic fracturing is available from Grande Prairie, Alberta (https://www.esgsolutions.com) where a 14-stage horizontal sliding sleeve hydraulic fracture program was implemented by the operators stimulating the target tight oil formation. During hydraulic fracturing stage the observation of fracture behavior in the formation was studied, in order to optimize well position and completion design for subsequent wells in the same field (Fig. 10.10). It is expected that fractures will create additional pathways perpendicular to and propagate parallel to the direction of principal maximum stress.

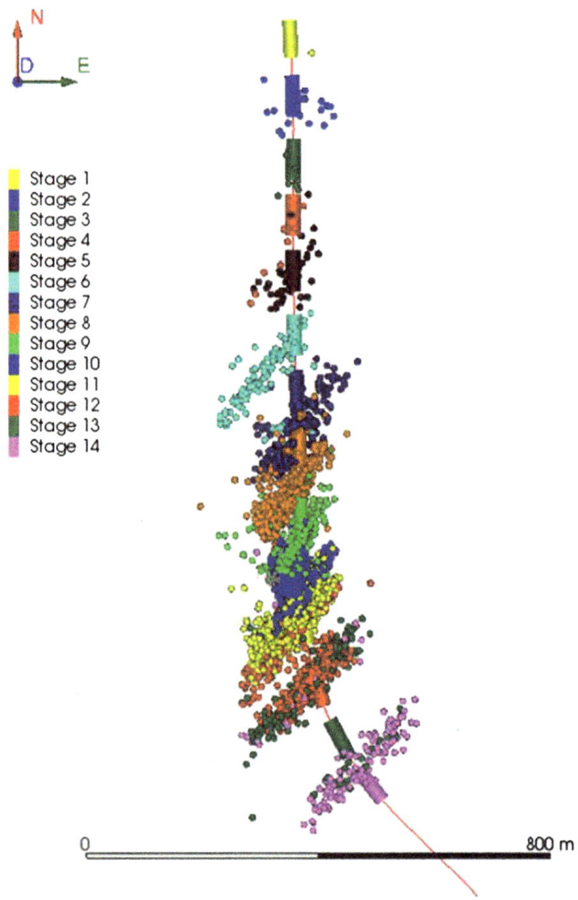

FIGURE 10.10

Microseismic events recorded during a 14-stage horizontal fracture stimulation. Events are colored by stage. (Source: https://www.esgsolutions.com)

Using a single monitoring well to capture microseismic activity associated with a 14-stage horizontal hydraulic fracture stimulation revealed information about local stress regimes and potential geological structures that may impact production efficiency in the tight oil field. During the fracture stimulation operations, it is very important to know where the fractures are being created in the reservoir. Measuring microseismic events that are occurring as the fracture stimulation takes place provides visual evidence that fractures are being developed both vertically and horizontally. The magnitude of seismic events created by hydraulic fracturing techniques is many times smaller than the events which can be felt at the surface. These models can be used to define the limit and reach of each fracture stimulation in the drilled wells. A schematic diagram illustrating fractures created during the multistage hydraulic stimulation process is shown in Fig. 10.11a; an illustration of the event is shown in 3D in Fig. 10.11b. The fracture stimulation example from Grande Prairie, Alberta, illustrated in Fig. 10.11c, shows opening of the fractures in a perpendicular direction and propagation of fractures in parallel to the direction of principal maximum stress. It is expected that fractures will open perpendicular to σ1, and propagate parallel to σ1. Knowledge of local stress conditions near the target formation suggested

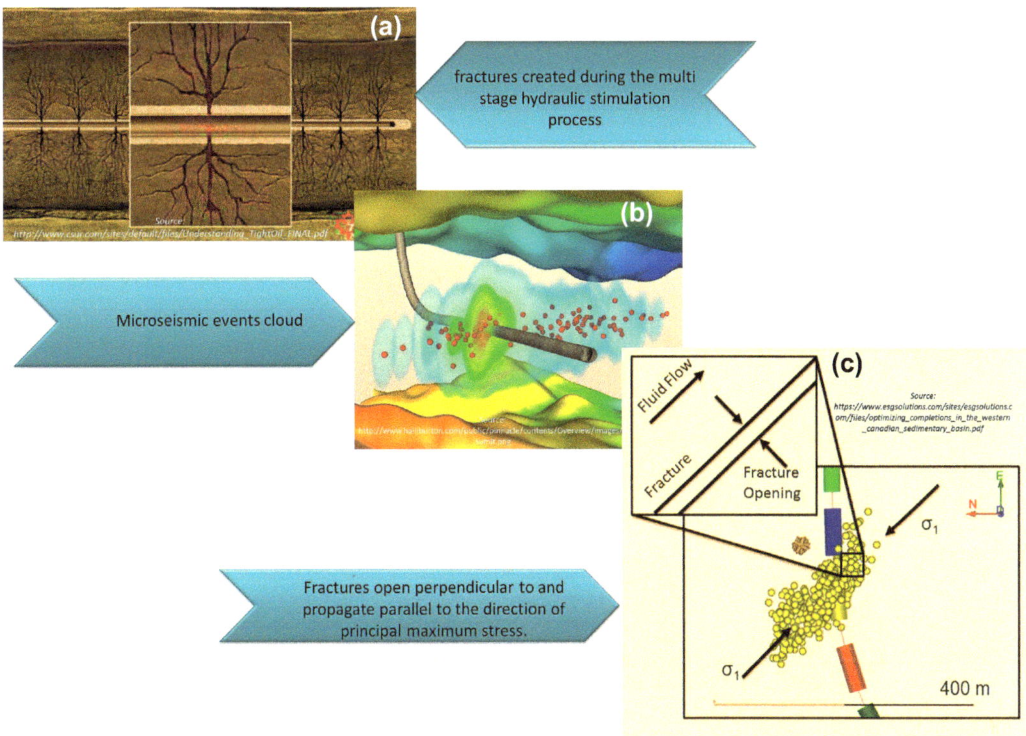

FIGURE 10.11

Multistage hydraulic stimulation, event cloud, and opening of the fractures. (a) A schematic diagram illustrating fractures created during the multistage hydraulic stimulation process. (b) 3D view of microseismic event cloud. (c) Opening of the fractures in perpendicular direction and propagation of fractures in parallel to the direction of principal maximum stress.

that σ1 typically trends with an azimuth of 50° E of N, therefore drilling future wells perpendicular to σ1 will maximize the fracture extent.

During the process of multistage hydraulic fracturing open pathways are created within the tight oil reservoir, allowing the oil to flow to the well-bore, and conventional methods are used to produce the well. If the objective of field development is to increase the overall recovery of oil from an existing field, an infill well should be the chosen option. These wells are located among the existing conventional wells with the purpose of extracting additional oil that has not been recovered using conventional production technology. The ultimate aim should be focused on expanding the boundaries of the "sweet spot" within the oil-bearing formation.

10.3.4 COAL BED METHANE OR COAL SEAM GAS

Coal bed methane (CBM) is also known as coal seam gas (CSG). CBM is an unconventional gas as it is contained in difficult-to-produce reservoirs, which require special completion, stimulation, and/or production techniques to achieve economic production. The first question that should be asked about this gas is "how is CBM or CSG created inside earth?" When plant materials such as roots, wood, and bark are deposited in swamps or swampy lakes, they undergo chemical and bacterial reactions and a deposit of peat is created. As the peat is buried deeper, under layers of sand and mud, over millions of years, it changes to brown coal, then bituminous coal, and eventually hard anthracite coal (coalification process) as shown in Fig. 10.12. CBM is generated by the conversion of plant material to coal through burial and heating. As "coalification" progresses, increasingly dense coal is formed, in order from low to high rank: lignite, sub-bituminous, bituminous, and anthracite. Biogenic methane (attributed to bacterial activity) is first to form, followed by thermogenic methane. Much of the methane generated by

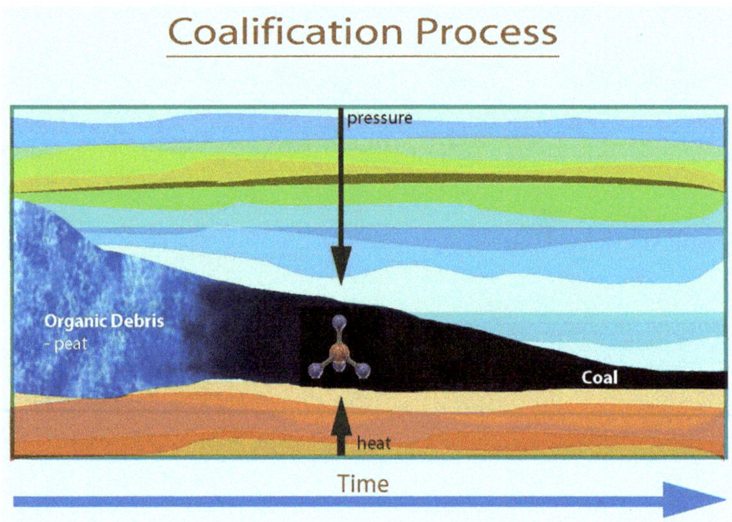

FIGURE 10.12

Coalification process. (Modified from http://www.energy.alberta.ca/OurBusiness/754.asp)

the coalification process escapes to the surface or migrates into adjacent reservoir or other rocks, but a significant volume remains trapped within the coal itself.

When the coal is formed, the decomposed organic material produces methane gas, as well as nitrogen, carbon dioxide, and additional gases. The pressure on coal increases which helps in storing the gas in the coal. CBM contains very little heavier hydrocarbons, such as propane or butane, and no natural gas condensate, which is often found in conventional natural gas. CBM may contain carbon dioxide, typically several percent or less.

Coal serves as both source and reservoir rocks. The principal trapping force for CBM is hydraulic pressure rather than a pressure seal or closed structure, which is common for conventional oil and gas. The methane content in coal seams generally increases the deeper the seam, and also with age. As the depth of the coal seam increases, so does the pressure level. This in turn reduces the level of permeability, causing the methane to be much more tightly bound to the coal and surrounding rock strata. Methane storage in coal beds is more complex than in most conventional reservoirs because of the unique properties of coal for gas storage. Methane storage in coal includes the "storage of methane as a free gas within the micropores and cleats", "as a dissolved gas in water which is present within coal"; as adsorbed gas held by molecular attraction on organic constituents that comprises the coal mass, micropores, and "cleat" in the coal; and as absorbed gas within the molecular structure of the coal molecules. Methane is highly combustible – its release can have serious implications for the safety of mine operations. It is also a potent greenhouse gas (GHG).

Cleats develop in the coal due to shrinking brought about by volatile loss during coalification, as well as due to tectonic stress. Primary cleats develop across coal layers and provide connections with noncoal interbeds. These cleat patterns are crucial for gas production because they allow for the release of adsorbed gas within coal beds and migration to the production well. Cleats can be destroyed by further burial in a process called healing or filled by secondary mineralization. Cleats are fractures that usually occur in two sets that are, in most instances, mutually perpendicular and also perpendicular to bedding (Laubach et al., 1998). Although pre-1990 geologic literature and current mining usage distinguishes these sets on the basis of factors such as "prominence" that are difficult to quantify, abutting relations between cleats generally show that one set predates the other as shown in Fig. 10.13.

The basic volumetric formula to determine original gas-in-place (OGIP) volumes for CBM is:

$$\mathrm{OGIP} = H \times (1 - (a + m)) \times \mathrm{GC} \times D \times A$$

where H = completable coal thickness, GC = gas content, a = ash content, D = coal density, m = moisture content, and A = drillable area.

CBM is extracted by drilling a well into a coal seam applying similar techniques used for other natural gas wells. The sides of the well are "cased" with cemented steel pipe. Usually, small holes, called perforations, are then made in the wall of the casing to let the CBM flow through into the well-bore and up the casing to the surface. To extract CBM a steel-encased hole is drilled into the coal seam, which may be 100 to 1500 m below ground. The well is equipped with a large water pump to lower the reservoir pressure within the coal seam. Water is produced to surface up the production tubing, whereas CBM gas is produced up the annulus. Produced gas is then sent to a compressor station and sold into natural gas pipelines (http://www.cbmasia.ca). Produced water is

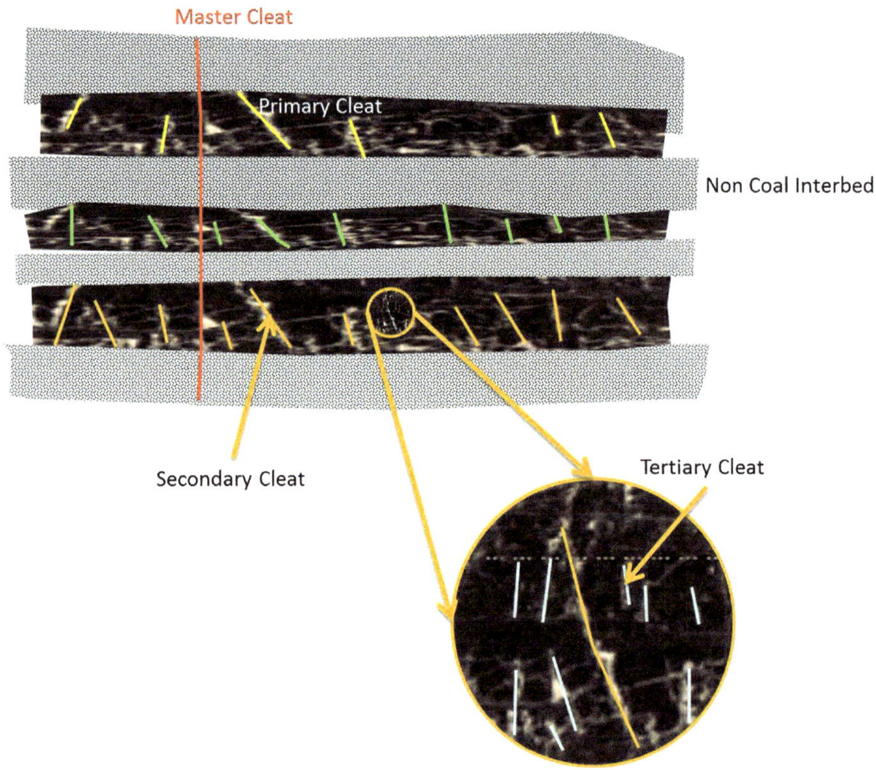

FIGURE 10.13

Cleat hierarchies in cross-section view. These conventions are used for cleat: length is dimension parallel to cleat surface and parallel to bedding; height is parallel to cleat surface and perpendicular to bedding; aperture is dimension perpendicular to fracture surface. Spacing between two cleats of same set is a distance between them at right angles to cleat surface. (Modified from Laubach and Tremain, 1991.)

either reinjected into isolated formations, released into streams, used for irrigation, or sent to evaporation ponds.

Methane is highly combustible – its release can have serious implications for the safety of mine operations. It is also a potent greenhouse gas (GHG). A range of technologies are available to recover methane from coal:

- Coal bed methane. Methane recovered from unmined coal seams. The coal seams may be mined in the future but this is largely dependent upon geological factors, such as coal depth and quality.
- Coal mine methane(CMM). Methane recovered during mining activities as the coal is in the process of being extracted and thus emitting significant quantities of the gas.
- Abandoned mine methane (AMM). Methane recovered from mines that have been abandoned following the completion of mining operations. Significant amounts of methane may remain trapped in the mine or may continue to be emitted from openings.

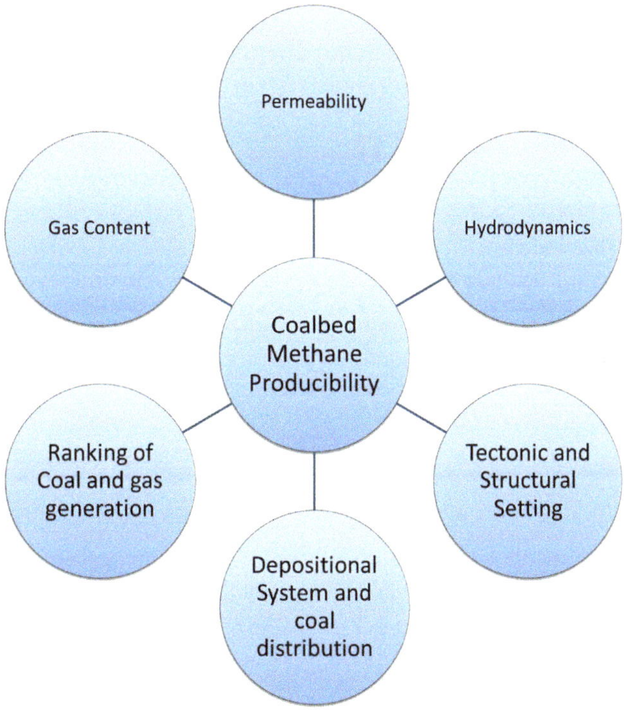

FIGURE 10.14

Development of a coal bed gas exploration and producibility model. (Modified from http://www.arcgroupllc.com/coalbed_methane.html)

Coal bed producibility model (http://www.arcgroupllc.com) can be applied to evaluation of coal bed gas exploration and producibility in coal basins throughout the world. The model indicates that tectonic/structural setting, depositional systems and coal distribution, coal rank, gas content, permeability, and hydrodynamics are controls critical to coal bed gas producibility (Fig. 10.14).

10.3.4.1 Key Parameters for Understanding CBM Plays

For successful reservoir characterization, the following characteristics need to be considered:

- Depth of coal seam
- Structural configuration
- Thickness of coal bed
- Areal extent of coals
- Organic matter type and richness (TOC)
- Maturation (R%)
- Density, porosity, and permeability
- Mineralogy
- Brittleness and cleat architecture
- Faults and fractures
- Pore pressure

- Stress and orientation
- Gas-in-place
- Aquifer strength
- Sorption isotherm characteristics
- Gas content
- Gas saturation

10.3.4.2 Seismic Attributes

Seismic response for a coal bed needs to be understood in detail. The following attributes prove useful for effective CBM exploration:

- Amplitude and frequency variation with thickness
- Spectral decomposition
- Seismic facies change
- Coherency, dip, and azimuth leads to mapping of fractures, faults, and cleats
- Quantitative seismic interpretation:
  - Poststack inversion
  - Prestack inversion
  - Stochastic inversion for detailed delineation of coal beds
  - Lithology prediction by seismic inversion derivatives
  - Geomechanical properties prediction

10.3.5 GAS HYDRATES

Methane hydrate is a class of clathrate, composed of water and low molecular weight gases, mainly methane, which forms under low temperature, high pressure, and suitable methane concentration. Hydrates with underlying free gas make a strong acoustic interface, which is often conspicuous in the seismic sections as bright reflections known as bottom-simulating reflectors (BSRs) and are interpreted as the primary geophysical indicator for the presence of gas hydrate. A BSR is identified on multichannel or multicomponent (4C) seismic sections as high-amplitude seismic reflection that mimics the seafloor with reverse polarity in comparison to the seafloor reflection, and being the thermodynamic phase boundary cross-cuts the bedding plane of sedimentary layers. A good example of a BSR is shown in Fig. 10.15.

Owing to very high temperatures below the BSR, the hydrates destabilize, so that in methane-saturated pore fluids, additional methane is in gas form, which is called free gas. Estimated seismic velocities may provide useful information about the probable model for the formation of BSR as well providing probable estimates of gas hydrates and free gas present in a region. The presence of hydrates in the pore spaces increases the seismic velocity and the presence of free gas in the sediment pores reduces the seismic velocity. However, such velocity estimations may be erroneous if the sedimentary rocks exhibit anisotropy (Crampin et al., 1984; Thomsen, 1986a,b). Transverse isotropy (TI) is the simplest form of anisotropy and is often described by Thomsen parameters ε, δ, and γ (Thomsen, 1986). Parameter ε is a measure of P-wave anisotropy, which is the normalized difference in horizontal and vertical velocities (Thomsen, 1986), whereas δ is more complex and has the strongest effect on P- and vertically polarized shear waves at intermediate angles. In this study, we disregard γ which only affects horizontally polarized shear waves. Seismic P-wave reflection data recorded on the surface are

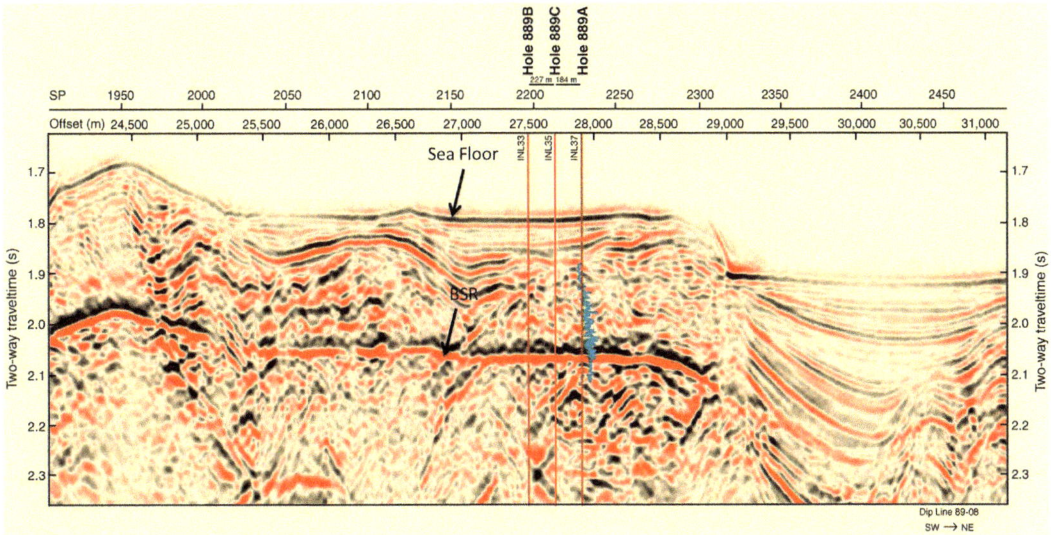

FIGURE 10.15

Multichannel and single-channel seismic lines showing BSR as a high amplitude and reverse polarity (with respect to seafloor) reflector. (Source: http://publications.iodp.org/scientific_prospectus/328/328_f3.htm)

not sufficient to estimate seismic anisotropy parameters, and as such vertical velocity and anisotropic parameters cannot be uniquely determined.

Many international gas hydrate research programs have been launched by different governments, including those in the United States, Southeast Asia, Japan, Australia, and Africa. India launched the National Gas Hydrate Program Expedition (NGHP), which is a consortium of National E&P companies (Oil and Natural Gas Corporation Ltd (ONGC), GAIL India Ltd, and Oil India Ltd (OIL)) and National Research Institutions (National Institute of Oceanography, National Geophysical Research Institute, and National Institute of Ocean Technology). Steered by the Ministry of Petroleum & Natural Gas and technically coordinated by the Directorate General of Hydrocarbons (DGH), the NGHP is given the responsibility to explore the gas hydrates in India. During its first expedition, the NGHP, through a consortium consisting of Overseas Drilling Ltd (ODL), Fugro McClelland Marine Geosciences, Geotek Ltd, Lamont-Doherthy Earth Observatory (LDEO) and scientists from numerous universities and national laboratories, struck rich deposits of massive gas hydrates present within the fractured shales in the Krishna Godavari (KG) Basin offshore area. The United States Geological Survey (USGS) was the main technical collaborator in this expedition.

The NGHP's efforts in gas hydrate exploration offshore India led to the following achievements:

- Conduction of comprehensive analyses of gas-hydrate-bearing marine sediments in both passive continental margins and marine accretionary wedge settings.
- Discovery of gas hydrates in numerous complex geological settings and collection of an unprecedented number of gas hydrate cores (more than 2800 m from 21 sites and 39 holes).
- Delineation and sampling of one of the richest marine gas hydrate accumulations discovered in the world to date (Krishna-Godovari basin).

- Discovery of one of the thickest and deepest gas hydrate occurrences yet known (Andaman Islands), which revealed gas-hydrate-bearing volcanic ash layers as deep as 600 m below the seafloor.
- Establishment of the existence of a fully developed gas hydrate system in the Mahanadi basin of the Bay of Bengal.

10.3.5.1 Recent Advances in Gas Hydrate Production

Japan has completed its first offshore methane hydrate production test. A test to dissolve methane hydrate and extract natural gas will be conducted off the coast of Japan, from Atsumi Peninsula to Shima Peninsula from January to March, 2013, as a part of the technical preparation that is required for commercializing methane hydrate. The production test site is located on the margin of the Daini Atsumi Knoll, off the coasts of Atsumi and Shima peninsulas, in the eastern Nankai Trough, Japan. The site was selected based on seismic and well data collected from 2001–2008 that indicate methane hydrate-rich sedimentary layers in this area. In the 21st century, the energy industry will witness a metamorphosis of current business models to incremental technology risk and a new way of "working together" models, which will lead to a "structural change" in the industry. The digital era of integration has arrived and there is a lot to learn from past mistakes. Collaboration and integration from all subsurface disciplines are key to optimizing project value while reducing uncertainties and associated risks. From past studies, five broad themes arise that are seen as the pillars necessary for consistently strong results with successful projects for conventional and conventional hydrocarbon reservoir characterization:

- Rock geometries. The best way to determine rock geometries is to establish depositional environment and focus on stratigraphic/structural patterns.
- Rock properties. Integration of wireline information and seismic interpretation offers rock property prediction by seismic inversion.
- Reservoir modeling. Integration of seismic inversion results with the geological and geocellular models provides predictive solutions.
- Reservoir simulation. This pillar calibrates the geological model against dynamic production scenarios.
- Strategy and management. Optimum results can be achieved with clarity of vision and goals which are clearly articulated throughout the life cycle of a field.

10.4 TOWARDS ADVANCED QUANTITATIVE INTERPRETATION METHODS

The progression from qualitative to quantitative interpretation methods is critical to establishing the long-term value of 3D and 4D seismic data integration with geological information together with engineering and petrophysical data. Qualitative seismic attributes such as seismic amplitude, envelope, root mean square (RMS) amplitude, spectral magnitude, relative impedance, and amplitude versus offset (AVO) measures are directly sensitive to the change in acoustic properties of sediments. Some qualitative attributes including peak-to-trough thickness, peak frequency, and bandwidth are sensitive to layer thicknesses. Attributes can be computed from prestack or from poststack data, before or after time migration. Prestack seismic attributes use seismic data to gather information on directional (azimuth) and offset. Poststack seismic attributes use stacked data and can be further classified into instantaneous, wavelet-based, physical attribute, geometry-based, reflective, and transmissive attributes. Future trends for qualitative seismic attributes include joint interpretation of combined attributes by using rule-based methods, statistical methods, or deterministic methods. Rule-based attributes are calculated by blending of physical and geometric attributes.

If the rock properties such as acoustic impedance (AI), compressional and shear velocity ratios (V_p/V_s), and density could be inverted without uncertainty so that a geologist can use these properties in building a static model, this information could facilitate the tie of seismic data to production data and help the industry to refine reservoir model history matching. Sophisticated software packages make the conversion of engineering measures to seismic data by using simulation-to-seismic modeling relatively easy, whereas the inversion problem of converting from seismic measures to engineering measures is nonunique (multiple acceptable solutions can exists for one problem), complex, and scale dependent. The trend is now for complete "seismic-to-simulator" workflows and this is equally applicable to conventional and unconventional reservoir exploration and development. Seismic measures of a reservoir include:

- AVO attributes
- Attenuation
- Anisotropy
- Elastic rock properties
- Travel time error
- Velocity change

Engineering measures of the reservoir include:

- Overburden stress and strain
- Saturation
- Pressure
- Compaction
- Fluid density
- Bulk modulus

The oil industry is moving to "earth-model-centric" workflows where the geological model is continuously updated throughout the life cycle of the field by incorporating any new data. This updated model is the key for future field development decisions. Seismic information is continuously used during inversion for prediction of elastic properties in the time domain. Inversion derivatives are then domain converted from time to depth and transferred into a static model framework, where seismic attributes are used to guide the interpolation of reservoir properties between the wells by using geostatistics. From an industry perspective, these techniques are well established and the question remains about the remaining challenges and future trends for seismic data integration into production data. Some of the future challenges and new directions are as follows:

- Use of full-waveform inversion for more accurate and higher resolution estimates of velocity, density, attenuation, and anisotropy changes.
- An important benchmark remains the integration of seismic inversion derivatives into the earth-model framework:
  - Seismic inversion results must be depth converted.
  - Consistent geometry with reservoir model:
    - Even if the inversion is performed in a stratigraphic grid, the grid geometry is always different from that of the reservoir model and seismic inversion results must be resampled into the reservoir model grid; this often proved to be a tedious and time-consuming process.

  - In future, the inversion should be performed directly into the depth domain and integrated directly with the reservoir model so much higher coupling between seismic inversion and geomodeling technologies is needed.
- Constraining a vertically detailed model with band-limited seismic data remains in the development stage. This is directly related to downscaling and is a focus of future research direction.
- Performing stochastic inversion in a meaningful manner still remains a challenge. Although many softwares are available that use geostatistical methods for inversion, the challenge to extract the rock properties directly in depth remains open. The promise has not been fulfilled so far. The workflows for stochastic inversion need to be developed and optimized for exploiting the multiple elastic realizations into the static model framework.
- Application of coil seismic acquisition (full azimuth), wide-azimuth, and multicomponent (4C) data into reservoir characterization, time-lapse changes monitoring, geomechanical properties prediction.
- Development in digital rock physics and statistical rock physics is fulfilling the promise, and the combination of rock physics workflows is successfully used with geostatistical methods to model the uncertainty. Future trends can be seen towards building rock physics-based seismic inversion softwares integrated with static modeling workflows.
- The challenges for unconventional reservoirs can be highlighted as follows:
  - Developing a geological understanding of unconventional reservoirs.
  - For unconventional reservoirs, predicting performance is challenging and fully integrated, scalable tools are needed to develop better understanding.
  - For effective exploration and development purposes, five pillars should be used to integrate the data from geology, geophysics, petrophysics, reservoir engineering, and production.
  - Designing hydraulic fracturing treatments to meet specific shale characteristics requires the precise application of a broad scope of expertise and technologies. Volumes of detailed formation data must applied to engineering the optimal fracture design. Service companies including three top providers (Schlumberger, Halliburtion, and Baker Hughes Incorporated) are continuously developing the sophisticated tools for precise execution of hydraulic fracturing, treatment, multistage fracturing design development, and optimized production.
  - Optimizing the location of horizontal wells and execution remain the primary challenges, and there are many failures. Future trends include minimization of failures and improving success rates by managing the horizontal boreholes with increased drilling efficiencies.

10.5 FINAL THOUGHTS AND SOME REMARKS

The oil industry is driven by its "return on investments". For every unconventional reservoir exploration and development project, the management asks, "What is the total cost and what is the expected return?" and in the current low oil price era it is becoming very difficult to get a blessing from higher management on investment decisions. The question should be rephrased by technical professionals integrated with commercial teams to, for example, "What is the value proposition if new 3D or 4D seismic is acquired and applied to the seismic-to-simulation workflow?" Therefore, the important word here is "integration". This needs to be applied in an effective way to maximize the value of technical and commercial workflows. Some workflows are available to help understand the investment strategy, e.g., a value of information exercise should provide information on cost.

To conclude, let us review the key factors that determine the success of exploration and development of an unconventional reservoir:

- Include multicomponent seismic projects as part of the field development and surveillance plans. Budgeting for planning of seismic and time constraints for the field plans offers tremendous help in optimizing the impact of field management.
- Define the business and technical objectives and expectations separately before the real field experiment.
- To fully understand the reservoir distribution, advanced quantitative seismic interpretation methods including pre- and poststack inversion, stochastic inversion integration with static model, and seismic measures integration with production data should be tested and implemented on a field scale.
- Develop knowledge of rock physics in order to understand reservoirs.
- Acquire and process seismic data with the highest resolution.
- A multidisciplinary team should be deployed with the understanding that interpretation of seismic data is the key.
- Conditioning of geological models with seismic data helps in reducing the uncertainties and making key decisions for field development purposes.

Gas hydrates should be considered as a future energy resource and the research should focus on extraction of methane with minimum uncertainty in production and in an environmentally friendly manner. Successful unconventional reservoir development requires time and input from dedicated and expert professionals, as well as support from all management and technical levels.

BIBLIOGRAPHY

Aguilera, R., May 1 2014. Flow Units: From Conventional to Tight-Gas to Shale-Gas to Tight-Oil to Shale-Oil Reservoirs. Society of Petroleum Engineers. http://dx.doi.org/10.2118/165360-PA.

Aller, R.C., 2001. Transport and reactions in the bioirrigated zone. In: Boudreau, B.P., Jorgensen, B.B. (Eds.), The Benthic Boundary Layer. Oxford University Press, Oxford, pp. 260–301.

Aller, R.C., Blair, N.E., 2004. Early diagenetic remineralization of sedimentary organic C in the Gulf of Papua deltaic complex (Papua New Guinea): net loss of terrestrial C and diagenetic fragmentation of C isotopes. Geochim. Coschim. Acta 68, 1815–1825.

Aller, R.C., Blair, N.E., 2006. Carbon remineralization in the Amazon-Guiana's tropical mobile mud belt: a sedimentary incinerator. Cont. Shelf Res. 26, 2241–2259.

Anderson, D.W., 1988. The effect of parent material and soil development on nutrient cycling in temperate ecosystems. Biogeochemistry 5, 71–97.

Archer, D.E., Morford, J.L., Emerson, S.R., 2002. A model for suboxic sedimentary diagenesis suitable for automatic tuning and gridded global domains. Glob. Biogeochem. Cycles 16. http://dx.doi.org/10.1029/2000GB001288.

Bastviken, D., Persson, L., Odham, G., Tranvin, L., 2004. Degradation of dissolved organic matter in oxic and anoxic lake waters. Limon. Oceanogr. 49, 109–116.

Benner, R., 1998. Cycling of dissolved organic matter in the oceans. In: Hessen, D.O., Tranvik, L.J. (Eds.), Aquatic Humic Substances: Ecology and Biogeochemistry. Springer, New York, pp. 317–331.

Benzing- Prudie, L., Ripmeester, J., Preston, C., 1983. Elucidation of nitrogen forms in the melanoidins and humic acid by nitrogen-15 cross polarization-majic angle spinning nuclear magnetic resonance spectroscopy. J. Agri. Food Chem. 31, 913–915.

Berner, R.A., 1978. Sulfate reduction and rate of deposition of marine sediments. Earth Planet. Sci. Lett. 37, 492–498.

Betts, J.N., Holland, N.D., 1991. The oxygen content of ocean bottom waters, the burial efficiency of organic carbon and regulation of atmospheric oxygen. Palaeogeogr. Palaeoclimatol. Palaeoecol. 97, 5–18.

Biscay, P.E., Eittreim, S.L., 1977. Suspended particle loads and transport in the Nepheloid layer of the abyssal Atlantic Ocean. Mar. Geol. 23, 155–172.

Boyd, P.W., Newton, P.P., 1999. Does plankton community structure determine downward particulate organic carbon flux in different oceanic provinces? Deep Sea Res. 46, 63–91.

Boyer, E.W., Swistock, B.R., Clark, J., Madden, M., Rizzo, D.E., 2011. The impact of Marcellus gas drilling on rural drinking water supplies. The Center for Rural Pennsylvania. http://www.rural.palegislature.us/documents/reports/Marcellus_and_drinking_water_2012.pdf.

Burdige, D.J., 2006. Geochemistry of Marine Sediments PhD thesis. Princeton University Press, Princeton, NJ.

Burdige, D.J., 2007. Preservation of organic matter in marine sediments: controls mechanisms, and an imbalance in sediment organic carbon budget? Chem. Rev. 107, 467–485.

Calvert, S.E., Pedersen, T.F., 1992. Organic carbon accumulation and preservation in marine sediments: how important is anoxia. In: Whelan, J.K., Farrington, J.W. (Eds.), Productivity, Accumulation, Preservation of Organic Matter in Recent and Ancient Sediments. Columbia University Press, New York, pp. 231–263.

Canfield, D.E., 1993. Organic matter oxidation in marine sediments. NATO ASI Ser. 14, 333–363.

Canfield, D.E., 1994. Factors influencing carbon preservation in marine sediments. Chem. Geol. 114, 315–319.

Chester, R., 1990. Marine Geochemistry. Unwin Hyman Ltd, London.

Davies, T.A., Gorsline, D.S., 1975. Oceanic sediments and sedimentary processes. In: Riley, J.P., Chester, R. (Eds.), Chemical Oceanography. vol. 5. Academic Press, New York, pp. 1–80.

Dean, W.E., Gardner, J.V., Anderson, R.Y., 1994. Geochemical evidence of enhanced preservation of organic material in the oxygen minimum zone of the continental margin of North California during the late Pleistocene. Paleoceanography 9, 47–61.

Donovaro, R., et al., 2008. Major viral impact on the functioning of benthic deep-sea ecosystems. Nature 454, 1084–1087.

Doyle, L.J., Garrels, R.M., 1985. What does percent organic carbon in sediments measure? Geol. Mar. Let. 5, 51–53.

Duce, R.E., Duursma, E.K., 1977. Inputs of organic matter to the ocean. Mar. Geochem. 5, 319–339.

Emery, O.K., 1969. The continental shelves. Sci. Am. 221, 106–122.

Eppley, R.W., Peterson, B.J., 1979. Particulate organic matter flux and planktonic new production in the deep ocean. Nature 282, 677–680.

Flood, R., 1978. Studies of Deep Sedimentary Microtopography in the North Atlantic Ocean PhD dissertation. Wood Hole Oceanographic Institute and Massachusetts Institute of Technology.

Folger, D.W., 1970. Wind transport of land derived mineral, biogenic and industrial matter over the North Atlantic. Deep Sea Res. 17, 337–352.

Gardner, W.D., 1978. Fluxes, Dynamics, and Chemistry of Particulates in the Ocean PhD thesis. Wood Hole Oceanographic Institute and Massachusetts Institute of Technology.

Garrels, R.M., Mackenzie, F.T., 1971. Evolution of Sedimentary Rocks. Norton, New York.

Harnett, H.E., Keil, R.G., Hedges, J.H., Devol, A.H., 1998. Oxygen exposure as a controlling factor in preservation of organic carbon in continental margin sediments. Nature 39, 572–574.

Hartmann, M., Müller, P.J., Suess, E., van der Weijden, C.H., 1976. Chemistry of Late Quaternary sediments and their interstitial waters from the NW African continental margin. Meteor Forsch. C 24, 1–67.

Harvey, G.R., Boram, D.A., Chesal, L.A., Tokar, J.M., 1983. The structure of marine fluvic and humic acid. Mar. Chem. 12, 119–132.

Hasse, L., 1983. Introductory meteorology and fluid mechanics. In: Liss, P.S., Slin, W.G.N. (Eds.), Air–sea Exchange of Gases and Particles. Reidel, Dordrecht, pp. 1–51.

Haszeldine, 2012. Shale gas, NW England earthquakes, and Uk regulation. Briefing note for Department of Energy and Climate Change Science Advisory Group. http://www.decc.gov.uk/assets/decc/11/about-us/science/5223-shale-gas-nw-england-earthquakesbriefing.pdf.

Hay, W., Southam, J., 1977. Modulation of marine sedimentation by continental shelf. In: Anderson, N.R., Malahoff, A. (Eds.), The Fate of Fossil Fuel CO_2 in the Ocean. Plenum Press, New York, pp. 569–604.

Hayes, D.E., et al., 1975. In: Initial Reports of the Deep Sea Drilling Project Leg. 28, vol. 28. US Government Printing Office, Washington, DC.

Hedges, J.I., Kiel, R.G., 1995. Sedimentary organic matter preservation: an assessment and speculative synthesis. Mar. Chem. 49, 81–115.

Hedges, J.I., Mann, D.C., 1979. The characterization of plant tissues by their lignin oxidation products. Geochim. Cosmochim. Acta 43, 1803–1807.

Hedges, J.I., Oades, J.M., 1997. Comparitive organic geochemistries of soils and marine sediments. Org. Geochem. 27, 319–361.

Hedges, J.I., Keil, R.G., Benner, R., 1981. What happens to land-derived organic matter in the ocean? Org. Geochem. 27, 195–212.

Hedges, J.I., et al., 2000. The molecularly uncharacterized component of nonliving organic matter in natural environments. Org. Geochem. 31, 945–958.

Hennings, et al., 2012. Relationship between fractures, fault zones, stress, and reservoir productivity in the Suban gas field, Sumatra, Indonesia. Am. Assoc. Pet. Geol. Bull. 96 (4), 753–772. http://doi.aapg.org/data/open/offer.do?target=%2Fbu lletns%2F2012%2F04apr%2FBLTN09084%2FIMAGE S%2FBLTN09084.PDF.

Hjulström, F., 1939. Transportation of detritus by moving water. In: Trask, P.D. (Ed.), Recent Marine Sediments. American Association of Petroleum Geologists, Tusla, OK, pp. 5–30.

HMSO, 1985. The Abbeystead Explosion: A Report of the Investigation by the Health & Safety Executive into the Explosion on 23 May 1984 at the Valve House of the Lunce/Wyre Water Transfer Scheme at Abbeystead. Publisher: HSE Books (December 1985). ISBN-13: 978-0118837958.

HoC, 2011. In: Shale Gas: Fifth Report of Session 2010–12. vol. I. House of Commons, London. http://www.scribd.com/doc/59299635/House-ofCommons-Shale-Gas.

Holeman, J.N., 1968. The sediment yield of major rivers. Water Resour. Res. 6, 737.

Holmes, E.H., Devalapally, H., Li, L., Perdue, M.L., Ostrander, G.K., 2013. Permeability Enhancers Dramatically Increase Zanamivir Absolute Bioavailability in Rats: Implications for an Orally Bioavailable Influenza Treatment. PLoS ONE 8 (4), e61853. http://dx.doi.org/10.1371/journal.pone.0061853.

Hunt, J.M., 1996. Petroleum Geochemistry and Geology. W.H. Freeman, New York.

Ingall, E.D., Cappellen, P.V., 1990. Relation between sedimentation rate burial of organic phosphorus and organic carbon in marine sediments. Geochim. Cosmochim. Acta 54, 373–386.

Ingall, R.D., Bustin, R.M., Van Cappallen, P., 1993. Influence of water-column anoxia on the burial and preservation of carbon and phosphorus in marine shales. Geochim. Cosmochim. Acta 57, 303–316.

ISRS, 2012. ISRS: for the health of your business. International Safety Rating System Det Norse Veritas http://viewer.zmags.com/publication/e94fa62a#/e94fa62a/1.

Jiao, N., et al., 2010. Microbial production of recalcitrant dissolved organic matter: long-term global carbon storage in the global ocean. Nat. Rev. Microbiol. http://dx.doi.org/10.1038/nrmicro2386.

Keil, R.G., Hu, F.H., Tsamaki, E.C., Hedges, R.I., 1994. Pollen in marine sediments as indicator of oxidation of organic matter. Nature 369, 639–641.

Keil, R.G., Cowie, G.L., 1999. Organic matter preservation through oxygen deficient zone of NE Arabian Sea as discerned by organic carbon: mineral surface area ratio. Mar. Geol. 161, 13–22.

Kennet, J., 1982. Marine Geology. Prentice-Hall, Inc, Eaglewood Cliff, NJ.

King, 2010. Thirty years of shale gas fracturing: what have we learned? Society of Petroleum Engineers Annual Technical Conference and Exhibition, Florence. http://www.spe.org/atce/2010/pages/schedule/tech_ program/documents/spe1334561.pdf.

King, 2012. Hydraulic fracturing 101: what every representative, environmentalist, regulator, reporter, investor, university researcher, neighbour and engineer should know about estimating frac risk and improving frac performance in unconventional gas and oil wells. Soc. Pet. Eng. http://www.kgs.ku.edu/PRS/Fracturing/Frac_Paper_ SPE_152596.pdf.

Knicker, H., Hatcher, P.G., 2001. Sequestration of organic nitrogen in the sapropel from Mangrove Lakes Bermuda. Org. Geochem. 32, 733–744.

Kristensen, E., Balckburn, T.H., 1987. The fate of carbon and nitrogen in experimental marine system: influence of bioturbation and anoxia. J. Mar. Res. 45, 231–257.

Kuuskraa, V.A., Steven, S., Van Leeuwen, T., Moodhe, K., 2011. World shale gas resources: An initial assessment of 14 regions outside united states. Washington DC US DOE EIA. http://www.adv-res.com/pdf/ARI%20EIA%20Intl%20Gas%20Shale%20APR%202011.pdf.

Lee, C., Wakeham, S., 1989. Organic matter in seawater biogeochemical process. In: Riley, J.P. (Ed.), Chemical Oceanography, Vol. IX. Academic Press, London, pp. 1–51.

Legendre, L., Rassoulzadegan, F., 2008. Biological Oceanography. In: Jacques, C.J., Chen, C.T.A. (Eds.), Oceanography. Encyclopedia of Life Support Systems (EOLSS), developed under the auspices of the UNESCO Eolss Publishers, Oxford, UK, pp. 1–35.

Levin, L.A., 2003. Oxygen minimum benthos: adaptation and community response to hypoxia. In: Gibson, R.N., Atkinson, R.J.A. (Eds.), Oceanography and Marine Biology: An Annual Review vol. 41, pp. 1–45. Taylor & Francis, UK.

Lisle and Srivastava, 2004. Test of the frictional reactivation theory for faults and validity of fault-slip analysis. Geology. 32, 569–572. http://geology.gsapubs.org/content/32/7/569.abstract.

Mayer, L.M., Macko, S.A., Cammen, L., 1988. Provenance concentration and nature of sedimentary organic nitrogen in Gulf of Maine. Mar. Chem. 25, 491–504.

McCarthy, M., Hedges, J.I., Benner, R., 1998. Major bacterial contribution to marine dissolved organic nitrogen. Science 281, 231–234.

McCave, I.N., 1975. Vertical flux of particles in the ocean. Deep Sea Res. 22, 297–310.

McKenzie, et al., 2012. Human health risks assessment of air emissions from development of unconventional natural gas resources. Sci. Total Environ. 424, 79–87. http://frackingfreeireland.org/wp-content/uploads/2011/08/Human_health_risk_assessment_ of_air_emissions1.pdf.

Menzel, D.W., 1974. Primary Productivity, Dissolved and Particulate Organic Matter and the Sites of Oxidation of Organic Matter. In: Goldberg, E.D. (Ed.), The sea. Wiley, New York, pp. 654–678.

Milliman, J.D., Meade, R.H., 1983. World-wide delivery of river sediments to the oceans. J. Geol. 91, 1–21.

Mills, K.W., 2012. Observations of ground movements within the overburden strata above longwall panels and implications for groundwater impacts. In: Proceedings of the 38th Symposium on the Advances in the Study of the Sydney Basin, May 10–11, 2012, Hunter Valley. Strata Control Technology.

Mincy, G., 2012. Effects of coal mining on freshwater mussels in the New River basin: a test using Corbicula as biomonitors. Abstr. Progr. Geol. Soc. Am. 44 (7), 465.

Mincy, G., McKinney, M.L., 2012. Effects of coal mining on survivorship and taphonomy of the Asian clam. Abstr. Progr. Geol. Soc. Am. 44 (4), 76.

Molnar, P., England, P., 1990. Late Cenozoic uplift of mountains ranges and glacial climate changes, chicken or egg. Nature 346, 29–34.

Molofsky, et al., 2011. Methane in Pennsylvania water wells unrelated to Marcellus shale fracturing. Oil Gas J. http://www.cabotog.com/pdfs/MethaneUnrelatedtoFracturing.pdf.

Mondal, S., Wickramasinghe, S.R., 2012. Photo-induced graft polymerization of N-isopropyl acrylamide on thin film composite membrane: produced water treatment and antifouling properties. Sep. Purif. Technol. 90, 231–238.

Moore, 2012. Gas works? Shale Gas and its Policy Implications. Policy Exchange, London. http://www.policyexchange.org.uk/images/publications/gas%20works%20-%20feb%2012.pdf.

Moore, C., Doherty, C., 2012. Upscaling Hydraulic Processes and Properties to Assess Impacts on Groundwater from Coal Seam Gas Abstractions. Hydrology and Water Resources Symposium 2012, November 19–22, 2012, Sydney.

Moser, A., Morris, R., 2012. New insights into the hydrodynamics of the Surat basin. Int. Geol. Congr. Abstr. 34.

Müller, P.J., 1975. Zur Diagenese stickstoffhaltiger Substazen in marine sedimenten unter oxydierenden und reduzierenden Bedingungen. Dissertation. Universitat Kiel.

Müller, P.J., Suess, E., 1979. Productivity, sedimentation rate and sedimentary organic matter in the oceans. I. Organic carbon preservation. Deep Sea Res. 26, 1347–1362.

Murray, J., 2012. Coal Seam Gas Water – River Discharge: Context Matters. SPE/APPEA International Conference on Health, Safety and Environment in Oil and Gas Exploration and Production 2012: Protecting People and the Environment – Evolving Challenges, September 11, 2012 to September 13, 2012. Origin Energy: Society of Petroleum Engineers, Perth, WA, Australia, pp. 2335–2342.

Namoi, C.M.A., 2013. Assessing the cumulative risks of mining and extractive industries to the Namoi Catchment's natural resource assets [Online]. Namoi Catchment Management Authority. Available at: http://www.namoi.cma.nsw.gov.au/41885.html?5.

Nguyen, R.T., Harvey, H.R., 2001. Preservation of protein in marine systems: hydrophobic and other noncovalent association of major stabilizing forces. Geochim. Cosmochim. Acta 69, 167–180.

Nosenzo, G., Whelan, B.E., Brunton, M., Kay, D., Buys, H., 2013. Continuous monitoring of mining induced strain in a road pavement using fiber Bragg grating sensors. Photonic Sens. 3 (2), 144–158.

Ogawa, H., Amagai, Y., Koke, T., Kaiser, K., Benner, R., 2001. Production of refractory dissolved organic matter by bacteria. Science 29, 217–220.

O'Leary, W.G., Nawrot, J.R., 2012. Stream reconstruction under SMCRA burning star #4: a case study. 29th Annual national Conference of the American Society of mining and Reclamation 2012, ASMR 2012, June 8, 2012 to June 15, 2012, Tupelo, MS, USA. Illinois Department of Natural Resources, Office of Mines and Minerals, Land Reclamation Division, Benton, IL, 62959, USA: American Society of Mining and Reclamation, pp. 406–423.

Ondrey, G., 2012. A treatment process for coal-bed-methane-extraction water. Chem. Eng. 119 (7), 10.

Orem, W.H., Tatu, C., Crosby, L.M., et al., 2012. Water chemistry in areas with surface mining of coal, West Virginia, Usa. Abstr. Progr. Geol. Soc. Am. 44 (7), 338.

Pagani, et al., 2010. High earth-system climate sensitivity determined from Pliocene carbon dioxide concentration. Nat. Geosci. 3, 27–30.

Pan, J., Zhang, X., Ju, Y., Zhao, Y., Bai, H., 2013. Stable isotope and water quality analysis of coal bed methane produced water in the southern Qinshui Basin, China. Membr. Water Treat. 4 (4), 265–275.

Paphitis, et al., 2002. A quantitative study of microbial decomposition of biopolymers in recent sediments from the Peru Margin. Mar. Geol. 113, 55–66.

Parkes, et al., 1993. A quantitative study of microbial decomposition of biopolymers in Recent sediments from Peru Margin. Mar. Geol. 113, 55–66.

Pederson, T.F., Calvert, S.E., 1990. Anoxia vs productivity: what controls the formation of organic carbon-rich sediments and sedimentary rocks? Bull. Am. Assoc. Pet. Geol. 74, 454–466.

Pratt, L.M., 1984. Influence of paleoenvironmental factors in the preservation of organic matter in Middle Cretaceous Greenhorn formation Pueblo Colorado. Bull. Am. Assoc. Pet. Geol. 68, 1148–1159.

Premuzic, E.T., Benovitz, C.M., Gaffney, J.S., Walsh, J.J., 1982. The nature and distribution of organic matter in near surface sediments of world oceans and Seas. Org. Geochem. 4, 63–77.

Procter, L.M., Fuhrman, J.A., 1990. Viral mortality of marine bacteria and cynobacteria. Nature 343, 60–62.

Ransom, B., Kim, D., Kastner, M., Wainwright, S., 1998. Organic matter preservation on continental slopes: importance of mineralogy and surface area. Geochim. Cosmochim. Acta 62, 1329–1345.

Reichart, G.L., Lourens, L.J., Zachariasse, W.J., 1998. Temporal variability in the northern Arabian Sea oxygen minimum zone (OMZ) during the last 225,000 years. Paleoceanography 13, 607–621.

Russel, R.D., 1939. Effects of transportation on sedimentary particles. In: Trask, P.D. (Ed.), Recent Marine Sediments. American Association of Petroleum Geologists, Tusla, OK, pp. 32–47.

Sackett, W.M., Eckelmann, W.R., Bender, M.L., Be, A.W.H., 1965. Temperature dependence of carbon composition in marine planktons and sediments. Science 148, 235–237.

Sarmiento, J.L., Herbert, T.D., Toggweiler, J.R., 1988. Causes of anoxia in the world ocean. Glob. Biogeochem. Cycles 2, 115–128.

Schlunz, B., Schneider, R.R., 2000. Transport of terrestrial organic carbon to oceans by rivers: re-estimating flux and burial rates. Int. J. Earth. Sci. 88, 599–660.

Schumm, S.A., 1963. The disparity between present rate of denudation and orogeny. U.S. Geol. Soc. Prof. Paper, 454-H 13.

Shaw, J., 1977. Till today morphology and structure related to glacier flow. Boreas 6, 189–201.

Stein, R., 1990. Organic carbon content/sedimentation rate relationship and its paleoenvironmetal significance for marine sediments. Geo. Mar. Lett. 10, 37–44.

Stevenson, F.J., Cheng, C.N., 1972. Organic geochemistry in Argentine Basin sediments: carbon-nitrogen relationship and Quaternary correlations. Geochim. Cosmochim. Acta 36, 653–671.

Strand, J., Freij-Ayoub, R., Ahmed, S., 2012. Simulating the impact of coal seam gas water production on aquifers. APPEA J. Conf. Proc. 52, 545–552.

Su, K.H., Liu, H., Wang, J., 2012. Borehole stability analysis and its application in Xujiahe formation of ZJ Block. In: Wu, J., Yang, J., Nakagoshi, N., Lu, X., Xu, H. (Eds.), Natural Resources and Sustainable Development, Ii, Pts 1–4. Trans Tech Publications Ltd, Stafa-Zurich, Switzerland, pp. 1359–1362.

Sun, Y.X., Xie, J.B., Zhao, J.Y., 2012. Study on borehole stability in Haita basin. In: Zhang, H.X., Han, Y., Chen, F., Wen, J. (Eds.), Materials and Computational Mechanics, Pts 1–3. Trans Tech Publications Ltd, Stafa-Zurich, Switzerland, pp. 1749–1753.

Sunborg, A., 1956. The river Klaralven, a study of fluvial processes. Geopgr. Ann. 38, 125–316.

Suttle, C.A., Chan, A.M., Cotrell, M.T., 1990. Infection of phytoplankton by viruses and reduction of primary productivity. Nat. Rev. Microbiol. 5, 801–812.

Swayne, 2012. Regulating coal seam gas in Queensland: lessons in an adaptive environmental management approach? Environ. Plan Law J. 29 (2).

Sweatman, R., Young, R., 2013. Modeling Reveals Hidden Conditions that Can Impair Wellbore Stability and Integrity. SPE/IADC Drilling Conference and Exhibition 2013, March 5, 2013 to March 7, 2013. Society of Petroleum Engineers (SPE), Amsterdam, Netherlands/Halliburton, United States, pp. 787–796.

Swift, G., 2013. Relationship between Joint Movement and Mining Subsidence. Springer Berlin Heidelberg, pp. 1–14. http://dx.doi.org/10.1007/s10064-013-0539-7.

Sydney Catchment Authority, 2012. Literature Review: Coal Seam Gas Impacts on Water Resources. http://www.waternsw.com.au/__data/assets/pdf_file/0011/58187/CSG-Impacts-on-Water-Resources-Draft-Revision-4.pdf.

Tan, Q., Deng, J.G., Yu, B.H., 2013. Study on Borehole Stability of Unconsolidated Sandstone in Depleted Reservior. Sustainable Development of Natural Resources, Pts 1–3. Trans Tech Publications Ltd, Stafa-Zurich, Switzerland, pp. 720–725.

Thiel, S., Peacock, J., Heinson, G., Hatch, M., Reid, P., 2012. Fracture delineation and monitoring of geothermal and coal seam gas areas using magnetotellurics. ASEG Ext. Abstr. 2012 (1), 3.

Thompson, I., Land, B., Holz, B., 2012. Rail corridor mine remediation – from investigation to verification. In: CORE 2012: Global Perspectives; Conference on Railway Engineering, September 10–12, 2012. Barton, A.C.T.: Engineers Australia, Brisbane, Australia, pp. 385–394.

Timms, W., Acworth, R.I., Hartland, A., Laurence, D., 2012. Leading practices for assessing the integrity of confining strata: application to mining and coal-seam gas extraction. International Mine Water Association Conference, September 30 to October 4, 2012, Bunbury.

Tissot, B.P., Welte, D.H., 1984. Petroleum Formation and Occurrence. Springer-Verlag, New York.

Tong, L., Liu, L., Yu, Q., 2013. Highway Construction across Heavily Mined Ground and Steep Topography in Southern China. pp. 1–18.

Turner, E.A., Kroeger, G.L., Arnold, M.C., et al., 2013. Assessing different mechanisms of toxicity in mountaintop removal/valley fill coal mining-affected watershed samples using Caenorhabditis elegans. PLoS One 8 (9), e75329.

Twenhofel, W.H., 1939. Principles of Sedimentation. McGraw-Hill Book Co., New York.

van der Elst, N.J., Savage, H.M., Keranen, K.M., Abers, G.A., 2013. Enhanced remote earthquake triggering at fluid-injection sites in the midwestern United States. Science 341 (6142), 164–167.

Varma, S., Michael, K., 2012. Impact of multi-purpose aquifer utilisation on a variable-density groundwater flow system in the Gippsland Basin, Australia. Hydrogeol. J. 20 (1), 119–134.

Versteegh, G.J.M., Blokker, P., Wood, G., et al., 2004. Oxidative polymerization of unsaturated fatty acids as a preservation pathway for microalgal organic matter. Org. Geochem. 35, 1129–1139.

Vink, S., 2013. Guidelines for Establishing Ecologically Sustainable Discharge Criteria in Seasonally Flowing Streams. Centre for Water in the Minerals Industry, University of Queensland.

von Balnckenburg, F., O'Nions, R.K., 1999. Response of beryllium and radio generic isotope ratios in North Atlantic Deep Water to the onset of northern hemispheric glaciation. Earth Planet. Sci. Lett. 167, 175–182.

Wakeham, S.G., Lee, C., Hedges, J.I., Hernes, P.J., Peterson, M.L., 1997. Molecular indicators of diagenetic status in marine organic matter. Geochim. Cosmochim. Acta 61, 5363–5369.

Walters, E.R., Pugh, C.W., Bender, K.S., Lefticariu, L., 2012. Geochemical and microbial community dynamics of a remediation system treating coal-derived acid mine drainage. Abstr. Progr. Geol. Soc. Am. 44 (7), 74.

Wangersky, P.J., 1978. Production of dissolved organic matter. In: Kinne, O. (Ed.), Marine Ecology, IV. Wiley and Sons, New York, pp. 115–220.

Weertman, J., 1961. Mechanism for the formation of inner moraines found near the edge of cold ice caps and ice sheets. J. Glaciol. 3, 965–978.

Wentworty, C.K., 1922. A scale of grade and class terms for classic sediments. J. Geol. 30, 377–392.

Westrich, J.T., Berner, R.A., 1984. The role of sedimentary organic matter in bacterial sulfate reduction: the G model tested. Limnol. Oceanogr. 29, 236–249.

Whitman, J.M., Davies, T.A., 1979. Cenozoic oceanic sedimentation rates: how good are the data? Mar. Geol. 30, 269–284.

Willenbring, J.N., von Blanckenburg, F., 2010. Long-term stability of glacial erosion rates and weathering during Late Cenozoic cooling. Nature. http://dx.doi.org/10.1038/nature09044.

William, P.M., 1968. Organic and inorganic constituents of Amazon river. Nature 218, 937–938.

Wyrtki, K., 1962. The oxygen minimum in relation to ocean circulation. Deep Sea. Res. 9, 11–23.

Zhang, P., Molnar, P., Downs, W.R., 2001. Increased sedimentation rate and grain sizes 2–4 Myr ago due to the influence of climate change and erosion rates. Nature 410, 891–897.

Zonnevede, et al., 2010. Selective preservation of origin matter in marine environments: processes and impact on the sedimentary record. Biogeosciences 7, 483–511.

Index

'Note: Page numbers followed by "f" indicate figures and "t" indicate tables.'

D

E

N

O

P

Q

R

S

T